AVIATION SPACE DICTIONARY

Editor
ERNEST J. GENTLE

President of Aero Publishers, Inc. Formerly: Senior Design Engineer and Senior Research Engineer for Lockheed Aircraft Corporation. Member of Tau Beta Pi, National Engineering Honorary, and Sigma Xi, National Science Honorary. Received Master of Science in Aeronautics from California Institute of Technology.

Co-Editor
LAWRENCE W. REITHMAIER

Free lance writer. Formerly Vice President of Aero Publishers, Inc.; Project Engineer at North American Rockwell Corporation on the Apollo and Skylab programs; Project Engineer at McDonnell Aircraft Corporation on military aircraft programs. FAA certificated Commercial Pilot / Instrument / Flight / Ground Instructor / A&P Mechanic. Received B.S. degree in Mechanical Engineering from the University of Illinois.

SIXTH EDITION
1980

AERO PUBLISHERS, INC.
329 W. Aviation Road   Fallbrook,   CA 92028

*Library of Congress Catalog Card Number
80-67567*

ISBN 0-8168-3002-9

Library of Congress Cataloging in Publication Data

Aviation & Space Dictionary-1st
Los Angeles, etc, Aero Publishers, etc, 1940

First ed. preceded by a "primary edition" with title Aviation Dic-
tionary, reference guide, published in 1939.
Title varies: 1st-2nd editions—Baughman's aviation dictionary and
reference guide. 3rd edition—Aviation Dictionary and reference guide.
The 1st-2nd editions prepared by H.E. Baughman; 3rd by E.J. Gentle.
Some editions issued in more than one printing.
1. Aeronautics-Dictionaries,   2. Astronautics-Dictionaries
I. Baughman, Harold Eugene.   II. Gentle, Ernest James
TL 509.A8   1980   629.1303   28287 Rev

Printed and Published in the United States

PREFACE TO THE SIXTH EDITION

In reviewing the fifth edition of this Dictionary for additions and changes to be incorporated into this sixth edition it was found to be amazingly up-to-date. The fifth edition had included all the new definitions resulting from the "space race" of the 1960s. This space race had been reduced to a walk during the 1970s with the space shuttle as the dominant program.

As is the case with all programs involving technology, an initial breakthrough results in a period of intense development, as in the 1960s with Apollo. Thereafter, development consists of continual refinement with less spectacular, but more efficient results as emphasized by the space shuttle. For this reason, the definitions involving space flight have changed little.

During the 1970s however, advancements in aeronautical technology dictated a complete review of definitions relating to all phases of aviation. The need to conserve fuel resulted in detail refinements in both powerplants and aircraft in order to increase efficiency. Advancements were also made in military aviation and changes took place in aircraft operations as well as air traffic control. Many new definitions were added in these areas and a few definitions considered obsolete were deleted.

This sixth edition of the Aviation/Space Dictionary would be an effective aid to aerospace communications throughout the decade of the 1980s.

PHOTO CREDITS

The basic sources of data for the *Aviation/Space Dictionary* were various dictionaries and glossaries issued by the United States Air Force, Nuclear Regulatory Commission, Federal Aviation Administration, National Weather Service and the National Aeronautics and Space Administration. Additional definitions were prepared by the editors.

ACKNOWLEDGEMENTS

Lockheed Aircraft Corporation, Pratt & Whitney Aircraft, General Electric Company-Aircraft Engine Groups, Delco Electronics Division of General Motors Corporation, Piper Aircraft Coporation, Cessna Aircraft Company, National Aeronautics and Space Administration, Rockwell International, LTV Aerospace Corporation, Bell Helicopter, McDonnell Douglas Corporation, Western Electric Company, Hughes Aircraft Company, Federal Aviation Administration, The Boeing Company, General Dynamics, Fairchild Industries and U.S. Department of Defense at Lawrence Livermore Laboratory (University of California).

ENDORSEMENTS

The American Society for Aerospace Education is pleased to endorse this major innovation in dictionary publishing. By now the grandeur of the achievement of landing men on the moon has taken its place in our language as a yardstick of human accomplishment. But what about the student or educator who until now had to research the technological spinoff of definitions and new terminology through scores of obscure publications? This 6th edition of the Aviation/Space Dictionary provides an invaluable glossary, in a single publication, that goes far beyond the specialized field of aerospace itself. Heretofore, unrelated fields such as computer technology, nucleonics, and meteorology are now included in this monumental work.

Though the Aviation/Space Dictionary will appeal to engineers, aviation and spacebuffs as well as technicians in every phase of the industry, it is inconceivable that the aerospace educator could possibly function efficiently without this excellent reference being readily available.

AMERICAN SOCIETY
For AEROSPACE EDUCATION

We here at National Headquarters, Civil Air Patrol, received the announcement of the planned Sixth Edition of the Aviation/Space Dictionary with great enthusiasm. Civil Air Patrol has as one of its missions that of implementing an understanding of aerospace power through education nationally. This is by no means an average task and in order to do this effectively, a vast amount and variety of instructional material and communiques must be produced. The foundation of this production is research communication and evaluation. Central to this foundation is, and has been for many years, your Aviation/Space Dictionary.

In announcing the Sixth Edition, with the expansion and improvements, is like telling us we will be improving our program in a direct ratio. Therefore we will not only endorse this Sixth Edition enthusiastically, we are telling you our work will be greatly enhanced and upgraded as a result. You have helped us with our future and we deeply appreciate that.

Accept this letter of endorsement as one of congratulations as well, because such a document should be in every library in America, on the desk of anyone who teaches in any classroom above the fifth grade, and should be an absolute source document for everyone in the nation who has anything to do with aviation and space.

U.S. AIR FORCE (CAP) DIVISION
of AEROSPACE EDUCATION

QUOTES from the late Dr. von Braun

The receding frontiers of space have resulted in a profusion of new concepts and technologies. The field of aviation has also expanded and reached a high degree of sophistication undreamed of even a decade ago. Out of this unprecedented progress has arisen a need for a comprehensive compilation of aviation and space terminology. In the process of learning, or during research, new words are often encountered, and consulting numerous reference books is time-consuming and often unproductive.

This edition of the Aviation/Space Dictionary brings together into one authoritative glossary, the terminology encompassed by aviation and space flight activities. The volume was compiled by experts and is based on the work of specialists in the many complex areas that constitute the aerospace sciences. The editors have succeeded in keeping the definitions concise and technically correct, yet easily understandable.

All of us engaged in any particular phase of this exciting field, so simply designated "aerospace", are in reality aviation and space enthusiasts. No facet of this magnificent jewel escapes our interest. This Aviation/Space Dictionary will unquestionably find a privileged place in the library of the managers, administrators, engineers, technicians, and flight personnel engaged in all phases of the industry. It should also appeal to aviation and space "buffs" as well as students, writers, and the curious and interested layman.

Dr. Wernher von Braun

Former Director—NASA George C. Marshall
Space Flight Center

MANNER OF PRESENTATION

The presentation emphasizes the technical content of the definitions and omits information on grammar usually given in conventional dictionaries. It is hoped that the definitions and examples will identify the parts of speech and indicate whether a verb is transitive or intransitive. If these points are not clear, it may be assumed that usage is not consistent. Terms are generally entered in the singular without mention of the plural unless only the plural is used or unless the plural is radically different from the singular. Variant spellings are entered in normal alphabetical order.

Bold face type is used for terms, the numeral and letters before definitions, for abbreviations immediately following terms, and for cross references. *Italic* is used in lieu of quotation marks to indicate that a term rather than a concept is being discussed. *Italic bold face* is used for symbols immediately following a term.

The definitions of different senses for a given term are separated by numerals and letters. The letters are used as subheadings under a numbered heading or are used to indicate a close relationship between the different headings.

All terms printed in **bold face** are defined in the Dictionary. Cross references are indicated by the use of **bold face** both in the definition and in specific instructions following the definition, such as "see" and "compare". The cross references following the definition should be used for full understanding of the term defined. The cross references within the definition are used to indicate key terms which are also defined in the Dictionary.

Synonyms for terms defined are entered in normal alphabetical order and a cross reference to the definition; as **autopilot** same as **automatic pilot**; or in some cases the words "same as" are omitted; as **Rnav. Area Navigation**. The inclusion of a definition for only one of a group of synonyms does *not* mean that the use of another synonym is incorrect. It does mean that the evidence examined indicated that the term defined was the most commonly used.

Acronyms, abbreviations and symbols are included in normal alphabetical order. Only the more commonly used acronyms, abbreviations and symbols are presented such as: **ATC** Air Traffic Control. More specialized items, as: **IMU** Inertial Measuring Unit are not included in the Dictionary. For consistency, all abbreviations are printed without periods between letters unless omission of periods may cause misunderstanding.

The U.S. Navy Lockheed carrier-based S-3A Viking is an **antisubmarine warfare** (**ASW**) *aircraft.*

The U.S. Air Force Fairchild A-10 is a close air support (**attack**) *aircraft.*

A

A. Military mission designation for **attack** aircraft.

A & P mechanic. Aircraft and powerplant mechanic. See **aviation mechanic** and **powerplant mechanic.**

AAM. Air-to-Air Missile.

abeam. Bearing of approximately 090 degrees or 270 degrees relative; at right angles to the longitudinal axis of a vehicle.

aberration. 1. In astronomy, the apparent angular displacement of the position of a celestial body in the direction of motion of the observer, caused by the combination of the velocity of the observer and the velocity of light. **2.** In optics, a specific deviation from perfect imagery, as, for example: spherical aberration, coma, astigmatism, curvature of field, and distortion.

ablate. To carry away; specifically, to carry away heat generated by **aerodynamic heating,** from a vital part, by arranging for its absorption in a nonvital part, which may melt or vaporize, then fall away taking the heat with it. See **heat shield, ablation.**

ablating material. A material, especially a coating material, designed to provide thermal protection to a body in a fluid stream through loss of mass.

ablating nose cone. A **nose cone** designed to reduce heat transfer to the internal structure by the use of an **ablating material.**

ablation. The removal of surface material from a body by vaporization, melting, chipping, or other erosive process; specifically, the intentional removal of material from a **nose cone** or **spacecraft** during high-speed movement through a planetary atmosphere to provide thermal protection to the underlying structure. See **ablating material.**

ablatively. By a process of ablation, as in *ablatively cooled.*

ablator. A material designed to provide thermal protection through **ablation.**

A-bomb. An **atomic bomb.**

abort. Failure of an aerospace vehicle to accomplish its purposes for any reason other than enemy action. An abort may occur at any point from start to count-down or takeoff to the destination. An abort can be caused by human, technical or meteorological errors, miscalculations or malfunction.

Abridged Nautical Almanac. See Nautical Almanac.

abscissa. The horizontal reference line of a curve or graph. Horizontal distances measured along this line from the ordinate line to points on a curve are called the abscissa values, e.g., when plotting an airfoil section, horizontal distances along the chord are plotted as abscissa.

absolute. 1. Pertaining to a measurement relative to a universal constant or natural **datum,** as *absolute coordinate system, absolute altitude, absolute temperature.* **2.** Complete, as in *absolute vacuum.*

absolute altimeter. Radio or similar apparatus that is designed to indicate the actual height of an aircraft above the terrain. See **radar altimeter.**

absolute angle of attack. The acute angle between the chord line of an airfoil at any instant in flight and its chord line at zero lift.

absolute altitude. Height above the surface. This height may be indicated directly on a radio/radar altimeter, which measures the time interval of a vertical signal bounced from the aircraft to the ground and back.

absolute ceiling. The maximum height above sea level at which a given airplane would be able to maintain horizontal flight under standard air conditions.

This is a theoretical value only since, at its absolute ceiling, an airplane's rate of climb is zero.

absolute coordinate system. An **inertial coordinate system** which is fixed with respect to the stars.

In theory, no absolute coordinate system can be established because the reference stars are themselves in motion. In practice, such a system can be established to meet the demands of the problem concerned by the selection of appropriate reference stars.

absolute humidity. See humidity.

absolute instrument. An instrument whose calibration can be determined by means of physical measurements on the instrument. Compare **secondary instrument.**

absolute magnitude. 1. A measure of the brightness of a star equal to the **magnitude** the star would have at a distance of 10 **parsecs** from the observer. **2.** The stellar magnitude any **meteor** would have if placed in the observer's zenith at a height of 100 kilometers.

absolute pressure. In engineering literature, a term used to indicate **pressure** above the absolute zero value of pressure that theoretically obtains in empty space or at the absolute zero of temperature as distinguished from *gage pressure.*

absolute system of units. 1. A system of units in which a smaller number of units are chosen as fundamental, and all other units are derived from them. **2.** Specifically, a system of electrical units put into effect by international agreement on 1 January 1948.

Prior to 1 January 1948 the international system was in effect; the two systems can be converted by the following relationships:

1 mean international ohm = 1.00049 absolute ohm.
1 mean international volt = 1.00034 absolute volt.

absolute temperature. Temperature measured from **absolute zero.**

absolute temperature scale. A temperature scale based upon the value zero as the lowest possible value. Thus, all obtainable temperatures are positive. The **Kelvin** and **Rankine** scales are absolute scales.

absolute vacuum. A void completely empty of matter. Also called *perfect vacuum.*

An absolute vacuum is not obtainable.

absolute zero. The theoretical temperature at which molecular motion vanishes and a body would have no heat energy; the zero point of the **Kelvin** and **Rankine** temperature scales.

Absolute zero may be interpreted as the temperature at which the volume of a perfect gas vanishes or, more generally, as the temperature of the cold source which would render a Carnot cycle 100 percent efficient. The value of absolute zero is now estimated to be —273.15° Celsius, —459.67° Fahrenheit, 0° Kelvin, and 0° Rankine.

absorbed dose. nucleonics: When **ionizing radiation** passes through **matter,** some of its energy is imparted to the matter. The amount absorbed per unit mass of irradiated material is called the absorbed dose, and is measured in **rems** and **rads.** (See **threshold dose.**)

absorber. nucleonics: Any material that absorbs or diminishes the intensity of ionizing **radiation.** Neutron absorbers, like boron, hafnium, and cadmium, are used in control rods for reactors. Concrete and steel absorb gamma rays and neutrons in reactor shields. A thin sheet of paper or metal will absorb or attenuate alpha particles and all except the most energetic beta particles. (Compare **moderator.**)

absorptance, absorbtance. The ratio of the radiant flux absorbed by a body to that incident upon it. Also called **absorption factor.** Compare **absorptivity.**

  Total absorptance refers to absorptance measured over all wavelengths.

  Spectral absorptance refers to absorptance measured at a specified wavelength.

absorption. 1.The process by which radiant energy is absorbed and converted into other forms of energy. **2.** In general, the taking up or assimilation of one substance by another. See **adsorption. 3.** In vacuum technology, gas entering into the interior of a solid.

absorption band. A range of wavelengths (or frequencies) in the **electromagnetic spectrum** within which radiant energy is absorbed by a susbstance. See **absorption spectrum.**

absorption cross section. In radar, the ratio of the amount of power removed from a beam by absorption of radio energy by a target to the power in the beam incident upon the target.

absorption line. A minute range of wavelength (or frequency) in the **electromagnetic spectrum** within which radiant energy is absorbed by the medium through which it is passing. Each line is associated with a particular mode of electronic excitation induced in the absorbing atoms by the incident radiation.

absorption spectrum. The array of **absorption lines** and **absorption bands** which results from the passage of radiant energy from a continous source through a selectively absorbing medium cooler than the source. See **electromagnetic spectrum.**

absorptivity. The capacity of a material to absorb incident radiant energy, measured as the **absorptance** of a specimen of the material thick enough to be completely opaque, and having an optically smooth surface.

absorptivity-emissivity ratio. In space applications, the ratio of absorptivity for **solar radiation** of a material to its infrared emissivity. Also called A/E ratio.

accelerate-stop distance. The distance required to accelerate an airplane to a specified speed and, assuming failure of the critical engine at the instant that speed is attained, to bring the airplane to a stop.

acceleration of gravity. Any acceleration due entirely to **gravity,** e.g., in a vacuum a freely falling body falls toward the earth's center at a velocity of 32.2 feet the first second, and an increased velocity of 32.2 feet per second the second second (velocity at the end of the second second is 64.4 feet per second, etc.). In air the rate of acceleration is modified by the force due to air resistance. A freely falling body will finally reach a speed at which the force of gravity and air resistance are equal and no greater speed is gained.

acceleration error. An error caused by the deflection of the vertical reference due to any change in velocity of the aircraft.

acceleration. The rate of change of **velocity** and expressed in feet per second per second or miles per hour per second, or any other unit of speed divided by any unit of time. Acceleration has direction as well as magnitude, and is forward when the velocity is increased or backward (deceleration) when the velocity is decreased.

accelerator. (nucleonics) A device for increasing the velocity and energy of charged **elementary particles,** for example, electrons or protons, through application of electrical and/or magnetic forces. Accelerators have made particles move in velocities approaching the speed of light. Types of accelerators include **betatrons, Cockcroft-Walton accelerators, cyclotrons, linear accelerators, synchrocyclotrons, synchrotons, and Van de Graaff generators.**

accelerometer. A **transducer** which measures acceleration or gravitational forces capable of imparting **acceleration.**

  An accelerometer usually uses a concentrated mass (seismic mass) which resists movement because of its inertia. The displacement of the seismic mass relative to its supporting frame or container is used as a measure of acceleration.

acceptance trials. Trials carried out by nominated representatives of the eventual military users of the weapon or equipment to determine if the specified performance and characteristics have been met.

accessory (aircraft). A mechanism or device employed in conjunction with a major item of equipment or installed as part of a system.

access to classified information. The ability and opportunity to obtain knowledge of classified information. A person has access to classified information if he is permitted to gain knowledge of the information or if he is in a place where he would be expected to gain such knowledge. A person does not have access to classified information by being in a place where classified information is kept if security measures prevent him from gaining knowledge of the information.

acclimatization. The adjustments of a human body or other organism to a new environment; the bodily changes which tend to increase efficiency and reduce energy loss.

accidental attack. An unintended attack which occurs without deliberate national design as a direct result of a random event, such as a mechanical failure, a simple human error, or an unauthorized action by a subordinate.

accumulator. 1.A device or apparatus that accumulates or stores up, as: (**a**) a contrivance in a hydraulic system that stores fluid under pressure; (**b**) a device sometimes incorporated in the fuel system of a gas-turbine engine to store up and release fuel under pressure as an aid in starting: (**c**) an electrical storage battery (British usage). **2.** In computer technology, a device which stores a number and upon receipt of another number adds it to the number already stored and stores the sum.

accuracy of fire. The measure of the deviation of fire from the target, expressed in terms of the distance between the center of mass of the target and the mean point of bursts.

acknowledgment. A message from the addressee informing the originator that his communication has been received and is understood.

aclinic line. The line through those points on the earth's surface at which **magnetic dip** is zero. The aclinic line is a particular case of an **isoclinic line.** Also called *dip equator, magnetic equator.* Compare **agonic line, geomagnetic equator.**

acoustic excitation. The process of inducing **vibration** in a structure by exposure to **sound waves.**

acoustic vibration. With respect to operational **environments, vibrations** transmitted through a gas. These vibrations **may** be **subsonic, sonic,** and **ultrasonic.**

acquire. 1. When applied to acquisition radars, the process of detecting the presence and location of a target in sufficient detail to permit identification. **2.** When applied to tracking radars, the process of positioning a radar beam so that a target is in that beam to permit the effective employment of weapons. See also **target acquisition.**

acquisition. 1. The process of locating the orbit of a satellite or trajectory of a space probe so that tracking or telemetry data can be gathered. **2.** The process of pointing an antenna or telescope so that it is properly oriented to allow gathering of tracking or telemetry data from a satellite or space probe.

acrobatic category. As applied to an aircraft **type certificate** issued by the **FAA,** acrobatic **category** is not limited to flight maneuver restrictions of the **normal category** or **utility category.** See **acrobatics.**

acrobatics. Unusual airplane maneuvers normally not necessary for air transportation or civil flight training but a requirement for military flying. Chandelles, lazy eights, steep turns (up to 60 degree bank angle) spins and stalls are normally not considered acrobatic maneuvers. Loops, rolls and variations thereof, are normally considered acrobatic maneuvers. See **limited acrobatics.**

actinic. Pertaining to **electromagnetic radiation** capable of initiating photochemical reactions, as in photography or the fading of pigments.

actinometer. An instrument used in determining the amount of insulation or protection received from the sun (it is a black bulb thermometer placed in a properly constructed vacuum chamber).

activate. To put into existence by official order a unit, post, camp, station, base, or shore activity which has previously been constituted and designated by name or number, or both, so that it can be organized to function in its assigned capacity.

active. 1. Transmitting a signal, as *active satellite.* Antonym of *passive.* **2.** = **radioactive,** as *active sample.* **3.** = **fissionable,** as *active material.* **4.** Receiving energy from some source other than a signal, as *active element.*

active aircraft. (military) Aircraft currently and actively engaged in supporting the flying missions either through direct assignment to operational units or in the preparation for such assignment or reassignment through any of the logistic processes of supply, maintenance, and modification. See also **aircraft.**

active controls. Automatic movement of airplane control surfaces in response to various sensors for load alleviation, stability augmentation or flutter load control. As applied to wing load alleviation, active control concept includes ailerons automatically deflected symmetrically, responding to a maneuver or gust load. With this system, which operates independently and in addition to the normal aileron control inputs, the center of the air load is forced to move inboard by unloading the outboard wing thus reducing the wing bending moments.

See **control configured vehicle.**

active air defense. Direct defensive action taken to destroy or reduce the effectiveness of any enemy air attack. It includes such measures as the use of aircraft, antiaircraft guns, electronic countermeasures, and surface-to-air guided missiles. See also **air defense.**

active communications satellite. A satellite which receives, regenerates, and retransmits signals between stations. See also **communications satellite.**

active front. A front which produces considerable cloudiness and precipitation. Also see **cold front, warm front.**

active homing. The **homing** of an aerodynamic or space vehicle in which energy waves (as radar) are transmitted from the vehicle to the target and reflected back to the vehicle to direct the vehicle toward the target. Compare **passive homing.**

active homing guidance. A system of homing guidance wherein both the source for illuminating the target, and the receiver for detecting the energy reflected from the target as the result of illuminating the target, are carried within the missile. See also **guidance.**

activity. (military) **1.** A unit, organization, or installation performing a function or mission, e. g., reception center, redistribution center, naval station, naval shipyard. **2.** A function or mission, e.g., recruiting, schooling. See also **establishment.**

active satellite. A **satellite** which transmits a signal, in contrast to *passive satellite.*

active tracking system. A system which requires addition of a **transponder,** or **transmitter** on board the vehicle to repeat, transmit, or retransmit information to the tracking equipment.

actuating cylinder. The purpose of an actuating cylinder is to transform energy in the form of fluid flow under pressure into mechanical force, or action, to perform some kind of work. Another definition is: A unit which receives fluid under pressure and which is connected to some movable part of the airplane in order to produce linear motion.

actuating system. A mechanical system that supplies and transmits energy for the operation of other mechanisms or systems.

adaptation luminance. The average **luminance** (or brightness) of those objects and surfaces in the immediate vicinity of an observer. Also called *adaptation brightness, adaptation level, adaptation illuminance.*

The adaptation luminance has a marked influence on an observer's estimate of the **visual range** because, along with the visual angle of the object under observation, it determines the observer's **threshold contrast.** High adaptation luminance tends to produce a high threshold contrast, thus reducing the estimated visual range. This effect of the adaptation luminance is to be distinguished from the influence of **background luminance.**

adapter skirt. A flange or extension of a **spacevehicle stage** or section that provides a ready means for fitting some object to the stage or section.

ADC. Aerospace Defense Command.

Adcock antenna. A pair of vertical antennas separated by a distance of one-half **wavelength** or less, and connected in **phase opposition** to produce a **radiation pattern** having the shape of a figure eight.

additive. Any material or substance added to something else. Specifically, a substance added to a **propellant** to achieve some purpose, such as a more even rate of combustion, or a substance added to fuels or lubricants to improve them or give them some desired quality, such as tetraethyl lead added to a fuel as an antidetonation agent, or graphite, talc, or other substances added to certain oils and greases to improve lubrication qualities.

address. 1. Of a **computer,** a location where information is stored. **2.** An expression, usually numerical, identifying an address (sense **1**).

ADF. Automatic Direction Finder.

adiabat. A line or curve on a temperature-pressure diagram along which a thermodynamic change takes place without the gain or loss of heat.

adiabatic. The word applied in the science of thermodynamics to a process during which no heat is communicated to or withdrawn from the body or system concerned. Adiabatic changes of temperature in a gas are those that occur only in consequence of compression or expansion accompanying an increase or a decrease of pressure. In the atmosphere such pressure changes take place most readily due to ascent (expansion) of air which produces cooling (see **adiabatic lapse rate**), or descent (compression), which produces heating.

adiabatic compression. See **adiabatic process.**

adiabatic efficiency. The efficiency with which work is done with respect to heat gains or losses. See **adiabatic process.**

adiabatic lapse rate. The rate at which ascending air cools, or descending air warms, when no heat is added or taken away. (See **adiabatic**.) The adiabatic rate for dry air is 5.4°Fahrenheit, per 1000 feet. However, in the case of ascending saturated air, the condensation of moisture releases **latent heat** of condensation, which lessens the decrease of temperature.

adiabatic process. A **thermodynamic** change of state of a system in which there is no transfer of heat or mass across the boundaries of the system. In an adiabatic process, compression always results in warming, expansion in cooling. See **diabatic process.**

adiabatic recovery temperature. 1. The temperature reached by a moving fluid when brought to rest through an **adiabatic process.** Also called *recovery temperature, stagnation temperature.* **2. adiabatic wall temperature. 3.** The final and initial temperature in an adiabatic, **Carnot cycle.**

adiabatic wall temperature. The temperature assumed by a wall in a moving fluid stream when there is no heat transfer between the wall and the stream.

ADP. Automatic Data Processing.

adsorption. The adhesion of a thin film of liquid or gas to the surface of a solid substance. The solid does not combine chemically with the adsorbed substance.

advanced landing field. (military) An airfield, usually having minimum facilities, in or near an objective area. See also **airfield.**

advection. The process of transfer by horizontal motion, particularly applied to the transfer of heat by horizontal motion of the air.

advectional currents. Horizontal air currents.

advection fog. Fog resulting from the transfer of warm, humid air over a cold surface, especially a cold ocean surface, or from the transport of humid air from an ocean to a cold land surface.

adverse weather. Weather in which flight operations are generally restricted or impeded.

adverse weather aerial delivery system. (military) The precise delivery of personnel, equipment, and supplies during adverse weather, utilizing a self-contained aircraft instrumentation system without artificial ground assistance, or use of ground navigational aids.

advisory service. Advice and information provided by a facility to assist pilots in the safe conduct of flight and aircraft movement.

aerial. 1. Antenna. **2.** Of or pertaining to the air, atmosphere, or aviation.

aerial combat tactics. Aerial tactics used to effectively employ fighter aircraft against hostile aircraft.

aerial delivery. The act or process of delivering cargo or personnel by airdrop or airland.

aerial delivery system. Any method by which personnel and/or material can be offloaded at a specific location by an aircraft in-flight.

aerial mining. The act or process of aerial mine-laying on land or sea routes of communication, facility accesses, and areas of enemy force concentrations.

aerial pickets. Aircraft disposed around a position, area, or formation, primarily to detect, report, and track approaching enemy aircraft.

aerial supply. (military) The act or process by which aerial delivery of supplies is made to ground units.

A/E ratio. Absorptivity-emissivity ratio.

aeroballistics. The study of the interaction of projectiles or high speed vehicles with the atmosphere. See **ballistics.**

The problem of the effect of reentry on the trajectory of a vehicle is a problem in aeroballistics.

aeroballistic missile. A wingless vehicle employing the boostglide and continuous roll technique for flight at hypersonic speeds within the earth's atmosphere. The trajectory is ballistic to apogee, after which the vehicle assumes an angle of attack (10 to 20 degrees) and descends partly ballistically and partly through aerodynamic lift to an altitude of about 60,000 feet, thereafter resuming a ballistic dive to the surface. A slow continuous roll is imparted to the wingless vehicle during the aerodynamic portion of flight to distribute frictional heat evenly over the airframe so as to preserve the structural integrity.

aerobiology. The study of the distribution of living organisms freely suspended in the atmosphere.

aerodrome. See **airfield, airport.**

aeroduct. A ramjet type of engine designed to scoop up ions and electrons freely available in the outer reaches of the atmosphere or in the atmospheres of other spatial bodies, and by a **metachemical** process within the duct of this engine, expel particles derived from the ions and electrons as a propulsive jetstream.

aerodynamic. Of or pertaining to **aerodynamics.**

aerodynamically balanced control surface. A control surface is balanced aerodynamically when a portion of the surface is ahead of the hingeline. Control surfaces are balanced aerodynamically to reduce hinge moments. Also see **balanced control surface.**

aerodynamic center of a wing section. A point located on the wing chord approximately one-quarter of the chord length back of the leading edge about which the **moment coefficient** is practically constant for all angles of attack.

aerodynamic coefficient. Any nondimensional **coefficient** relating to aerodynamic forces or moments, such as a coefficient of drag, a coefficient of lift, etc.

aerodynamic force. The force exerted by a moving gaseous fluid upon a body completely immersed in it.

aerodynamic heating. The heating of a body produced by passage of air or other gases over the body; caused by friction and by compression processes and significant chiefly at high speeds.

aerodynamic missile. A missile which uses aerodynamic forces to maintain its flight path, generally employing propulsion guidance. See also **ballistic missile; guided missile.**

aerodynamics. 1. The science that deals with the motion of air and other gaseous fluids, and of the forces acting on bodies when the bodies move through such fluids, or when such fluids move against or around the bodies, as *his research in aerodynamics.* 2. (**a**) The actions and forces resulting from the movement or flow of gaseous fluids against or around bodies, as, *the aerodynamics of a wing in supersonic flight.* (**b**) The properties of a body or bodies with respect to these actions or forces, as, *the aerodynamics of a turrent or of a configuration.* 3. The application of the principles of gaseous fluid flows and of their actions against and around bodies to the design and construction of bodies intended to move through such fluids, as *a design used in aerodynamics.*

aerodynamic trail. A condensation trail formed by **adiabatic** cooling to saturation (or slight supersaturation) of air passing over the surfaces of high-speed aircraft.

Aerodynamic trails form off the tips of wings and propellers and other points of maximum pressure decrease. They are relatively rare and of short duration compared to **exhaust trails.**

aerodynamic twist. Variation of the zero-lift line along the span of a wing or other airfoil. See **washin, washout.**

aerodynamic vehicle. A device, such as an airplane, glider, etc., capable of flight only within a **sensible atmosphere** and relying on aerodynamic forces to maintain flight.

The term is used when the context calls for discrimination from *space vehicle*

aerodynamicist. A person trained in the science of aerodynamics and persues it as a vocation.

aeroelasticity. The study of the response of structurally elastic bodies to **aerodynamic** loads.

aeroembolism. 1. The formation or liberation of gases in the blood vessels of the body, as brought on by a too-rapid change from a high, or relatively high, atmospheric pressure to a lower one. 2. The disease or condition caused by the formation of gas bubbles (mostly nitrogen) in the body fluids. The disease is characterized principally by neuralgic pains, cramps, and swelling, and sometimes results in death. Also called *decompression sickness.*

aerolite. A **meteorite** composed principally of stony material.

aerology. 1. As officially used in the U. S. Navy until early 1957, same as **meteorology;** this usage was more administrative than scientific. 2. As a subdivision of meteorology, the study of the **free atmosphere** throughout its vertical extent, as distinguished from studies confined to the layer of the atmosphere adjacent to the earth's surface.

aeromedical evacuation. (military) The movement of patients to and between medical treatment facilities by air transportation.

aeronautical chart. A specialized representation of mapped features of the earth, or some part of it, produced to show selected terrain, cultural and hydrographic features, and supplemental information required for air navigation, pilotage, or for planning air operations.

aeronautical light beacon. The aeronautical beacon is a visual navaid displaying flashes of white and/or colored light, which is used to indicate the location of airports, landmarks, and certain points of the Federal airways in mountainous terrain and to mark hazards. The principal light so used is a rotating beacon of relatively high intensity, which is often supplemented by nonrotating flashing lights of lesser intensity.

aeronautics. A general term applied to everything associated with or used in any way in the study or design, construction and operation of an aircraft.

aeronomy. The study of the upper regions of the atmosphere where ionization, dissociation, and chemical reactions take place.

aeropause. A region of indeterminate limits in the upper atmosphere, considered as a boundary or transition region between the denser portion of the **atmosphere** and **space.**

aeroplane. British; **airplane,** U.S.

aerospace. From *aeronautics* and *space*). 1. Of or pertaining to both the earth's **atmosphere** and **space,** as in *aerospace industries.* 2. Earth's envelope of air and space above it; the two considered as a single realm for activity in the flight of air vehicles and in the launching, guidance, and control of ballistic missiles, earth satellites, dirigible space vehicles, and the like.

Aerospace in sense **2** is used primarily by the U.S. Air Force.

The term *aerospace* first appeared in print in the *Interim Glossary; Aero-Space Terms* (edited by Woodford Agee Heflin) published in February 1958 at the Air University, Maxwell Air Force Base, Alabama.

aerospace defense. All measures designed to reduce or nullify the effectiveness of hostile acts by aircraft, missiles, and space vehicles after they leave the earth's surface; an inclusive term encompassing air defense and space defense.

Aerospace Defense Command (ADC). See **major command.**

aerospace ground equipment (AGE). All equipments required on the ground to make a weapon system, command and control system, support system, advanced objective, subsystem, or end item of equipment operational in its intended environment. This includes all equipment required to install, launch, arrest, guide, control, direct, inspect, test, adjust, calibrate, appraise, gage, measure, assemble, disassemble, handle, transport, safeguard, store, activate, service, repair, overhaul, maintain, or operate the system, subsystem, end item, or component. This definition applies regardless of the method of development, funding or procurement.

aerospace medicine. That branch of medicine dealing with effects of flight through the atmosphere or in space upon the human body and with the prevention or cure

A General Electric J-79 turbojet engine running on a test stand. Its **afterburner** *provides a large increase in thrust. The variable area nozzle is in the full open position during afterburning.*

of physiological or psychological malfunctions arising from these effects.

aerospace vehicle. A vehicle capable of flight within and outside the **sensible atmosphere.**

aerostat. A generic term for aircraft whose support is chiefly due to buoyancy derived from aerostatic forces. The immersed body consists of one or more containers, filled with a gas which is lighter than air. The classification includes lighter-than-air-craft, i.e., airships and balloons.

aerothermodynamic border. An altitude at about 100 miles, above which the atmosphere is so rarefied that the skin of an object moving through it at high speeds generates no significant heat.

aerothermodynamic duct. The full term for *athodyd.*

aerothermodynamics. The study of **aerodynamic** phenomena at sufficiently high gas velocities that **thermodynamic** properties of the gas are important.

aft. Backward or in the rear of any designated position.

after body. 1. A companion body that trails a satellite. **2.** An unprotected section or piece of a ballistic missile that re-enters the atmosphere behind the nose cone or other body protected for re-entry.

afterburner. A device for augmenting the thrust of a **jet engine** by burning additional fuel in the uncombined oxygen in the gases from the turbine.

afterburning. 1. Irregular burning of fuel left in the **firing chamber** of a rocket after **fuel cutoff. 2.** The function of an **afterburner,** a device for augmenting the thrust of a jet engine by burning additional fuel in the uncombined oxygen in the gases from the turbine.

aftercooling. 1. The cooling of a gas after compression. **2.** The necessary cooling of **reactor core** after its shutdown by pumping a liquid or gas through it to carry off the excess heat generated by continuing radioactive decay of fission products within the core.

afterglow. 1. A broad, high arch of radiance or glow seen occasionally in the western sky above the highest clouds in deepening twilight, caused by the scattering effect of very fine particles of dust suspended in the upper atmosphere. **2.** The transient decay of a **plasma** after the power has been turned off.

The decay time involved is a direct consequence of the charged particle loss mechanisms, such as **diffusion** and **recombination.** The magnitude of these quantities is determined by measuring the decay time under controlled conditions.

afterheat. The heat generated in a **reactor core** after shutdown by continuing radioactive decay of fission products.

AGE. aerospace ground equipment. See **GSE.**

age of the moon. The elapsed time, usually expressed in days, since the last new moon. See **phases of the moon.**

aging. In a metal or alloy, a change in properties that generally occurs slowly at room temperature and more rapidly at higher temperatures.

agl. Above ground level.

agonic line. A line drawn on a map or chart joining points of zero magnetic variation for a specified epoch.

agravic. The condition of no gravity. Theoretically, absolute agravic conditions in the true sense of the word do not exist in the universe. See also **zero gravity.**

ailerons. Pairs of control surfaces, normally hinged along the wing span, designed to control an aircraft in roll by their differential movement. They produce primarily rolling moment.

aileron roll. An airplane acrobatic maneuver consisting of a **roll** in one or more complete revolutions executed as a single quick maneuver principally, if not entirely by use of ailerons. Sometimes call slow roll.

air. 1. The mixture of gases comprising the earth's atmosphere.

The percent by volume of those gases found in relatively constant amount in dry air near sea level is very nearly as follows:

	%
nitrogen (N_2)	78.084
oxygen (O_2)	20.9476
argon (A)	0.934
carbon dioxide (CO_2)	0.0314 (variable)
neon (Ne)	0.001818
helium (He)	0.000524
methane (CH_4)	0.000524
krypton (Kr)	0.0002 (variable)
hydrogen (H_2)	0.000114
nitrous oxide (N_2O)	0.00005
xenon (Xe)	0.0000087

In addition to the above constituents there are many variable constituents. Chief of these is water vapor, which may vary from zero to volume percentages close to 4 percent. Ozone, sulfur dioxide, ammonia, carbon monoxide, iodine, and other trace gases occur in small and varying amounts.

The above composition of dry air is true to about 90 kilometers. See **upper atmosphere.**

2. The realm or medium in which **aircraft** operate.

air alert. (military) The operational status of aircraft in the air that are ready for the immediate accomplishment of a mission.

air alert mission. Aircraft airborne in the battle area to answer calls for immediate air support from the ground forces.

Air Almanac. A joint publication of the Nautical Almanac Office of the United States Naval Observatory and Her Majesty's Nautical Almanac Office. It covers a four month period and it contains tabulated values of the greenwich hour angle and declination of selected celestial bodies, plus additional celestial data used in navigation.

air attack. 1. (**coordinated**) A combination of two or more types of air attack (dive, glide, low-level) in one strike, using one or more types of aircraft. **2.** (**deferred**) A procedure in which attack groups rendezvous as a single unit. It is used when attack groups are launched from more than one station with their departure on the mission being delayed pending further orders. **3.** (**divided**) A method of delivering a coordinated air attack which consists of holding the units in close tactical concentration up to a point, then splitting them to attack an objective from different directions.

airborne. 1. Applied to personnel, equipment, etc., transported by air, e.g., airborne infantry. **2.** Applied to materiel being or designed to be transported by aircraft, as distinguished from equipment installed in and remaining a part of the aircraft. **3.** Applied to an aircraft from the instant it becomes entirely sustained by air until it ceases to be so sustained. A lighter-than-air aircraft is not considered to be airborne when it is attached to the ground, except that moored balloons are airborne whenever sent aloft.

airborne alert. A state of aircraft readiness wherein

combat-equipped aircraft are airborne and ready for immediate action. It is designed to reduce reaction time and to increase the survivability factor. See also **combat air patrol; fighter cover.**

airborne command post. A suitably equipped aircraft used by the commander for the control of his forces.

airborne early warning. The detection of enemy air or surface units by radar or other equipment carried in an airborne vehicle and the transmitting of a warning to friendly units.

airborne early warning and control. Air surveillance and control provided by airborne early warning vehicles which are equipped with search and height-finding radar and communications equipment for controlling weapons. See also **air pickets.**

airborne force. A force composed primarily of ground and air units organized, equipped, and trained for airborne operations. See also **force(s).**

airborne intercept equipment. A fire control system, including radar equipment, installed in interceptor aircraft used to effect air interception.

airborne launch control center. A suitably equipped aircraft used to monitor status and command launch of strategic missiles based in the ground.

airborne launch control system. A system consisting of an airborne launch control center used in conjunction with launch and service facility operating ground equipment for airborne launch control of stragic missiles.

airborne operation. An operation involving the air movement into an objective area, of combat forces and their logistic support for execution of a tactical or a strategic mission. The means employed may be any combination of airborne units, air transportable units and types of transport aircraft, depending on the mission and the overall situation.

airborne stationkeeping. (military) The use of air inplane radar capability to maintain a specific position in close formation during night and instrument flight conditions.

airborne warning and control system. An aircraft suitably equipped to provide an airborne control, surveillance and communications capability for strategic defense and/or tactical air operations.

air brake. Any device primarily used to increase the drag of an aircraft at will.

air breakup. The breakup of a test **reentry body** after reentry into the atmosphere.

airbreather. An **aerodynamic vehicle** propelled by fuel oxidized by intake from the atmosphere; an airbreathing vehicle.

airbreathing. Of an engine or **aerodynamic vehicle,** required to take in air for the purpose of combustion.

air-breathing missile. A missile with an engine requiring the intake of air for combustion of its fuel, as in a ramjet or turbojet. To be contrasted with the rocket missile, which carries its own oxidizer and can operate beyond the atmosphere.

airburst. An explosion of a bomb or projectile above the surface, as distinguished from an explosion on contact with the surface or after penetration.

air carrier. A person who undertakes directly by lease, or other arrangement, to engage in air transportation.

air carrier airport: an airport (or runway) designated by design and/or use for air carrier operations.

air carrier—all cargo: a certificated route air carrier authorized to perform scheduled air freight, express, and mail transportation service as well as the conduct of nonscheduled operations (which may include passengers over specified routes).

air carrier—certificated route: an air carrier holding a Certificate of Public Convenience and Necessity issued by the Civil Aeronautics Board to conduct scheduled services over specified routes and a limited amount of nonscheduled operations.

air carrier—commuter: an air taxi operator which: (1) performs at least five round trips per week between two or more points and publishes flight schedules which specify the times, days of the week and places between which such flights are performed; or (2) transports mail by air pursuant to a current contract with the U.S. Postal Service.

air carrier—intrastate: an air carrier licensed by a state to operate wholly within its borders but not permitted to carry interline passengers from out of state. Intrastate air carriers are not regulated by the Civil Aeronautics Board.

air carrier—supplemental: an air carrier holding a Certificate of Public Convenience and Necessity issued by the Civil Aeronautics Board; authorizing it to perform passenger charter and/or cargo charter service supplementing the scheduled services of the certificated route air carriers.

air command. A major subdivision of the Air Force; for operational purposes it normally consists of two or more air forces. See also **command.**

air commerce. Interstate, overseas, or foreign air commerce or the transportation of mail by aircraft or any operation or navigation of aircraft within the limits of any Federal airway or any operation or navigation of aircraft which directly affects, or which may endanger safety in, interstate, overseas, or foreign air commerce.

air controller. An individual especially trained for and assigned the duty of the control (by use of radio, radar, or other means) of such aircraft as may be allotted to him for operation within his area. See also **air traffic controller; air weapons controller; tactical air controller; fighter controller; fighter direction.**

air corridors. (military) Restricted air routes of travel specified for use by friendly aircraft and established for the purpose of preventing friendly aircraft from being fired on by friendly forces.

aircraft. Any structure, machine, or contrivance, especially a **vehicle,** designed to be supported by the air, being borne up either by the dynamic action of the air upon the surfaces of the structure or object, or by its own buoyancy; such structures, machines, or vehicles collectively, as, *fifty aircraft.*

Aircraft, in its broadest meaning, includes fixed-wing airplanes, helicopters, gliders, airships, free and captive balloons, ornithopters, flying model aircraft, kites, etc., but since the term carries a strong vehicular suggestion, it is more often applied, or recognized to apply, only to such of these craft as are designed to support or convey a burden in or through the air.

aircraft accident. An occurrence associated with the operation of an aircraft which takes place between the time any person boards the aircraft with the intention of flight until such time as all such persons have disembarked, in which

a) any person suffers death or serious injury as a result of being in or upon the aircraft or by direct contact with the aircraft or anything attached thereto, or
b) the aircraft receives substantial damage.

aircraft arresting barrier. A device not dependent on special modification to an aircraft, used to engage and absorb the forward momentum of an emergency landing (or aborted takeoff). See also **aircraft arresting system.**

aircraft arresting gear. A device dependent on special modification to an aircraft, used to engage and absorb the forward momentum of a routine or emergency landing. See also **aircraft arresting system.**

aircraft arresting hook. A device fitted to an aircraft to engage arresting gear. See also **aircraft arresting system.**

aircraft arresting system. A series of components used to engage an aircraft and absorb the forward momentum of a routine or emergency landing (or aborted takeoff). See also **aircraft arresting barrier; aircraft arresting gear; aircraft arresting hook.**

aircraft block speed. True air speed in knots under zero wind conditions adjusted in relation to length of sortie to compensate for take-off, climb-out, let-down, instrument approach and landing.

aircraft cable. A flexible stranded wire cable used for operating controls.

aircraft carrier. A ship designed to carry and maintain aircraft and permit their take-off and landing.

aircraft class. See **class.**

aircraft climb corridor. Positive controlled airspaces of defined vertical and horizontal dimensions extending from an airfield.

aircraft commander. (military) See **captain/aircraft commander.**

aircraft dispatcher. A person, usually ground based, who exercises with the **pilot in command,** of the operational control of a flight. With regards to U. S. **Civil aircraft** engaged in **air commerce**, an aircraft dispatcher must have in his possession a current **aircraft dispatcher certificate** issued by the **FAA.** Also see **airman certificates.**

aircraft dispatcher certificate. A certificate of competency issued by the **FAA** to a person meeting the requirements of the applicable **Federal Aviation Regulations.** Also see **aircraft dispatcher, airman certificates.**

aircraft dispersal area. An area on a military installation designed primarily for the dispersal of parked aircraft, whereby such aircraft will be less vulnerable in the event of an enemy air raid.

aircraft engine. An engine that is used or intended to be used in propelling aircraft. It includes engine appurtenances and accessories necessary for its functioning, but does not include propellers.

Aircraft Flight Manual: a document containing the limitations, procedures, information, and data necessary for the safe operation of aircraft. See **airplane flight manual.**

aircraft handover. The process of transferring control of aircraft from one controlling authority to another.

aircraft hook wire. Any type of arresting device to prevent the overrun of aircraft in which a wire is engaged by an aircraft hook.

aircraft inspection. The process of systematically examing, checking and testing aircraft structural members, components and systems, to detect actual or potential unservicable conditions.

aircraft log. A log of an aircraft containing a complete operating history of the aircraft including flight time, reports of inspection, repairs, and alterations.

aircraft mechanic. A person who repairs and/or maintains an aircraft. When applied to U. S. civil aviation, aircraft mechanic may be a **rating** to an **aviation mechanic certificate.**

aircraft modification. A change in the physical characteristics of aircraft, accomplished either by a change in production specifications or by alternation of items already produced.

aircraft operating weight. The basic weight of the aircraft plus the weight of the crew, equipment, and oil.

aircraft operations: the airborne movement of aircraft in controlled or noncontrolled airport terminal areas and about given en route fixes or at other points where counts can be made. There are two types of operations—local and itinerant.
1. Local operations are performed by aircraft which:
  a) Operate in the local traffic patterns or within sight of the airport.
  (b) Are known to be departing for, or arriving from, flight in local practice areas within a 20 mile radius of the airport.
  (c) Execute simulated instrument approaches or low passes at the airport.
2. Itinerant operations are all aircraft operations other than local operations.

aircraft repair. The process of restoring aircraft or aircraft material after damage or wear to a serviceable condition.

aircraft rocket. A rocket-powered missile carried by, and launched from, an aircraft. It may be guided or unguided.

aircraft scrambling. (military) Directing the immediate takeoff of aircraft from a ground alert condition of readiness.

aircraft tiedown. Securing aircraft when parked in the open to restrain movement due to the weather or condition of the parking area. See **tie-down point.**

aircraft turn around. The process of replenishing an aircraft with consumable or expendable stores and equipment so as to render it fit for immediate operational readiness.

aircraft type. See **type.**

aircraft utilization. Average number of hours during each 24-hour period that an aircraft is actually in flight.

aircraft vectoring. The directional control of in-flight aircraft through transmission of azimuth headings.

air cushion vehicle. See **hovercraft.**

air data computer (ADC). See **central air data computer (CADC).**

air defense. All measures designed to reduce or nullify the effectiveness of hostile acts by vehicles (including missiles) in the earth's envelope of atmosphere. See also **active air defense; passive air defense.**

air defense artillery. Weapons and equipment for actively combating air targets from the ground. Weapons are classed as:
  light—20-57mm
  medium—58-99mm
  heavy—100mm or greater.

Air Defense Command (ADC): a United States Air Force command charged with the issuance of North American

Air Defense Command policies for United States Air Force participation in air defense.

air defense control center. The principal information, communications, and operations center from which all aircraft, antiaircraft operations, air defense artillery, guided missiles, and air warning functions of a specific area of air defense responsibility are supervised and coordinated. See also **combat information center.**

air defense early warning. Early notification of approach of enemy airborne weapons or weapons carriers obtained by electronic or visual means.

air defense emergency. An emergency condition, declared or confirmed by either the Commander in Chief, North American Air Defense Command or Commander in Chief, Continental Air Defense Command, or higher authority, which exists when attack upon the continental United States, Alaska, Canada, or United States installations in Greenland by hostile aircraft or missiles is considered probable, is imminent, or is taking place.

air defense identification zone (ADIZ). The area of airspace over land or water within which the ready identification, the location, and the control of aircraft are required in the interest of national security.

air defense operations area. An area and the airspace above it within which procedures are established to minimize mutual interference between air defense and other operations and which may include designation of one or more of the following: air defense action area, air defense identification zone, firepower umbrella, and/or positive identification and radar advisory zone. See also **air defense action area; air defense identification zone; firepower umbrella.**

air defense readiness. An operational status requiring air defense forces to maintain higher than ordinary preparedness for short periods of time.

air defense warning conditions. A degree of air raid probability according to the following code. The term air defense division/sector referred to herein may include forces and units afloat and/or deployed to forward areas, as applicable. **Air defense warning yellow**—attack by hostile aircraft and/or missiles is probable. This means that hostile aircraft and/or missiles are en route toward an air defense division/sector, or unknown aircraft and/or missiles suspected to be hostile are en route toward or are within an air defense division/sector. **Air defense warning red**—attack by hostile aircraft and/or missiles is imminent or is in progress. This means that hostile aircraft and/or missiles are within an air defense division/sector or are in the immediate vicinity of an air defense division/sector with high probability of entering the division/sector. **Air defense warning white**—attack by hostile aircraft and/or missiles is improbable. May be called either before or after air defense warning yellow or red. The initial declaration of air defense emergency will automatically establish a condition of air defense warning other than white for purposes of security control of air traffic.

air delivery. (military) See **airdrop; air landed; air movement; air supply.**

air delivery equipment. (military) Special items of equipment, such as parachutes, air delivery containers, platforms, tie downs, and related items used in air delivery of personnel supplies, and equipment.

air division. A unit or its headquarters, on a level of command above wing level, composed of two or more combat wings, but sometimes adapted to other organizational structures.

air drag. The drag exerted by air particles upon a moving object.

airdrop. (military) The unloading of personnel or materiel from aircraft in flight. See also **air movement; free drop; free fall.**

air evacuation. (military) Evacuation by aircraft of personnel and cargo.

air facility. (military) An installation from which air operations may be or are being conducted. See also **facility.**

airfield. An area prepared for the accommodation, (including any buildings, installations, and equipment), landing and take-off of aircraft. See also **alternative airfield; departure airfield; main airfield; redeployment airfield.** The term airfield is normally used to describe a U. S. military facility whereas airport is a U.S. civil facility. Aerodrome is an international term.

airflow. A flow or stream of air. An airflow may take place in a wind tunnel, in the induction system of an engine, etc., or a relative airflow can occur, as past the wing or other parts of a moving craft; a rate of flow, measured by mass or volume per unit of time. See **flow.**

airfoil. A structure, piece, or body, originally likened to a foil or leaf in being wide and thin, designed to obtain a useful reaction of itself in its motion through the air.

Air Force. See **military department.**

Air Force base. (AFB) An air base for support of Air Force units consisting of landing strips and all components or related facilities for which the Air Force has operating responsibility, together with interior lines of communications and the minimum surrounding area required for local security. (Normally, not greater than an area of 20 square miles.) See also **base complex.**

airframe. The assembled structural and aerodynamic components of an **aircraft** or **rocket vehicle** that support the different systems and subsystems integral to the vehicle.

The word *airframe,* a carryover from aviation usage, remains appropriate for rocket vehicles since a major function of the airframe is performed during flight within the atmosphere.

air freight terminal. A facility which provides administrative functions and space for intransit storage: the receipt and processing of originating, terminating, and in-transit air cargo; and the marshalling, manifesting, and forwarding of air cargo to destination of either domestic or overseas bases.

airhead. 1. A designated area in a hostile or threatened territory which, when seized and held, insures the continuous air landing of troops and materiel and provides maneuver space necessary for projected operations. Normally, it is the area seized in the assault phase of an airborne operation. **2.** A designated location in an area of operations used as a base for supply and evacuation by air.

air intercept control common. A tactical air-to-ground radio frequency monitored by all air intercept control facilities within an area, which is used as a backup for other discrete tactical control frequencies.

air interception. To effect visual or electronic contact by a friendly aircraft with another aircraft. Normally the air intercept is conducted in the following five phases:

a. climb phase—Airborne to cruising altitude;

b. maneuver phase—Receipt of initial vector to target until beginning transition to attack speed and altitude;

c. transition phase—Increase or decrease of speed and altitude required for the attack;

d. attack phase—Turn to attack heading, acquisition of target, completion of attack and turn to breakaway heading; and

e. recovery phase—Breakaway to landing. See also **broadcast controlled air interception; close controlled air interception.**

air intercept missile (AIM). A missile launched from an aircraft against an airborne target.

air intercept rocket (AIR). A rocket launched from an aircraft against an airborne target.

air intercept zone. A subdivided part of the destruction area in which it is planned to destroy or defeat the enemy airborne threat with interceptor aircraft. See also **destruction area.**

air interdiction. Air operations conducted to destroy, neutralize, or delay the enemy's military potential before it can be brought to bear effectively against friendly forces, at such distance from friendly forces that detailed integration of each air mission with the fire and movement of friendly forces is not required.

air landed. (military) Moved by air and disembarked, or unloaded, after the aircraft has landed or while a helicopter is hovering.

air-launch. To launch from an aircraft in the air, as, to air-launch a guided missile, an aerial target, an aircraft, etc.

air-launched ballistic missile. An air-launched ballistic missile launched from an airborne vehicle.

air liaison officer. An officer (aviator/pilot) attached to a ground unit who functions as the primary advisor to the ground commander on air operation matters.

airlift. (military) **1.** The total weight of personnel and/or cargo that is, or can be, carried by air, or that is offered for carriage by air. **2.** To transport passengers and cargo by use of aircraft. **3.** The carriage of personnel and/or cargo by air.

airlift force. Includes military strategic and tactical airlift aircraft augmented by the Civil Reserve Air Fleet.

airlift service. The performance or procurement of air transportation and services incident thereto required for the movement of persons, cargo, mail, or other goods.

airline. Same as **air carrier.**

airliner. A large transport type airplane.

airline transport pilot. An airline transport pilot has the privileges of a **commercial pilot** with an **instrument rating.** See **airline transport pilot certificate.**

airline transport pilot certificate. A certificate of competency issued by the FAA to a pilot meeting the requirements of the applicable Federal Aviation Regulations.

air lock. 1. A stoppage or diminution of flow in a fuel system, hydraulic system, or the like, caused by a pocket of air or vapor. **2.** A chamber capable of being hermetically sealed that provides for passage between two places of different pressure, as between an altitude chamber and the outside atmosphere.

air logistic support. Support by air landing or airdrop including air supply, movement of personnel, evacuation of casualties and prisoners of war, and recovery of equipment and vehicles.

air mass. A meteorological term for a body of air which may extend hundreds of miles horizontally and have a great height, extending sometimes to the stratosphere, and maintaining practically uniform temperature and humidity at any given level during its movement over a continent or an ocean. Examples are a **tropical air mass** and a **polar air mass.**

airman certificates. The **FAA** issues a **pilot certificate** to those who are able to pass appropriate medical, written and flight examinations. There are different classes of examinations, depending on whether a pilot wants to fly by referring to the ground (under **Visual Flight Rules,** or **VFR** as it is called), to fly by instruments (**Instrument Flight Rules** or IFR), to be a flight instructor or to fly for hire (a crop duster, business pilot or airline pilot, for example). The pilot of an airliner must pass the most demanding written, flight and physical examinations.

The law requires that other airmen (such as aviation mechanics, flight engineers and parachute riggers) also meet strict standards and be able to pass practical and written examinations required by the **FAA.**

As defined by the **FAA** an airman is a **flight crew member (pilot flight engineer and flight navigator), aviation mechanic, repairman, parachute rigger, air traffic control tower operator, aircraft dispatcher.** See also **medical certificate.**

air mass weather. Weather which is typical of a given **air mass** but which is not typical of a given **front.**

AIRMET. See **SIGMET** and **AIRMET.**

air mission. See **mission.**

air movement. (military) Air transport of units, personnel, supplies, and equipment including airdrops and air landings. See also **airdrop; free drop.**

air munitions. All munitions such as bombs, dispenser munitions, rockets, missiles, flares, incendiary bombs, ammunition, aerial land mines, etc., which are delivered to the target by an aerospace vehicle.

Air National Guard of the United States. All Federally recognized units, organizations, and members of the Air National Guard. In addition to their status as members of the Air National Guard, these persons are Reserves of the Air Force in the same commissioned, warrant officer, or airman grades they hold in the Air National Guard of the several states, Commonwealth of Puerto Rico or District of Columbia.

air navigation. See **navigation.**

air navigation facility (NAVAID). Any facility used in, available for use in, or designed for use in aid of air navigation, including landing areas, lights, any apparatus or equipment for disseminating weather information, for signaling, for radio direction-finding, or for radio or other electronic communication, and any other structure or mechanism having a similar purpose for guiding or controlling flight in the air or the landing or takeoff of aircraft. Also see **beacon, consolan, DME, ILS, loran, NDB, tacan, VOR, VORTAC.**

air observer. An individual whose primary mission is to observe or take photographs from an aircraft in order to adjust artillery fire or obtain military information.

air offensive. Sustained operations by strategic and/or tactical air weapon systems against hostile air forces or surface targets.

airpark. A little used term for an **airport** at which flight operations of small aircraft are conducted.

air photographic reconnaissance. The obtaining of infor-

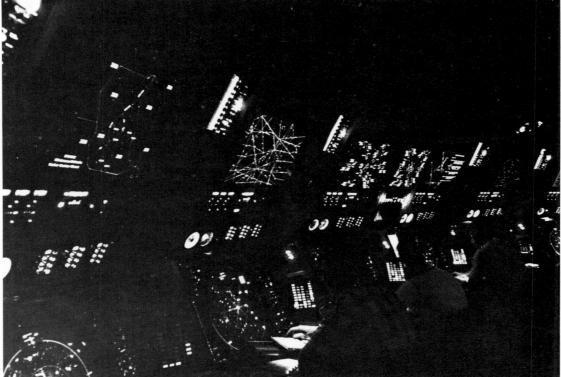

Airport surveillance radar (ASR) *is used for arrival and departure air traffic control in the vicinity of a busy terminal airport.*

mation by air photography—divided into three types: **a.** strategic photographic reconnaissance; **b.** tactical photographic reconnaissance; and **c.** survey/cartographic photography—air photography taken for survey/cartographic purposes and to survey/cartographic standards of accuracy. It may be strategic or tactical.

air pickets. Airborne early warning aircraft disposed around a position, area, or formation primarily to detect, report, and track approaching enemy aircraft and to control intercepts. See also **airborne early warning and control.**

airplane. A mechanically driven fixed-wing aircraft, heavier than air, which is supported by the dynamic reaction of the air over its wing surfaces.

airplane flight manual. A manual, associated with the **certificate of airworthiness,** containing limitations within which the airplane is to be considered airworthy, and instructions and information necessary to the flight crew members for the safe operation of the airplane.

air plot. 1. A continuous plot used in air navigation of a graphic representation of true headings steered and air distances flown. **2.** A continuous plot of the position of an airborne object represented graphically to show true headings steered and air distances flown. **3.** Within ships, a display which shows the positions and movements of an airborne object relative to the plotting ship.

airport. An area of land or water that is used or intended to be used for the landing and takeoff of aircraft, and includes its buildings and facilities, if any. The term airport is normally used to describe a U.S. civil facility whereas airfield is a U.S. military facility. Aerodrome is an international term.

airport advisory area. The area within five statute miles of an uncontrolled airport on which is located a Flight Service Station so depicted on the appropriate Sectional Aeronautical Chart.

airport advisory service. A service provided by a Flight Service Station to enhance the safety of terminal operations of airports where a station is operating but where there is no control tower.

airport beacon. See **aeronautical light beacon.**

airport of entry: See **international airport.**

airport operation: a landing or a takeoff at the airport at which the facility is located. (A low approach below traffic pattern altitudes or a touch-and-go operation shall be counted as both a landing and a takeoff; i.e., two operations)

airport surface detection equipment (ASDE). Short range radar displaying the airport surface. Aircraft and vehicular traffic operating on runways, taxiways, and ramps, moving or stationary, may be observed with a high degree of resolution.

airport surveillance radar (ASR). Radar providing position of aircraft by azimuth and range data without elevation data. It is designed for a range of 50 miles. Used for Terminal approach and departure control.

airport traffic area. Unless otherwise specifically designated, that airspace within a horizontal radius of five statute miles from the geographical center of any airport at which a control tower is operating, extending from the surface up to, but not including, 3000 feet above the surface.

airport traffic control service. Air traffic control service provided by an airport traffic control tower for aircraft operating on the movement area and in the vicinity of an airport.

airport traffic control tower (TOWER). A facility providing airport traffic control service.

air position. The calculated position of an aircraft assuming no wind effect.

air position indicator. An airborne computing system which presents a continuous indication of the aircraft position on the basis of aircraft heading, airspeed, and elapsed time.

air reconnaissance. The acquisition of intelligence information employing visual observation and/or sensors in air vehicles.

Air Reserve Forces. All units, organizations, and members of the Air National Guard of the United States and the United States Air Force Reserve.

Air Reservist. A member of the Air Reserve Forces, i.e., Air National Guard of the United States or Air Force Reserve.

air route. The navigable airspace between two points, identified to the extent necessary for the application of flight rules.

air route surveillance radar (ARSR). Long range radar which increases the capability of ATC for handling heavy en route traffic. An ARSR site is usually located at some distance from the ARTCC it serves. Range, approximately 200 NM.

air route traffic control center (ARTCC) (CENTER). A facility established to provide air traffic control service to aircraft operating on an IFR flight plan within controlled airspace and principally during the en route phase of flight.

air-sea rescue. See **search and rescue.**

airscoop. A hood or open end of an air duct or a similar structure, projecting into the **airstream** about a vehicle in such a way as to utilize the motion of the vehicle in capturing air to be conducted to an engine, a ventilator, etc.

airship (Z), (military designation). A self-propelled lighter-than-air aircraft. See **military aircraft types.**

airship. An aerostat provided with a propelling system and with means of controlling the direction of motion.

air shower. A grouping of cosmic-ray particles observed in the atmosphere; a **cascade shower** in the **atmosphere.** Also called *shower.*

airsickness. Motion sickness occurring in flight.

air sounding. The act of measuring atmospheric phenomena or determining atmospheric conditions at altitude, especially by means of apparatus carried by balloons or rockets. See **sounding.**

airspace. Specifically, the **atmosphere** above a particular portion of the earth, usually defined by the boundaries of an area on the surface projected upward.

Airspace is sometimes particularized by altitude, as the airspace above 20,000 feet.

airspace reservation. The airspace located above an area on the surface of the land or water, designated and set apart by Executive Order of the President or by a state, commonwealth, or territory, over which the flight of aircraft is prohibited or restricted for the purpose of national defense or for other governmental purposes.

airspeed. The speed of an aircraft relative to its surrounding air mass. The unqualified term "airspeed" can mean any one of the following:

    a. calibrated airspeed—Indicated airspeed corrected

for instrument installation error. Also called true in-dicated airspeed.

    b. equivalent airspeed—Calibrated airspeed corrected for compressibility error.

    c. indicated airspeed—The airspeed shown by an air-speed indicator.

    d. true airspeed—Equivalent airspeed corrected for error due to air density (altitude and temperature).

air spot. The correcting adjustment of gunfire based on air observation.

air staging unit. (military) A unit situated at an airfield and concerned with the reception, handling, servicing, and preparation for departure of aircraft and control o personnel and cargo.

airstart. An act or instance of starting an aircraft's engine while in flight, especially a **jet engine** after flameout. Compare **in-flight start, ground start.**

airstream. airflow.

air strike. An attack on specific objectives by fighter, bomber, or attack aircraft on an offensive mission. May consist of several air organizations under a single command in the air.

air strike coordinator. The air representative of the force commander in a target area. He is responsible for direc-ting all aircraft in the target area and coordinating their efforts to achieve the most effective use of air striking power.

air strip. An unimproved surface which has been adapted for takeoff or landing of aircraft, usually having minimum facilities. See also **airfield, airport.**

air superiority. That degree of dominance in the air battle of one force over another which permits the conduct of operations by the former and its related land, sea, and air forces at a given time and place without prohibitive interference by the opposing force.

air supply. (military) The delivery of cargo by airdrop or air landing.

air support. All forms of support given by air forces to forces on land or sea. See also **close air support; tactical air support.**

air support radar team. A subordinate operational com-ponent of a tactical air control system which provides ground controlled precision flight path guidance and weapons release.

air supremacy. That degree of air superiority wherein the opposing air force is incapable of effective interference.

air surveillance. (military) The systematic observation of airspace by electronic, visual, or other means primarily for the purpose of identifying and determining the movements of aircraft and missiles, friendly and enemy, in the airspace under observation. See also **sur-veillance.**

air survey/cartographic photography. The taking and processing of air photographs for mapping and charting purposes.

air terminal. An installation provided with the facilities for loading and unloading aircraft and the intransit handling of traffic (passengers, cargo, and mail) which is moved by aircraft.

air-to-air missile (AAM). A missile launched from an air-borne carrier at a target above the surface.

air-to-surface missile (ASM). A missile launched from an airborne carrier to impact on a surface target.

air traffic. Aircraft operating in the air or on an airport surface, exclusive of loading ramps and parking areas.

air traffic clearance (CLEARANCE). An authorization by air traffic control for the purpose of preventing colli-sion between known aircraft, for an aircraft to proceed under specified traffic conditions within controlled air-space.

air traffic control (ATC). A service provided by the FAA to promote the safe, orderly, and expeditious flow of air traffic along the nation's airways and within all other controlled airspace.

air traffic control radar beacon system (ATCRBS). The form of **secondary radar** used in the air traffic control system for control purposes. It consists of coded signal initiated from the ground coupled with a similar reply from the aircraft to provide accurate identification and position information. Also see **radar, transponder, IFF, mode, interrogator.**

air traffic control service (CONTROL). A service provid-ed for the purpose of promoting the safe, orderly, and expeditious flow of air traffic, including airport, ap-proach, and enroute air traffic control service.

air traffic controller. An air controller especially trained for and assigned to the duty of airspace management and traffic control of airborne objects. See also **air con-troller, airport control tower operator.**

air traffic control tower operator. A person assigned to an **airport traffic control tower** to provide for the safe, orderly and expeditious flow of traffic in accordance with the appropriate regulations and procedures. See **air traffic control tower operator certificate, air traffic controller.**

air traffic control tower operator certificate. A certificate of competency issued by the FAA to a person meeting the requirements of the applicable **Federal Aviation Regulations.** Also see **air traffic control tower operator, airmen certificates.**

air traffic control specialist. A person responsible for providing air traffic control services. Commonly referred to as a controller. See **air controller.**

air transportation. Interstate, overseas, or foreign air transportation or the transportation of mail by aircraft.

air vane. A vane that acts in the air, as contrasted to a **jet vane** which acts within **a jetstream.** See **control vane.**

air vehicle. aircraft.

airway. A control area or portion thereof established in the form of a corridor marked with radio navigation aids. Also see **Federal Airways System.**

airworthiness certificate. An airworthiness certificate is issued by a representative of the **Federal Aviation Ad-ministration** after the aircraft has been inspected and it is found that it meets the requirements of the **Federal Aviation Regulations (FARs),** and is found to be in a condition for safe operation. The certificate must be displayed in the aircraft according to applicable regulations.

airworthy. The status of being in condition suitable for safe flight.

Aitken dust counter. An instrument developed by John Aitken for determining the dust content of the at-mosphere. In operation, a sample of air is mixed, in an expandable chamber, with a larger volume of dust-free air containing water vapor. Upon a sudden expansion, the chamber cools adiabatically below its dewpoint, and droplets form with the dust particles as nuclei (**Aitken nuclei**). A portion of these droplets settle on a ruled plate in the instrument and are counted with the aid of a microscope. Also called *Aitken nucleus counter.*

Aitken nuclei. The microscopic particles in the atmosphere which serve as **condensation nuclei** for droplet growth during the rapid **adiabatic** expansion produced by an **Aitken dust counter.** These nuclei are both solid and liquid particles whose diameters are of the order of tenths of microns or even smaller.

The Aitken nuclei play an important role in atmospheric electrical processes, for they are the particles which capture (by absorption or other surface electrical processes) small ions and thereby form large ions. In air containing large numbers of Aitken nuclei, the **small ion** population is small, the **large ion** population is large, and the air conductivity is low.

albedo. The ratio of the amount of **electromagnetic radiation** reflected by a body to the amount incident upon it, often expressed as a percentage, as, *the albedo of the earth is 34%.*

albedometer. An instrument used for the measurement of the reflecting power, the **albedo,** of a surface.

A **pyranometer** adapted for the measurement of radiation reflected from the earth's surface is sometimes employed as an albedometer.

alert. 1. Readiness for action, defense, or protection. **2.** A warning signal of a real or threatened danger, such as an air attack. **3.** The period of time during which troops stand by in response to an alarm. **4.** To forewarn; to prepare for action. See also **airborne alert; air defense warning conditions; ground alert.**

alert area. Airspace which may contain a high volume of pilot training activities or an unusual type of aerial activity.

Alford loop. A multielement antenna, having approximately equal and in-phase currents uniformly distributed along each of its peripheral elements, producing a substantially circular radiation pattern in the plane of polarization.

alga (*plural,* **algae**). Any plants of a group of unicellular and multicellular primitive organisms that include the Chlorella, Scenedesmus, and other genera.

The green algae and blue-green algae, for example, provide a possible means of photosynthesis in a **closed ecological system,** also a source of food.

algorism. The art or system of calculating with any species of notation, as in arithmetic with nine figures and a zero. Also called *algorithm.*

Different algorisms have been used in the design of computing machines.

algorithm. 1. A special mathematical procedure for solving a particular type of problem. **2. algorism.**

alkali metal. A metal in group IA of the periodic system; namely, lithium, sodium, potassium, rubidium, cesium, and francium.

Alkali metals are being considered as coolants (in liquid state) for nuclear reactors for spacecraft. See **liquid-metal corrosion.**

all burnt. The time at which a rocket consumes its propellants. See **burnout,** note.

all-inertial guidance. The guidance of a rocket vehicle entirely by use of inertial devices; the equipment used for this.

allocation. (military) The designation of specific numbers and types of aircraft sorties for use during a specified time period or for carrying out an assigned task.

all out war. Not to be used. See **general war.**

alloying element. An element added to a metal to affect changes in properties and which remains within the metal.

all weather fighter. A **fighter aircraft** with radar devices and other special equipment which enable it to intercept its target in dark or daylight weather conditions which do not permit visual interception.

alphanumeric (*alphabet* and *numeric*). Including letters and digits.

alphanumeric display. Use of letters of the alphabet and numerals to show altitude, beacon code, and other information about a target on a radar display.

alpha particle. A positively charged particle emitted by certain radioactive materials. It is made up of two neutrons and two protons bound together, hence is identical with the nucleus of a helium atom. It is the least penetrating of the three common types of *radiation* (alpha, beta, gamma) emitted by radioactive material, being stopped by a sheet of paper. It is not dangerous to plants, animals or man unless the alpha-emitting substance has entered the body. (See *decay, radioactive.*)

alpha ray. A stream of alpha particles. Loosely, a synonym for *alpha particle.*

alternative airfield. (military) An airfield with minimum essential facilities for use as an emergency landing ground, or when main or redeployment airfields are out of action, or as required for tactical flexibility. See also **airfield.**

alternate airport. An airport specified on a flight plan to which a flight may proceed when a landing at the point of first intended landing becomes inadvisable.

alternate configuration. The configuration of a weapon system which gives it a capability to perform other than its primary mission.

alternate fuels. As alternate fuels to petroleum based Jet-A fuel (conventional kerosene), there are a number of synthetic fuels which can be produced from United States energy resources. Based on NASA studies, a number of synthetic fuels appear suitable for application to aviation. These are: synthetic Jet-A produced from coal and shale, liquid methane (LCH_4) from coal, and liquid hydrogen (LH_2) from coal. The studies have been limited to production from coal since it is the largest fossil fuel resource in the United States.

altimeter. An instrument for measuring height above a reference datum; specifically, an instrument similar to an **aneroid** barometer that utilizes the change of atmospheric pressure with altitude to indicate the approximate elevation above a given point or plane used as a reference.

altimeter, classifications.
    absolute,
    aneroid
    ground avoidance radar
    pressure
    radar
    radio

altimeter setting. Pressure in inches of mercury converted to make the altimeter read zero elevation at an altitude of 10 feet (average cockpit height) above mean sea level, or to read field elevation 10 feet above the runway.

altitude. 1. In astronomy, angular displacement above the horizon; the arc of a verticle circle between the horizon and a point on the **celestial sphere,** measured upward from the horizon.

Angular displacement below the horizon is called *negative altitude* or *dip*. See **horizon system.**

2. Height, especially radial distance as measured above a given datum, as average sea level.

In space navigation *altitude* designates distance from the mean surface of the reference body as contrasted to *distance,* which designates distance from the center of the reference body.

See **absolute altitude, density altitude, indicated altitude, pressure altitude, true altitude.**

altitude acclimatization. A physiological adaptation to reduced atmospheric and oxygen pressure.

altitude chamber. A chamber within which the air pressure, temperature, etc., can be adjusted to simulate conditions at different altitudes; used for experimentation and testing.

altitude datum. The arbitrary level from which vertical displacement is measured. The datum for height measurement is the terrain directly below the aircraft or some specified datum other than mean sea level; for pressure altitude, the level at which the atmospheric pressure is 29.92 inches of mercury (1013.2 m.bs); and for true altitude, mean seal level.

altitude difference. In navigation, the difference between computed and observed altitudes, or between precomputed and sextant altitudes.

It is labeled **T** (toward) or **A** (away) as the observed (or sextant) altitude is greater or smaller than the computed (or precomputed) altitude. Also called *altitude intercept, intercept.*

altitude intercept. altitude difference. Often shortened to *intercept.*

altitude sickness. In general, any sickness brought on by exposure to reduced oxygen tension and barometric pressure.

altitude wind tunnel. A **wind tunnel** in which the air pressure, temperature, and humidity can be varied to simulate conditions at different altitudes.

In an altitude wind tunnel for testing engines, provision is made for exchanging fresh air for exhaust-laden air during operation.

altocumulus. A form of middle cloud. Also see **cumuliform clouds.**

altostratus. A form of middle cloud. Also see **stratiform clouds.**

AM. amplitude modulation.

ambient. Surrounding; especially, of or pertaining to the environment about a flying aircraft or other body but undisturbed or unaffected by it, as in *ambient* air, or *ambient temperature.*

ambient noise. The pervasive **noise** associated with a given environment, being usually a composite of sounds from sources both near and distant.

ambiguity. In navigation, the condition obtaining when a given set of observations defines more than one point, direction, line of position, or surface of position.

American Ephemeris and Nautical Almanac. An annual publication of the U. S. Naval Observatory, containing elaborate tables of the predicted positions of various celestial bodies and other data of use to astronomers and navigators.

Beginning with the editions for 1960, The American Ephemeris and Nautical Almanac issued by the Nautical Almanac Office, United States Naval Observatory, and The Astronomical Ephemeris issued by H. M. Nautical Almanac Office, Royal Greenwich Observatory, were unified. With the exception of a few introductory pages, the two publications are identical; they are printed separately in the two countries, from reproducible material prepared partly in the United States of America and partly in the United Kingdom.

American Nautical Almanac. See **Nautical Almanac.**

ammunition. A contrivance charged with explosives, propellants, pyrotechnics, initiating composition, or nuclear, biological, or chemical material for use in connection with defense or offense including demolitions. Certain ammunition can be used for training, ceremonial, or nonoperational purposes.

amphibian. An airplane designed to rise from and alight on either land or water.

amplidyne. A special type of direct current generator used as a power **amplifier** in which the output voltage responds to changes in field excitation; used extensively in **servo systems.**

amplifier. A device which enables an input signal to control a source of power, and this is capable of delivering at its output an enlarged reproduction of the essential characteristics of the signal.

Typical amplifying elements are electron tubes, transistors, and magnetic circuits.

amplitude. 1. The maximum value of the displacement of a **wave** or other periodic phenomenon from a reference position. **2.** Angular distance north or south of the **prime vertical;** the arc of the horizon, or the angle at the **zenith** between the prime vertical and a **vertical circle,** measured north or south from the prime vertical to the vertical circle.

The term is customarily used only with reference to bodies whose centers are on the celestial horizon, and is prefixed E or W, as the body is rising or setting, respectively; and suffixed N or S to agree with the declination. The prefix indicates the origin, and the suffix indicates the direction of measurement. Amplitude is designated as *true, magnetic, compass,* or *grid* as the reference direction is true, magnetic, compass, or grid east or west, respectively.

amplitude-modulated indicator. One of two general classes of **radar indicators,** in which the sweep of the electron beam is deflected vertically or horizontally from a base line to indicate the existence of an **echo** from a target. The amount of deflection is usually a function of the echo signal strength. Also called *deflection-modulated indicator.* Compare **intensity modulated indicator.**

amplitude modulation. (A.M.) The usual form of radio broadcasting in which the amplitude change of a carrier wave corresponds to power variations of the transmitted signal.

AMR. Atlantic Missile Range.

anacoustic zone. The zone of silence in space; the region above 100 miles altitude where the distance between the rarefied air molecules is greater than the wave length of sound, and sound waves can no longer be propagated.

analog. In computers, pertaining to the use of physical variables such as voltage, distance, rotation, etc. To represent numerical variables as in *analog computer, analog output.* Compare **digital.**

analog computer. A computing machine working on the principle of measuring, as distinguished from counting, in which the input data is analogous to a measurement continuum, such as linear lengths, voltages, resistances, etc., which can be manipulated by the computer.

Analog computers range in complexity from a slide rule to electrical computers used for solving mathematical problems.

analog output. Transducer output in which the amplitude is continuously proportional to a function of the stimulus. Distinguished from *digital output.*

analog to digital conversion. A process by which a sample of **analog** information is transformed into a **digital** code.

analog to digital converter. A device which will convert an **analog** voltage sample to an equivalent **digital** code of some finite resolution. Also called *digitizer, encoder.*

analysis. A stage in the intelligence cycle in which information is subjected to review in order to identify significant facts and derive conclusions therefrom.

analytical photography. Photography, either motion picture or still, accomplished to determine (by qualitative, quantitative, or any other means) whether a particular phenomenon does or does not occur. See **technical photography.**

Differs from **metric photography** in that measurements are not a prime requisite.

AND. In **Boolean algebra,** the operation of **intersection.**

AND gate, and gate. A circuit or device used in computers whose output is energized only when every input is in its prescribed state. It performs the logical function of the *AND,* the Boolean operation of **intersection.** Also called *intersector, AND circuit.*

anemometer. An instrument which measures and indicates the speed of the wind, in common use at airports.

aneroid. A thin, disk-shaped box or capsule, usually metallic, partially evacuated of air and sealed, which expands and contracts with changes in atmospheric or gaseous **pressure.**

The aneroid is the sensing and actuating element in various meters or gages, such as barometers, altimeters, manifold-pressure gages, etc; it is also the triggering or operating element in various automatic mechanisms.

A device similar to an aneroid, but open to outside pressures, such as the capsule in an airspeed indicator, is not commonly called an *aneroid.*

aneroid altimeter. A common type **altimeter,** the indications of which depend on the deflection of a pressure-sensitive element (**aneroid barometer**). The graduations of the dial correspond to an empirical or arbitrary pressure-temperature-altitude formula.

aneroid barometer. A barometer constructed around an evacuated cell or chamber. The difference between atmospheric pressure and that contained in the cell tends to collapse the cell. An increase or decrease of this pressure difference will tend to collapse or expand the cell. Through proper linkage a needle attached to the side of this cell is actuated by the movement. The needle, moving past a properly calibrated dial, indicates changes in pressure.

angel. A radar **echo** caused by a physical phenomenon not discernible to the eye.

angels (military). Aircraft altitude (in thousands of feet).

angle. The inclination to each other of two intersecting lines, measured by the arc of a circle intercepted between the two lines forming the angle, the center of the circle being the point of intersection.

An acute angle is less than 90°; a right angle, 90°; an obtuse angle, more than 90° but less than 180°; a straight angle, 180°; a reflex angle, more than 180° but less than 360°; a perigon, 360°. Any angle not a multiple of 90° is an oblique angle. If the sum of two angles is 90°, they are complementary angles; if 180°, supplementary angles; if 360°, explementary angles. Two adjacent angles have a common vertex and lie on opposite sides of a common side. A dihedral angle is the angle between two intersecting planes. A spherical angle is the angle between two intersecting great circles.

angled deck. A flight deck arranged for the landing of aircraft at an angle, in the horizontal plane, to the fore-and-aft axis of an aircraft carrier.

angle of attack. The angle between the chord of a wing or the reference line in a body and the direction of the undisturbed flow or **relative wind** in the absence of sideslip.

angle-of-attack indicator. A device for displaying the angle-of-attack of the wing of an airplane to the pilot.

angle of climb. The angle between the **flight path** of a climbing vehicle and the local horizontal.

angle of depression. The angle in a vertical plane between the local horizontal and a descending line. Also called *depression angle.* See **angle of elevation.**

angle of descent. The angle between the **flightpath** of a descending vehicle and the local horizontal.

angle of elevation. The angle in a vertical plane between the local horizontal and an ascending line, as from an observer to an object. Also called *elevation angle.*

A negative angle of elevation is usually called an *angle of depression.*

angle of pitch. 1. The angle, as seen from the side, between the longitudinal **body axis** of an aircraft or similar body and a chosen reference line or plane, usually the horizontal plane. This angle is positive when the forward part of the longitudinal axis is directed above the reference line. **2.** Same as **blade angle** (in all senses).

angle of incidence. 1. The angle at which a ray of energy impinges upon a surface, usually measured between the direction of propagation of the energy and a perpendicular to the surface at the point of impingement, or incidence. Compare **angle of arrival.** See also **angle of reflection, angle of refraction.**

In some cases involving radio waves, the angle of incidence is measured relative to the surface.

2. A fixed angle between the plane of the wing chord and the line of thrust or any other longitudinal line which is level when the fuselage is level longitudinally. The angle of incidence is always a fixed angle except in a few rare cases where variable incidence wings have been tried.

angle of reflection. The angle at which a reflected ray of energy leaves a reflecting surface, measured between the direction of the outgoing ray and a perpendicular to the surface at the point of reflection. Compare **angle of incidence.**

In some cases involving radio waves, the angle of reflection is measured relative to the surface.

angle of refraction. The angle at which a refracted ray of energy leaves the interface at which the **refraction** occurred, measured between the direction of the refracted ray and a perpendicular to the **interface** at the point of refraction.

angle of roll. The angle that the lateral body axis of an aircraft or similar body makes with a chosen reference plane in rolling; usually, the angle between the lateral axis and a horizontal plane. The angle of roll is considered positive if the roll is to starboard.

angle of yaw. The angle, as seen from above, between the longitudinal **body axis** of an aircraft, rocket, or the like and a chosen reference direction. This angle is positive when the forward part of the longitudinal axis is directed to starboard. Also called *yaw angle.*

angstrom. A unit of length, used in measuring electromagnetic radiation, equal to 10-ˣcentimeter. Named for A. J. Angstrom, Swedish spectroscopist.

angular acceleration. The rate of change of **angular velocity** per unit of time.

angular distance. 1. The angular difference between two directions, numerically equal to the angle between two lines extending in the given directions. **2.** The arc of the **great circle** joining two points, expressed in angular units. **3.** Distance between two points, expressed in **wave lengths** at a specified **frequency.**

angular resolution. Specifically, the ability of a radar to distinguish between two targets solely by the measurement of angles.

It is generally expressed in terms of the minimum angle by which targets must be spaced to be separately distinguishable. See **resolution.**

angular velocity. Angular velocity is the rate of changing direction or the angle through which any radius of a rotating body turns in a unit of time. It may be expressed in revolutions per minute (R.P.M.), degrees per second, or **radians** per second.

annular. Pertaining to an annulus or ring; ring shaped.

annular eclipse. An **eclipse** in which a thin ring of the source of light appears around the obscuring body.

anode. The **positive pole** or terminal of an electrical unit. An **electrode** to which an **electron** stream flows.

anomalistic month. The average period of revolution of the moon from **perigee** to perigee, a period of 27 days 13 hours 18 minutes 33.2 seconds.

anomalistic period. The interval between two successive **perigee** passages of a **satellite** in orbit about a primary. Also called *perigee-to-perigee period.*

anomalistic year. The period of one revolution of the earth about the sun from **perihelion** to perhihelion; 365 days 6 hours 13 minutes 53.0 seconds in 1900 and increasing at the rate of 0.26 second per century.

anomaly. 1. In general, a deviation from the norm. **2.** In geodesy, a deviation of an observed value from a theoretical value, due to an abnormality in the observed quantity. **3.** In celestial mechanics, the angle between the **radius vector** to an orbiting body from its primary (the focus of the orbital ellipse) and the **line of apsides** of the orbit, measured in the direction of travel, from the point of closest approach to the primary (perifocus).

anoxia. An absence of oxygen in the blood cells or tissues of the body. (Often confused with hypoxia.)

antenna. A conductor or system of conductors for radiating or receiving **radio waves.**

antenna array. A system of two or more antennas, often similar, excited by the same source for the purpose of obtaining directional effects.

antenna pattern. Same as **radiation pattern.**

antiaircraft operations center. The tactical headquarters of an antiaircraft commander. The agency provided to collect and evaluate information, and disseminate intelligence for the antiaircraft defense, and through which operational control over subordinate units is exercised.

antiaircraft weapon. See **Duster (antiaircraft weapon).**

antiair warfare. A United States Navy/United States Marine Corps term to indicate that action required to destroy or reduce to an acceptable level the enemy air and missile threat. It includes such measures as the use of interceptors, bombers, antiaircraft guns, surface-to-air and air-to-air missiles, electronic countermeasures, and destruction of the air or missile threat both before and after it is launched. Other measures which are taken to minimize the effects of hostile air action are: cover, concealment, dispersion, deception (including electronic), and mobility. See also **counter air.**

anti-collision light. A high intensity, flashing or condenser discharge light or group of lights mounted on an aircraft to allow it to be seen more easily.

anticrop agent. A living organism or chemical used to cause disease or damage to selected food or industrial crops.

anticrop operations. The employment of anticrop agents in military operations to destroy the enemy's source of selected food or industrial crops.

anticyclone. An area of high barometric pressure and its attendant system of winds. (Winds are clockwise in the Northern Hemisphere and counter-clockwise in the Southern Hemisphere.)

anti-G suit. A device worn by aircrew to counteract the effects on the human body of positive acceleration.

anti-icing. The protection of aircraft against icing by preventing ice formation (e. g. by continuous heating). Compare **de-icing.**

antimateriel agent. A living organism or chemical used to cause deterioration of or damage to selected materiel.

antimateriel operations. The employment of antimateriel weapons or agents in military operations.

antimatter (antiparticles). Matter in which the ordinary nuclear *particles* (neutrons, protons, electrons, etc.) are conceived of as being replaced by their corresponding antiparticles (antineutrons, antiprotons, positrons, etc.). An antihydrogen atom, for example, would consist of a negatively charged antiproton with an orbital positron. Normal matter and antimatter would mutually annihilate each other upon contact, being converted totally into energy. (Compare *matter.*)

antinode. 1. Either of the two points on an **orbit** where a line in the orbit plane, perpendicular to the **line of nodes,** and passing through the **focus,** intersects the orbit. **2.** A point, line, or surface in a standing wave where some characteristic of the wave field has maximum amplitude. Also called *loop.*

In sense **2,** the appropriate modifier should be used before the word *antinode* to signify the type that is intended; e.g., *displacement* antinode, *velocity* antinode, *pressure* antinode.

antiplant agent. A microorganism or chemical which will kill, disease or damage plants. See also **anticrop agent.**

antipode. Anything exactly opposite to something else. Particularly, that point on the earth 180° from a given place.

antiradiation missile. A missile which homes passively on a radiation source.

anti-spin parachute. A parachute attached to an aircraft to assist in its recovery from a spin. Normally used for flight test purposes only.

antisolar point. That point on the **celestial sphere** 180° from the sun.

antisubmarine (S), (military designation). Aircraft designed to search out, detect, indentify, attack, and destroy enemy submarines. See **military aircraft types.**

antisubmarine action. An operation by one or more antisubmarine ships or aircraft, or a combination of the two, against a particular enemy submarine. It begins when contact has been gained by any ship or aircraft of the unit. Any number of antisubmarine attacks may be carried out as part of the action. The action ends when the submarine has been destroyed or when contact has been lost and cannot be regained.

antisubmarine air escort and close support. The provision of air protection to a particular convoy or force threatened by imminent submarine attack. Aircraft provide increased defense in depth and are under the tactical control of the officer in tactical command.

antisubmarine air offensive operations. Carrier-based and shore-based aircraft operated singly and in coordination with other aircraft, ships, or both, to conduct offensive operations. While the purpose of such operations differs fundamentally from that of operations in distant support, the search localization and attack tactics are similar to those in the conduct of antisubmarine air distant support.

antisubmarine operation. Operation contributing to the conduct of antisubmarine warfare.

antisubmarine patrol. The systematic and continuing investigation of an area or along a line to detect or hamper submarines, used when the direction of submarine movement can be established.

antisubmarine search. Systematic investigation of a particular area for the purpose of locating a submarine known or suspected to be somewhere in the area. Some types of search are also used in locating the position of a distress incident.

antisubmarine support aircraft carrier. A ship primarily designed to support and operate aircraft and for sustained antisubmarine warfare and escort convoys. It also may be used to provide close air support. Designated as CVS. These are former CVAs which have been redesignated.

antisubmarine warfare (ASW). Operations conducted with the intention of denying the enemy the effective use of his submarines.

antisubmarine warfare forces. Forces organized primarily for antisubmarine action. May be comprised of surface ships, aircraft, submarines, or any combination of these, and their supporting systems.

anvil cloud. A heavy **cumulus** or **cumulonimbus** with an anvil-like form in its upper portions. The anvil formation usually indicates a thunderstorm in the dissipating stage. Also see **cloud, thunderstorm.**

apareon. The point on a Mars-centered **orbit** where a satellite is at its greatest distance from Mars. *Apareon* is analogous to *apogee.* See **geo.**

apastron. That point of the **orbit** of one member of a **binary star** system at which the stars are farthest apart. That point at which they are closest together is called *periastron.*

aperiodic. Without a period; not cyclic; completely **damped.**

aperture. 1. An opening; particularly, that opening in the front of a camera through which light rays pass when a picture is taken. **2.** The diameter of the **objective** of a telescope or other optical instrument, usually expressed in inches, but sometimes as the angle between lines from the principal focus to opposite ends of a diameter of the objective. **3.** Of a unidirectional **antenna,** that portion of a plane surface near the antenna, perpendicular to the direction of maximum radiation, through which the major part of the radiation passes.

aperture ratio. The ratio of the useful diameter of a lens to its focal length. It is the reciprocal of the **f-number.**

In application to an optical instrument, rather than to a lens, *numerical aperture* is more commonly used. The aperture ratio is then twice the tangent of the angle whose sine is the numerical aperture.

apex of the sun's motion. Same as **solar apex.**

apex of the sun's way. Same as **solar apex.**

aphelion. Point on an elliptical orbit around the Sun which is farthest from the Sun. (The Earth's aphelion is about 94,500,000 miles from the Sun.)

apoapis. That point in an orbit farthest from the center of attraction.

apocynthion. That point in the **orbit** of a moon satellite which is farthest from the moon.

apogee. 1. That point in a **geocentric orbit** which is most distant from the earth. That orbital point nearest the earth is called *perigee.* See **geo.**

By extension, *apogee* and *perigee* are also used in reference to orbits about other planets and natural satellites.

2. Of a satellite or rocket: To reach its apogee (sense **1**), as in *the Vanguard apogees at 2,560 miles.*

apparent position. The position on the **celestial sphere** at which a heavenly body (or a space vehicle) would be seen from the center of the earth at a particular time. Compare **astrometric position.**

The apparent position of a body is displaced from the true position at the time of observation by the motion of the body during the time it takes light to travel from the body to the earth (see **planetary aberration**) and by **aberration.**

Most ephemerides tabulate apparent position of the sun, moon, and planets.

apparent solar day. The duration of one rotation of the earth on its axis, with respect to the **apparent sun.** It is measured by successive transits of the apparent sun over the lower branch of a meridian. The length of the apparent solar day is 24 hours of apparent time and averages the length of the **mean solar day,** but varies somewhat from day to day.

apparent sun. The actual sun as it appears in the sky. Also called *true sun.* See **mean sun.**

apparent time. Time based upon the rotation of the earth relative to the **apparent** or **true sun.** This is the time shown by a sundial. See **equation of time.**

Apparent time may be designated as either local or Greenwich, as the local or Greenwich meridian is used as the reference.

apparent wander. Apparent change in the direction of the axis of rotation of a spinning body, as a gyro, due to rotation of the earth. Often shortened to *wander.* See **precession.**

The horizontal component of apparent wander is called *drift,* and the vertical component is called *topple.*

appliance. Any instrument, mechanism, equipment, part,

apparatus, appurtenance, or accessory, including communications equipment, that is used or intended to be used in operating or controlling an aircraft in flight, is installed in or attached to the aircraft, and not part of an airframe, engine, or propeller.

applied research. Research concerned with the practical application of knowledge, material, and/or techniques directed toward a solution to an existent or anticipated requirement. See also **basic research; research.**

apportionment. A commander's decision on division of the total tactical air capability among air strike tasks to be performed for a specified period.

approach. That part of an airplane flight just preceeding, and in preparation for a landing.

approach clearance. Authorization for a pilot conducting flight in accordance with instrument flight rules to commence an approach to an airport.

approach control (departure control). A service established to control IFR flights arriving at or departing from, or operating within the vicinity of an airport. The service utilizes direct radio communication, radar and effects all IFR and occasionaly VFR, traffic in the designated area.

approach control service. Air traffic control service, provided by a terminal area traffic control facility, for arriving and departing IFR aircraft and, on occasion, VFR aircraft. Also called "Approach Control" and "Departure Control".

approach fix. The fix from or over which final approach (IFR) to an airport is executed.

approach gate. That point on the final approach course which is 1 mile from the approach fix on the side away from the airport or 5 miles from the landing threshold, whichever is farther from the landing threshold.

approach lights. A system of lights so arranged as to assist a pilot in aligning his aircraft with a runway and in following a straight path when descending preppreparatory to landing. Approach lights are normally associated with an instrument runway.

approach sequence. The order in which aircraft are positioned while awaiting approach clearance or while on approach.

approximate absolute temperature scale. A temperature scale with the ice point at 273° and boiling point of water at 373°. It is intended to approximate the Kelvin temperature scale with sufficient accuracy for many sciences, notably meteorology, and is widely used in the meteorological literature. Also called *tercentesimal thermometric scale.*

appulse. 1. The near approach of one celestial body to another on the **celestial sphere,** as in **occultation, conjunction,** etc.

apron (airfield). A paved, surfaced, or prepared area where aircraft stand for purposes of loading or unloading passengers or cargo, refueling, parking, or servicing.

approved. Unless used with reference to another person, means approved by the Administrator (FAA).

apsis (*plural* **apsides**). In celestial mechanics, either of the two orbital points nearest or farthest from the center of attraction. Also called *apse.*

　　The apsides are the **perihelion** and **aphelion** in the case of an orbit about the sun, and the **perigee** and **apogee** in the case of an orbit about the earth. The line connecting these two points is called *line of apsides.* The nearest point is the lower apsis while the farthest point

is the higher apsis.

APU. auxiliary power unit.

arc. The track over the ground of an aircraft flying at a constant distance from a navaid by reference to distance measuring equipment.

arc. 1.A part of a curved line, as a circle. **2.** A luminous glow which appears when an electric current passes through ionized air or gas. **3.** An auroral arc. See **aurora.** See **arc discharge.**

arc discharge. A luminous, gaseous, electrical discharge in which the charge transfer occurs continuously along a narrow channel of high **ion density.** An arc discharge requires a continuous source of electric potential difference across the terminals of the arc.

　　Arc discharge is to be distinguished from **corona discharge, point discharge,** and **spark discharge.**

arc tunnel. A wind tunnel employing high temperature air heated by an electric arc. Used to simulate the environment of hypersonic flight.

ARDC model atmosphere. See standard atmosphere.

area air defense commander. Within an overseas unified command, subordinate unified command, or joint task force, the commander will assign overall responsibility for air defense to a single commander. Normally, this will be the Air Force component commander. Representation from the other Service components involved will be provided, as appropriate, to the area air defense commander's headquarters.

area bombing. Bombing of a target which is in effect a general area rather than a small or pinpoint target.

area damage control. Measures taken before, during, or after hostile action or natural or man-made disasters, to reduce the probability of damage and minimize its effects.

area defense. Area defense involves the concept of locating defense units to intercept enemy attacks remote from and without reference to individual vital installations, industrial complexes, or population centers.

area forecast. An aviation weather forecast on a regional basis. Also see **terminal forecast, weather forecast.**

area navigation (Rnav). A method of navigation that permits aircraft operations on any desired course within the coverage of station-referenced navigation signals or within the limits of self-contained system capability.

area of interest. That area of concern to the commander, including the area of influence, areas adjacent thereto, and extending into enemy territory to the objectives of current or planned operations. This area also includes areas occupied by enemy forces who could jeopardize the accomplishment of the mission.

area of militarily significant fallout. The area in which radioactive fallout affects the ability of military units to carry out their normal missions.

area of northern operations. (military) A region of variable width in the Northern Hemisphere that lies north of the 50 degrees isotherm—a line along which the average temperature of the warmest four month period of the year does not exceed 50 degrees Fahrenheit. Mountain regions located outside of this area are included in this category of operations provided these same temperature conditions exist.

area of operations. That portion of an area of conflict necessary for military operations, either offensive or defensive, pursuant to an assigned mission, and for the administration incident to such military operations.

area rule. A prescribed method of design for obtaining minimum zero-lift **drag** for a given aerodynamic configuration, such as a wing-body configuration, at a given speed.

For a transonic body, the area rule is applied by subtracting from, or adding to, its *cross-sectional* area distribution normal to the airstream at various stations so as to make its cross-sectional area distribution approach that of an ideal body of minimum drag; for a supersonic body, the sectional areas are frontal projections of areas intercepted by planes inclined at the Mach angle.

area search (military). Visual reconnaissance of limited or defined areas.

areodesy. That branch of mathematics which determines by observation and measurement, the exact positions of points and figures and areas of large portions of the surface of the planet Mars, or the shape and size of the planet Mars.

argument. In astronomy, an angle or arc, as in *argument of perigee.*

argument of latitude. In celestial mechanics, the **angular distance** measured in the **orbit plane** from the **ascending node** to the orbiting object; the sum of the argument of perigee and the true **anomaly.**

argument of perigee. In celestial mechanics, the angle or arc, as seen from a **focus** of an elliptical **orbit,** from the **ascending node** to the closest approach of the orbiting body to the focus. The angle is measured in the orbital plane in the direction of motion of the orbiting body.

Ariel. A satellite of Uranus orbiting at a mean distance of 192,000 kilometers.

arithmetic mean. One of several accepted measures of central tendency, physically analogous to *center of gravity.* Also called *mean, average, simple average.*

Since the word *mean* is also applied to other measures of central tendency, such as weighted means, geometric means, harmonic means, the adjective *arithmetic* is used for clarity. However, when used without further qualification, the term *mean* is understood as *arithmetic mean.*

armed conflict. Conflict between nations or other contestants entailing the physical destruction of, or injury to, one another's armed forces. Armed conflict exists if the lives or safety of members of the armed services of a nation, belligerent power, coalition, or faction are endangered as a direct result of the use of physical force.

armed forces. The military forces of a nation or a group of nations. See also **force(s).**

Armed Forces of the United States. The Army, Navy, Air Force, Marine Corps, and Coast Guard, including their regular and reserve components and members serving without component status.

armed reconnaissance. A mission with the primary purpose of locating and attacking targets of opportunity, i.e., enemy materiel, personnel, and facilities, in assigned general areas or along assigned ground communications routes, and not for the purpose of attacking specific briefed targets.

arming. As applied to weapons and ammunition, the changing from a safe condition to a state of readiness for initiation.

arming system. That portion of a weapon which serves to ready (arm), safe, or re-safe (disarm) the firing system and fuzing system and which may actuate devices in the nuclear system.

arms control. A concept which connotes: **a.** any plan, arrangement, or process, resting upon explicit or implicit international agreement, governing any aspect of the following: the numbers, types, and performance characteristics of weapon systems (including their command and control, logistics support arrangements, and any related intelligence-gathering mechanisms); and the numerical strength, organization, equipment, deployment or employment of the armed forces retained by the parties. (It encompasses "disarmament".) and **b.** on some occasions, those measures taken for the purpose of reducing instability of the military environment.

arms control agreement. The written or unwritten embodiment of the acceptance of one or more arms control measures by two or more nations.

Army air ground system. The Army system which provides for interface between Army and tactical air support agencies of other Services in planning, evaluating, processing, and coordinating of air support requirements and operations. It is composed of appropriate staff members, including G-2 air and G-3 air personnel, and necessary communications equipment.

arresting barrier. See **aircraft arresting system.**

arresting gear. See **aircraft arresting system.**

arrival. Any aircraft arriving at an airport is referred to as an arrival.

arrow wing. A wing plan form configuration under study for a supersonic-transport aircraft which consists of a highly swept (about 60 degrees) forward wing section combined with an aft wing section of approximately 45 degrees sweepback. The trailing edge of the aft section is also swept approximately 45 degrees. The basic arrow wing configuration consists of the wing as described and a long fuselage including a tail. The complete configuration is therefore in the shape of an arrow with an arrowhead, a slender shaft and tail feathers.

ARSR. Air Route Surveillance Radar

ARTCC. Air Route Traffic Control Center.

artificial antenna. A device which has the equivalent impedance characteristics of an **antenna** and the necessary power-handling capabilities, but which does not radiate nor intercept radiofrequency energy. Also called *dummy antenna.*

artificial asteroid. A manmade object placed in orbit about the sun. See **asteroid.**

artificial earth satellite. A manmade Earth satellite, as distinguished from the Moon. Usually called "Earth satellite".

artificial feel. A **control feel** simulated by mechanisms incorporated in the control system of an aircraft or spacecraft where the forces acting on the control surfaces are not transmitted to the cockpit controls, as in the case of an irreversible control system or a power-boosted system.

artificial gravity. A simulated **gravity** established within a space vehicle by rotation or acceleration.

artificial horizon. **1.** A gyro-operated flight instrument that shows the pitching and banking attitudes of an aircraft or spacecraft with respect to a reference line horizon, within limited degrees of movement, by means of the relative position of lines or marks on the face of the instrument representing the aircraft and the **horizon.** See **attitude gyro. 2.** A device, such as a spirit level, pendulum, etc., that establishes a horizontal reference in a navigation instrument.

ascending node. That point at which a planet, planetoid, or comet crosses to the north side of the **ecliptic;** that point at which a **satellite** crosses to the north side of the equatorial plane of its primary. Also called *northbound node.* The opposite is *descending node* or *southbound node.*

ASDE. Airport Surface Detection Equipment.

asl. Above sea level.

ASM. Air-to-Surface Missile.

aspect ratio, of an airfoil wing. The ratio of span to mean chord of an airfoil, i.e., the ratio of the square of the maximum span to the total area of an airfoil. In a simple rectangular airfoil it is the ratio of the span to the chord.

aspects. The **apparent positions** of celestial bodies relative to one another; particularly, the apparent positions of the moon or a planet relative to the sun.

ASR. Airport Surveillance Radar.

assault. 1. The climax of an attack; closing with the enemy in hand-to-hand fighting. **2.** In an amphibious operation, the period of time between the arrival of the major assault forces of the amphibious task force in the objective area and accomplishment of the amphibious task force mission. **3.** To make a short, violent, but well-ordered attack against a local objective, such as a gun emplacement, a fort, or a machine gun nest. **4.** A phase of an airborne operation beginning with delivery by air of the assault echelon of the force into the objective area and extending through attack of assault objectives and consolidation of the initial airhead.

assault aircraft. Powered aircraft, including helicopters, which move assault troops and cargo into an objective area and which provide for their resupply.

assault airdrop. The delivery of an airborne force into an objective area under combat conditions by unloading men and materiel while in flight.

assault airland. The delivery of an airborne force by unloading men and materiel after landing in the objective area under combat conditions.

assault echelon (air transport). The element of a force which is scheduled for initial assault on the objective area.

assemble. In computer terminology, to organize the **subroutines** into a complete **program.**

assisted take-off. A take-off of an aircraft using a supplementary source of power, usually rockets. See RATO.

associate contractor. A prime contractor to the Procuring Government Agency for the development or production of subsystems, equipments, or components meeting specifications furnished or approved by the Agency. One member of a group of contractors which together is developing and producing a complete weapon system.

A-station. In **loran,** the designation applied to the transmitting station of a pair, the signal of which always occurs less than half a repetition period after the next preceding signal and more than half a repetition period before the next succeeding signal of the other station of the pair, designated as *B-station.*

asteroid. One of many thousands of minor planets which revolve around the Sun mostly between the orbits of Mars and Jupiter. All are very small compared with the major planets. Ceres, the largest, is 480 miles in diameter; the majority are less than 50 miles; and some are about one mile in diameter.

astro. A prefix meaning *star* or *stars* and, by extension, sometimes used as the equivalent of *celestial,* as in *astro*nautics.

astroballistics. The study of the phenomena arising out of the motion of a solid through a gas at speeds high enough to cause **ablation;** for example, the interaction of a meteoroid with the atmosphere.

Astroballistics uses the data and methods of astronomy, aerodynamics, ballistics, and physical chemistry.

astrobiology. A branch of biology concerned with the discovery or study of life on planets.

astro compass. An instrument used primarily to obtain true heading by reference to celestial bodies.

astrodome. A transparent bubble calibrated for refraction mounted in the top of an aircraft fuselage through which celestial observations are taken.

astrodynamics. The practical application of **celestial mechanics, astroballistics,** propulsion theory, and allied fields to the problem of planning and directing the **trajectories** of space vehicles.

Astrodynamics is sometimes used as a synonym for *celestial mechanics.* This usage should be discouraged.

astrolabe. 1. In general, any instrument designed to measure the **altitudes** of celestial bodies. **2.** Specifically, an instrument designed for very accurate celestial altitude measurements, as in survey work.

astrometric position. The position of a heavenly body (or space vehicle) on the **celestial sphere** corrected for **aberration** but not for **planetary aberration.** Compare **apparent position.**

Astrometric positions are used in photographic observations where the position of the observed body can be measured in reference to the positions of comparison stars in the field of the photograph.

astrometry. The branch of **astronomy** dealing with the geometrical relations of the celestial bodies and their real and **apparent motions.**

The techniques of astrometry, especially the determination of accurate position by photographic means, are used in tracking satellites and space probes.

astronaut. 1. A person who rides in a space vehicle. **2.** Specifically, one of the test pilots selected to participate in Project Mercury, Project Gemini, Project Apollo, or any other United States program for manned space flight.

astronautics. 1. The art, skill, or activity of operating **spacecraft. 2.** In a broader sense the science of space flight.

astronavigation. The plotting and directing of the movement of a **spacecraft** from within the craft by means of observations on celestial bodies. Sometimes contracted to *astrogation* or called *celestial navigation.*

astronomical. Of or pertaining to **astronomy** or to observations of the celestial bodies. Also called *astronomic.*

Astronomers have long preferred *astronomical,* Geodesists usually use *astronomic* as an intended parallel to *geodetic.* The Coast and Geodetic Astronomers have long preferred *astronomical.* Geodesists usally use *astronomic* as an intended parallel to *geodetic.* The Coast and Geodetic Survey uses *astronomic* in their publications insofar as is compatible with established practice.

astronomical constants. The elements of the orbits of the bodies of the **solar system,** their masses relative to the sun, their size, shape, orientation, rotation, and inner constitution, and the velocity of light.

astronomical coordinates. Coordinates defining a point on the surface of the earth, or of the **geoid,** in which the local direction of gravity is used as a reference. Sometimes called *geographic coordinates,* which see. See **astronomical latitude, astronomical longitude.**

astronomical day. A **mean solar day** beginning at mean noon, 12 hours later than the beginning of the civil day of the same date. Astronomers now generally use the civil day. See **Julian day, astronomical time.**

astronomical equator. A line on the surface of the earth connecting points having 0° **astronomical latitude.** Sometimes called *terrestrial equator.*

When the astronomical equator is corrected for station error, it becomes the *geodetic equator.*

astronomical latitude. Angular distance between the direction of gravity and the plane of the **celestial equator.** Sometimes called *geographic latitude.*

Astronomical latitude corrected for the *meridional* component of station error becomes *geodetic latitude.*

astronomical longitude. The angle between the plane of the reference meridian and the plane of the **celestial meridian.** Sometimes called *geographic longitude.*

Astronomical longitude corrected for the prime-vertical component of station error divided by the cosine of the latitude becomes *geodetic longitude.*

astronomical meridian. A line connecting points having the same **astronomical longitude.** Also called *terrestial meridian.*

Because the deflection of the vertical varies from point to point, the astronomical meridian is an irregular line. When the astronomical meridian is corrected for station error, it becomes the *geodetic meridian.*

astronomical parallel. A line connecting points having the same **astronomical latitude.**

Because the deflection of the vertical varies from point to point, the astronomical parallel is an irregular line. When the astronomical parallel is corrected for station error, it becomes the *geodetic parallel.*

astronomical position. 1. A point on the earth whose coordinates have been determined as a result of observation of celestial bodies.

The expression is usually used in connection with positions on land determined with great accuracy for **survey** purposes.

2. A point on the earth, defined in terms of **astronomical latitude** and **longitude.**

astronomical refraction. 1. The angular difference between the apparent **zenith distance** of a celestial body and its true zenith distance, produced by **refraction** effects as the light from the body penetrates the atmosphere. Also called *atmospheric refraction, astronomical refraction error.*

For bodies near zenith the astronomical refraction is only about 0.1 minute, but for bodies near the horizon it becomes about 30 minutes or more and contributes measurably to the length of the apparent day.

2. Any **refraction** phenomenon observed in the light originating from a source outside of the earth's atmosphere; as contrasted with terrestrial refraction. This is applied only to refraction cuased by inhomogeneities of the atmosphere itself, and not to that caused by ice crystals suspended in the atmosphere.

astronomical scintillation. Any **scintillation** phenomena, such as irregular oscillatory motion, variation of intensity, and color fluctuation observed in the light emanating from an extraterrestrial source; to be distinguished from **terrestrial scintillation** primarily in that the light source for the latter lies somewhere within the earth's atmosphere. Also called *stellar scintillation.* See **seeing.**

Astronomical scintillation is typically strongest for celestial objects lying at large zenith distances and is not easily observed by eye for objects whose zenith distances are under 30°. Nonperiodic vibratory motions of stellar images with frequencies of the order of 1 to 10 cycles per second create a troublesome problem of **seeing** in astronomical work. The size of the **schlieren** producing vibratory scintillations has been estimated to be of the order of centimeters, and chromatic scintillations of celestial objects appear to be produced by parcels whose dimensions are of the order of decimeters or, perhaps, meters. Hence, astronomical scintillation is primarily a consequence of the high-frequency, short-wavelength type of atmospheric turbulence.

astronomical triangle. The **navigational triangle,** either terrestrial or celestial, used in the solution of **celestial observations.**

astronomical unit (*abbr* **AU**). **1.** A unit of length, usually defined as the distance from the earth to the sun, 149,-599,000 kilometers (approximately 93,000,000 miles).

This value for the AU was derived from radar observations of the distance of Venus. The value given in astronomical ephermerides, 149,500,000 kilometers, was derived from observations of the minor planet Eros.

2. The unit of distance in terms of which, in the Kepler Third Law, $n^2a^3 = k^2(1 + m)$, the semimajor axis a of an elliptical orbit must be expressed in order that the numerical value of the Gaussian constant k may be exactly 0.01720209895 when the unit of time is the ephemeris day.

In astronomical units, the mean distance of the earth from the sun, calculated by the Kepler law from the observed mean motion n and adopted mass $m,$ is 1.00000003.

astronomy. The science that treats of the location, magnitudes, motions, and constitution of **celestial bodies** and structures.

astrophysics. A branch of **astronomy** that treats of the physical properties of celestial bodies, such as luminosity, size, mass, density, temperature, and chemical composition.

astro-tracker. A navigation equipment which automatically acquires and continuously tracks a celestial body in azimuth and altitude. Also see **star tracker.**

ASW. antisubmarine warfare.

ATA. Actual time of arrival. A term used in navigation.

ATC. Air Traffic Control

ATCRBS. Air Traffic Control Radar Beacon System.

athodyd. A type of **jet engine** consisting essentially of a duct or tube of varying diameter and open at both ends, which admits air at one end, compresses it by the forward motion of the engine, adds heat to it by the combustion of fuel, and discharges the resulting gases at the other end to produce thrust.

The **ramjet** is an athodyd; the **pulsejet,** especially the earlier type, is usually not considered an athodyd.

ATIS. Automatic Terminal Information Service.

atmosphere. 1. The envelope of **air** surrounding the earth; also the body of gases surrounding or comprising any **planet** or other celestial body. Compare **biosphere,**

geosphere, hydrosphere, lithosphere. See **atmospheric shell. 2.** = **standard atmosphere. 3.** (*abbr* **atm**) A unit of pressure equal to 14.7 pounds per square inch.

atmospheric braking. The action of atmospheric drag in decelerating a body that is approaching a planet; can be deliberately used, where sufficient atmosphere exists, to lose much of the vehicle velocity before landing.

atmospheric entry. The penetration of any planetary **atmosphere** by any object from outer space; specifically, the penetration of the earth's atmosphere by a manned or unmanned capsule or **spacecraft.**

atmospheric interference. See **atmospherics.**

atmospheric pressure. The pressure at any point in an **atmosphere** due solely to the weight of the atmospheric gases above the point concerned. See **station pressure, sea-level pressure.**

atmospheric radiation. Infrared radiation emitted by or being propagated through the atmosphere. See **insolation.**

Atmospheric radiation, lying almost entirely within the wavelength interval of from 3 to 80 microns, provides one of the most important mechanisms by which the heat balance of the earth-atmosphere system is maintained. Infrared radiation emitted by the earth's surface (terrestial radiation) is partially absorbed by the water vapor of the atmosphere which in turn reemits it, partly upward, partly downward. This secondarily emitted radiation is then, in general, repeatedly absorbed and reemitted, as the radiant energy progresses through the atmosphere. The downward flux, or counterradiation, is of basic importance in the **greenhouse effect;** the upward flux is essential to the radiative balance of the planet.

atmospheric shell. Any one of a number of strata or *layers* of the earth's atmosphere. Also called *atmospheric layer, atmospheric region.*

Temperature distribution is the most common criterion used for denoting the various shells. The *troposphere* (the *region of change*) is the lowest 10 or 20 kilometers of the atmosphere, characterized by decreasing temperature with height. The top of the troposphere is called the *tropopause.* Above the tropopause, the *stratosphere,* a region in which the temperature generally increases with altitude, extends to the *stratopause,* the top of the inversion layer, at about 50 to 55 kilometers. Above the stratosphere, the *mesosphere,* a region of generally decreasing temperatures with height extends to the *mesopause,* the base of an inversion layer at about 80 to 85 kilometers. The region above the mesopause, in which temperature generally increases with height, is the *thermosphere.*

The distribution of various physicochemical processes is another criterion. The *ozonosphere,* lying roughly between 10 and 50 kilometers, is the general region of the upper atmosphere in which there is an appreciable ozone concentration and in which ozone plays an important part in the radiative balance of the atmosphere; the *ionosphere,* starting at about 70 to 80 kilometers, is the region in which ionization of one or more of the atmospheric constituents is significant; the *neutrosphere* is the shell below this which is, by contrast, relatively unionized; and the *chemosphere,* with no very definite height limits, is the region in which photochemical reactions take place.

Dynamic and kinetic processes are a third criterion. The *exosphere* is the region at the *top* of the at-

mosphere, above the critical level of escape, in which atmospheric particles can move in free orbits, subject only to the earth's gravitation.

Composition is a fourth criterion. The *homosphere* is the shell in which there is so little photodissociation or gravitational separation that the mean molecular weight of the atmosphere is sensibly constant; the *heterosphere* is the region above this, where the atmospheric composition and mean molecular weight are not constant. The boundary between the two is probably at the level at which molecular oxygen begins to be dissociated, and this occurs in the vicinity of 80 or 90 kilometers.

The term *mesosphere* has been given another definition which does not fit into any logical set of criteria, i.e., the shell between the exosphere and the ionosphere. This use of *mesosphere* has not been widely accepted.

atmospherics. The **radiofrequency** electromagnetic radiations originating, principally, in the irregular surges of charge in thunderstorm lightning discharges. Atmospherics are heard as a quasi-steady background of crackling noise (static) in ordinary amplitude-modulated radio receivers. Also called *atmospheric interference, strays.*

atom. A particle of matter indivisible by chemical means. It is the fundamental building block of the chemical elements. The elements, such as iron, lead, and sulfur, differ from each other because they contain different kinds of atoms. There are about six sextillion (6 followed by 21 zeros, or 6×10^{21}) atoms in an ordinary drop of water. According to present-day theory, an atom contains a dense inner core (the *nucleus*) and a much less dense outer domain consisting of *electrons* in motion around the nucleus. Atoms are electrically neutral. (Compare *element, ion, molecule;* see *matter.*)

atom smasher. An *accelerator.*

atomic battery. A *radioisotopic generator.*

atomic bomb. A bomb whose energy comes from the *fission* of heavy elements, such as uranium or plutonium. (Compare *hydrogen bomb.*)

atomic clock. A device that uses the extremely fast vibrations of molecules or atomic nuclei to measure time. These vibrations remain constant with time, consequently short intervals can be measured with much higher precision than by mechanical or electrical clocks. (Compare *radioactive dating.*)

atomic cloud. The cloud of hot gases, smoke, dust, and other matter that is carried aloft after the explosion of a nuclear weapon in the air or near the surface. The cloud frequently has a mushroom shape. (See *fireball, radioactive cloud.*)

atomic energy. See **nuclear energy.**

atomic mass. See **atomic weight, mass.**

atomic number. The number of protons in the *nucleus* of an atom, and also its positive charge. Each chemical element has its characteristic atomic number, and the atomic numbers of the known elements form a complete series from 1 (hydrogen) to 103 (lawrencium). (Compare **atomic weight, mass number;** see **element, isotope, Periodic Table.**)

atomic reactor. See **nuclear reactor.**

atomic weapon. An explosive weapon in which the energy is produced by nuclear *fission* or *fusion.* (Compare *device, nuclear.*)

atomic weight. The mass of an atom relative to other atoms. The present-day basis of the scale of atomic weights is carbon; the commonest isotope of this element has arbitrarily been assigned an atomic weight of 12. The unit of the scale is 1/12 the weight of the carbon-12 atom, or roughly the mass of one proton or one neutron. The atomic weight of any element is approximately equal to the total number of protons and neutrons in its *nucleus.* (Compare *atomic number;* see *atomic mass unit, Periodic Table.*)

attached shock. Same as **attached shock wave.**

attached shock wave. An oblique or conical shock wave that appears to be in contact with the leading edge of an airfoil or the nose of a body in a supersonic flow field. Also called *attached shock.*

attack (A), (military designation). Aircraft designed to search out, attack, and destroy enemy land or sea targets, using conventional or special weapons. Also used for interdiction and close air support missions. See **military aircraft types.**

attack aircraft carrier. A warship designed to support and operate aircraft, engage in attacks on targets afloat or ashore, and engage in sustained operations in support of other forces. Designated as CVA and CVAN. CVAN is nuclear powered.

attack altitude. The altitude at which the interceptor will maneuver during the attack phase of an air intercept.

attack carrier striking forces. Naval forces, the primary offensive weapon of which is carrier-based aircraft. Ships, other than carriers, act primarily to support and screen against submarine and air threat, and secondarily against surface threat.

attack heading. 1. The interceptor heading during the attack phase which will achieve the desired track-crossing angle. **2.** The assigned magnetic compass heading to be flown by aircraft during the delivery phase of an air strike.

attack speed. The speed at which the interceptor will maneuver during the attack phase of an air intercept.

attack sortie. A combat air sortie intended to inflict damage against an enemy surface target.

attitude. The position or orientation of an aircraft, spacecraft, etc., either in motion or at rest, as determined by the relationship between **its axes** and some reference line or plane or some fixed system of reference axes.

attitude control. 1. The regulation of the **attitude** of an aircraft, spacecraft, etc. **2.** A device or system that automatically regulates and corrects attitude, especially of a pilotless vehicle.

attitude gyro. 1. A **gyro**-operated flight instrument that indicates the **attitude** of an aircraft or spacecraft with respect to a reference coordinate system throughout 360° of rotation about each axis of the craft.

This instrument is similar to the **artificial horizon,** but has greater angular indication.

2. Broadly, any gyro-operated instrument that indicates attitude.

attitude jets. Fixed or movable gas nozzles on a rocket missile or satellite, operated continuously or intermittently to change the attitude or position in aerospace. Sometimes called steering jets, attitude control jets, or roll, pitch, and yaw jets.

attribute. A characteristic of a thing which can be appraised only in terms of whether it does or does not ex-

ist. See **method of attributes.**

attributes testing. A **reliability** test procedure where the items under test are classified according to qualitative rather than quantitative characteristics.

attrition. (military) The reduction of the effectiveness of a force caused by loss of personnel and materiel.

attrition rate. A factor, normally expressed as a percentage, reflecting the degree of losses of personnel or materiel due to various causes within a specified period of time.

attrition reserve aircraft. Aircraft procured for the specific purpose of replacing the anticipated losses of aircraft due to peacetime and/or wartime attrition.

AU (*abbr*). **astronomical unit.**

audible sound. Sound containing **frequency** components lying between about 15 and 20,000 cycles per second.

audio. Pertaining to **audiofrequency range.**

The word **audio** may be used as a modifier to indicate a device or system intended to operate at audiofrequencies, e.g. **audioamplifier.**

audiofrequency range. The range of frequencies to which the human ear is sensitive, approximately 15 cycles per second to 20,000 cycles per second. Also called *audiorange.*

auditory sensation area. In acoustics, the frequency region enclosed by the curves defining the **threshold of pain** and the **threshold of audibility.**

auger shower. A very large **cosmic-ray** shower. Also called *extensive air shower.*

augmentation. The apparent increase in the **semidiameter** of a celestial body, as observed from the earth, as its altitude increases, due to the reduced distance from the observer.

The term is used principally in reference to the moon.

augmentation correction. A correction due to **augmentation,** particularly that sextant altitude correction due to the apparent increase in the **semidiameter** of a celestial body as its altitude increases.

augmenter tube. A tube or pipe, usually one of several, through which the exhaust gases from an aircraft **reciprocating engine** are directed especially to provide additional thrust.

AUM. Air-to-Underwater Missile.

aurora. The sporadic radiant emission from the **upper atmosphere** over middle and high latitudes. It is believed to be due primarily to the **emission** from atomic molecular and ionic nitrogen; atomic oxygen; atomic sodium; the hydroxyl radical; and hydrogen. Compare **airglow.**

According to various theories, auroras seem definitely to be related to magnetic storms and the influx of charged particles from the sun. The exact details of the nature of the mechanisms involved are still being investigated, but release of trapped particles from the Van Allen belt apparently plays an important part. The aurora is most intense at times of magnetic storms (when it is also observed farthest equatorward), and shows a periodicity which is related to the sun's 27-day rotation period and the 11-year sunspot cycle. The distribution with height shows a pronounced maximum near 100 kilometers. The lower limit is probably near 80 kilometers.

The aurora can often be clearly seen, and it assumes a variety of shapes and colors which are characteristic patterns of auroral emission.

aurora borealis. The aurora of northern latitudes. Also

called *aurora polaris, northern lights.*

auroral zone. A roughly circular band around either **geomagnetic pole** above which there is a maximum of auroral activity. It lies about 10° to 15° of **geomagnetic latitude** from the geomagnetic poles.

The auroral zone broadens and extends equatorward during intense auroral displays.

The northern auroral zone is centered along a line passing near Point Barrow, Alaska, through the lower half of Hudson Bay, slightly off the southern tip of Greenland, through Iceland, northern Norway and northern Siberia. Along this line auroras are seen on an average of 240 nights a year. The frequency of auroras falls off both to the north and to the south of this line but more rapidly to the south. The most severe blackouts occur in the auroral zone.

authority. The power to direct action or to use resources to accomplish assigned responsibility. Authority must be commensurate with responsibility. Authority resides in an individual and is received from one having greater authority.

autogiro. A gyroplane.

autokinetic illusion. The illusion of a fixed object or light moving when gazed at steadily.

automated data processing system. The total of the automatic data processing equipment, automated data systems, software, maintenance services, and other items required to achieve desired performance.

automated radar terminal systems (ARTS). computer-aided radar display subsystems capable of associating alphanumeric data with radar returns. Systems of varying functional capability, determined by the type of automation equipment and software, are denoted by a number/letter suffix following the name abbreviation.

automatic approach and landing. A control mode in which the aircraft's speed and flight path are automatically controlled for approach, flareout, and landing.

automatic control. Control of devices and equipment, including aerospace vehicles, by automatic means.

automatic data processing. Data processing performed by a system of electronic or electrical machines so interconnected and interacting as to reduce to a minimum the need for human assistance or intervention. See also **automated data processing system, electronic data processing.**

automatic data processing equipment. General purpose, commercially available, mass produced automatic data processing components and the equipment configurations created from them regardless of use, size, capacity, or price, that are designed to be applied to the solution or processing of a variety of problems or applications, and that are not specially designed, as opposed to configured, for any specific applications. It includes:

a. Digital, analog, or hybrid computer equipment;

b. Auxiliary or accessorial equipment, such as plotters, communications terminals, tape cleaners, tape testers, source data automation recording equipment (optical character recognition equipment, paper tape typewriters, magnetic tape cartridge typewriters, and other data acquisition devices), etc., to be used solely in support of digital, analog, or hybrid computer equipment, either cable-connected, wire-connected, or self-standing, and whether selected or acquired with a computer or separately; and/or

c. Punched card accounting machines (PCAM) used in conjunction with or independently of digital, analog, or hybrid computers.

automatic direction finder (*abbr* **ADF**). A radio **direction finder** which automatically and continuously provides a measure of the direction of arrival of the received signal. Data are usually displayed visually. Also see **non-directional radio beacon, radio compass, radio direction finder, radio direction finder, automatic.**

automatic flight control system. A system which includes all equipment to automatically control the flight of an aircraft or missile to a path or attitude described by references internal or external to the aircraft or missile. Also see **automatic pilot.**

automatic frequency control. An arrangement whereby the **frequency** of an **oscillator** is automatically maintained within specified limits.

automatic landing system. See **automatic approach and landing system.**

automatic navigation system. An operable data processing system for the collation, processing and display of navigational data derived from sensors, including the inertial sensor systems, VOR and DME receivers, magnetic compass system and air data sensors. The system provides steering signals to the **automatic flight control system.**

automatic pilot. Equipment which automatically stabilizes the attitude of a vehicle about its **pitch, roll,** and **yaw** axes. Also called *auto-pilot.*

automatic radio direction finder. See **automatic direction finder.**

automatic terminal information service (ATIS). The continuous broadcast of recorded noncontrol information in selected high activity terminal areas. Its purpose is to improve controller effectiveness and to relieve frequency congestion by automating the repetitive transmission of essential but routine information.

automatic tracking. Tracking in which a **servomechanism** automatically follows some characteristic of the signal; specifically, a process by which tracking or data acquisition systems are enabled to keep their antennas continually directed at a moving target without manual operation.

automation. The technique of improving human productivity in the processing of materials, energy, and information, by utilizing in various degrees, elements of automatic control, and of automatically executed product programming.

autonomous operation. In air defense, the mode of operation assumed by a unit after it has lost all communications with higher echelons. The unit commander assumes full responsibility for control of weapons and engagement of hostile targets.

autopilot. Same as **automatic pilot.**

autoradiograph (nucleonics). A photographic record of *radiation* from radioactive material in an object, made by placing the object very close to a photographic film or emulsion. The process is called autoradiography. It is used, for instance, to locate radioactive atoms or *tracers* in metallic or biological samples. (Compare **radiography.**)

autorotation. Continuous rotation of a body about any axis in a uniform air stream due solely to aerodynamic

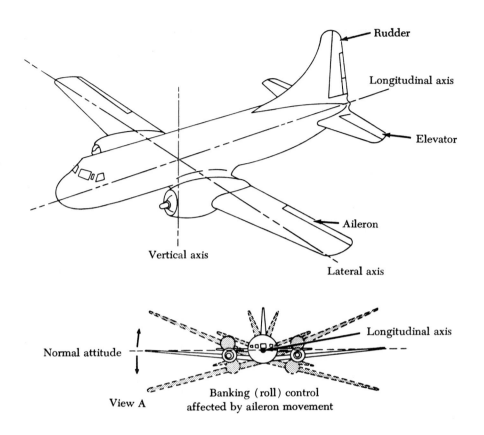

Rudder

Longitudinal axis

Elevator

Aileron

Vertical axis

Lateral axis

Normal attitude

Longitudinal axis

View A

Banking (roll) control
affected by aileron movement

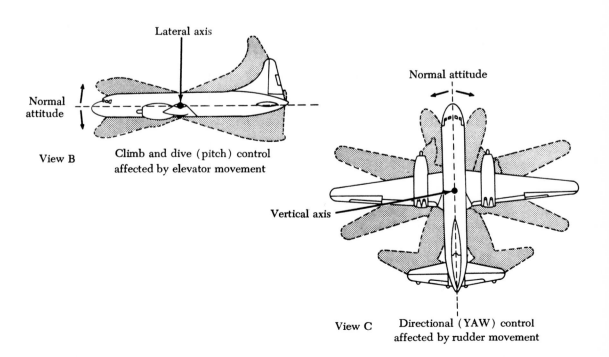

Lateral axis

Normal
attitude

View B

Climb and dive (pitch) control
affected by elevator movement

Normal attitude

Vertical axis

View C

Directional (YAW) control
affected by rudder movement

Motion of an aircraft about its axes.

moments. See **spin.**

autorotation (rotor). That condition of flight of a rotorcraft wherein there is free and continuous rotation of the rotor when it is not power driven.

autosyn. (A trade name, from *autosynchronous,* often capitalized.) A remote-indicating instrument or system based upon the synchronousmotor principle, in which the angular position of the rotor of one motor at the measuring source is duplicated by the rotor of the indicator motor, used, e.g., in fuel-quantity or fuel-flow measuring systems, position-indicating systems, etc.

autumnal equinox. 1. That point of intersection on the **celestial sphere** of the **ecliptic** and the **celestial equator** occupied by the sun as it changes from north to south **declination,** on or about September 23. Also called *September equinox, first point of Libra.* **2.** That instant the sun reaches the point of zero declination when crossing the celestial equator from north to south.

auxiliary circle. In celestial mechanics, a circumscribing circle to an orbital **ellipse** with a radius *a*, the **semimajor axis.**

auxiliary fluid ignition. A method of ignition of a liquid-propellant **rocket engine** in which a liquid which is **hypergolic** with either the **fuel** or the **oxidizer** is injected into the combustion chamber to initiate combustion.

Aniline is used as an auxiliary fluid with nitric acid and some organic fuels to initiate combustion.

auxiliary landing gear. That part or parts of a **landing gear,** as an outboard wheel, which is intended to stabilize the craft on the surface but which bears no significant part of the weight.

auxiliary power unit (*abbr* **APU**). A power unit carried on an aircraft or spacecraft which can be used in addition to the main sources of power of the craft.

auxiliary rotor. A rotor that serves either to counteract the effect of the main rotor torque on a rotorcraft or to maneuver the rotorcraft about one or more of its three principal axes.

available payload. The passenger and/or cargo capacity expressed in weight and/or space available to the user.

average deviation. In statistics, the average or arithmetic mean of the deviations, taken without regard to sign, from some fixed value, usually the **arithmetic mean** of the data. Also called *mean deviation.* See **standard deviation.**

avgas. An aviation gasoline.

aviation.

   a. The operation of heavier-than-air aircraft.

   b. Synonym for Aeronautics.

aviation mechanic. A person who repairs, alters and/or maintains an aircraft and/or its powerplant in accordance with applicable regulations and procedures. A U.S. civil aviation mechanic must meet applicable requirements of the **Federal Aviation Regulations (FAR).** Also see **aviation mechanic certificate, airman certificates.**

aviation mechanic certificate. A certificate of competency issued by the **FAA** to a person meeting the requirements of the applicable **Federal Aviation Regulations.** Also see **aircraft mechanic, powerplant mechanic, airman certificates.**

aviation weather report. See **sequence report, National Weather Service.**

aviator. The pilot of an aircraft. This term is seldom used. **Pilot** is preferred.

aviatrix. Female pilot of an aircraft. This term is seldom used. **Pilot** is preferred.

avigation. Navigation as applied to aircraft. The term has gained only limited acceptance. See **navigation.**

avionics. The application of electronics to aviation and astronautics.

axes, of an aircraft. Three fixed lines of reference, usually centroidal and mutually perpendicular. The first, the **longitudinal axis** in the plane of symmetry, usually parallel to the axis of the propeller, or **thrust line.** The second, about which the plane rotates in **yawing** is known as the vertical axis, and third, the axis perpendicular to the other two is called the lateral axis. In mathematical discussions the first of these axis, drawn from front to rear, is called the X axis; the second, drawn upward, the Z axis; and the third, running from left to right, the Y axis.

axial flow compressor. A rotary **compressor** having interdigitated rows or stages of rotary and of stationary blades through which the flow of fluid is substantially parallel to the rotor's axis of rotation. Compare **centrifugal compressor.**

axis (*plural* **axes**). **1.** A straight line about which a body rotates, or along which its center of gravity moves (axis of translation). **2.** A straight line around which a plane figure may rotate to produce a solid; a line of symmetry. **3.** One of a set of reference lines for a **coordinate** system.

axis of freedom. Of a gyro, an axis about which a **gimbal** provides a **degree of freedom.**

azimuth. 1. Horizontal direction or **bearing.** Compare **azimuth angle. 2.** In navigation, the horizontal direction of a **celestial** point from a **terrestrial** point, expressed as the angular distance from a reference direction, usually measured from 0° at the reference direction clockwise through 360°.

An azimuth is often designated as *true, magnetic, compass, grid,* or *relative* as the reference direction is true, magnetic, compass, grid north, or heading, respectively. Unless otherwise specified the term is generally understood to apply to *true azimuth,* which may be further defined as the arc of the horizon, or the angle at the zenith, between the north part of the celestial **meridian** or principal vertical circle and a vertical circle, measured from 0° at the north part of the principal vertical circle clockwise through 360°.

3. In astronomy, the direction of a celestial point from a terrestrial point measured clockwise from the north or the south point of **meridian** plane. See **horizon system.**

4. In surveying, the horizontal direction of an object measured clockwise from the south point of the meridian plane.

In surveying, an azimuth of a celestial body is called an *astronomic azimuth.*

azimuth angle. 1. Azimuth measured from 0° at the north or south reference direction clockwise or counterclockwise through 90° or 180°.

Azimuth angle is labeled with the reference direction as a prefix and the direction of measurement from the reference direction as a suffix. Thus, azimuth angle S 144° W is 144° west of south, or azimuth 324°. When azimuth angle is measured through 180°, it is labeled N or S to agree with the latitude and E or W to agree with the meridian angle.

2. In surveying, an angle in triangulation or in traverse through which the computation of azimuth is carried.

azimuth error. An error in the indicated **azimuth** of target detected by radar, resulting from horizontal **refraction.** Compare **range error.**

Inasmuch as significant horizontal gradients of **index of refraction** are very uncommon in the atmosphere, these errors almost invariably are negligible. Seacoast areas may give rise on occasion to appreciable horizontal bending of radio waves because of the contrast of refractive index values between the air over land and the air over water.

azimuth marker. 1. A scale encircling the plan position indicator (PPI) scope of a **radar** on which the **azimuth** of a target from the radar may be measured. **2.** Reference limits inserted electronically at 10° or 15° intervals which extend radially from the relative position of the radar on an offcenter PPI scope. These are employed for target azimuth determination when the radar position is not at the center of the PPI scope and hence the fixed azimuth scale on the edge of the scope cannot be employed.

On such markers north is usually 0°, east 90°, etc. Occasionally, on ship or airborne radars. 0° is used to indicate the direction in which the craft is heading, in which cases the relative bearing, not azimuth, of the target is indicated.

azimuth stick. See **cyclic pitch (rotorcraft).**

azran. Azimuth and range. This term was coined in the field of radar, and has since been extended in application to the locating of any object (or target) by means of polar coordinates.

azusa. A short-baseline, continuous-wave, phase comparison, single-station, tracking system operating at C-band and giving two **direction cosines** and **slant range** which can be used to determine space position and velocity.

*The U.S. Air Force Rockwell International B-1 strategic **bomber** is equipped with a variable geometry or swing wing. Wings are in the fully swept or high speed position in the view shown. The B-1 is not in production, as of 1980.*

B

B. Military mission designation for **bomber** aircraft.

background luminance. In visual-range theory the **luminance** (brightness) of the background against which a target is viewed. Compare **adaptation luminance.**

background noise. 1. In recording and reproducing, the total system **noise** independent of whether or not a signal is present. The signal is not to be included as part of the noise. **2.** In receivers, the noise in the absence of signal modulation on the carrier.

Ambient noise detected, measured, or recorded with the signal becomes part of the background noise.

Included in this definition is the interference resulting from primary power supplies, that separately is commonly described as **hum.**

background radiation. The radiation in man's natural environment, including cosmic rays and radiation from the naturally radio-active elements, both outside and inside the bodies of men and animals. It is also called *natural radiation.* The term may also mean radiation that is unrelated to a specific experiment. (See **cosmic rays.**)

background return. See **clutter.**

backlash. Dead space or unwanted movement in a **control system.**

backout. An undoing of things already done during a **countdown,** usually in reverse order.

back pressure. Pressure exerted backward; in a field of fluid flow, a pressure exerted contrary to the pressure producing the main flow.

back scatter. Same as **backward scatter.**

backup. 1. An item kept available to replace an item which fails to perform satisfactorily. **2.** An item under development intended to perform the same general functions of another item also under development performs.

backward scatter. The **scattering** of radiant energy into the hemisphere of space bounded by a plane normal to the direction of the incident **radiation** and lying on the same side as the incident ray; the opposite of **forward scatter.** Also called *back scattering.*

Atmospheric backward scatter depletes 6 to 9 percent of the incident solar beam before it reaches the earth's surface.

In radar usage, *backward scatter* refers only to that radiation scattered at 180° to the direction of the incident wave.

bail out. To jump from an aircraft and descend by means of a parachute.

bailout bottle. A personal supply of oxygen usually contained in a cylinder under pressure and utilized when the individual has left the central oxygen system as in a parachute jump.

bakeout. The **degassing** of surfaces of a vacuum system by heating during the pumping process.

balance. 1. The equilibrium attained by an aircraft, rocket, or the like when forces and **moments** are acting upon it so as to produce steady flight, especially without rotation about its axes; also used with reference to equilibrium about any specified **axis** as, *an airplane in balance about its longitudinal axis.* **2.** A weight that

counterbalances something, especially on an aircraft control surface, a weight installed forward of the hinge axis to counterbalance the surface aft of the hinge axis.

balance tab. A **tab** designed to reduce the effort required to operate a control surface.

balanced circuit. A **circuit,** the two sides of which are electrically alike and symmetrical with respect to a common reference point, usually ground.

balanced control surface. 1. A control surface which is aerodynamically balanced to reduce the force necessary to displace it by providing some area ahead of the hinge line, or **2.** a control surface which is balanced by weights ahead of its hinge line to prevent possible flutter. Also see **aerodynamic balance.**

balanced detector. A **demodulator** for frequency-modulation systems. In one form the output consists of the rectified difference of the two voltages produced across two resonant circuits, one circuit being tuned slightly above the **carrier frequency** and the other slightly below.

balanced modulator. A device in which the **carrier** and modulating signal are so introduced that, after modulation takes place, the output contains the two **sidebands** without the carrier.

balanced runway concept. A runway length design concept wherein the length of prepared runway is such that the **accelerate-stop** distance is equal to the takeoff distance for the aircraft for which the runway is designed.

ballistic camera. A ground-based camera using multiple exposures on the same plate to record the **trajectory** of a rocket.

ballistic coefficient. A design parameter indicating the relative magnitude of inertial and aerodynamic affects (weight to a function of drag), used in performance analysis of objects which move through the atmosphere.

ballistic missile. Any missile which does not rely upon aerodynamic surfaces to produce lift and consequently follows a ballistic trajectory when thrust is terminated. See also **aerodynamic missile; guided missile.**

ballistic missile early warning system. An electronic system for providing detection and early warning of attack by enemy intercontinental ballistic missiles.

ballistic temperature. That temperature (in °F) which, when regarded as a surface temperature and used in conjunction with the lapse rate of the standard artillery atmosphere, would produce the same effect on a projectile as the actual temperature distribution encountered by the projectile in flight.

ballistic trajectory. The **trajectory** followed by a body being acted upon only by gravitational forces and the resistance of the medium through which it passes.

A rocket without lifting surfaces is in a *ballistic trajectory* after its engines cease operating.

ballistic vehicle. A nonlifting vehicle; a vehicle that follows a **ballistic trajectory.**

ballistic wind. The constant wind which would produce the same effect upon the **trajectory** of a projectile as the actual wind encountered in flight. Ballistic winds can be regarded as made up of **range wind** and **crosswind** components.

ballistics. The science that deals with the motion, behavior, and effects of projectiles, especially bullets, aerial bombs, rockets, or the like; the science or art of designing and hurling projectiles so as to achieve a desired performance.

balloon. An **aerostat** without a propelling system.

balloon barrage. See **barrage,** part **2.**

balloon-type rocket. A liquid-fuel **rocket,** such as Atlas, that requires the pressure of its **propellants** (or other gases) within it to give it structural integrity.

ballute. A cross between a balloon and a parachute, used to brake the free fall of sounding rockets.

band. 1. frequency band. **2.** absorption band. **3.** A group of tracks on a **magnetic drum. 4.** auroral band. See aurora.

band-elimination filter. A **wave filter** that attenuates one **frequency band,** neither the critical nor cutoff frequencies being zero or infinite.

bandwidth. 1. In an **antenna,** the range of **frequencies** within which its performance, in respect to some characteristic, conforms to a specified standard. **2.** In a **wave,** the least frequency interval outside of which the **power spectrum** of a time-varying quantity is everywhere less than some specified fraction of its value at a reference frequency. **3.** The number of cycles per second between the limits of a **frequency band.**

Sense **2** permits the spectrum to be less than the specified fraction within the interval. Unless otherwise stated, the reference frequency is that at which the spectrum has its maximum value.

4. In information theory, the information-carrying capacity of a communications channel.

bang-bang control. Flicker control, especially as applied to rockets.

Bang-bang in this term is imitative, arising from the noise made by control mechanisms slamming first to one side, then to the other, in this sort of control.

bank, to bank an airplane. To incline the airplane so that the **lateral axis** of the airplane makes an angle with the horizontal. Banks are incident to all properly executed turns.

bank indicator. See **turn-and-bank indicator.**

baralyme. A commercial trade name for a type of carbon dioxide absorber, a mixture of calcium hydroxide and barium hydroxide.

Barany chair. (After Robert Barany, 1876-1936, Swedish physician.) A kind of chair in which a person is revolved to test his susceptibility to **vertigo.**

bare base. (military) A base having a runway, taxiways, and parking areas which are adequate for the deployed force and possesses an adequate source of water that can be made potable.

barn. A unit of area for measuring a **nuclear cross section.** One barn equals 10^{-24} square centimeter.

barometer. An instrument for measuring the pressure of the atmosphere. The two principal types are the mercurial and the aneroid. Also see **aneroid barometer.**

barometric tendency. The change of barometric pressure within a specified time (usually three hours), before the observation.

barosphere. The atmosphere below the **critical level** of escape.

baroswitch. (from *barometric switch).* **1.**Specifically, a pressure-operated switching device used in a **radiosonde.** In operation, the expansion of an aneroid capsule causes an electrical contact to scan a radiosonde commutator composed of conductors separated by insulators. Each switching operation corresponds to a particular pressure level. The contact of an insulator or a conductor determines whether temperature, humidity, or reference signals will be transmitted. **2.** Any switch operated by a change in **atmospheric pressure.**

barrage. 1. A prearranged barrier of fire, except that delivered by small arms, designed to protect friendly troops and installations by impeding enemy movements across defensive lines or areas. See also **fire. 2.** A protective screen of balloons that are moored to the ground and kept at given heights to prevent or hinder operations by enemy aircraft. This meaning also called **balloon barrage. 3.** A type of electronic countermeasures intended for simultaneous jamming over a wide area of frequency spectrum. See also **barrage jamming; electronic countermeasures; electronic jamming.**

barrage jamming. Simultaneous electronic jaming over a broad bank of frequencies. See also **jamming.**

barrier combat air patrol. One or more divisions or elements of a fighter aircraft employed between a force and an objective area as a barrier across the probable direction of enemy attack. It is used as far from the force as control conditions permit, giving added protection against raids that use the most direct routes of approach. See also **combat air patrol.**

barycenter. 1. The center of gravity of the Earth-Moon system. **2.** The center of gravity of any set of revolving masses.

barye. The pressure unit of the **centimeter-gram-second system** of physical units; equal to one dyne per square centimeter. (0.001 millibar). Sometimes called *bar* or *microbar.*

base. 1. A locality from which operations are projected or supported. **2.** An area or locality containing installations which provide logistic or other support. **3.** Home airfield, or home carrier.

base. A quantity, the powers of which are assigned as the unit value of columns in a **numeric** system; for example, *two* is the base in **binary notation,** and *ten* in **decimal notation.** Also called *radix.* See **logarithm, binary notation.**

base command. An area containing a military base or a group of such bases organized under one commander. See also **command.**

base complex. See **Air Force base.**

base drag. Drag owing to a **base pressure** lower than the ambient pressure. It is a part of the **pressure drag.**

base leg. A flight path in the **traffic pattern** at right angles to the landing runway off its approach end and extending from the downwind leg of the extended runway centerline.

base line. 1. Any line which serves as the basis for measurement of other lines, as in a surveying triangulation, measurement of auroral heights, etc. **2.** The **geodesic line** between two stations operating in conjunction for the determination of a line of position, as the two stations constituting a loran rate. **3.** In radar, the line traced on **amplitude-modulated indicators** which corresponds to the power level of the weakest echo detected by the radar. It is retraced with every pulse transmitted by the radar, but appears as a nearly continuous display on the scope.

Target signals show up as perpendicular deviations from the base line; range is measured along the base line; signal strength is indicated by the magnitude of the deviations; and the type of target usually can be determined by the appearance of the deviations.

base of operations. An area or facility from which a military force begins its offensive operations, to which it falls back in case of reverse, and in which supply facilities are organized.

base period. That period of time for which factors were determined for use in current planning and programming.

base point. In computer terminology, the character, or the location of an implied symbol, which separates the integral part of an expression in **positional notation** from the fractional part; the point which marks the place between the zero and negative powers of the **base.** Also called *radix point.*

base pressure. In aerodynamics, the **pressure** exerted on the base, or extreme aft end, of a body, as of a cylindrical or boattailed body or of a blunt-trailing-edge wing, in a fluid flow.

base-timing sequencing (abbr **BT sequencing**). The control of the time sharing of a single **transponder** between several ground transmitters through the use of suitable coded timing signals.

basic cover (photogrammetry). Air coverage of any installation or area of a permanent nature with which later cover can be compared to discover any changes that have taken place.

basic encyclopedia. (military) A compilation of identified installations and physical areas of potential significance as objectives for attack.

basic intelligence. General intelligence concerning the capabilities, vulnerabilities, and intentions of foreign nations; used as a base for a variety of intelligence products for the support of planning, policy making, and military operations. See also **intelligence.**

basic load (ammunition). That quantity of nonnuclear ammunition which is autorized and required by each Service to be on hand within a unit or formation at all times. It is expressed in rounds, units, or units of weight as appropriate.

basic research. Research directed toward the increase of knowledge, the primary aim being a greater knowledge or understanding of the subject under study. See also **applied research; research.**

basic tactical organization. The conventional organization of landing force units for combat, involving combinations of infantry, supporting ground arms, and aviation for accomplishment of missions ashore. This organizational form is employed as soon as possible following the landing of the various assault components of the landing force.

basic thermal radiation. Thermal radiation from a **quiet sun.**

battle group. Army tactical and administrative infantry or airborne unit, on a command level below a division or brigade whose next lower echelons are companies, the entire organization of which is prescribed by a table of organization.

battle map. A map showing ground features in sufficient detail for tactical use by all forces, usually at a scale of 1:25,000. See also **map.**

baud. A unit of signaling speed. The speed in bauds is the number of **code elements** per second.

Baume scale. Either of two scales sometimes used to graduate **hydrometers;** one scale is for liquids heavier than water, the other for liquids lighter than water.

Bayard-Alpert ionization gage. A type of **ionization vacuum gage** using a tube with an electrode structure designed to minimize X-ray induced electron emission from the ion collector.

beachhead. A designated area on a hostile shore which, when seized and held, insures the continuous landing of troops and materiel, and provides maneuver space requisite for subsequent projected operations ashore. It is the physical objective of an amphibious operation.

beacon. A light, or group of lights or electronic source which emits a distinctive or characteristic signal used for the determination of bearings, courses, or location. See **aeronautical light beacon, crash locator beacon, fan marker beacon, homing beacon, non-directional radio beacon (NBD), marker beacon, personnel locator beacon, radar beacon, radio beacon (RBN), z-marker beacon.**

beacon delay. The amount of inherent delay within a **beacon,** i.e., the time between the arrival of a signal and the response of the beacon.

beacon skipping. A condition where **transponder** return pulses from a **beacon** are missing at the interrogating **radar.**

Beacon skipping can be caused by interference, over-interrogation of beacon, **antenna nulls,** or pattern minimums.

beacon stealing. Loss of **beacon tracking** by one **radar** due to (interfering) **interrogation** signals from another radar.

beacon tracking. The tracking of a moving object by means of signals emitted from a transmitter or **transponder** within or attached to the object.

beam. 1. A ray or collection of focused rays of radiated energy. See **beam width, radiation pattern. 2.** A beam (sense **1**) of radio waves used as a navigation aid. **3.** electron beam. **4.** A body, one of whose dimensions is large compared with the others, whose function is to carry lateral loads (perpendicular to the long dimension) and bending movements.

beam attack. In air interception, an attack by an interceptor aircraft which terminates with a heading crossing angle greater than 45 degrees but less than 135 degrees. See also **heading crossing angle.**

beam hole. nucleonics: An opening through a reactor shield and, generally, through the reactor reflector, which permits a beam of radioactive particles or radiation to be used for experiments outside the reactor.

beam rider. A missile guided by an electronic beam.

beam-rider guidance. A system for guiding aircraft or spacecraft in which a craft follows a radar beam, light beam, or other kind of beam along the desired path. Also see *beam-climber guidance.* See **guidance.**

beam width. A measure of the concentration of power of a **directional antenna.** It is the angle in degrees subtended at the antenna by arbitrary power-level points across the axis of the **beam.** This power level is usually the point where the **power density** is one-half that which is present in the axis of the beam at the same distance from the antenna (half-power points). Also called *beam angle.*

The beam width of a radar determines the minimum angular separation which two targets can have and still be resolved. Roughly speaking, two targets at the same range whose angular separations at the radar antenna exceeds one-half of the beam width between half-power points will be resolved or distinguishable as two individual targets. The smaller the beam width, the greater the angular resolving power. Beam width may be at different locations through the axis depending upon the shape of the antenna reflector.

bearing. (navig.) The horizontal angle at a given point, measured from a specific reference datum, to a second point. The direction of one point relative to another as measured from a specific reference datum. In air navigation, the same as azimuth. "Azimuth" is preferred for use in celestial navigation; "bearing" is preferred in all other forms of air navigation. Also see **direction, magnetic bearing, relative bearing, true bearing.**

beat. 1. One complete cycle of the variations in the **amplitude** of two or more periodic phenomena of different frequency which mutually react. See **beat frequency. 2.** To produce beating.

beat frequency. The frequency obtained when two simple **harmonic** quantities of different frequencies are superimposed.

beating. A wave phenomenon in which two or more periodic quantities of different **frequencies** produce a resultant having pulsations of **amplitude.**

This process may be controlled to produce a desired **beat frequency.** See **heterodyne.**

Beaufort scale of wind force. A numerical scale, ranging from 0 (Calm) to 12 (Hurricane), originated by Rear-Admiral Sir Francis Beaufort for the estimation of wind force by observing its effects on common objects. It is largely used in marine operations but occasionally appears in aviation operations.

beavertail antenna. A type of radar antenna which forms a beam having a greater beam width in azimuth than in elevation, or vice versa. In physical dimensions, its long axis lies in the plane of smaller beam width.

behavior. The way in which an organism, organ, body, or substance acts in an **environment** or responds to excitation, as the *behavior of steel under stress,* or *the behavior of an animal in a test.*

bench check. A work shop or servicing bay check which includes the typical check or actual functional test of an item to ascertain what is to be done to return the item to a serviceable condition or ascertain the item's temporary or permanent disposition.

bends. The acute pain and discomfort in the arms, legs, and joints resulting from the formation of nitrogen bubbles, together with other biological gases, in body tissues and fluids, caused by exposure to reduced barometric pressure. The incidence of bends at high altitudes can be greatly reduced by denitrogenation (breathing pure oxygen) at ground level before ascent. See **decompression sickness, dysbarison.**

Bernoulli law or **Bernoulli theorem.** After Daniel Bernoulli, 1700-1782, Swiss scientist.) **1.** In aeronautics, a law or theorem stating that in a flow of incompressible fluid the sum of the **static pressure** and the **dynamic pressure** along a **streamline** is constant if gravity and frictional effects are disregarded.

From this law it follows that where there is a velocity increase in a fluid flow there must be a corresponding pressure decrease. Thus an airfoil, by increasing the velocity of the flow over its upper surface, derives lift from the decreased pressure.

2. As originally formulated, a statement of **conservation of energy** (per unit mass) for a nonviscous fluid in steady motion. The specific energy is composed of the kinetic energy; the potential energy; and the work done by the pressure forces of a compressible fluid.

best angle-of-climb airspeed. (airplane). The airspeed which results in the greatest increase in altitude in a given distance.

best rate-of-climb airspeed. (airplane). The airspeed which results in the greatest increase in altitude in a unit of time.

beta particle. [Symbol β (beta)] An *elementary particle* emitted from a nucleus during radioactive decay, with a single electrical charge and a mass equal to $1/1837$ that of a proton. A negatively charged beta particle is identical to an *electron.* A positively charged beta particle is called a *positron.* Beta radiation may cause skin burns, and beta-emitters are harmful if they enter the body. Beta particles are easily stopped by a thin sheet of metal, however. (See **decay, radioactive. radioactive.**)

betatron. A doughnut-shaped accelerator in which electrons, traveling in an orbit of constant radius, are accelerated by a changing magnetic field. Energies as high as 340 Mev have been attained. (See **accelerator.**)

bias error. A measurement error that remains constant in magnitude for all observations. A kind of **systematic error.**

An example is an incorrectly set **zero** adjustment.

bimetallic strip gage. A thermal conductivity **vacuum gage** in which deflection of a bimetallic strip with changing temperature indicates the changes in pressure.

binary. 1. Involving the integer two (2). See **binary notation. 2.** binary cell. **3.** binary star.

binary cell. Any device or circuit that can be placed in either of two stable states to store a **bit** of binary information. Often called a **binary.**

binary chain. A cascaded series of **binary cells.**

binary code. A code composed of a combination of entities each of which can assume one of two possible states. Each entity must be identifiable in time or space.

binary counter. A **counter** with two distinguishable states.

binary digit. A **digit** (0 or 1) in binary **notation.** See **bit.**

binary magnetic core. A ferromagnetic material which can be caused to assume either of two stable magnetic states and thus can be used in a **binary cell.**

binary notation. A system of **positional notation** in which the digits are coefficients of powers of the base 2 in same way as the digits in the conventional decimal system are coefficients of powers of the base 10.

Binary notation employs only two digits, 1 and 0, therefore is used extensively in computers where the *on* and *off* positions of a switch or storage device can represent the two digits.

In decimal notation $111 = (1 \times 10^2) + (1 \times 10^1) + (1 \times 10^0) = 100 + 10 + 1 =$ one hundred and eleven. In binary notation $111 = (1 \times 2^2) + (1 \times 2^1) + (1 \times 2^0) = 4 + 2 + 1 =$ seven.

binary number system. See **binary notation.**

binary point. The **base point** in binary notation.

binary star. Two stars revolving around a common center of gravity.

binding energy. (nucleonics). The binding energy of a *nucleus* is the minimum energy required to dissociate it into its component neutrons and protons. Neutron or proton binding energies are those required to remove a neutron or a proton, respectively, from a nucleus. Electron binding energy is that required to remove an electron from an atom or a molecule. (Compare **fission, ionization.**)

binding energy. 1.The force which holds molecules, atoms, or atomic particles together; specifically, the force which holds an atomic nucleus together. **2.** The energy required to break chemical, atomic, or molecular bonds.

bingo. (military.)**1.**(when orginated by controlling activity)—Proceed to alternate airfleld or carrier as specified. **2.**(when orginated by pilot)—I have reached minimum fuel for safe return to base or to designated alternate.

bingo field. (military) Alternate airfield.

bioastronautics. The study of biological, behavioral, and medical problems pertaining to **astronautics.** This includes systems functioning in the **environments** expected to be found in space, vehicles designed to travel in space, and the conditions on celestial bodies other than on earth.

biochemistry. Chemistry dealing with the chemical processes and compounds of living organisms.

bioclimatology. The study of the relations of climate and life, especially the effects of climate on the health and activity of human beings (human bioclimatology) and on animals and plants.

biodynamics. The study of the effects of dynamic processes (motion, **acceleration, weightlessness,** etc.) on living organisms.

biological agent. A microorganism which causes disease in man, plants, or animals or causes the deterioration of materiel. See also **chemical agent.**

biological dose. The radiation dose absorbed in biological material. Measured in *rems*. (See **absorbed dose.**)

biological half-life. The time required for a biological system, such as a man or an animal, to eliminate, by natural processes, half the amount of a substance (such as a radioactive material) that has entered it. (Compare **half-life;** see **half-life, effective.**)

biological shield. A mass of absorbing material placed around a reactor or radioactive source to reduce the radiation to a level that is safe for human beings. (See **absorber**)

biological warfare. Employment of living organisms, toxic biological products, and plant growth regulators to produce death or casualties in man, animals or plants; or defense against such action.

biological weapon. An item of materiel which projects, disperses, or disseminates a biological agent including arthropod vectors.

bionics. The study of systems, particularly **electronic** systems, which function after the manner of, or in a manner characteristic of, or resembling, living systems.

bio-pak. A container for housing a biological organism in a habitable environment and to record biological functions during space flight.

biosatellite. An artificial **satellite** which is specifically designed to contain and support man, animals, or other living material in a reasonably normal manner for an adequate period of time and which, particularly for man and animals, possesses the proper means for safe return to the earth. See **ecological system.**

biosensor. A **sensor** used to provide information about a life process.

biosphere. That transition zone between earth and at-mosphere within which most forms of terrestrial life are commonly found; the outer portion of the **geosphere** and inner or lower portion of the atmosphere. See **hydrosphere.**

biotechnology. The application of engineering and technological principles to the life sciences.

biotelementry. The remote measuring and evaluation of life functions, as, e.g., in **spacecraft** and artificial **satellites.**

biotron. A **test chamber** used for biological research within which the environmental conditions can be completely controlled, thus allowing observations of the effect of variations in **environment** on living organisms.

biplane. An airplane with two main supporting surfaces (wings) placed one above the other.

bipropellant. A rocket **propellant** consisting of two unmixed or uncombined chemicals (**fuel and oxidizer**) fed to the combustion chamber separately.

bipropellant rocket. A rocket using two separate **propellants** which are kept separate until mixing in the **combustion chamber.**

biquinary notation. A numerical system in which each **decimal digit** is represented by a pair of digits consisting of a coefflcient of five followed by a coefflcient of one.

For example, the decimal digit 7 is represented in biquinary notation by 12[(1 x 5) + (2 x 1)], and the decimal quantity 3648 is represented by 03 11 04 13. The abacus is based on biquinary notation.

bistable elements. In computer terminology, a device which can remain indefinitely in either of two stable states.

bit. 1. An abbreviation of *b*inary dig*it*. **2.** A single character of a language employing only two distinct kinds of characters. **3.** A quantity of intelligence which is carried by an identifiable entity and which can exist in either of two states. **4.** A unit of **storage capacity;** the capacity in bits of a storage device is the logarithm to the base two of the number of possible states of the device. **5.** A **quantum** of information. **6.** Loosely, a mark.

bit rate. The frequency derived from the period of time required to transmit one **bit.**

black body, blackbody. 1. An ideal emitter which radiates energy at the maximum possible rate per unit area at each wavelength for any given temperature. A black body also absorbs all the radiant energy in the near visible spectrum incident upon it.

No actual substance behaves as a true black body, although platinum black and other soots rather closely approximate this ideal. However, one does speak of a black body with respect to a particular wavelength interval. This concept is fundamental to all the **radiation laws,** and is to be compared with the similarly idealized concepts of the **white body** and the **gray body.** In accordance with the **Kirchoff law,** a black body not only absorbs all wavelengths, but emits at all wavelengths and does so with maximum possible intensity for any given temperature.

2. A laboratory device which simulates the characteristics of a **black body** (sense 1).

black-body radiation. The **electromagnetic radiation** emitted by an ideal **black body;** it is the theoretical maximum amount of radiant energy of all wavelengths which can be emitted by a body at a given temperature.

The spectral distribution of black-body radiation is

described by **Planck law** and the related **radiation laws.** If a very tiny opening is made into an otherwise completely enclosed space (hohlraum), the radiation passing out through this hole when the walls of the enclosure have come to thermal equilibrium at some temperature will closely approximate ideal black-body radiation for that temperature.

black box. 1. In engineering design, a unit whose **output** is a specified function of the **input,** but for which the method of converting input to output is not necessarily specified. **2.**Colloquially, any unit, usually an electronic device such as an amplifier, which can be mounted in, or removed from, a rocket, spacecraft, or the like as a single package.

black list. An official counterintelligence listing of actual or potential enemy collaborators, sympathizers, intelligence suspects, and other persons whose presence menaces the security of friendly forces.

blackout. 1. A **fadeout** of radio communications due to ionospheric disturbances.

Blackouts are most common in, but are not restricted to, the arctic. An arctic blackout may last for days or even weeks during periods of intense auroral activity.

Recent experiments with high-altitude nuclear detonations have produced blackouts and artificial auroras over the subtropics.

2.A fadeout of radio and telemetry transmission between ground stations and vehicles travelling at high speeds in the atmosphere caused by signal attenuation in passing through ionized boundary layer (**plasma sheath**) and shock wave regions generated by the vehicle.

3.A vacuum tube characteristic which results from the formation of a **dielectric** film on the surface of the control grid.

A negative charge, accumulated on the film when the grid is driven positive with respect to the cathode, affects the operating characteristics of the tube.

4.A condition in which vision is temporarily obscured by a blackness, accompanied by a dullness of certain of the other senses, brought on by decreased blood pressure in the eye and a consequent lack of oxygen, as may occur, e.g., in pulling out of a high-speed dive in an airplane. Compare **grayout, redout.**

blade. 1. (a) An arm of a propeller; a rotating wing. (**b**)Specifically, restrictive, that part of a propeller arm or of a rotating wing from the shank outward, i.e., that part having an efficient airfoil shape and that cleaves the air. **2.** A **vane** (in sense **2**), such as a rotating vane or stationary vane in a rotary air compressor, or a vane of a turbine wheel.

blade damper (rotor). A device for damping the motion of a rotor blade about the lag hinge.

blanket. To blank out or obscure weak radio signals by a stronger signal.

blast. 1. The brief and rapid movement of air or other fluid away from a center of outward pressure, as in an explosion. **2.**The characteristic instantaneous rise in **pressure,** followed by a sudden decrease, that results from this movement, differentiated from less rapid pressure changes. **3.**To **take off** from a launching pad or stand. Said of a rocket in reference to the blast effects caused by rapid combustion of fuel as the rocket starts to move upward. (Popular.)

This term is commonly used for *explosion,* but the two terms should be distinguished. In space, an explo-

sion could take place, but no blast would follow

blast chamber. A **combustion chamber,** especially a combustion chamber in a gas-turbine engine, jet engine, or rocket engine.

blast deflector. A type of deflector used to divert the blast of a rocket fired from a vertical position.

blast effect. Destruction of or damage to structures and personnel by the force of an explosion on or above the surface of the ground. Blast effect may be contrasted with the cratering and ground-shock effects of a projectile or charge which goes off beneath the surface.

blastoff. A missile **launch.** (Slang.)

blast wave. A pulse of air, propagated from an explosion, in which the pressure increases sharply at the front of a moving air mass, accompanied by strong, transient winds.

bleed. To let a fluid, such as air or liquid oxygen, escape from a pipe, tank, or the like.

bleed off. To take off a part or all of a fluid from a tank or line, normally through an escape valve or outlet, as in *to bleed off excess oxygen from a tank.*

blimp. A colloquial term and refers to a small, nonrigid airship.

blind toss. A weapon delivery maneuver, executed under instrument conditons using radar to identify the target. The aircraft is pulled up and the weapon released in such a manner that it is tossed onto the target.

blip. A spot of light or deflection of the **trace** on a radarscope, loran indicator, or the like caused by the received signal, as from a reflecting object. Also called *a pip or echo.*

blob. A fairly small-scale temperature and moisture inhomogeneity produced by **turbulence** within the atmosphere.

The abnormal gradient of the **index of refraction** resulting from a *blob* can produce a radar echo of the type known as *angels.*

block. In computer operations, a group of **machine words** considered as a unit.

blockhouse, block house. 1.A reinforced concrete structure, often built underground or half underground, and sometimes dome shaped, to provide protection against blast, heat, or explosion during rocket launchings or related activities; specifically, such a structure at a **launch site** that houses electronic control instruments used in launching a rocket. **2.**The activity that works in such a structure.

blocking oscillator. A regenerative circuit which generates **pulses** of short duration.

Blocking oscillators are used in **digital computers.**

block time. The period from the moment the chocks are withdrawn and brakes released, or moorings dropped, to the return to rest or take up of moorings after the flight.

blood chit. A small cloth chart depicting an American Flag and a statement in several languages to the effect that anyone assisting the bearer to safety will be rewarded.

blowdown tunnel. A type of **wind tunnel** in which stored compressed gas is allowed to expand through a test section to provide a stream of gas or air for model testing.

The downstream side may or may not be reduced in pressure to provide greater expansion potential.

blowoff. Separation of an instrument section or package from the remainder of the rocket vehicle by application of explosive force, to retrieve the instruments after they have collected the required information. See **air breakup, fallaway section.**

bluff body. A body having a broad, flattened front, as in some **reentry vehicles.**

bluntness. A parameter of a **conic** related to the **eccentricity** of the conic.

boattail. The rear portion of an elongated body, as a rocket, having decreasing cross-sectional area toward the rear.

bobbing. Fluctuation of the strength of a radar **echo,** or its indication on a **radarscope,** due to alternate interference and reinforcement of returning reflected waves.

body. 1. The main part or main central portion of an airplane, airship, rocket, or the like; a fuselage or hull. **2.** In a general sense, any fabrication, structure, or other material form, especially one aerodynamically or ballistically designed, as, *an airfoil is a body designed to produce an aerodynamic reaction.*

body burden. (nucleonics) The amount of radioactive material present in the body of a man or an animal. (See **background radiation, whole body counter.**)

body of revolution. A symmetrical body having the form described by rotating a plane curve about an **axis** in its plane.

bogey. An air contact which is unidentified but assumed to be enemy. (Not to be confused with "unknown.")

bogie. 1. A supporting and aligning wheel or roller on the inside of an endless track, used, e.g., in certain types of landing gear. **2.** A type of landing-gear unit consisting of two sets of wheels in tandem with a central strut.

boilerplate model. A metal copy of a flight vehicle, the structure or components of which are heavier than the flight model.

boiling point. The temperature at which the **equilibrium vapor pressure** between a liquid and its vapor is equal to the external pressure on the liquid. Compare **ice point.**

boiling water reactor. (nucleonics) A reactor in which water, used as both coolant and moderator, is allowed to boil in the core. The resulting steam can be used directly to drive a turbine.

boiloff. The vaporization of a liquid, such as liquid oxygen or liquid hydrogen, as its temperature reaches its **boiling point** under conditions of exposure, as in the tank of a rocket being readied for launch.

bolide. A brilliant **meteor,** especially one which explodes; a detonating **fireball.**

bolometer. An instrument which measures the intensity of radiant energy by employing a thermally sensitive electrical resistor; a type of **actinometer.** Also called *actinic balance. Compare* **radiometer.**

Two identical, blackened, thermally sensitive electrical resistors are used in a Wheatstone bridge circuit. Radiation is allowed to fall on one of the elements, causing a change in its resistance. The change is a measure of the intensity of the radiation.

bolometric magnitude. 1. The **magnitude** of a star for the entire **electromagnetic spectrum** without atmospheric absorption.

The magnitude measured within the earth's atmosphere by a bolometer is the *radiometric magnitude.* **2.** Loosely, **radiometric magnitude.**

Boltzmann constant. The ratio of the **universal gas constant** to **Avogadro number;** equal to 1.38054×10^{-16} erg/°K. Sometimes called *gas constant per molecule, Boltzmann universal conversion factor.*

bombardier. A member of a bomber crew who operates the bomb sight and release mechanism.

bomb alarm system. A fully automatic system of detectors ringing key target areas in North America for transmitting to display centers reports of nuclear bursts. See also **nuclear detonation detection and reporting system.**

bomb damage assessment. The determination of the effect of all air attacks on targets (e.g., bombs, rockets, or strafe).

bomber (B). 1. light: A bomber designed for a tactical operating radius of under 1,000 nautical miles at design gross weight and design bomb load. **2. medium:** A bomber designed for a tactical operating radius of between 1,000 to 2,500 nautical miles at design gross weight and design bomb load. **3. heavy:** A bomber designed for a tactical operating radius over 2,500 nautical miles at design gross weight and design bomb load. See **military aircraft types.**

bomb impact plot. A graphic representation of target area, usually a prestrike air photograph, on which prominent dots are plotted to mark the impact or detonation points of bombs dropped on a specific bombing attack.

bomb release point. The point in space at which bombs must be released to reach the desired point of detonation.

Bond albedo. The ratio of the amount of light reflected from a sphere exposed to parallel light to the amount of light incident upon it. Sometimes shortened to *albedo.*

The Bond albedo is used in planetary astronomy.

bonding. 1. Specifically, a system of connections between all metal parts of an aircraft or other structure forming a continuous electrical unit and preventing jumping or arcing of static electricity. **2.** Glueing or cementing together for structural strength.

bone seeker. (nucleonics). A *radioisotope* that tends to accumulate in the bones when it is introduced into the body. An example is strontium-90, which behaves chemically like calcium.

Boolean algebra. The study of the manipulation of symbols representing operations according to the rules of logic.

Boolean algebra corresponds to an algebra using only the numbers 0 and 1, therefore can be used in programming digital computers which operate on the **binary** principle.

boost. 1. Additional power, pressure, or force supplied by a **booster,** as, hydraulic boost, or extra propulsion given a flying vehicle during lift-off, climb, or other part of its flight as with a booster engine. **2. Boost pressure. 3.** To supercharge. **4.** To **launch** or to push along during a portion of flight, as *to boost a ramjet to flight speed by means of a rocket, or a rocket boosted to altitude with another rocket.*

booster. 1. Short for **booster engine** or **booster rocket. 2. launch vehicle.**

booster engine. An engine, especially **a booster rocket,** that adds its thrust to the thrust of the **sustainer engine.**

booster pump. A pump in a fuel system, oil system, or the like, used to provide additional or auxiliary pressure when needed or to provide an initial pressure differential before entering a main pump, as in pumping hydrogen near the boiling point.

booster rocket. 1. A **rocket motor,** either solid or liquid, that assists the normal propulsive system or **sustainer**

engine of a rocket or aeronautical vehicle in some phase of its flight. **2.** A rocket used to set a vehicle in motion before another engine takes over.

In sense **2** the term *launch vehicle* is preferred.

booster vehicle. launch vehicle.

boostglide vehicle. A **vehicle** designed to glide in the atmosphere following a rocket-powered phase. Portions of the flight may be **ballistic,** out of the atmosphere.

boost pressure. Manifold pressure greater than the ambient **atmospheric** pressure, obtained by supercharging. Often called *boost.*

bootstrap. 1. Referring to a self-generating or self-sustaining process; specifically, the operation of liquid-propellant rocket engines in which, during main-stage operation, the gas generator is fed by the main propellants pumped by the turbopump, and the turbopump in turn is driven by hot gases from the gas generator system.

Such a system must be started in its operation by outside power or propellants. When its operation is no longer dependent on outside power or propellant the system is said to be in *bootstrap* operation.
2. In computer operations, the coded instructions at the beginning of an input tape which together with manually inserted instructions, initiate a **routine. 3. leap-frog.** See, **leapfrog test.**

boresight camera. A camera mounted in the optical axis of a **tracking radar** to photograph rockets being tracked while in camera range and thus provide a correction for the alinement of the radar.

boresighting. In radio the process of alining a **directional antenna** system by an optical procedure.

bounce table. A testing device which subjects devices and **components** to impacts such as might be encountered in accidental dropping.

boundary conditions. A set of mathematical conditions to be satisfied, in the solution of a differential equation, at the edges or physical boundaries (including fluid boundaries) of the region in which the solution is sought. The nature of these conditions usually is determined by the physical nature of the problem. See **boundary-value problem.**

boundary layer. The layer of **fluid** in the immediate vicinity of a bounding surface; in fluid mechanics, the layer affected by viscosity of the fluid, referring ambiguously to the *laminar boundary layer, turbulent boundary layer, planetary boundary layer,* or *surface boundary layer.*

In aerodynamics the boundary-layer thickness is measured from the surface to an arbitrarily chosen point, e.g., where the velocity is 99 percent of the stream velocity. Thus, in aerodynamics, *boundary layer* by selection of the reference point, can include only the laminar boundary layer or the laminar boundary layer plus all, or a portion of, the turbulent boundary layer.

boundary-layer control (of a wing). Control by artificial means of the development of the boundary layer with the object of affecting transition or separation, for example by withdrawing air from the boundary layer through the surface (suction) or be injecting air or other gas into the boundary layer (blowing). Use of boundary layer control provides a greater range of useable lift coefficients.

boundary-value problem. A physical problem completely specified by a differential equation in an unknown, valid in a certain region of space, and certain informa-

tion (**boundary condition**) about the unknown, given on the boundaries of that region. The information required to determine the solution depends completely and uniquely on the particular problem.

bow wave. A **shock wave** in front of a body, such as an airfoil, or apparently attached to the forward tip of the body.

Boyle law. Boyle-Mariotte law.

Boyle-Mariotte law. The empirical generalization that for many so-called perfect gases, the product of pressure p and volume V is constant in an isothermal process:
$$pV = F(T)$$
where the function F of the temperature T cannot be specified without reference to other laws (e.g., **Charles-Gay-Lussac law**). Also called *Boyle law, Mariotte law.*

brake horse-power—BHP. The power delivered at the propeller shaft of the engine.

brake parachute. deceleration parachute.

braking ellipses. A series of ellipses, decreasing in size due to **aerodynamic drag,** followed by a **spacecraft** in entering a planetary atmosphere.

In theory, this maneuver will allow a spacecraft to dissipate the heat generated in entry without burning up.

braking rocket. retrorocket.

branch. 1. A subdivision of any organization. **2.** A geographically separate unit of an activity which performs all or part of the primary functions of the parent activity on a smaller scale. Unlike an annex, a branch is not merely an overflow addition. **3.** An arm or service of the Army.

branch. 1. In an electrical circuit, a portion of **a network** consisting of one or more two-terminal elements in series. **2.** The point in **a computer** program at which the machine will proceed with one of two or more possible **routines** according to existing conditions and instructions.

Brayton cycle. (After George B. Brayton, American engineer.) Same as **Joule cycle.**

brazing. Joining metals by flowing a thin-layer capillary thickness of nonferrous filler metal into the space between them.

Bonding results from the intimate contact produced by the dissolution of a small amount of base metal in the molten filler metal, without fusion of the base metal. Sometimes, the filler metal is put in place as a thin solid sheet or as *cladding* and the composite is heated as in *furnace brazing.*

The term *brazing* is used where the temperature exceeds some arbitrary value, such as 800°F; the term *soldering* is used for temperatures lower than the arbitrary value.

breadboard. 1. An assembly of preliminary circuits or parts used to prove the feasibility of a device, circuit, system, or principle without regard to the final configuration or packaging of the parts. **2.** To prepare a *breadboard,* sense **1.**

breakaway. The action of a **boundary layer** separating from a surface.

breakaway. (military). After completion of attack, turn to heading as directed.

breakdown potential. dielectric strength.

break off. A command utilized to immediately terminate an attack.

breakoff phenomenon. The feeling which sometimes occurs during high-altitude flight of being totally separated and detached from the earth and human society. Also called the *breakaway phenomenon.*

break point. In computer operations, a point at which a **break-point instruction** inserted in the **routine** will cause the machine to stop, upon a command from the operator, for a check of progress.

break-point instruction. In computer operations, an instruction which, in conjunction with a manually operated control, causes the machine to stop.

breeder reactor. A reactor that produces fissionable fuel as well as consuming it, especially one that creates more than it consumes. The new *fissionable material* is created by capture in fertile materials of neutrons from fission. The process by which this occurs is known as *breeding.* (Compare **converter reactor; see fertile material.**)

bremsstrahlung. *Electromagnetic radiation* emitted by a fast-moving charged particle (usually an electron) when it is slowed down (or accelerated) and deflected by the electric field surrounding a positively charged atomic nucleus. X rays produced in ordinary X-ray machines are bremsstrahlung. (In German, the term means "braking radiation".) (See **X ray.**)

briefing. The act of giving in advance specific instructions or information.

broadcast. A transmission of information relating to air navigation that is not addressed to a specific station or stations.

broadcast controlled air interception. An interception in which the interceptor is given a continuous broadcast of information concerning an enemy raid and effects interception without further control. See also **air interception; close controlled air interception.**

broadside array. An **antenna array** whose direction of maximum radiation is perpendicular to the line or plane of the array according as the elements lie on a line or plane. A uniform broadside array is a **linear array** whose elements contribute fields of equal **amplitude** and **phase.**

brush discharge. corona discharge.

B-station. In **loran,** the designation applied to the transmitting station of a pair, the signal of which always occurs more than half a repetition period after the next succeeding signal and less than half a repetition period before the next preceding signal from the other station of the pair, designated an **A-station.**

B-trace. The second trace of an **oscilloscope** having more than one, as the lower trace of a loran indicator.

Btu *(abbr.).* **British** *t*hermal *u*nit.

bubble horizon. An artificial horizon parallel to the celestial horizon, established by means of a bubble level.

buckling. 1.An unstable state of equilibrium of a thin-walled body stemming from compressive stresses in the walls. 2.The lateral deflection of a thin-walled body resulting from such instability.

buffer. In computers: 1.An isolating **circuit** used to avoid reaction of a driven circuit on the corresponding driving circuit. 2.A **storage device** used to compensate for a difference in rate of flow of information or time or occurrence of events when transmitting information from one device to another.

buffer distance (nuclear). The horizontal distance which, when added to the radius of safety will give the desired

assurance that the specific degree of risk will not be exceeded. The buffer distance is normally expressed quantitatively in multiples of delivery error. 2.The vertical distance which is added to the fallout safe height of burst in order to determine a desired height of burst which will provide the desired assurance that fallout will not occur. It is normally expressed quantitatively in multiples of the vertical error.

buffer storage. In computer operations, **storage** used to compensate for a difference in rate of flow or time of occurrence when transferring information from one device to another.

buffeting. The beating of an aerodynamic structure or surfaces by unsteady flow, gusts, etc.; the irregular shaking or oscillation of a vehicle component owing to turbulent air or separated flow.

bug. 1.A concealed microphone or listening device or other audiosurveillance device. 2. To install means for audiosurveillance.

bugged. Room or object which contains a concealed listening device.

build. Of a radiant energy signal, to increase, often temporarily, in received **signal strength** without a change of receiver controls.

The opposite of *fade.*

bulk cargo. General cargo capable of being stacked on the floor of the aircraft.

bulkhead. A wall, partition, or similar member in a rocket, spacecraft, airplane fuselage, or similar structure, at right angles to the **longitudinal axis** of the structure, and serving to strengthen, divide, or help give shape to the structure.

bulk petroleum products. Liquid petroleum products which are normally transported by pipeline, rail tank car, road tank truck, road tank trailer, barge, harbor or coastal tanker and oceangoing tanker and stored in a tank or container having a fill capacity greater than 55 United States gallons (45 Imperial gallons).

burble. A separation or breakdown of the **laminar flow** past a body; the eddying or **turbulent flow** resulting from this.

Burble occurs over an airfoil operating at an angle of attack greater than the angle of maximum lift, resulting in a loss of lift and an increase of drag. See **compressibility burble.**

burble point. A point reached in an increasing **angle of attack** at which **burble** begins. Also called *burble angle.*

burn (military). 1.Deliberately expose the true status of a person under cover. 2. The legitimate destruction and burning of classified material, usually accomplished by the custodian as prescribed in regulations.

burnable poison. (nucleonics). A neutron absorber (or poison), such as boron, which, when purposely incorporated in the fuel or fuel cladding of a *nuclear reactor,* gradually "burns up" (is changed into nonabsorbing material) under neutron irradiation. This process compensates for the loss of reactivity that occurs as fuel is consumed and fission-product poisons accumulate, and keeps the overall characteristics of the reactor nearly constant during its use. (See **reactivity.**)

burner. combustion chamber.

burn-in. debug.

burn notice. An official statement by one intelligence agency to other agencies, domestic or foreign, that an individual or group is unreliable for any of a variety of reasons.

burnout. 1. An act or instance of fuel or oxidant depletion or, ideally, the simultaneous depletion of both; the time at which this occurs. Compare **cutoff.**

In the United Kingdom *all burnt* is preferred to *burnout.*

2. An act or instance of something burning out or of overheating; specifically, an act or instance of a rocket combustion chamber, nozzle, or other part overheating so as to result in damage or destruction.

burn-through range. The distance at which a specific radar can discern targets through the external interference being received.

bus. In computer operations, a main circuit, channel, or path for the transfer of information. Also called *trunk*

buster (military). Fly at maximum continuous speed (or power).

buzz. 1. In **supersonic diffuser** aerodynamics, a nonsteady shock motion and airflow associated with the shock system ahead of the inlet, very rapid pressure pulsations are produced which can affect downstream operation in the burner, nozzle, etc. **2.** Sustained oscillation of an aerodynamic control surface caused by intermittent **flow** separation on the surface, or by a motion of **shock waves** across the surface, or by a combination of flow separation and shock-wave motion on the surface.

bypass ratio. In a **turbofan,** the ratio of secondary or non-combusted air flow to the primary air flow which is combusted. The bypass ratio is approximately 5 to 1 for most turbofans or five times as much secondary air flows along the engine axis between the inner and outer casing as flows through the combustion chambers and turbines.

by-product material (nucleonics). Any radioactive material (except source material or fissionable material) obtained during the production or use of source material or fissionable material. It includes *fission products* and many other *radioisotopes* produced in nuclear reactors. (Compare **fissionable material, source material.**)

C

C. Military mission designation for **cargo/transport** aircraft.

cabin. In an aircraft, all the compartments used for the carriage of passengers or cargo.

cabin supercharger. In pressurized aircraft, a compressor for maintaining the pressure in the cabin above the ambient air pressure.

cage. To lock a **gyro** in a fixed position in its case.

caging. The process of orienting and mechanically locking the spin axis of a **gyro** to an internal reference position.

caisson disease. Those conditions including collapse, neurological changes, and pain, associated with relatively rapid reduction of ambient pressure from levels appreciably higher than 1 atmosphere to 1 atmosphere; and due to the release of inert gases in the body. Also called *compressed air illness, bends.*

calculating punch. A punched-card machine in which information is read from cards, and the results of sequential operations are punched on cards as they pass through the machine.

calendar year. The year of the **Gregorian calendar,** common years having 365 days and leap years 366 days.

Each year exactly divisible by 4 is a leap year, except century years (1800, 1900, etc.), which must be exactly divisible by 400 (2000, 2400, etc.) to be leap years. The calendar year is based on the **tropical year.** Also called *civil year.*

calibrated airspeed. See **airspeed.**

calibration card. A card mounted near an instrument indicating the corrections for instrument and installation errors.

call for fire. (military). A request for fire containing data necessary for obtaining the required fire on a target.

call mission. (military) A type of air support mission which is not requested sufficiently in advance of the desired time of execution to permit detailed planning and briefing of pilots prior to takeoff. Aircraft scheduled for this type of mission are on air, ground, or carrier alert, and are armed with prescribed load.

call number. In computer operations, a set of characters identifying a **subroutine** and containing (**a**) information concerning parameters to be inserted in the subroutine, (**b**) information to be used in generating the subroutine, or (**c**) information related to the **operands.**

call sign. Any combination of characters or pronounceable words which identifies a communication facility, a command, an authority, an activity, or a unit; used primarily for establishing and maintaining communications.

calorimeter. An instrument designed to measure **heat** evolved or absorbed.

Calorimeters are used in some **pyrheliometers.**

camber.

1. Curvature of the median line of an air foil section; more generally, the curvature of a surface.

2. The ratio of the maximum height of the median line above the chord to the chord length.

camouflage. The use of concealment and disguise to minimize the possibility of detection and/or identification of troops, materiel, equipment and installations. It includes taking advantage of the natural environment as well as the application of natural and artificial materials.

camouflage detection photography. Photography utilizing a special type of film (usually infrared) designed for the detection of camouflage.

canard. Pertaining to an **aerodynamic vehicle** in which horizontal surfaces used for trim and control are forward of the main lifting surface; the horizontal trim and control surfaces in such an arrangement.

candela. The unit of luminous intensity in the International System of Units, 1960; equal to one-sixtieth of the luminous intensity from 1 square centimeter of a **black body** at 2046°K (the temperature of solidification of platinum). Also called *candle.*

candle. candela.

cannibalize. To remove serviceable parts from one item of equipment in order to install them on another item of equipment.

canonical time unit. For geocentric orbits, the time required by a hypothetical satellite to move one radian in a circular **orbit** of the earth's equatorial radius; 13.447052 minutes.

canopy. **1.** The fabric body of a parachute, which provides high air-drag when inflated. **2.** A transparent, bubble-like enclosure for the flight crew of an aircraft.

cantelever wing. A wing built on the principle of a cantelever beam with no external struts or bracing.

CAP. Civil Air Patrol.

capability. A power or capacity to do something. Compare **characteristic.**

Capabilities belong to people, organized forces, or things.

capacity. In computer operations, (**a**) the largest quantity which can be stored, processed, or transferred; (**b**) the largest number of **digits** or characters which may regularly be processed; (**c**) the upper and lower limits of the quantities which may be processed.

capsule. **1.** A boxlike component or unit, often sealed. See **aneroid.** **2.** A small, sealed, pressurized cabin with an internal **environment** which will support life in a man or animal during extremely high altitude flight, space flight, or emergency escape. See **ejection capsule.**

The term *spacecraft* is preferred to *capsule* for any man-carrying vehicle.

3. A container carried on a rocket or spacecraft, as an instrument *capsule* holding instruments intended to be recovered after a flight.

captain/aircraft commander. The crew member authorized to command an aircraft.

captive test. A **holddown test** of a propulsion subsystem, rocket engine or motor. Distinguished from a *flight test.*

capturing. The use of a **torquer** to restrain the spin axis of a **gyro** to a specified position relative to the spin reference axis.

carbide. A compound of carbon with one or more metallic elements.

carbon cycle. A sequence of atomic **nuclear** reactions and spontaneous **radioactive decay** which serves to convert

The Lockheed, C-5A Galaxy **cargo** / **transport** *aircraft cruises at speeds above 500 miles per hour and hauls payloads up to 130 tons. It is powered by four General Electric turbofan engines.*

matter into energy in the form of radiation and high-speed particles, and which is regarded as one of the principal sources of the energy of the sun and other similar stars.

This cycle, first suggested by Bethe in 1938, gets its name from the fact that carbon plays the role of a kind of catalyst in that it is both used by and produced by the reaction, but is not consumed itself. Four protons are, in net, converted into an alpha particle and two positrons (with accompanying neutrinos); and three gamma-ray emissions are emitted directly in addition to the two gamma emissions that ensue from annihilation of the positrons by ambient electrons. This cycle sets in at stellar interior temperatures of the order of 5 million degrees Kelvin.

An even simpler reaction, the **proton-proton** reaction, is also believed to occur within the sun and may be of equal or greater importance.

carburetor icing. Ice formation either in the carburetor venturi or air scoop.

card. 1. A punched card, used in **computer** operations for the storage of information in the form of holes punched through the card material.

Standard punched cards are 7.375 X 3.250 X 0.007 inches, containing either 80 columns in each of which any of 12 positions may be punched or 90 columns in each of which any combination of 6 places may be punched. **2.** Any card adapted for the storage of information. **3.** A **printed-circuit** board, usually before other parts are mounted therein. See **module, package.**

cardinal altitudes or flight levels. "Odd" or "even" thousand-foot altitudes or flight levels. Examples: 5000, 6000, 7000, FL 250, FL 260, FL 270.

cardinal points. The directions: north, south, east, and west.

card punch. A mechanism which punches holes in **cards** used in **computer** operations.

An automatic card punch punches cards according to a stored program.

card reader. A mechanism that reproduces the information on punched **cards** in another form, usually electrical signals.

cargo transport (C), (military designation). Aircraft designed for carrying cargo and/or passengers. See **military aircraft types.**

carpet bombing. The progressive distribution of a mass bomb load upon an area defined by designated boundaries, in such a manner as to inflict damage to all portions thereof.

Carnot cycle. An idealized reversible **thermodynamic** cycle. The Carnot cycle consists of four stages: (**a**) an **isothermal** expansion of the gas at temperature T_1; (**b**) an **adiabetic** expansion to temperature T_2; (**c**) an isothermal compression at temperature T_2; (**d**) an adiabatic compression to the original state of the gas to complete the cycle. See **Carnot engine, thermodynamic efficiency.**

In a Carnot cycle, the net work done is the difference between the heat input Q_1 at higher temperature T_1 and the heat extracted Q_2 at the lower temperature T_2

Carnot efficiency. thermodynamic efficiency.

Carnot engine. An idealized reversible **heat engine** working in a **Carnot cycle.** It is the most efficient engine that can operate between two specified temperatures; its efficiency is equivalent to the **thermodynamic efficiency.**

The Carnot engine is capable of being run either as a conventional engine or as a refrigerator.

carrier, aircraft. See **aircraft carrier.**

carrier; nucleonic: A stable *isotope,* or a normal element, to which radioactive atoms of the same element can be added to obtain a quantity of radioactive mixture sufficient for handling, or to produce a radioactive mixture that will undergo the same chemical or biological reaction as the stable isotope. A substance in weighable amount which, when associated with a trace of another substance, will carry the trace through a chemical, physical or biological process. (See **tracer, isotopic.**)

carrier; elec. or rdo. A general term designating **carrier waves,** carrier current, or carrier voltage.

carrier air group. Two or more aircraft squadrons formed under one command for administrative and tactical control of operations from a carrier.

carrier striking force. A naval task force composed of aircraft carriers and supporting combatant ships capable of conducting strike operations.

carry time. In computer operations, the time required for a **binary chain** to complete its response to an input pulse.

carrier wave; rdo. A **wave,** modulated by a signal and which enables the signal to be transmitted through a specific physical system.

Cartesian coordinates. A **coordinate system** in which the locations of points in space are expressed by reference to three planes, called coordinate planes, no two of which are parallel. Compare **curvilinear coordinates.**

The three planes intersect in three straight lines, called coordinate axes. The coordinate planes and coordinate axes intersect in a common point, called the origin. From any point P in space three straight lines may be drawn, each of which is parallel to one of the three coordinate planes. If A, B, C denote these points of intersection, the Cartesian coordinates of P are the distances PA, PB, and PC. If the coordinate axes are mutually perpendicular, the coordinate system is rectangular; otherwise, oblique.

CAS. Calibrated Air Speed.

Cassegrain telescope. A reflecting telescope in which a small hyperboloidal mirror reflects the convergent beam from the paraboloidal primary mirror through a hole in the primary mirror to an eyepiece in back of the primary mirror. Also called *Cassegrainian telescope, Cassegrain.* See **Newtonian telescope.**

cassette. In photography, a reloadable container for either unexposed or exposed sensitized materials which may be removed from the camera or darkroom equipment under lightened conditions.

casualty (military). Any person who is lost to his organization by reason of having been declared dead, wounded, injured, diseased, interned, captured, missing; or a person whose whereabouts or status has not been determined.

CAT. clear air turbulence.

catalogue number. The designation of a star by the name of a particular **star catalogue** and the number of the star in that catalogue.

catalytic attack. An attack designed to bring about a war between major powers through the disguised machinations of a third power.

catapult. A power-actuated machine or device for hurling forth something, as an airplane or missile, at a high initial speed; also, a device, usually explosive, for ejecting a person from an aircraft. Compare **launcher,** senses **1** and **2.**

category. (**1**) As used with respect to the FAA certification, ratings, privileges, and limitations of airmen, means a broad classification of aircraft. Examples include: **airplane; rotorcraft; glider;** and **lighter-than-air;** and (**2**) As used with respect to the FAA certification of aircraft, means a grouping of aircraft based upon intended use or operating limitations. Examples include: **transport; normal; utility; acrobatic; limited; restricted;** and **provisional.**

cathode. The **electrode** from which the **electron** stream flows.

cathode-ray oscilloscope. An instrument which displays visually on the face of a **cathode-ray tube** instantaneous voltages of electrical signals. Either the intensity or the displacement of the trace may be controlled by the signal voltage. More commonly called *oscilloscope.* Also called *cathode-ray oscillograph.* See **radarscope.**

cathode rays. A stream of *electrons* emitted by the cathode, or negative electrode, of a gas-discharge tube or by a hot filament in a vacuum tube, such as a television tube.

cathode-ray tube (*abbr* **CRT**). A vacuum tube consisting essentially of an **electron gun** producing a concentrated electron beam (or cathode ray) which impinges on a phosphorescent coating on the back of a viewing face (or screen). The **excitation** of the **phosphor** produces light, the intensity of which is controlled by regulating the flow of electrons. Deflection of the beam is achieved either electromagnetically by currents in coils around the tube, or electrostatically by voltages on internal deflection plates.

catoptric light. A light concentrated into a parallel beam by means of a reflector.

A light so concentrated by means of refracting lenses or prisms is a dioptric light.

cat whisker. A fine wire **pickoff,** specifically a **gyro pickoff.**

cavitation. The formation of bubbles in a liquid, occurring whenever the **static pressure** at any point in the fluid flow becomes less than the fluid vapor pressure.

cavu. Ceiling and visibility unlimited.

C-band. See **frequency band.**

C-day (military). The unnamed day on which a deployment operation commences or is to commence. The deployment may be movement of troops, cargo, weapon systems, or a combination of these elements utilizing any or all types of transport. The letter "C" will be the only one used to denote the above. The highest command or headquarters responsible for coordinating the planning will specify the exact meaning of C-day within the aforementioned definition. The command or headquarters directly responsible for the execution of the operation, if other than the one coordinating the planning, will do so in light of the meaning specified by the highest command or headquarters coordinating the planning.

cease engagement. An order that weapons will disengage a particular target or targets and prepare to engage another target. Missiles in flight will continue to intercept. The order terminates engagement on a particular target.

cease fire. A command normally given to air defense artillery units to refrain from firing on, but to continue to track, an airborne object. Missiles already in flight will

be permitted to continue to intercept.

ceiling. (aircraft) The maximum altitude the airplane is capable of attaining under standard conditions.

ceiling. (meteorology). The height above the earth's surface of the lowest layer of clouds or obscuring phenomena that is reported as "broken," "overcast," or "obscuration" and not classified as "thin" or "partial."

celestial. 1. Of or pertaining to the heavens. 2. Short for *celestial navigation.*

celestial body. Any aggregation of matter in space constituting a unit for astronomical study, as the sun, moon, a planet, comet, star, nebula, etc. Also called *heavenly body.*

celestial coordinates. Any set of **coordinates** used to define a point on the **celestial sphere.**

The horizon, celestial equator, ecliptic, and galactic systems of celestial coordinates are based on the celestial horizon, celestial equator, ecliptic, and galactic equator, respectively, as the **primary great circle.** See **coordinate,** table |I, for a comparison of the systems.

celestial equator or equinoctial equator. The **great circle** of the **celestial sphere** formed by extending the plane of the earth's equator through the **celestial sphere.**

celestial guidance. The process of directing movements of an aircraft or spacecraft, especially in the selection of a **flight path,** by reference to celestial bodies. Also called *automatic celestial navigation.* See **guidance, celestial navigation.**

celestial horizon. The **great circle** of the **celestial sphere** created by passing a plane through the center of the earth **normal** to the vertical straight line connecting the **Zenith and Nadir.**

celestial-inertial guidance. The process of directing the movements of an aircraft or spacecraft, especially in the selection of a flight path, by an **inertial guidance** system which also receives inputs from observations of celestial bodies.

celestial latitude. Angular distance north or south of the **ecliptic;** the arc of a **circle of latitude** between the ecliptic and a point on the **celestial sphere,** measured northward or southward from the ecliptic through 90°, and labeled N or S to indicate the direction of measurement. See **ecliptic system of coordinates.**

celestial line of position. A **line of position** determined by observation of one (or more) celestial bodies.

celestial longitude. Angular distance east of the **vernal equinox,** along the **ecliptic;** the arc of the ecliptic or the angle at the ecliptic pole between the **circle of latitude** of the vernal equinox and the circle of latitude of a point on the celestial sphere, measured eastward from the circle of latitude of the vernal equinox, through 360°. See **ecliptic system of coordinates.**

celestial mechanics. The study of the theory of the motions of **celestial bodies** under the influence of gravitational fields. See **gravitation.**

celestial meridian. A **great circle** of the **celestial sphere,** through the celestial poles and the **zenith.**

The expression usually refers to the upper branch, that half of the great circle from pole to pole which passes through the zenith; the other half being the lower branch. The celestial meridian coincides with the **hour circle** through the zenith and the **vertical circle** through the elevated pole.

celestial navigation. The process of directing a craft from one point to another by reference to **celestial bodies** of known coordinates.

Celestial navigation usually refers to the process as accomplished by a human operator. The same process accomplished automatically by a machine is usually termed *celestial guidance* or sometimes *automatic celestial navigation.*

celestial observation. In navigation, the measurement of the **altitude** of a **celestial body,** or the measurement of **azimuth,** or measurement of both altitude and azimuth. Also called *sight.*

The expression may also be applied to the data obtained by such measurement.

celestial sphere. An imaginary sphere of infinite radius concentric with the earth, on which all **celestial bodies** except the earth are assumed to be projected.

celestial triangle. A spherical triangle on the **celestial sphere,** especially the **navigational triangle.**

cell. In computers, an elementary unit of **storage,** as *binary cell, decimal cell.*

Celsius temperature scale (*abb* **C**). Same as **centigrade temperature scale.**

The Ninth General Conference on Weights and Measures (1948) replaced the designation *degree centigrade* by *degree Celsius.*

center. See **Air Route Traffic Control Center.**

center frequency. The assigned **carrier frequency** of a **frequency-modulation** (**FM**) station; the unmodulated frequency of an FM system.

center of gravity. The point within an aerospace vehicle through which for balance purposes, the total force due to gravity is considered to act.

center of gravity limits. The limits within which an aircraft's center of gravity must lie to ensure safe flight. The center of gravity of the loading aircraft must be within these limits at take-off, in the air, and on landing. In some cases take-off and landing limits may also be specified.

center of lift. The mean of all the centers of pressure on an airfoil.

center of pressure (**aerodyn**). The point on some reference line (e.g. the chord of an airfoil) about which the pitching moment is zero.

center of thrust. thrust axis.

center section. The middle or center section of a wing, to which the outer wing penals are attached. Where a wing has no clearly defined central section, the center section is considered to lie between points of attachment of the wing to the fuselage or fuselage struts.

centigrade temperature scale (*abbr* **C**). A temperature scale with the ice point at 0° and the boiling point of water at 100°. Now called *Celsius temperature scale.*

Conversion to the Fahrenheit temperature scale is according to the formula

$$°C = 5/9 \, (°F—32)$$

centimeter. A metric unit of measurement equal to one hundredth of a meter. (Equal to .39 of an inch.)

centimeter-gram-second system (*abbr* **cgs**). A system of units based on the centimeter as the unit of length, the gram as the unit of mass, and the second as the unit of time.

centipoise. A unit of **viscosity.** See **poise.**

central air data computer (**CADC**). The central air data computer receives pressure inputs from the pitot-static system and temperature inputs from total temperature probes. The computer applies appropriate correction factors and computations and provides outputs of

altitude, vertical speed, airspeed, Mach number and temperature for use by interfacing systems such as the pilots' instruments, altitude reporting, warning and navigation systems.

central control. 1. Control exercised over an extensive and complicated system from a single center. **2.** Usually capitalized. The place, facility, or activity from which this control is exercised; specifically, at Cape Kennedy or at Vandenberg AFB, the place, facility, or activity at which the whole action incident to a test launch and flight is coordinated and controlled, from the make-ready at the launch site and on the range, to the end of the rocket flight downrange.

For a few seconds during the actual launch, control of a missile is exercised from the **blockhouse,** but it almost immediately reverts to Central Control for guidance and tracking, with two men in essential control. One of these is the supervisor of range operations, the other is the range safety officer.

centrifugal acceleration. centrifugal force.

centrifugal compressor. A **compressor** having one or more vaned rotary **impellers** which accelerate the incoming fluid radially outward into a **diffuser,** compressing by centrifugal force. Sometimes called a *centrifugal-flow compressor.* Compare **axial-flow compressor.**

centrifugal force. The **apparent force** in a rotating system, deflecting masses radially outward from the axis of rotation, with magnitude per unit mass $\omega^2 R$, where ω is the angular speed of rotation; and R is the radius of curvature of the path. This magnitude may also be written as V^2/R, in terms of the linear speed V. This force (per unit mass) is equal and opposite to the **centripetal acceleration.** Also called *centrifugal acceleration.*

The centrifugal force on the earth and atmosphere due to rotation about the earth's axis is incorporated with field of gravitation to form the field of gravity.

centrifuge. A large motor-driven apparatus with a long rotating arm at the end of which human and animal subjects or equipment can be revolved at various speeds to simulate very closely the prolonged accelerations encountered in high-performance aircraft, rockets, and spacecraft.

centripetal acceleration. The **acceleration** on a particle moving in a curved path, directed toward the instantaneous center of curvature of the path, with magnitude v^2/R, where v is the speed of the particle and R the radius of curvature of the path. This acceleration is equal and opposite to the **centrifugal force** per unit mass.

ceramic. An inorganic compound or mixture requiring heat treatment to fuse it into a homogeneous mass usually possessing high temperature strength but low ductility.

Types and uses range from china for dishes to **refractory** liner for **nozzles.**

Cerenkov radiation. The **radiation** from a charged particle whose velocity is greater than the **phase velocity** that an electromagnetic wave would have if it were propagating in the medium. The particle will continue to lose energy by radiation until its velocity is less than this phase velocity.

This phenomenon is analogous to the generation of a shock wave when an object is traveling faster than the sound velocity of the medium. A bow wave is set up which radiates energy into the medium and slows down the object.

The angle that the cone of **luminescence** makes with the direction of motion of the particle can be used to measure the velocity of the particle.

cermet (*cera*mic + *met*al). A body consisting of ceramic particles bonded with a metal; used in aircraft, rockets, and spacecraft for high strength, high temperature applications. Also called *ceramal* (*ceram*ic + *al*loy).

certificated airman. See **airman certificates.**

certificate of airworthiness. See **airworthiness certificate.**

certificate of registration. See **registration certificate.**

CGS system. A system of units based on the centimeter, the gram, and the second.

chad. The piece removed when punching a hole, as in a card. See **chadless.**

chadless. A type of punching in which the chad is left attached by about 25 percent of the circumference of the hole, at the leading edge.

Chadless punching is used where it is undesirable to mutilate information written or printed on the punched medium.

chaff. Radar confusion reflectors which consists of thin, narrow metallic strips of various lengths and frequency responses, used to reflect echoes for confusion purposes. See also **rope; window.**

chain of command. The succession of commanding officers from a superior to a subordinate through which command is exercised. Also called command channel.

chain radar beacon. A radar beacon with a very fast recovery time.

This recovery time provides the possibility of simultaneously interrogating and tracking the beacon by as many radars as required so long as they are phased, synchronized, or the sum total pulse recurrence frequency does not exceed the maximum pulse recurrence frequency characteristics of the beacon.

chain reaction. nucleonics: A reaction that stimulates its own repetition. In a fission chain reaction a fissionable *nucleus* absorbs a neutron and fissions, releasing additional neutrons. These in turn can be absorbed by other fissionable nuclei, releasing still more neutrons. A fission chain reaction is self-sustaining when the number of neutrons released in a given time equals or exceeds the number of neutrons lost by absorption in non-fissioning material or by escape from the system. (See **criticality, fission.**)

chamber. combustion chamber.

chamber pressure. The pressure of gases within the **combustion chamber** of a **rocket** engine.

chamber volume. The volume of the rocket **combustion chamber** including the convergent portion of the **nozzle** up to the throat.

chandelle. An airplane flight maneuver consisting of an abrupt climbing turn to approximately a stall in which the momentum of the airplane is used to obtain a higher rate of climb than would be possible in unaccelerated flight.

channel. 1. Short for *frequency channel.* **2.** In computer operations: (**a**) That portion of a **storage** medium which is accessible to a given reading station. See **track.** (**b**) A path of flow, usually including one or more operations.

Chapman region. A hypothetical region in the **upper atmosphere** in which the distribution of electron density with height can be described by a theoretical equation derived by Sydney Chapman.

Some of the basic assumptions used to develop the equation were that the ionizing radiation from the sun is essentially monochromatic, that the ionized constituent is distributed exponentially (with a constant **scale height**), and that there is an equilibrium condition between the creation of free electrons and their loss by recombination.

character. One of a set of elementary marks or events which may be combined to express information.

For example, a decimal digit (0 to 9), a letter (A to Z), or a symbol (comma, plus, minus, etc.).

characteristic. Specifically, a distinguishing quality, property, feature, or **capability** of a machine or piece of equipment, or of a component part.

The characteristics of an aircraft are (**1**) qualities such as stability, maneuverability, and strength; (**2**) features such as number, kind, or power of engines, and size, shape, or number of wings; and (**3**) capabilities such as range, speed, and payload.

characteristic equation. 1. An equation defining the characteristics of a set of partial differential equations. **2.** A linear algebraic equation determining the **eigenvalues** or free waves of a boundary-value problem. See **characteristic-value problem.**

characteristic length. A convenient reference length (usually constant) of a given configuaration, such as overall length of an aircraft, the maximum diameter or radius of a body of revolution, a chord or span of a lifting surface, etc.

characteristics. Lines or surfaces associated with a partial differential equation, or with a set of such equations, which are at all points tangent to characteristic directions, determined by certain specified linear combinations of the equations.

The use of these lines or surfaces may facilitate the solution of the equations and is known as the *method of characteristics.* The method has been particularly successful, for example, in the problem of finite-amplitude expansion and shock waves.

characteristic-value problem. A problem in which an undetermined **parameter** is involved in the coefficients of a differential equation, and in which the solution of the differential equation, with associated boundary conditions, exists only for certain discrete values of the parameter, called *eigenvalues,* or *characteristic values,* sometimes *principal values.*

An important example of a physical problem which leads to a characteristic-value problem is the determination of the modes and frequencies of a vibrating system. In this case the dependent variable of the differential equation represents the displacements of the system and the parameter represents the frequencies of vibration.

charge. The propellant of semifixed or separate loading ammunition.

charged particle. An *ion;* an *elementary particle* that carries a positive or negative electric charge. (See **plasma.**)

charge neutrality. The approximate equality of positive and negative particles in high-density **plasmas.**

This phenomenon, which is sometimes called *electrical neutrality,* is a result of the extremely large electric space charge fields that would arise if the densities were not equal. Although the positive and negative charge densities are seldom exactly equal, their percentage difference is so small as to be negligible. It is not difficult to maintain this condition in an active plasma since **ionization** or **recombination** always produces or destroys an ion pair together.

charge spectrum. The range and magnitude of electric charges with reference to **cosmic rays** at a specific altitude.

charging point. A connection on an aircraft, or aircraft component, through which the aircraft or aircraft component can be replenished with a specific commodity, e.g., oxygen, air, or hydraulic fluid, etc.

Charles-Gay-Lussac law. An empirical generalization that in a gaseous system at constant pressure, the temperature increase and the relative volume increase stand in approximately the same proportion for all so-called perfect gases. Also called *Charles law, Gay-Lussac law.*

charring ablator. An **ablation material** characterized by the formation of a carbonaceous layer at the heated surface which impedes heat flow into the material by its insulating and reradiating characteristics.

chart. A graphic representation, specifically designed for navigational purposes, of a section of the earth's surface. A chart may also be referred to as a map, although a chart is usually specifically designed as a plotting medium for marine or aerial navigation. Also see **aeronautical chart.**

chase pilot. A pilot who flies an escort airplane advising a pilot who is making a check, training, or research flight in another craft.

chaser. The vehicle that maneuvers in order to effect a **rendezvous** with an orbiting object.

check flight. 1. A flight made to check or test the performance of an aircraft, rocket, or spacecraft, or a piece of equipment or component, or to obtain measurements or other data on performance; a **test flight. 2.** A familiarization flight in an aircraft, or a flight in which a pilot or other aircrew member or members are tested or examined for proficiency.

checking. Presence of a network of fine hairline cracks on the surface of a structure usually induced by poor machining technique.

checklist. A list of items to be checked, written out to guard against memory failure. Specifically a list carried in an airplane setting forth procedures and/or instructions to be followed before and during various flight phases such as preflight, takeoff, landing, etc.

checkout. 1. A sequence of actions taken to test or examine a thing as to its readiness for incorporation into a new phase of use, or for the performance of its intended function. **2.** The sequence of steps taken to familiarize a person with the operation of an airplane or other piece of equipment.

In sense **1,** a checkout is usually taken at a transition point between one phase of action and another. To shorten the time of checkout, automation is frequently employed.

check pilot. An instructor pilot who accompanies a pilot during a **check flight.**

check point. 1. A predetermined point on the earth's surface used as a means of controlling movement, a registration target for fire adjustment, or a reference for location. **2.** Center of impact; a burst center. **3.** Geographical location on land or water above which the position of an aircraft in flight may be determined by observation or by electronic means.

checkout GSE. Ground support equipment used to make a

checkout, which see, sense **1.**

chemical agent. A solid, liquid, or gas which through its chemical properties produces lethal or damaging effects on man, animals, plants, or material, or produces a screening or signaling smoke.

chemical defense. The methods, plans, and procedures involved in establishing and executing defensive measures against attack by chemical agents.

chemical energy. Energy produced or absorbed in the process of a chemical reaction. In such a reaction, energy losses or gains usually involve only the outermost **electrons** of the atoms or ions of the system undergoing change; here a chemical bond of some type is established or broken without disrupting the original atomic or ionic identities of the constituents.

Chemical changes, according to the nature of the materials entering into the change, may be induced by heat (*thermochemical*), light (*photochemical*), and electric (*electrochemical*) energies.

chemical fuel. 1. A fuel that depends upon an oxidizer for combustion or for development of thrust, such as liquid or solid rocket fuel, jet fuel, or internal-combustion engine fuel. Distinguished from nuclear fuel. **2.** An exotic fuel that uses special chemicals.

chemical munitions. Ammunition such as bombs, projectiles, grenades, or the like, containing a chemical agent(s). Such agents include war gases, smokes, and incendiaries.

chemical pressurization. The **pressurization** of propellant tanks in a **rocket** by means of high-pressure gases developed by the combustion of a fuel and oxidizer or by the decomposition of a substance.

chemical warfare. Employment of chemical products to produce death or casualties in man, to create a military advantage, or to defend against such action.

chemisorption. The binding of a liquid or gas on the surface or in the interior of a solid by chemical bonds or forces.

chemosphere. The vaguely defined region of the upper atmosphere in which **photochemical reactions** take place. It is generally considered to include the stratosphere (or the top thereof) and the mesosphere, and sometimes the lower part of the thermosphere. See **atmospheric shell.**

This entire region is the seat of a number of important photochemical reactions involving atomic oxygen O, molecular oxygen O_2, ozone O_3, hydroxyl OH, nitrogen N_2, sodium Na, and other constituents to a lesser degree.

chicks. Friendly fighter aircraft.

chief of staff. The senior or principal member or head of a staff, or the principal assistant in a staff capacity to a person in a command capacity; the head or controlling member of a staff, for purposes of the coordination of its work; a position, which in itself is without inherent power of command by reason of assignment, except that which is invested in such a position by delegation to excerise command in another's name. In the Army and Marine Corps, the title is applied only to the staff on a brigade or division level or higher. In lower units, the corresponding title is executive officer. In the Air Force, the title is applied normally in the staff on an Air Force level and above. In the Navy, the title is applied only on the staff of a commander with rank of rear admiral or above. The corresponding title on the staff of a commander of rank lower than rear admiral is chief staff officer, and in the organization of a single ship, executive officer.

chine. As applied to aircraft, the intersection of the upper and lower surfaces of a highly swept wing of a supersonic airplane configuration.

chirp. An all-encompassing term for the various techniques of pulse expansion-pulse compression applied to **pulse radar;** a technique to expand narrow pulses to wide pulses for transmission, and compress wide received pulses to the original narrow pulse width and wave shape, to gain improvement in **signal-to-noise ratio** without degradation to range resolution and range discrimination.

chlorate candles. Usually, a mixture of solid chemical compounds which, when ignited, liberates oxygen.

Chlorella. A genus of unicellular green **algae,** considered to be adapted to converting carbon dioxide into oxygen in a closed ecological system. See **closed ecological system.**

choked flow. Flow in a duct or passage such that the flow upstream of a certain critical section cannot be increased by a reduction of downstream pressure.

chokes. Pain and irritation in the chest and throat as a result of reduced ambient pressure.

choking Mach number. The **Mach number** at some reference point in a duct or passage (e.g., at the inlet) at which the flow in the passage becomes choked. See **choked flow.**

chord. 1.A straight line intersecting a circle or other curve, or a straight line connecting the ends of an arc. **2.** (*symbol c*). In aeronautics, a straight line intersecting or touching an airfoil profile at two points; specifically, that part of such a line between two points of intersection.

This line is usually a **datum line** joining the leading and trailing edges of an airfoil, joining the ends of the mean line of an airfoil profile, or running along the lower surface or line of an airfoil profile, from which the ordinates and angles of the airfoil are measured. As such a datum line, it is sometimes called the *geometric chord,* to distinguish it from a chord established on the basis of any other considerations.

3. chord length.

In sense **3,** points or stations along a chord are designated in percentages or fractions of the chord or chord length from the leading edge, as, a point at 25 percent, or one-quarter, chord.

chord length. The length of the **chord** of an **airfoil** section between the extremities of the section.

For many airfoils, the chord is established intersecting the airfoil profile at its extremities, and tye chord length is equal to the length of the chord between the points of intersection; for airfoils where the chord is established by a point or points of tangency or intersection not at the extremities, however, the chord length is considered to extend beyond either or both points, as necessary, to equal the maximum length of the profile. See **chord,** senses **2** and **3.**

cnromatography. The separation of chemical substances by making use of differences in the rates at which the substances travel through or along a stationary medium.

chromosphere. A thin layer of relatively transparent gases above the photosphere of the sun.

chugging. A form of **combustion instability** in a **rocket engine,** characterized by a pulsing operation at a fairly low frequency, sometimes defined as occuring between particular frequency limits; the noise made in this kind of combustion. Also called *chuffing, bumping.*

chute. Colloquial for **parachute.**

cine-theodolite. A photographic **tracking** instrument which records on each film frame the target and the azimuth and elevation angles of the optical axis of the instrument. Also called *Askania.*

circle of equal altitude. Same as **parallel of altitude.**

circle of equal probability. A measure of the accuracy with which a rocket or missile can be guided; the radius of the circle at a specific distance in which 50 percent of the reliable shots land. Also called *circular error probable, circle of probable error.*

circle of latitude. A **great circle** of the **celestial sphere** through the **ecliptic poles,** and hence perpendicular to the plane of the **ecliptic.**

circle of longitude. A circle of the **celestial sphere,** parallel to the **ecliptic.** Also called *parallel of latitude.*

circuit. A **network** providing one or more closed paths.

circuitry. A complex of circuits describing interconnection within or between systems.

circular dispersion. In rocketry, the diameter of a circle within which 75 percent of the events under study occur.

Circular dispersion is most often used as a measure of error of the accuracy with which rockets reach their intended target.

circular error. An error associated with delivery of munitions on a target. It is the distance measured between the desired and actual point of impact of a munition.

circular velocity. At any specific distance from the primary, the orbital velocity required to maintain a constant-radius **orbit.**

circumlunar. Around the moon, generally applied to **trajectories.**

cirrocumulus. A form of **high cloud.** Also see **cloud, cumuliform clouds.**

cirrostratus. A form of high cloud. Also see **cloud, stratiform clouds.**

cirrus. A form of **high cloud.** Also see **cloud.**

cislunar. (Latin cis, *on this side*). Of or pertaining to phenomena, projects, or activity in the space between the earth and moon, or between the earth and the moon's orbit. Compare **translunar, circumlunar.**

civil aircraft. Aircraft other than public aircraft.

Civil Air Patrol. A federally chartered, non-frofit corporation which was designated by Congress in 1948 as a volunteer civilian auxiliary of the Air Force. Its mission is to provide public service during local and national emergencies, motivate youth to high ideals of leadership, and further United States air and space supremacy through aerospace education and training.

civil day. See **mean solar day.**

civil defense. All those activities and measures designed or undertaken to: **a.** minimize the effects upon the civilian population caused or which would be caused by an enemy attack upon the United States; **b.** deal with the immediate emergency conditions which would be created by any such attack; and **c.** effectuate emergency repairs to, or the emergency restoration of, vital utilities and facilities destroyed or damaged by any such attack.

civil reserve air fleet. A group of commercial aircraft with crews which is allocated in time of emergency for exclusive military use in both international and domestic service.

civil time. See **mean time,** note.

civil twilight. See **twilight,** note.

civil year. Same as **calendar year.**

cladding. A coating placed on the surface of a material and usually bonded to the material. Also called *clad.*

Cladding is used extensively in nuclear reactor cores to prevent corrosion of the fissionable material by the coolant.

cladding. nucleonics: The jacket of nuclear *fuel elements.* It prevents corrosion of the fuel ind the release of fission products into the coolant. Aluminum or its alloys, stainless steel and zirconium alloys are common cladding materials.

clamping circuit. 1. A **circuit** which maintains either extremity of a **waveform** at a prescribed potential. **2.** A **network** for adjusting the absolute voltage level of a waveform.

class. (1) As used with respect to the FAA certification, rating, privileges, and limitations of airmen, means a classification of aircraft within a category having similar operating characteristics. Examples include: single engine; multiengine; land; water; gyroplane; helicopter; airship; and free balloon; and (2) As used with respect to the FAA certification of aircraft, means a broad grouping of aircraft having similar characteristics of propulsion, flight, or landing. Examples include: airplane; rotorcraft; glider; balloon; landplane; and seaplane.

classification. The determination that official information requires, in the interests of national defense, a specific degree of protection against unauthorized disclosure, coupled with a designation signifying that such a determination has been made. See also **defense classification.**

classified information. Official information which has been determined to require, in the interests of national defense, protection against unauthorized disclosure and which been so designated.

clean aircraft. 1. An aircraft in flight configuration, versus landing configuration, i.e., landing gear and flaps retracted, etc. **2.** An aircraft that does not have external stores.

clean bomb. A nuclear bomb that produces relatively little radioactive *fallout.* A *fusion bomb.* (Compare **dirty bomb.**)

cleanup. 1. The process of removing gas from a vacuum system or device by **sorption** or ion pumping. **2.** In aeronautics, the process of improving external shape and smoothness of an aircraft to reduce its **drag.**

clear. 1. To approve or authorize, or to obtain approval or authorization for: **a.** a person or persons with regard to their actions, movements, duties, etc.; **b.** an object or group of objects, as equipment or supplies, with regard to quality, quantity, purpose, movement, disposition, etc.; and **c.** a request, with regard to correctness of form, validity, etc. **2.** Specifically, to give one or more aircraft a clearance. **3.** To give a person a security clearance. **4.** To fly over an obstacle without touching it. **5.** To pass a designated point, line, or object. The end of a column must pass the designated feature before the latter is cleared. **6. a.** To operate a gun so as to unload it or make certain no ammunition remains; and **b.** to free a gun of stoppages. **7.** To clear an engine; to open the throttle of an idling engine to free it from carbon. **8.** To clear the air to gain either temporary or permanent air superiority or control in a given section. **9.** To restore a **storage** or **memory** device to a prescribed state, usually that denoting zero.

clear air turbulence (CAT). Turbulence encountered in air where no clouds are present; more popularly applied to high-level turbulence associated with wind shear; often encountered in the vicinity of the jet stream.

clear ice. Clear ice is transparent with a glassy surface identical to the glaze which forms on trees and other objects during a freezing rain. It is smooth and streamlined when deposited from large droplets without solid precipitation. But if mixed with snow or sleet it is rough, irregular, and whitish. The deposit then becomes very bluntnosed.

Conditions most favorable for clear ice formation are high water content, large droplet size, temperature only slightly below freezing, high airspeed, and thin airfoils. Encountered most frequently in cumuliform clouds, clear ice also accumulates rapidly on aircraft flying in freezing rain or drizzle.

clearance limit. The fix to which an aircraft is issued an air traffic clearance.

clearway. For turbine engine powered airplanes, an area beyond the runway, not less than 500 feet wide, centrally located about the extended centerline of the runway, and under the control of the airport authorities. The clearway is expressed in terms of a clearway plane, extending from the end of the runway with an upward slope not exceeding 1.25 percent, above which no object nor any terrain protrudes. However, threshold lights may protrude above the plane if their height above the end of the runway is 26 inches or less if they are located to each side of the runway.

clo. The amount of insulation which will maintain normal skin temperature of the human body when heat production is 50 kilogram-calorie per meter squared per hour, air temperature is 70 F, and the air is still.

One clo is roughly equivalent to the amount of insulation provided by the average businessman's suit in a temperate climate.

clock code position. The position of a target in relation to an aircraft or ship with dead-ahead position considered as twelve o'clock.

clock frequency. The master **frequency** of periodic **pulses** which schedule the operation of a machine, as a computer.

clock pulse. A **pulse** used for timing purposes.

In pulse-code-modulation systems, a timing pulse which occurs at the **bit rate.**

close air support. Air attacks against hostile targets which are in close proximity to friendly forces and which require detailed integration of each air mission with the fire and movement of those forces. See also **air interdiction; air support.**

close-controlled air interception. An interception in which the interceptor is continuously controlled into a position from which the target is within visual range of radar contact. See also **air interception; broadcast-controlled air interception.**

closed area. A designated area in or over which passage of any kind is prohibited.

closed-cycle reactor system. nucleonics: A reactor design in which the primary heat of fission is transferred outside the reactor core to do useful work by means of a *coolant* circulating in a completely closed system that includes a *heat exchanger.* (Compare **direct-cycle reactor system, indirect-cycle reactor system, open-cycle reactor system.**)

closed ecological system. A system that provides for the maintenance of life in an isolated living chamber through complete reutilization of the material available, in particular, by means of a cycle wherein exhaled carbon dioxide, urine, and other waste matter are converted chemically or by photosynthesis into oxygen, water, and food. Compare **controlled-leakage system, open system.**

closed-loop system. A system in which the **output** is used to control the **input.** See **feedback control loop.**

closed-loop telemetry. 1. A **telemetry** system which is used as the indicating portion of a remote-control system. **2.** A system used to check out test vehicle or telemetry performance without radiation of radio-frequency energy.

closed system. 1. In thermodynamics, a system so chosen that no transfer of mass takes place across its boundaries; for example, a fluid parcel undergoing a saturation-adiabatic process, as opposed to a pseudoadiabatic expansion. See **open system. 2.** In mathematics, a system of differential equations and supplementary conditions such that the values of all the unknowns (dependent variables) of the system are mathematically determined for all values of the independent variables (usually space and time) to which the system applies. **3. closed ecological system. 4.** A system which constitutes a **feedback** loop so that the inputs and controls depend on the resulting output. For example, an automatic radar-controlled tracking system.

closest approach. The place or time at which two planets are nearest each other as they orbit about the Sun.

closing rate. The speed at which two bodies approach each other.

cloud. A visible cluster of minute water and/or ice particles in the atmosphere above the earth's surface. Also see **anvil, cumuliform, high clouds, lenticular cloud, middle alto clouds, noctilucent clouds, low clouds, roll cloud, stratiform clouds, vertical development clouds.**

cloud absorption. The **absorption** of **electromagnetic radiation** by the water drops and water vapor within a cloud. Compare **cloud attenuation.**

For **insolation** (incoming solar radiation), clouds absorb rather small fractions, particularly of the shorter wavelengths. Even for depths of clouds of the order of 20,000 feet, measurements suggest absorptions of less than 30 percent, while layers only 1000 or 2000 feet thick may absorb only about 5 percent. However, for longwave **terrestrial radiation,** even very thin layers of cloud act as almost complete black-body absorbers.

cloud bank. A well-defined mass of clouds observed at a distance. It covers a considerable portion of the horizon sky, but it is not overhead.

cloud attenuation. Usually, the reduction in intensity of **microwave radiation** by clouds in the earth's atmosphere. For the centimeter wavelength band, clouds produce **Rayleigh scattering.** The attenuation is due largely to scattering, rather than to absorption, for both ice and water clouds. See **precipitation attenuation.** Compare **cloud absorption.**

cloud chamber. A device for observing the paths of **ionizing particles,** based on the principle that supersaturated vapor condenses more readily on ions than on neutral molecules.

cloud physics. A subdivision of **physical meteorology** concerned with physical properties of clouds in the atmosphere and the processes occurring therein.

Cloud physics, broadly considered, embraces not only the study of condensation and precipitation processes in clouds, but also radiative transfer, optical phenomena, electrical phenomena, and a wide variety of hydrodynamic and thermodynamic processes peculiar to natural clouds.

cloud seeding. The art or activity of sowing clouds from an aircraft with an inoculant, such as dry ice particles or silver iodide crystals, so as to induce precipitation.

cluster. 1. Fireworks signal in which a group of stars burns at the same time. **2.** Groups of bombs released together. A cluster usually consists of fragmentation or incendiary bombs. **3.** Two or more engines coupled together so as to function as one power unit.

clutter. 1. Atmospheric noise, extraneous signals, etc., which tend to obscure the reception of a desired signal in a radio receiver, radarscope, etc.

As compared with *interference, clutter* refers more particularly to unwanted reflections on a radar **plan position indicator,** such as ground return, but the terms are often used interchangeably. **2. window.**

co. 1. A prefix meaning 90° minus the value with which it is used. Thus, if the *latitude* is 30°, the *colatitude* is 90°—30° = 60°. **2.** A prefix meaning in common, as in *coaxial,* having a common axis.

coast. A memory feature on a **radar** which, when activated, causes the range and angle systems to continue to move in the same direction and at the same speed as that required to track an original target.

Coast is used to prevent **lock-on** to a stronger target if approached by the target being tracked.

coated optics. Optical elements (lenses, prisms, etc.) which have their surfaces covered with a thin transparent film to minimize reflection and loss of light in the system.

coaxial cable. A form of **waveguide** consisting of two concentric conductors insulated from each other.

coaxial propellers. Two propellers mounted on concentric shafts having independent drives and normally rotating in opposite directions.

cobalt bomb. If a nuclear weapon were encased in colbalt, large amounts of radioactive cobalt-60 could be produced when it was detonated. Such a weapon (only theoretical today) could add to the explosive force of the bomb the danger of the highly penetrating and long-lasting gamma radiation emitted by cobalt-60.

cockpit. A compartment housing the pilot(s).

cocooning. The spraying or coating of an aircraft or equipment with a substance, e.g., a plastic, to form a cocoon-like seal against the effects of the atmosphere.

codan (*abbr*). **carrier operated device, antinoise.** A device which silences a receiver except when a carrier signal is being received.

code. 1. A system of symbols or **signals** for representing **information,** and the rules for associating them. **2.** The set of characters resulting from the use of a code as defined in sense **1. 3.** Specifically, to translate a problem to a **routine** expressed in machine language for a specific **computer. 4.** To express given information by means of a code, to encode.

coded decimal digit. In computer operation, a **decimal digit** expressed in a code, usually a four-digit **binary code.**

codes. The numbers assigned to the multiple pulse reply signals transmitted by **ATCRBS** and military transponders.

coding. The arrangement in a coded form, usually acceptable to a specific **computer,** of the instructions for the operations necessary to solve a problem.

coefficient. 1.A number indicating the amount of some change under certain specified conditions, often expressed as a ratio.

For example, the coefficient of linear expansion of a substance is the ratio of its change in length to the original length for a unit change of temperature from a standard.

2.A constant in an algebraic equation. **3.**One of several parts which combine to make a whole, as the maximum deviation produced by each of several causes.

coherent. 1. Of electromagnetic radiation, being **in phase,** so that waves at various points in space act in unison, as in a *laser producing coherent light.* **2.** Having a fixed relation between frequency and phase of input and output signal.

coherent echo. A radar **echo** whose **phase** and **amplitude** at a given range remain relatively constant.

Hills, buildings, and slowly moving point targets such as ships are examples of objects which produce coherent radar echoes. Volume targets (such as clouds and precipitation) give noncoherent echoes. The classification of an echo as coherent or noncoherent is closely related to the spatial resolution (beam width) of the radar or the volume occupied by the radar pulse. Thus, small atmospheric inhomogeneities which give rise to noncoherent echoes, would give coherent echoes if the radar volume were reduced in size to the order of magnitude of the inhomogeneities themselves.

coherent oscillator. An **oscillator** which provides a reference by which the radiofrequency **phase** difference of successive received pulses may be recognized. See **coherent reference.**

coherent radar. A type of **radar** that employs circuitry which permits comparison of the **phase** of successive received target signals.

coherent reference. The reference **signal,** usually of stable frequency, to which other signals are **phase-locked** to establish **coherence** throughout a system.

coherent transponder. A **transponder,** the output signal of which is **coherent** with the input signal.

coincidence circuit. An electronic **circuit** that produces a usable **output** pulse only when each of two or more **input** circuits receive pulses simultaneously or within an assignable time interval.

coincidence counter. A device using **ionization counters** and **coincidence circuits** to count and determine the direction of travel of **ionizing particles,** particularly cosmic rays.

coincident-current magnetic core. A **binary magnetic core** in which information is stored as the result of current flowing simultaneously in two or more independent windings.

Usually a number of cores are arranged in the form of a matrix.

cold cathode. A **cathode** whose operation does not depend on its temperature being above the ambient temperature.

cold-cathode ionization gage. An **ionization gage** (vacuum gage) in which the ions are produced by a discharge between two electrodes, both near room temperature. The discharge usually takes place in the presence of a magnetic field which lengthens the path of the electrons between cathode and anode.

cold front. The boundary line at the Earth's surface between advancing cold air and the warmer air under which it replaces. Also see **front.**

cold-flow test. A test of a liquid rocket, especially to check or verify the efficiency of a propulsion subsystem that provides for the conditioning and flow of propellants, including tank pressurization, propellant loading, and propellant feeding.

cold war. A state of international tension, wherein political, economic, technological, sociological, psychological, paramilitary, and military measures short of overt armed conflict involving regular military forces are employed to achieve national objectives.

coleopter. An **aircraft** having an annular (barrel-shaped) wing, the engine and body being mounted within the circle of the wing.

collective pitch control (rotorcraft). A control by which an equal change of blade angle is imposed on all the blades independently of their azimuth position.

collector ring. An exhaust manifold in the form of a ring, used in **radial engines.**

collimate. 1. To render parallel, as rays of light. **2.** To adjust the line of sight of an optical instrument, such as a theodolite, in proper relation to other parts of the instrument.

collimation tower. A tower on which are mounted a visual and a radio target for use in checking the electrical **axis** of an **antenna.**

collimator. A device for focusing or confining a *beam* of particles or radiation, such as X rays.

collision. nucleonics. A close approach of two or more *particles, photons,* atoms or nuclei, during which such quantities as energy, momentum and charge may be exchanged.

collision course interception. Any course whereby the interception is accomplished by the constant heading of both aircraft.

collision parameter. 1. In orbit computation, the distance between a center of attraction of a **central force field** and the extension of the velocity vector of a moving object at a great distance from the center. **2.** In gas dynamics and atomic physics, any of several parameters, such as **cross section, collision rate, mean free path,** etc., which provide a measure of the probability of collision.

collision rate. The average number of **collisions** per second suffered by a molecule or other **particle** moving through a gas. Also called *collision frequency.*

colloid. See **colloidal system.**

colloidal system. An intimate mixture of two substances one of which, called the *dispersed phase* (or *colloid*) is uniformly distributed in a finely divided state through the second substance, called the *dispersion medium* (or *dispersing medium*). The dispersion medium may be a gas, a liquid, or a solid, and the dispersed phase may also be any of these, with the exception that one does not speak of a colloidal system of one gas in another. Also called *colloidal dispersion, colloidal suspension.*

A system of liquid or solid particles colloidally dispersed in a gas is called an *aerosol.* A system of solid substance or water-insoluble liquid colloidally dispersed in liquid water is called a *hydrosol.* There is no sharp line of demarcation between true solutions and colloidal systems on the one hand, or between mere suspensions and colloidal systems on the other. When the particles of the dispersed phase are smaller than about 1 millimicron in diameter, the system begins to assume the properties of a true solution; when the particles dispersed are much greater than 1 micron, separation of the dispersed phase from the dispersing medium becomes so rapid that the system is best regarded as a suspension.

colorblind. Applied to a photographic emulsion sensitive only to blue, violet, and ultraviolet light.

color excess. The difference between the **apparent color index** of a star and its **true color index** as computed for its stellar type.

Color excess is a measure of **space reddening.**

color sensitive. Referring to a photographic emulsion which is not **colorblind.**

An emulsion sensitive not only to blue, violet, and ultraviolet, but also to yellow and green, is called *orthochromatic;* if sensitive to red as well, it is called *panchromatic.*

color temperature. 1.An estimate of the **temperature** of an incandescent body, determined by observing the **wavelength** at which it is emitting with peak intensity (its color) and using that wavelength in **Wien law.**

If such a body were an ideal black body, the temperature so estimated would be its true temperature and would also agree with its **effective temperature;** but for actual bodies, the color temperature is generally only an approximate value. Thus, the sun's color temperature is about 6100°K, a few hundred degrees hotter than most approximations of its effective temperature.

2. The temperature to which a **black body** radiator must be raised in order that the light it emits may match a given light source in color. (Usually expressed in degrees Kelvin (°K).)

coma. 1.The gaseous envelope that surrounds the nucleus of a comet. 2.In an optical system, a result of spherical aberration in which a point source of light, not on the axis, has a blurred, comet-shaped image.

combat air patrol. An aircraft patrol provided over an objective area, over the force protected, over the critical area of a combat zone, or over an air defense area, for the purpose of intercepting and destroying hostile aircraft before they reach their target.

combat area. A restricted area (air, land, or sea) which is established to prevent or minimize mutual interference between friendly forces engaged in combat operations. See also **combat zone.**

combat center. A major facility of the North American Air Defense Command from which the region commander and his staff exercise operational control over assigned forces. Computers, electronic displays, and internal and external communications may be utilized to provide the information necessary to the exercising of operational control over assigned forces.

combat control team. A team of Air Force personnel organized, trained, and equipped to establish and operate navigational or terminal guidance aids, communications, and aircraft control facilities within the objective area of an airborne operation.

combat forces. Those forces whose primary missions are to participate in combat.

combat load. The total warlike stores carried by an aircraft.

combat readiness. Synonymous with "operational readiness" with respect to missions or functions performed in combat.

combat survival. Those measures to be taken by Service personnel when involuntarily separated from friendly forces in combat, including procedures relating to individual survival, evasion, escape and conduct after capture.

combat trail. Interceptors in trail formation. Each interceptor behind the leader maintains position visually or with airborne radar.

combat zone. 1. That area required by combat forces for the conduct of operations. 2. The territory forward of the Army rear boundary. See also **combat area; communications zone.**

combined error. A term used to specify the largest possible error of an instrument in the presence of adding or interacting effects.

Generally applied to the largest error due to the combined effect of nonlinearity and hysteresis.

combined force. A military force composed of elements of two or more allied nations. See also **force(s).**

combustion efficiency. The efficiency with which **fuel** is burned, expressed as the ratio of the actual energy released by the combustion to the potential **chemical energy** of the fuel.

combustion instability. Unsteadiness or abnormality in the **combustion** of fuel, as may occur, e.g., in a **rocket engine.**

combustion starter. A device in which the firing of a charge provides the energy to rotate the engine for starting.

combustor. A name generally assigned to the combination of flame holder or stabilizer, igniter, combustion chamber, and injection system of a ramjet or gas turbine.

comet. A loose body of gases and solid matter revolving around the Sun.

command. A **signal** which initiates or triggers an action in the device which receives the signal. In computer operations also called *instruction.*

command (military). 1. The authority which a commander in the military Service lawfully exercises over his subordinates by virtue of rank or assignment. Command includes the authority and responsibility for effectively using available resources and for planning the employment of, organizing, directing, coordinating, and controlling military forces for the accomplishment of assigned missions. It also includes responsibility for health, welfare, morale, and discipline of assigned personnel. 2. An order given by a commander; that is, the will of the commander expressed for the purpose of bringing about a particular action. 3. A unit or units, an organization, or an area under the command of one individual. 4. To dominate by a field of weapon fire or by observation from a superior osition. See also **air command; area command; base command.**

command altitude. Altitude which must be assumed and/or maintained by the interceptor.

command control. A system whereby functions are performed as the result of a transmitted **signal.**

command destruct. A **command control** system that destroys a flightborne test **rocket,** actuated on command of the range safety officer whenever the rocket performance indicates a safety hazard.

command guidance. The **guidance** of a spacecraft or rocket by means of electronic **signals** sent to receiving devices in the vehicle.

command heading (military). Heading that the controlled aircraft is directed to assume by the control station.

The INTELSAT IV-A communications satellite, was developed by Hughes Aircraft for the International Telecommunications Satellite Organization. The new spacecraft has a capacity for 11,000 telephone circuits or 20 simultaneous TV channels from its 22,300 mile high synchronous orbit.

command speed (military). The speed at which the controlled aircraft is directed to fly.

commercial items. Articles of supply readily available from established commercial distribution sources, which the Department of Defense or inventory managers in the military services have designated to be obtained directly or indirectly from such source.

commercial pilot. A **pilot** authorized by the **FAA** to accept payment for his services and carry passengers for hire. See **commercial pilot certificate.**

commercial pilot certificate. A certificate of competency issued by the **FAA** to a person meeting the requirements of the applicable **Federal Aviation Regulations.** See **commercial pilot, instrument rating, category, class, type, airmen certificates.**

commercial operator. A person who, for compensation or hire, engages in the carriage by aircraft in air commerce of persons or property, other than as an air carrier or foreign air carrier. Where it is doubtful that an operation is for "compensation or hire", the test applied is whether the carriage by air is merely incidental to the person's other business or is, in itself, a major enterprise for profit.

commission (military). 1. To put in or make ready for service or use, as to commission an aircraft or a ship. **2.** A written order giving a person rank and authority as an officer in the armed forces. **3.** The rank and the authority given by such an order. See also **activate.**

commit. The process of committing one or more air interceptors or surface-to-air missiles for interception against a target track.

commonality. The term applied to equipment or systems which possess like and interchangeable characteristics. Equipment and systems are common when: they possess compatibility; each can be operated and maintained by personnel trained on the others without additional specialized training; repair parts (components and/or subassemblies) are interchangeable; and consumable items are interchangeable between them.

common user airlift service. In military transport service usage, the airlift service provided on a common basis for all Department of Defense agencies and, as authorized, for other agencies of the United States Government.

communication deception. (military) Use of devices, operations, and techniques with the intent of confusing or misleading the user of a communications link or a navigation system.

communications intelligence. Technical and intelligence information derived from foreign communications by other than the intended recipients. Also called **COMINT.**

communications satellite. An orbiting vehicle, which relays signals between communications stations. They are of two types: **a.** Active Communications Satellite-A satellite which receives, regenerates, and retransmits signals between stations. **b.** Passive Communications Satellite—A Satellite which reflects communications signals between stations.

commutation. Sequential sampling, on a repetitive timesharing basis, of multiple data sources for transmitting or recording, or both, on a single channel.

commutator. A device used to accomplish **time division multiplexing** by repetitive sequential switching.

companion body. A nose cone, last-stage rocket, or other body that orbits along with an Earth satellite. See **after-body, fallaway section.**

compartment marking. In an aircraft, a system of marking a cabin into compartments for the positioning of loads in accordance with the weight and balance requirements.

compass. 1. An instrument for indicating a horizontal reference direction, specifically a **magnetic compass. 2.** Referring to or measured from **compass north.**

compass acceleration error. The error induced in a magnetic compass by vertical magnetic components when acceleration deflects the detecting element from its normal position.

compass compensation. The systematic reduction of compass deviation by inserting or adjusting small magnets incorporated in a magnetic compass for that purpose. Sometimes called compass calibration.

compass course. A compass course is equal to a magnetic course + or - deviation. Also see **course.**

compass direction. The horizontal direction expressed as an angular distance measured clockwise from **compass north.**

compass locator. Compass locator transmitters are often situated at the middle and outer marker sites of an ILS. The transmitters have a power of less than 25 watts, a range of at least 15 miles and operate between 200 and 415 kHz. At some locations, higher-powered radio beacons, up to 400 watts, are used as outer marker compass locators. These generally carry Transcribed Weather Broadcast information.

The following abbreviations and terms describe COMPASS LOCATORS:

LMM—COMPASS LOCATOR combined with MIDDLE MARKER of ILS.

LOM—COMPASS LOCATOR combined with OUTER MARKER of ILS.

Also see **non-directional radio beacon (NDB).**

compass, magnetic. An instrument indicating magnetic directions by means of a freely suspended compass card; the primary means of indicating the heading or direction of flight of an aircraft. Also see **compass.**

compass north. The direction indicated by the northseeking end of a compass needle.

compass points. The thirty-two divisions which the circumference of a compass and consequently the horizon is divided into. The **cardinal** compass points are north, east, south and west, and each space between is divided into eight equal parts.

compass rose. A large circle on the ground graduated clockwise from 0° to 360° for use as a reference in ground swinging aircraft compasses. It is oriented with 0 toward magnetic or true north.

compass swing. A procedure for determining compass deviation on various aircraft headings for use in compensating or calibrating the compass. This can be done either on the ground or in the air. A compass rose is utilized to swing the compass on the ground. A deviation card is prepared from this operation.

compatibility. 1. A characteristic ascribed to a major subsystem that indicates it functions well in the overall **system. 2.** Also applied to the overall system with reference to how well its various subsystems work together, as in *the vehicle has good compatibility.* **3.** Also applied to materials which can be used in conjunction with other materials and not react with each other under normal operating conditions.

compensation signals. In telemetry, a **signal** recorded on a tape, along with the data and in the same track as the data, used during the playback of data to correct electrically the effects of tape-speed errors.

compile. In computer terminology, to **assemble** the necessary subroutines into a main **routine** for a specific problem.

complement. 1. An angle equal to 90° minus a given angle.

Thus, 50° is the complement of 40°, and the two are said to be *complementary*. See **explement.**

2. The true complement of any quantity which, when added to the first quantity, gives the least quantity containing one more place. **3.** The base-minus-one complement of any quantity in positional notation; i.e., the quantity which, when added to the first quantity, gives the largest quantity containing the same number of places.

In many computing machines a negative quantity is represented as a complement of the corresponding positive quantity.

complete round. A term applied to an assemblage of explosive and nonexplosive components designed to perform a specific function at the time and under the conditions desired. Examples of complete rounds of ammunition are:

a. (separate loading)—consisting of a primer, propelling charge and except for blank ammunition, a projectile and a fuze;

b. (fixed or semifixed)—consisting of a primer, propelling charge, cartridge case, a projectile and a fuze except when solid projectiles are used;

c. (bomb)—consisting of all component parts required to drop and function the bomb once;

d. (missile)—consisting of a complete warhead section and a missile body with its associated components and propellants; and

e. (rocket)—consisting of all components necessary for it to function.

complex. 1. Short for **launch complex,** as in *Complex 25B at Cape Kennedy.* **2.** Pertaining to a magnitude composed of a real number and an imaginary number.

complexity units. In reliability studies of electronic devices, an approximate figure of merit for complexity based on the sum of the number of tubes plus the number of relays in a unit or system. The total number of parts is roughly 10 times the number of complexity units.

component. An article which is a self-contained element of a complete operating *unit* and performs a function necessary to the operation of that unit.

component life. The period of acceptable usage after which the likelihood of failure sharply increases and before which the components are removed in the interests of reliability of operation.

composite air photography. Air photographs made with a camera having one principal lens and two or more surrounding the oblique lenses. The several resulting photographs are correct or transformed in printing to permit assembly as verticals with the same scale.

Composite Air Strike Force. A group of selected United States Air Force units composed of appropriate elements of tactical air power (tactical fighters, tactical reconnaissance, tankers, airlift, and command and control elements) capable of employing a spectrum of nuclear and nonnuclear weapons. Composite Air Strike Force forces are held in readiness for immediate deployment from the continental United States to all areas of the world to meet national emergency contingency plans.

composite materials. Structural materials of metals, **ceramics,** or plastics with built-in strengthening agents which may be in the form of filaments, foils, powders, or flakes of a different compatible material.

composite propellant. A **solid rocket propellant** consisting of a fuel and an oxidizer neither of which would burn without the presence of the other.

compressed-air starter. A device for starting an engine by utilizing the expansive energy of compressed air, in the cylinders or otherwise.

compressibility. The property of a substance, as air, by virtue of which its density increases with increase in pressure.

In aerodynamics, this property of the air is manifested expecially at high speeds. (speeds approaching that of sound and higher speeds). Compressibility of the air about an aircraft may give rise to buffeting, aileron buzz, shifts in trim, and other phenomena not ordinarily encountered at low speeds, known generally as *compressibility effects.*

compressibility burble. A region of disturbed flow, produced by, and rearward of, a shock wave. See **burble.**

compressibility drag. The increase in drag arising from the compressibility of the air which occurs at high speeds.

compressible flow. In aerodynamics, **flow** at speeds sufficiently high that density changes in the fluid cannot be neglected.

compression. 1. ellipticity. See **flattening. 2.** More generally, the act of compressing, pressing together; as in *compression waves, compression ratio.*

compression-ignition engine. An engine in which ignition of the charge in the cylinder is produced by the heat of compression alone.

compressor. A machine for compressing air or other fluid.

Compressors are distinguished (**1**) by the manner in which fluid is handled or compressed, as the **axial-flow, centrifugal, double-entry, free-vortex, mixed-flow, single-entry,** and **supersonic compressor;** or (**2**) by the number of stages, as the **multistage** or **single-stage compressor.** See individual entries on the different types.

computer. 1. A machine for carrying out calculations and performing specified transformations on information. Also called *computing machinery.* **.2.** One who computes, or who operates a computer.

conceptual phase. That period prior to beginning engineering development during which comprehensive system studies and experimental hardware efforts are accomplished.

concourse. A passageway for passengers and public between the principal terminal building waiting area and the fingers and/or aircraft loading positions.

condensation. The process of cooling and/or coagulating through which vapor becomes a liquid.

condensation level; metgy. The height at which a rising column of air reaches saturation, and clouds form.

condensation nucleus. 1. A **particle,** either liquid or solid, upon which **condensation** of vapor begins. **2.** Specifically, in meteorology, a particle upon which condensation of water begins in the atmosphere.

condensation shock wave. A sheet of discontinuity associated with a sudden **condensation** and fog formation in a field of **flow.** It occurs, e.g., on a wing, where a rapid drop in pressure causes the temperature to drop considerably below the dew point. Also called *condensation shock.*

condensation trail. A visible trail of condensed water vapor or ice particles left behind an aircraft, an airfoil, etc. in motion through the air. Also called a *contrail* or *vapor trail.*

There are three kinds of condensation trails: the aerodynamic type, caused by reduced pressure of the air in certain areas as it flows past the aircraft; the convection type, caused by the rising of air warmed by an engine; and the engine-exhaust, or exhaust-moisture, type, formed by the ejection of water vapor from an engine into a cold atmosphere.

condenser discharge light. A lamp in which the high brightness flashes of extemely short duration are produced by the discharge of electricity at high voltage through a gas enclosed in a tube.

conditional stability (or conditional instability). (meteorology). The state of a column of air when its vertical distribution of temperature is such that the layer is stable for dry air, but unstable for saturated air.

conductance. 1. In electricity, the ratio of the current flowing through an electric circuit to the difference of potential between the ends of the circuit, the reciprocal of **resistance.** See **conductivity. 2.** In vacuum systems, the **throughput** Q under steady-state conservative conditions divided by the measured difference in pressure p between two specified cross sections inside a pumping system.

conduction. The transfer of energy within and through a conductor by means of internal particle or molecular activity and without any net external motion.

Conduction is to be distinguished from **convection** (of heat) and **radiation** (of all electromagnetic energy).

conductivity. 1. The ability to transmit, as electricity, heat, sound, etc. **2.** A unit measure of electrical conduction; the facility with which a substance conducts electricity, as represented by the current density per unit electrical-potential gradient in the direction of flow.

Electrical conductivity is the reciprocal of electrical **resistivity** and is expressed in units such as mhos (reciprocal ohms) per centimeter. It is an intrinsic property of a given type of material under given physical conditions (dependent mostly upon temperature).

Conductance, on the other hand, varies with the dimensions of the conducting system, and is the reciprocal of the electrical resistance.

conductor. A substance or entity which transmits electricity, heat, sound, etc.

cone of escape. A hypothetical cone in the **exosphere,** directed vertically upward, through which an atom or molecule would theoretically be able to pass to outer space without a collision, that is, in which the **mean free path** is infinite.

Such a cone would open wider with increasing altitude above the **critical level of escape,** and would be nonexistent below the critical level of escape.

cone of silence. An inverted cone-shaped space directly over the aerial towers of some forms of radio beacons in which signals are unheard or greatly reduced in volume. See also **Z marker beacon.**

confidence interval. In statistics, a range of values which is believed to include, with a preassigned degree of confidence (**confidence level**), the true characteristic of the lot or universe a given percentage of the time.

For example: 95-percent confidence limits for a sample of 10 with a ratio of successes to total number tested of 0.9 (9 successes and 1 failure) would be 0.54 to 1.0. That is, even with an observed success ratio of 0.9 (90 percent) the best that can be said is that the true ratio lies between 0.54 (54 percent) and 1.0 (100 percent) an estimated 95 percent of the time.

confidence level. In statistics, the degree of desired trust or assurance in a given result.

A confidence level is always associated with some assertion and measures the **probability** that a given assertion is true. For example, it could be the probability that a particular characteristic will fall within specified limits, i.e., the chance that the true value of **P** lies between **P** = a and **p** = b. See **confidence interval.**

confidential. See **defense classification.**

configuration. 1. Relative position , disposition of various things, or the figure or pattern so formed. **2.** A geometric figure, usually consisting principally of points and connecting lines. **3. planetary configuration. 4.** A particular type of a specific aircraft, rocket, etc., which differs from others of the same model by virtue of the arrangement of its components or by the addition or omission of auxiliary equipment as *long-range configuration, cargo configuration.*

conformal projection. A map or chart projection on which all angles and distances at every point are correctly represented.

conic. 1. A curve formed by the intersection of a plane and a right circular cone. Originally called *conic section.*

The conic sections are the **ellipse,** the **parabola,** and the **hyperbola,** curves that are used to describe the **paths** of bodies moving in space.

The circle is a special case of the ellipse, an ellipse with an eccentricity of zero.

The conic is the locus of all points the ratio of whose distances from a fixed point, called the *focus,* and a fixed line, called the *directrix,* is constant.

2. In reference to satellite orbital parameters, without consideration of the perturbing effects of the actual shape or distribution of mass of the **primary.**

Thus, *conic perigee* is the perigee the satellite would have if all of the mass of the primary were concentrated at its center.

conical beam. The **radar beam** produced by **conical scanning** methods.

This type of beam has an advantage over that produced by a single radiating element placed at the focus of a parabolic reflector in that much greater angular accuracy is possible in locating targets.

conical scanning. *Scanning* in which the direction of maximum radiation generates a cone whose vertex angle is of the order of the **beam width.** Such scanning may be either rotating or nutating, according as the direction of polarization rotates or remains unchanged.

conic section. The original name for **conic.**

coning angle (rotor). The angle between the longitudinal axis of a blade and the tip-path plane.

conjunction. 1. The situation of two **celestial bodies** having either the same **celestial longitude** or the same sidereal hour angle. Compare **opposition, quadrature.**

A planet is at superior conjunction if the sun is between it and the earth; at inferior conjunction if it is between the sun and the earth.

2. The time at which conjunction, as defined in sense **1** , takes place.

conservation of energy. The principle that the total **energy** of an isolated system remains constant if no interconversion of **mass** and energy takes place.

This principle takes into account all forms of energy in the system; it therefore provides a constraint on the conversions from one form to another. See **energy equation, conservation of energy.**

conservation of mass. The principle in **Newtonian mechanics** which states that **mass** cannot be created or destroyed but only transferred from one volume to another. See **continuity equation.**

conservation of momentum. The principle that in the absence of forces absolute **momentum** is a property which cannot be created or destroyed. See **Newton laws of motion.**

consolan. A low frequency, long-distance navaid used principally for transoceanic navigation.

console. Term applied to a grouping of controls, indicators and similar electrical or mechanical equipment which is used to monitor readiness of and/or control specific functions such as missile checkout, count-down, or launch operations.

constant-level balloon. A balloon designed to float at a constant-pressure level. Also called *constant-pressure balloon.* See **skyhook balloon.**

constant pressure chart (meteorology). A chart which usually contains plotted data and analysis of the distribution of heights of any selected isobaric surface, and the analyses of the wind, temperature, and humidity existing at each height of the selected isobaric surface.

constant speed propeller. A propeller that tends to maintain the rotational speed (R.P.M.) of an engine at a constant, predetermined value. This is accomplished by pitch angle changes controlled by a governor.

containment (nucleonics). The provision of a gastight shell or other enclosure around a *reactor* to confine fission products that otherwise might be released to the atmosphere in the event of an accident.

constellation. Originally a conspicuous configuration of stars; now a region of the celestial sphere marked by arbitrary boundary lines.

constituent day. The duration of one rotation of the earth on its axis, with respect to an astre fictif, a fictitious star representing one of the periodic elements in the tidal forces. It approximates the length of a **lunar** or **solar day.**

construction weight. The weight of a **rocket** exclusive of propellant, load, and crew, if any. Also called *structural weight.*

contact (military). Unit has an unevaluated target.

contact approach. An approach wherein an aircraft on an IFR flight plan, operating clear of clouds with at least one mile flight visibility and having received an air traffic control authorization, may deviate from the prescribed instrument approach procedure and proceed to the airport of destination by visual reference to the surface.

contact (flight) conditions. Weather conditions permitting a pilot to guide and control an aircraft by visual reference to the horizon, terrain features or clouds.

contact lost. A target tracking term used to signify that a target believed to be still within the area of visual, sonar, or radar coverage is temporarily lost but the termination of track plotting is not warranted.

contact point. 1. In land warfare, a point on the terrain, easily identifiable where two or more units are required to make contact. **2.** In air operations, the position at which a flight leader makes radio contact with an air control agency. See also **forward air controller; pull-up point; turn-in point.**

contact print. A print made from a negative or a diapositive in direct contact with sensitized material.

contact report. A report of visual, radio, sonar, or radar contact with the enemy. The first report, giving the information immediately available when the contact is first made, is known as an initial contact report. Subsequent reports containing additional information are referred to as amplifying reports. See also **sighting.**

container delivery system (miliary). An aerial delivery system which provides for the air drop of from one to sixteen 2200 lb. bundles into small drop zones.

contamination. The deposit and/or absorption of radioactive material, biological, or chemical agents on and by structures, areas, personnel, or objects.

conterminous U.S. Forty-eight states and the District of Columbia.

continental control area. See **controlled airspace.**

continental United States (CONUS). United States territory, including the adjacent territoral waters, located within the North American continent between Canada and Mexico.

Continental United States Commands (United States Air Force). Those major commands with units in the Continental United States, i.e., Aerospace Defense Command, United States Air Force Academy, Air Force Communications Service, Air Force Logistics Command, Air Force Systems Command, Air Training Command, Air University, Continental Air Command, Headquarters Command, Military Airlift Command, Strategic Air Command, Tactical Air Command, United States Air Force Security Service.

contingency. An event or happening which may occur in relation to a course of action or an event.

contingent effects. The effects, both desirable and undesirable, which are in addition to the primary effects associated with nuclear detonation.

continuity equation. In a **steady-flow** process, the mathematical statement of the principle of the **conservation of mass** by equating the flow at any section x, to the flow at any section y.

continuous-flow system. An **oxygen system** in which the oxygen flows during both inspiration and expiration by the individual.

continuous-pressure breathing. A kind of **pressure breathing** in which a minimum amount of pressure variation exists inside the mask.

continuous spectrum. 1. A **spectrum** in which wavelengths, wave numbers, and frequencies are represented by the continuum of real numbers or a portion thereof, rather than by a discrete sequence of numbers. See **discrete spectrum. 2.** For **electromagnetic radiation,** a spectrum that exhibits no detailed structure and represents a gradual variation of intensity with wavelength from one end to the other, as the spectrum from an incandescent solid. Also called *continuum, continuum radiation.* **3.** For **particles,** a spectrum that exhibits a continous variation of the momentum or energy.

continuous strip camera. A camera in which the film moves continuously past a slit in the focal plane, producing a photograph in one unbroken length by virtue of the continuous forward motion of the aircraft.

continuous-wave radar. A general species of **radar** transmitting continuous waves, either modulated or unmodulated. The simplest form transmits a single frequency and detects only moving targets by the Doppler effect. This type of radar determines direction but usually not range. Also called *CW radar.* Compare **pulse radar.**

 Two advantages of CW radar are the narrow bandwidth and low power required. Range information may be obtained by some form of modulation, e.g., frequency modulation, pulse modulation.

continuous waves (*abbr* **CW**). Waves, the successive oscillations of which are identical under steady-state conditions.

continuum. 1. Something which is continuous, which has no discrete parts, as the *continuum of real numbers* as opposed to the sequence of discrete integers, as the *background continuum* of a **spectrogram** due to thermal radiation. 2. **continuous spectrum.**

contour interval. Difference in elevation between two adjacent contour lines.

contour line. A line on a map or chart connecting points of equal elevation.

contract termination. As used in Department of Defense procurement, refers to the cessation or cancellation in whole or in part, or work under a prime contract, or a subcontract thereunder , for the convenience of, or at the option of, the government, or due to failure of the contractor to perform in accordance with the terms of the contract (default).

contrail. Same as **condensation trail.**

contra-rotating propellers. Two propellers mounted on concentric shafts having a common drive and rotating in opposite directions.

contrast. In general, the degree of differentiation between different tones in an image.

contravane. A vane that reverses or neutralizes rotation of a **flow.** Also called a *countervane.*

control. 1. A lever, switch, cable, knob, push-button, or other device or apparatus by means of which direction, regulation, or restraint is exercised over something. 2. In plural (**a**) A system or assembly of levers, gears, wheels, cables, boosters, valves, etc., used to control the attitude, direction, movement, power, and speed of an aircraft, rocket, spacecraft, etc. (**b**)Control surfaces or devices. 3. Sometimes capitalized. An activity or organization that directs or regulates an activity. See **central control.** 4. Specifically, to direct the movements of an áircraft or rocket with particular references to changes in attitude and speed. Compare **guidance.**

control column (airplane). The lever, or pillar, supporting a handwheel or its equivalent, by which the longitudinal and lateral controls are operated.

control configured vehicle (CCV). Control configured vehicles utilize **active controls** technology which imply the use of automatic **feedback control systems** that sense aircraft control motion and provide signals to a control surface, to reduce vehicle weight or to improve performance.

control and reporting center. An element of the United States Air Force tactical air control system, subordinate to the Tactical Air Control Center, from which radar control and warning operations are conducted within its area of responsibility.

control and reporting system. An organization set up for: **a.** early warning, tracking, and recognition of aircraft and tracking of surface craft; and **b.** control of all active air defenses. It consists primarily of a chain of radar reporting stations and control centers and an observer organization, together with the necessary communications network.

control area. See **controlled airspace.**

control feel. The ımpression of the stability and control of an aircraft that a pilot receives through the cockpit **controls,** either from the **aerodynamic forces** acting on the control surfaces or from forces simulating these aerodynamic forces. See **artificial feel, feel.**

controllability. The capability of an aircraft rocket, or other vehicle to respond to **control,** especially in direction or attitude.

controlled airspace. Airspace designated as continental control area, control area, control zone, or transition area, within which some or all aircraft may be subject to air traffic control.

 Continental control area. Airspace at and above 14,500 feet MSL of the 48 contiguous states, the District of Columbia, and Alaska south of Lat. 68°00'00"N, excluding the Alaska peninsula west of Long. 160°00'00"W; but does not include airspace less than 1,500 feet above the surface of the earth, and prohibited and restricted areas (except certain specified restricted areas).

 Control area. Unless otherwise provided, airspace extending upward from 700 feet above the surface (until designated 1200 feet above the surface or from at least 300 feet below the MEA, whichever is higher) to the base of the Continental Control Area.

 Control zone. Controlled airspace which extends upward from the surface of the earth and terminates at the base of the Continental Control Area. Control zones that do not underlie the Continental Control Area have no upper limit. A control zone may include one or more airports and is normally a circular area of five statute miles in radius with extensions where necessary to include instrument approach and departure paths.

 Transition area. Airspace extending upward from 700 feet or more above the surface of the earth when designated in conjunction with an airport for which an approved instrument approach procedure has been prescribed, or from 1,200 feet or more above the surface of the earth when designated in conjunction with airway route structures or segments. Unless otherwise limited, transition areas terminate at the base of the overlying controlled airspace.

controlled environment. The environment of any object, such as an instrument, a man, or an unlaunched missile, in which matters such as humidity, pressure, temperature, etc. are under control.

controlled fusion experiment. See **controlled thermonuclear reaction.**

controlled general war. A war in which the homeland of one or both of the combatants is subjected to attack but in which efforts are made to limit the target system, hold collateral damage to a minimum, and achieve a satisfactory termination without reaching the level of counter-value war.

controlled interception. An aircraft intercept action wherein the friendly aircraft are controlled from a

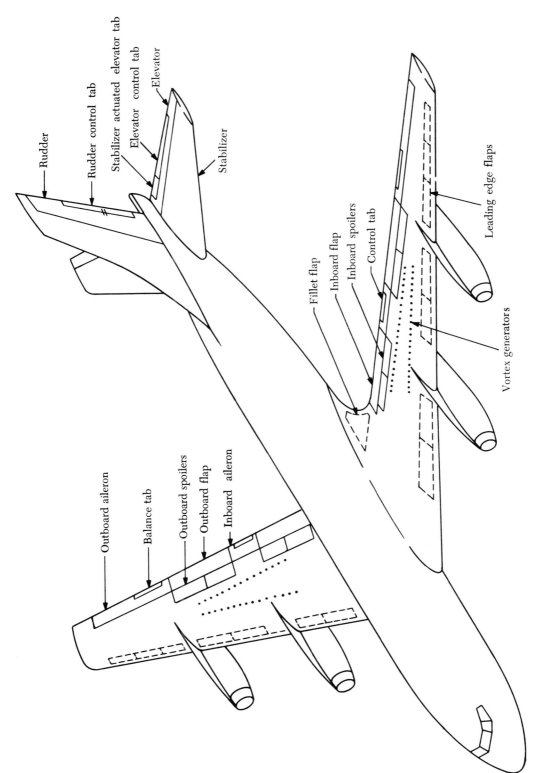

Rudder

Rudder control tab

Stabilizer actuated elevator tab

Elevator control tab

Elevator

Stabilizer

Leading edge flaps

Fillet flap

Inboard flap

Inboard spoilers

Control tab

Vortex generators

Outboard aileron

Balance tab

Outboard spoilers

Outboard flap

Inboard aileron

Control surfaces *on a large turbojet aircraft.*

ground, ship, or airborne station. See also **air interception.**

controlled leakage system. A system that provides for the body's metabolism in an aircraft or spacecraft cabin by a controlled escape of carbon dioxide and other waste from the cabin, with replenishment provided by stored oxygen and food. See **closed ecological system.**

controlled thermonuclear reaction. Controlled *fusion*, that is, fusion produced under research conditions, or for production of useful power.

control of electromagnetic radiation. A national operational plan to minimize the use of electromagnetic radiation in the United States, its possessions and the Panama Canal Zone in the event of attack or imminent threat thereof, as an aid to the navigation of hostile aircraft, guided missiles, or other devices.

control rocket. A vernier rocket, retro-rocket, or other such rocket, used to guide, accelerate, or decelerate a ballistic missile, spacecraft, or the like.

control rod (nucleonics). A rod, plate, or tube containing a material that readily absorbs *neutrons* [hafnium, boron, etc.), used to control the power of a *nuclear reactor.* By absorbing neutrons, a control rod prevents the neutrons from causing further *fission.*

control stick. A form of lever used by the **pilot** in some aircraft to operate the flight controls. See **control, control column, control wheel.**

control surface. An airfoil or part thereof which moves to produce changes in the forces and/or moments acting on an aircraft in order to control it.

control unit. That part of a **computer** which causes the **arithemetic unit, storage,** and transfer of a computer to operate in proper sequence.

control vane. A movable **vane** used for **control,** especially a movable **air vane** or **jet vane** on a rocket, used to control flight attitude.

control wheel. A wheel or the like, on the control column of an airplane used by the pilot to actuate the lateral control system (normally ailerons.) Also see **control, control column.**

control zone. See **controlled airspace.**

convection. 1. In general, mass motions within a **fluid** resulting in transport and mixing of the properties of that fluid. Compare **conduction, radiation. 2.** Specifically, in meteorology, atmospheric motions that are predominantly vertical. Compare **advection.**

conventional weapons. Nonnuclear weapons. Excludes all biological weapons, and generally excludes chemical weapons except for existing smoke and incendiary agents, and agents of the riot-control type.

converge (military). A request or command used in a call for fire to indicate that the observer or spotter desires a sheaf in which the planes of fire intersect at a point.

convergence. 1. The contraction of a vector field; also, a precise measure thereof.

Mathematically, convergence is negative divergence, and the latter term is used for both. (For mathematical treatment, see **divergence.)**

2.The property of a sequence or series of numbers or functions which ensures that it will approach a definite limit.

A series representation of a mathematical function exhibits convergence if the sum of the terms of the series approaches the value of the function more closely as more terms of the series are taken, the two agreeing in the limit of an infinite number of terms.

3.Decrease in area or volume.

convert. In computer terminology, (**a**) to change the manner of representing information, e.g., from analog to digital; (**b**) to translate the medium of conveying or storing information, e.g., from punched cards to magnetic tape; (**c**)to change **numeric** information from one **notation** to another.

converter. 1. A rotary device for changing alternating current to direct current.

A static device for this purpose is called a *rectifier.* A device for changing direct current to alternating current is called an *inverter.*

2. A transducer whose output is a different frequency from its input. **3.** In computer terminology same as **conversion device.**

converter reactor. A reactor that produces some *fissionable material,* but less than it consumes. In some usages, a reactor that produces a fissionable material different from the fuel burned, regardless of the ratio. In both usages the process is known as *conversion.* (Compare **breeder reactor.)**

convertiplane. A hybrid form of heavier-than-air aircraft that is capable, by virtue of one or more horizontal rotors or units acting as rotors, of taking off, hovering, and landing as, or in a fashion similar to, a helicopter, and once aloft, and moving forward, capable, by means of a mechanical conversion of one sort or another, of flying purely as a fixed-wing aircraft, especially in its higher speed ranges.

convoy. 1. A number of merchant ships or naval auxiliaries, or both, usually escorted by warships and/or aircraft, or a single merchant ship or naval auxiliary under surface escort, assembled and organized for the purpose of passage together. **2.** A group of vehicles organized for the purpose of control and orderly movement with or without escort protection.

coolant. A **liquid** or **gas** used to cool something, as a rocket combustion chamber.

This word is used in many self-explanatory compounds, which include: *coolant* chamber, *coolant* gallery, *coolant* hose, *coolant* jacket, *coolant* passage, *coolant* pump, *coolant* radiator.

cooling drag. Drag associated with the cooling of the power plant.

coordinate. One of a set of measures defining a point in space.

If the point is known to be on a given line, only one coordinate is needed; if on a surface, two are required; if in space, three. Cartesian coordinates define a point relative to two intersecting lines, called *axes.* If the axes are perpendicular, the coordinates are rectangular; if not perpendicular, they are oblique coordinates. A three-dimensional system of Cartesian coordinates is called space coordinates. Polar coordinates define a point by its distance and direction from a fixed point called the *pole.* Direction is given as the angle between a reference **radius vector** and a radius vector to the point. If three dimensions are involved, two angles are used to locate the radius vector. Space-polar coordinates define a point on the surface of a sphere by **(1)** its distance from a fixed point at the center, the pole; **(2)** the colatitude or angle between the polar axis (a reference line through the pole) and the radius vector (a straight line connecting the pole and the point); and **(3)** the longitude or angle between a reference plane through the polar axis and a plane through the radius vector and the polar axis. Spherical coordinates define a point on a sphere or spheroid by its angular distances from a primary great circle and from a reference secondary great circle. Geographical or terrestrial coordinates define a point on the surface of the earth. Celestial coordinates define a point on the celestial sphere.

coordinate system. Any scheme for the unique identification of each point of a given **continuum.** The geometry of the system is a matter of convenience determined by the other boundaries of the continuum or by other considerations. Also called *reference frame.*

coordination; aircraft pilot. The ability to use and move the hands and feet simultaneously and subconsciously and in the proper relationship to produce desired **flight attitudes** and functioning of the airplane.

co-pilot. A pilot, responsible for assisting the first pilot to fly the aircraft.

core. The central portion of a *nuclear reactor* containing the *fuel elements* and usually the *moderator,* but not the *reflector.*

coriolis effects. The physiological effects (nausea, vertigo, dizziness, etc.) felt by a person moving radially in a rotating system, as a rotating space station.

coriolis force. An inertial force on a moving body, or particles, produced by the movement of the masses involved, perpendicular to the axis of the primary rotating system. Also called *compound centrifugal force, deflecting force.* See **inertial force.**

Such a force is required if Newton laws are to be applied in the system.

coriolis force (meteorology). A deflecting force normal to the velocity, to the right of motion in the Northern Hemisphere and to the left in the Southern Hemisphere. It cannot alter the speed of the particle. Coriolis force tends to balance the pressure gradient between highs and lows and to cause the air to move parallel to the isobars above ground friction levels.

corner reflector. In radar, three conducting surfaces mutually intersecting at right angles designed to return

electromagnetic radiations toward their sources and used to render a position more conspicuous to **radar** observations.

corona. 1. The outer visible envelope of the sun. Also called *solar corona.*

It is observed at solar eclipse or with the **coronagraph.** The shape of the corona varies during the sunspot cycle. At sunspot minimum the corona has large extensions along the sun's equator, with short brushlike tufts near the poles. At sunspot maximum the equatorial extensions are much smaller and the corona is more regular in shape. The temperature of the corona appears to be in the vicinity of 1,000,000° K.

2. The extremely tenuous outer atmosphere of the sun now known to extend past the earth's orbit. 3. A set of one or more prismatically colored rings of small radii, concentrically surrounding the disk of the sun, moon, or other luminary when veiled by a thin cloud.

The corona is due to **diffraction** by numerous water drops. It can be distinguished from the relatively common halo of 22° by the much smaller angular diameter of the corona, which is often only a few degrees, and by its color sequence, which is from blue inside to red outside, the reverse of that in the 22° halo.

corona discharge. A luminous, and often audible, **electric discharge** that is intermediate in nature between a **spark discharge** (with, usually, its single discharge channel) and a **point discharge** (with its diffuse, quiescent, and nonluminous character). Also called *brush discharge, St. Elmo's fire, corposant.*

coronagraph. An instrument for photographing the **corona** and **prominences** of the sun at times other than at solar eclipse. An occulting disk is used to block out the image of the body of the sun in the focal plane of the objective lens. The light of the corona passes the occulting disk and is focused on a photographic film.

Great care must be taken to avoid light scattered from the atmosphere and the lenses, and from reflections in the tube of the instrument. The coronagraph is

TABLE I.—TERRESTRIAL AND CELESTIAL COORDINATE SYSTEMS[a]

Terrestrial (*long, lat*)	Celestial equator (α, δ)	Horizon (*h, Zn*)	Ecliptic (λ, β)	Galactic (*l, b*)
equator	celestial equator	horizon	ecliptic	galactic equator
poles	celestial poles	zenith, nadir	ecliptic poles	galactic poles
meridians	hour, circles, celestial meridians	vertical circles	circles of latitude	
prime meridian	hour circle through ♈[b], Greenwich celestial meridian, local celestial meridian	principal vertical circle, prime vertical circle	circle of latitude through ♈[b]	great circle through galactic poles so that angle θ to celestial north pole is 123°
parallels	parallels of declination	parallels of altitude	parallels of latitude	
latitude (*L, lat*)	declination (δ, *d*)	altitude (*h*)	celestial latitude (β)	galactic latitude (*b*)
colatitude	polar distance	zenith distance	celestial colatitude	galactic colatitude
longitude (λ, *long*)	sidereal hour angle (*SHA*) right ascension (*RA*, α) Greenwich hour angle (*GHA*) local hour angle (*LHA*)	azimuth (*Zn*) azimuth angle	celestial longitude (λ)	galactic longitude (*l*)

[a] For definitions of terms, see individual entries.
[b] ♈ = first point of Aries

used with a narrow-band polarizing filter or with a spectroscope.

corpuscular. Consisting of **particles,** specifically atomic
particles.

corpuscular cosmic rays. Primary cosmic rays from outer
space which consist of particles, mainly atomic nuclei
(protons) of hydrogen and helium, positively charged
and possessing extemely high kinetic energy. About 1%
of the primary cosmic rays consists of atomic nuclei of
elements heavier than hydrogen or helium.

correction. A quantity, equal in absolute magnitude to
the **error,** added to a calculated or observed value to obtain the true value.

correlation. 1. In statistics, a relationship between two
occurrences which is expressed as a number between
minus one (-1) and plus one (+1).

correlation. In air defense, the determination that an aircraft appearing on a radar scope, on a plotting board,
or visually is the same vehicle as that on which information is being received from another source.

correlation detection. A method of **detection** in which a
signal is compared, point-to-point, with an internally
generated reference. Also called *cross correlation detection.*

correlation tracking system. A **trajectory measuring
system** utilizing correlation techniques where signals
derived from the same source are correlated to derive
the **phase difference** between the signals.

This phase difference contains the system data.

corrosion. Deterioration of metals due to electrochemical
or chemical attack resulting from exposure to natural
or induced environmental conditions, or from the
destructive attack of fungi or bacteria.

cosinophils. A type of white blood cell or leukocyte which
stains a red color with eosin stain; normally about 2 to 3
percent of white cells in the blood but tending to
decrease during stressful situations and thus usable as
an index for **stress.**

cosmic. Of or pertaining to the universe, especially that
part of it outside the earth's atmosphere. Used by the
USSR as equivalent to space, as in *cosmic rocket,
cosmic ship.*

cosmic dust. Finely divided solid matter with particle
sizes smaller than a **micrometeorite,** thus with
diameters much smaller than a millimeter, moving in
interplanetary space. See **dust.**

Cosmic dust in the solar system is thought to be concentrated in the plane of the ecliptic, thus causing the
zodiacal light.

cosmic noise. Interference caused by cosmic **radio waves.**

cosmic radio waves. Radio **waves** emanating from extraterrestrial sources.

They are *galactic radio waves* if their origin is within
our galaxy and *extragalactic radio waves* if their origin
is outside our galaxy. *Solar radio waves* emanate from
the sun.

cosmic-ray burst. An extensive production of ionization
from a common origin by **cosmic rays** in a recording
device such as a **cloud chamber.**

cosmic rays. The aggregate of extremely high-energy subatomic **particles** which travel the solar system and bom-
bard the earth from all directions. Cosmic-ray
primaries seem to be mostly protons, hydrogen nuclei,
but also contain heavier nuclei. On colliding with atmospheric particles they produce many different kinds
of lower energy **secondary cosmic radiation** (see

cascade shower). Also called *cosmic radiation.*

Cosmic rays thought to originate outside the solar
system are called *galactic cosmic rays.* Those thought
to originate in the sun are called *solar cosmic rays.*

In the earth's atmosphere, the maximum flux of
cosmic rays, both primary and secondary, is at an
altitude of 20 km, and below this the absorption of the
atmosphere reduces the flux, though the rays are still
readily detectable at sea level. Intensity of cosmic-ray
showers has also been observed to vary with latitude,
being more intense at the poles.

cosmonaut. A Soviet astronaut.

cost contract. 1.A contract which provides for payment
to the contractor of allowable costs, to the extent
prescribed in the contract, incurred in performance of
the contract. **2.** A cost reimbursement type contract under which the contractor receives no fee.

cost effectiveness. A comparative evaluation derived from
analyses of alternatives (actions, methods, approaches,
equipment, weapon systems, support systems, force
combinations, etc.) in terms of the interrelated influences of cost and effectiveness in accomplishing a
specific mission.

cost plus a fixed-fee contract. A cost reimbursement type
contract which provides for the payment of a fixed fee
to the contractor. The fixed fee, once negotiated, does
not vary with actual cost, but may be adjusted as a
result of any subsequent changes in the scope of work or
services to be performed under the contract.

cost plus incentive fee contract. A contract, with specified
cost limitations, in which the government agrees to
reimburse the contractor for costs incurred in producing the articles or performing the services covered by
the contract and, in addition, pay a target fee which is
subject to subsequent incentive adjustment dependent
upon prescribed contract performance and/or cost factors.

cost-sharing contract. A cost reimbursement type contract under which the contractor receives no fee but is
reimbursed only for an agreed portion of its allowable
costs.

count. To proceed from one point to another in a countdown or countup, normally by calling a number to
signify the point reached; to proceed in a count-down,
as in "**T** minus 90 and counting".

countdown. 1.A step-by-step process that culminates in a
climactic event, each step being performed in accordance with a schedule marked by a count in inverse
numerical order; specifically, this process is used in
leading up to the launch of a large or complicated
rocket vehicle, or in leading up to a captive test, a
readiness firing, a mock firing, or other firing test. **2.**
The act of counting inversely during this process.

In sense **2,** the countdown ends with T-time; thus, *T
minus 60 minutes* indicates there are 60 minutes to go,
excepting for holds and recycling. The countdown may
be hours, minutes, or seconds. At the end, it narrows
down to seconds, 4—3—2—1—0. See **plus count.**

counter (nucleonics). A general designation applied to
radiation detection instruments or *survey meters* that
detect and measure radiation in terms of individual
ionizations, displaying them either as the accumulated
total of their rate of occurrence. (See **Geiger-Muller
counter.)**

counter air. United States Air Force term for air
operations conducted to attain and maintain a desired

degree of air superiority by the destruction or neutralization of enemy forces. Both air offensive and air defensive actions are involved. The former range throughout enemy territory and are generally conducted at the initiative of the friendly forces. The latter are conducted near to or over friendly territory and are generally reactive to the initiative of the enemy air forces. See also **antiair warfare.**

counterattack. Attack by a part or all of a defending force against an enemy attacking force, for such specific purposes as regaining ground lost or cutting off or destroying enemy advance units, and with the general objective of denying to the enemy the attainment of his purpose in attacking. In sustained defensive operations, it is undertaken to restore the battle position and is directed at limited objectives.

counterespionage. That aspect of counterintelligence designed to detect, destroy, neutralize, exploit or prevent espionage activities through identification, penetration, manipulation, deception and repression of individuals, groups or organizations conducting or suspected of conducting espionage activities.

counterforce. The employment of strategic air and missile forces in an effort to destroy, or render impotent, selected military capabilities of an enemy force under any of the circumstances by which hostilities may be initiated.

countermeasures. That form of military science which by the employment of devices and/or techniques has as its objective the impairment of the operation effectiveness of enemy activity. See also **electronic countermeasures.**

counterradiation. The downward **flux** of **atmospheric radiation** passing through a given level surface, usually taken as the earth's surface. Also called *back radiation.*

This result of infrared (long-wave) absorption and reemission by the atmosphere is the principal factor in the **greenhouse effect.**

countervalue. The employment of air and missile forces against the urban industrial areas of an enemy.

coupled modes. Modes of **vibration** that are not independent but which influence one another because of energy transfer from one mode to the other.

coupling. 1. A device or contrivance for joining adjacent ends or parts of anything. **2.** A device permitting transfer of **energy** from one electrical circuit to another, or from one mechanical device to another.

course. 1. A predetermined or intended route or direction to be followed, measured with respect to a geographic reference direction; a line on a chart representing a course. **2.** A line of flight taken by an aircraft, rocket, etc. **3.** A **radio beam** in a radio range.

course, compass. See **compass course.**

course deviation indicator (CDI). See **VOR receiver.**

course, great circle. See **great circle course.**

course, magnetic. See **magnetic course.**

course of action. 1. Any sequence of acts which an individual or a unit may follow. **2.** A possible plan open to an individual or commander which would accomplish or is related to the accomplishment of his mission. **3.** The scheme adopted to accomplish a job or mission. **4.** A line of conduct in an engagement.

course, omnirange. One of an infinite number of radials sent out from the VOR (Very-high-frequency radio omnirange) station. Radial information is translated by the aircraft receiving equipment into designated compass bearings (courses) of the aircraft to the VOR sta-

tion or from the VOR station.

course selector. See **VOR receiver.**

course, true. See **true course.**

cover. 1. The action by land, air, or sea forces to protect by offense, defense, or threat of either or both. **2.** Shelter or protection, either natural or artificial. See also **concealment. 3.** To maintain a continuous receiver watch with transmitter calibrated and available, but not necessarily available for immediate use. **4.** Photographs or other recorded images which show a particular area of ground. See also **comparative cover. 5.** Keep fighters between force/base and contact designated at distance stated from force/base (e.g. "cover bogey" twenty-seven to thirty miles). **6.** Protective guise used by a person, organization, or installation to prevent identification with clandestine activities.

Cowell method. A method of **orbit** computation using direct step-by-step integration in rectangular coordinates of the total acceleration of the orbiting body.

cowling. A cover surrounding the whole or part of a power unit when installed in an aircraft.

crab. A correction of aircraft heading, into the wind, to make good a given track; correction for wind drift.

craft. 1. An aircraft, or aircraft collectively. **2.** Any vehicle or machine designed to fly through air or space.

crash locator beacon. An electronic device attached to the aircraft structure as far aft as practicable in the fuselage, or in the tail surface, in such a manner that damage to the beacon will be minimized in the event of crash impact. It may be automatically ejectable or be permanently mounted. If it is automatically ejectable it will also have provision for manual removal and operation. The beacon operates from its own power source on 121.5 MHz and/or 243 MHz, preferably on both emergency frequencies, transmitting a distinctive downward swept audio tone for homing purposes, and is designed to function without human action after an accident.

crater. 1. lunar crater. **2.** The depression resulting from high speed solid particle impacts on a rigid material as a **meteoroid** impact on the skin of a spacecraft.

creep. The slow but continuous deformation of a material under constant load or prolonged **stress** (usually critically encountered at elevated temperatures).

crew duty time. The total time an aircrew is on duty before resting. Time begins when an aircrew reports to a designated place of duty to initially begin preparation for a mission, and ends when post-flight crew duties are completed (i.e., refueling, parking and securing of aircraft, debriefing, equipment turn-in, or other required post-flight activities).

crewmember. A person assigned to perform duty in an aircraft during flight time.

crew ratio (military) The number of aircrews authorized per each unit equipment aircraft. By multiplying the crew ratio by the number of authorized aircraft, the total aircrew manpower requirements for a unit can be determined.

critical (nucleonics). Capable of sustaining a *chain reaction.* (See **criticality.**)

critical altitude. The maximum altitude at which, in standard atmosphere, it is possible to maintain, at a specified rotational speed, a specified power or a specified manifold pressure. Unless otherwise stated, the critical altitude is the maximum altitude at which it is possible to maintain, at the maximum continuous rotational speed, one of the following:

(1) The maximum continuous power, in the case of engines for which this power rating is the same at sea level and at the rated altitude.

(2) The maximum continuous rated manifold pressure, in the case of engines, the maximum continuous power of which, is governed by a constant manifold pressure.

critical assembly (nucleonics). An assembly of sufficient fissionable material and moderator to sustain a fission *chain reaction* at a very low power level. This permits study of the behavior of the components of the assembly for various fissionable materials in different geometrical arrangements. (Compare **nuclear reactor.**)

critical damping. The minimum **damping** that will allow a displaced system to return to its intitial position without **oscillation.**

critical engine. The engine whose failure would most adversely affect the performance or handling qualities of a multi-engine aircraft.

critical engine failure speed. That speed to which the aircraft can be accelerated, lose an engine, and then continue the takeoff or stop in the computed critical field length.

critical field length. The length of runway required to accelerate on all engines to the critical engine failure speed, experience an engine failure or a malfunction, then continue to takeoff or abort.

criticality. The state of a nuclear reactor when it is sustaining a *chain reaction.*

critical Mach number. The **free-stream** Mach number at which a local **Mach number** of 1.0 is attained at any point on the body under consideration.

For example, an airplane traveling at a Mach number of 0.8 with respect to the undisturbed flow might attain a Mach number of 1 in the flow about the wing; the critical Mach number would thus be 0.8.

critical mass. The smallest mass of *fissionable material* that will support a self-sustaining *chain reaction* under stated conditions.

critical point. The **thermodynamic** state in which liquid and gas **phases** of a substance coexist in equilibrium at the highest possible temperature. At higher temperatures than the critical no liquid phase can exist.

critical pressure. 1. In rocketry, the pressure in the **nozzle throat** for which the isentropic **weight flow rate** is a maximum. **2.** The pressure of a gas at the **critical point,** which is the highest pressure under which a liquid can exist in equilibrium with its vapor.

critical Reynolds number. The **Reynolds number** at which some significant change occurs, e.g., the Reynolds number at which a transition from **laminar** to **turbulent flow** begins, or at which the **drag** of a cylinder or sphere drops sharply.

critical speed. A speed of a rotating system that corresponds to a **resonance frequency** of the system.

critical temperature. 1. The **temperature** above which a substance cannot exist in the **liquid** state, regardless of the pressure. **2.** As applied to reactor overheat or **afterheat,** the temperature at which the least resistant component of the **reactor core** begins to melt down. **3.** As applied to materials, the temperature at which a change in phase takes place causing an appreciable change in the properties of the material.

critical velocity. In rocketry, the **speed of sound** at the conditions prevailing at the **nozzle throat.** Also called *throat velocity, critical throat velocity.*

critical zone. The area over which a bombing plane engaged in horizontal or glide bombing must maintain straight flight so that the bomb sight can be operated properly and bombs dropped accurately.

cross-country flying. Flying an airplane from one geographical location to another especially over distances great enough from the point of take off to require some form of **navigation.**

cross feed. A system in a large airplane by which fuel, or oil, may be transferred from engine to engine, or from tank to tank.

crossflow. A **flow** going across another flow, as a spanwise flow over a wing.

crosshair. A hair, thread, or wire constituting part of a **reticle.**

cross over point. That range in the air warfare area at which a target ceases to be an air intercept target and becomes a surface-to-air missile target.

crosstalk. Electrical disturbances in a communication channel as a result of **coupling** with other communication channels.

crosswind. That wind vector component which is perpendicular to the course of an exposed moving object. Compare **range wind.**

cruise. To move at a speed fit for sustained travel.

cruise. (air traffic control). A word used instead of the word "maintian" in an ATC clearance to indicate to a pilot that climb and descent from the assigned altitude may be made at his discretion, and is authorization for the pilot to proceed to and make an approach at the destination airport.

cruise control. The procedure for the operation of an airplane, and its power plants, to obtain the maximum efficiency on extended flights.

cruise missile. Guided missile, the major portion of whose flight path to its target is conducted at approximately constant velocity; depends on the dynamic reaction of air for lift and upon propulsion forces to balance drag.

cruising altitude. A level determined by vertical measurement from mean sea level, maintained during a flight or portion thereof.

cryogenic liquid. Liquefied gas at very low temperature, such as liquid oxygen, nitrogen, argon.

cryogenic materials. Those metals and alloys which are usable in structures operating at very low temperature, and usually possess improved strength properties at these temperatures.

cryogenic propellant. A rocket **fuel, oxidizer,** or propulsion fluid which is liquid only at very low temperatures.

cryogenic pump. A type of pump which uses **cryopumping** to attain a vacuum.

cryogenics. 1. The study of the methods of producing very low temperatures. **2.**The study of the behavior of materials and processes at cryogenic temperatures.

cryogenic temperature. In general, a temperature range below the boiling point of nitrogen (—195° C); more particularly, temperatures within a few degrees of absolute zero.

cryopump. 1. An exposed surface refrigerated to **cryogenic temperature** for the purpose of pumping gases in a vacuum chamber by condensing the gas and maintaining the condensate at a temperature such that the equilibrium vapor pressure is equal to or less than the desired ultimate pressure in the chamber. **2.** The act of removing gases from an enclosure by condensing the gases on surfaces at cryogenic temperature.

    Also referred to as a *cryogenic pump* and not to be confused with a *cryogenic fluid pump* for circulating *cryogenic propellants.*

cryopumping. The process of removing gas from a system by condensing it on a surface maintained at very low temperatures.

cryotron. A device based upon the principle that **super-conductivity** established at temperatures near absolute zero is destroyed by the application of a magnetic field.

cryptanalysis. The study of encrypted texts. The steps or processes involved in converting encrypted text into plain text without initial knowledge of the key employed in the encryption.

crystal lattice. The three-dimensional, recurring pattern in which the atoms of a crystal are arranged.

crystal transducer. A **transducer** in which the method of transduction is accomplished by means of the **piezoelectric** properties of certain crystals or salts. Also called *crystal.*

culture (cartographic). Man-made or artificial features of the terrain.

cumuliform clouds. Cumuliform clouds are formed by rising air currents in unstable air. They are cauliflower-like in appearance with appreciable vertical development and dome-shaped upper surfaces. Usually cumuliform clouds are separate and distinct from each other. They also have flat bases and rarely cover the entire sky. Precipitation from cumuliform is usually of a shower nature. They also produce turbulent flying conditions. Also see **cloud, cumulonimbus, cumulus, cirrocumulus, stratocumulus.**

cumulonimbus. A form of cloud with extensive **vertical development.** Also see **cloud, cumuliform, nimbus, thunderstorm.**

cumulus. A form of cloud with less vertical development than **cumulonimbus.** Also see **cloud, cumuliform clouds, vertical development clouds.**

curie. The basic unit to describe the intensity of *radioactivity* in a sample of material. The curie is equal to 37 billion disintegrations per second, which is approximately the rate of decay of 1 gram of *radium.* A curie is also a quantity of any nuclide having 1 curie of radioactivity. Named for Marie and Pierre Curie, who discovered radium in 1898. (Compare **roentgen.**)

Curie point. The temperature in a **ferromagnetic** material above which the material becomes substantially nonmagnetic.

cursor. A device used with an instrument to provide a movable reference, as the runner of a slide rule or a rotatable plastic disk with inscribed crosslines, used in reading bearings on a **plan position indicator.**

curve of pursuit. The curved path described by a fighter plane making an attack on a moving target while holding the proper aiming allowance.

curvilinear coordinates. Any linear **coordinates** which are not **Cartesian coordinates.** Examples of frequently used curvilinear coordinates are **polar coordinates** and **cylindrical coordinates.** See **natural coordinates, spherical coordinates.**

cutoff or **cut-off. 1.** An act or instance of shutting something off; specifically, in rocketry, an act or instance of shutting off the **propellant** flow in a rocket, or

of stopping the **combustion** of the propellant. Compare **burnout. 2.** Something that shuts off, or is used to shut off. See **fuel shutoff. 3.** Limiting or bounding as in *cutoff frequency.*

cutoff attack. An attack that provides a direct vector from the interceptor position to an intercept point with the target track.

cutoff velocity. The velocity attained by a missile at the point of cutoff.

CW system. A **trajectory measuring system** that utilizes a **continous-wave** signal to obtain information on the trajectory of a target.

cyanometry. The study and measurement of the blueness of the sky.

    The characteristic blue color of clear skies is due to preferential **scattering** of the short wavelength components of visible sunlight by air molecules. Presence of foreign particles in the atmosphere alters the scattering processes in such a way as to reduce the blueness. Hence spectral analysis of **diffuse sky radiation** provides useful information concerning the scattering particles.

cybernetics. The study of methods of **control** and communication which are common to living organisms and machines.

cycle *(symbol* **c). 1.** The complete sequence of values of a **periodic quantity** that occur during a period. **2.** One complete wave, a **frequency** of 1 **wave** per second. **3.** Any repetitive series of operations or events.

cycle efficiency. The efficiency of a given **cycle** in an internal combustion engine, in producing work, expressed as the useful work output divided by the work input. For a gas-turbine engine, the cycle efficiency is the useful work energy less the work required for compression divided by the heat energy in the fuel used; for a reciprocating engine, it is the energy of the indicated horsepower divided by the heat energy of the fuel.

cyclic. Of or pertaining to a **cycle or cycles.**

cyclic control stick. See **cyclic pitch (rotorcraft).**

cyclic pitch (rotorcraft). Repetitive once-around-the-circle change in the pitch angle of each rotor blade as it turns around the axis. Cyclic pitch control is also known as "azimuth" control and its purpose is to tilt the direction of lift force of the rotor rather than to change its magnitude. Azimuth usually refers to the cyclic control stick, which can tilt the rotor in any direction of the azimuth; hence "azimuth stick".

cyclic pitch control (rotorcraft). A control by which the blade pitch angle is varied sinusoidally with the blade azimuth position.

cyclone. An area of low barometric pressure with its attendant system of winds. Also see **cyclonic.**

cyclonic. Having a sense of **rotation** about the **local vertical** the same as that of the earth's rotation: that is, as viewed from above, counter-clockwise in the Northern Hemisphere, clockwise in the Southern Hemisphere, undefined at the equator; the opposite of **anticyclonic.**

cyclotron. A particle *accelerator* in which charged particles receive repeated synchronized accelerations by electrical fields as the particles spiral outward from their source. The particles are kept in the spiral by a powerful magnetic field. (Compare **synchrocyclotron.**)

D

D. Military mission designation for **director** aircraft.

dadcap. Dawn and dusk combat air patrol.

Dalton law. The empirical generalization that for many so-called **perfect gases,** a mixture of these gases will have a pressure equal to the sum of the partial pressures that each of the gases would have as sole component with the same volume and temperature, provided there is no chemical interaction.

damp. To level out or retard. Also to reduce sound and deaden vibration, etc.

damped wave. Any **wave** whose **amplitude** decreases with time or whose total energy decreases by transfer to other portions of the wave spectrum.

danger area. A specified area within, below, or over which there may exist activities constituting potential danger to aircraft flying over it, or to persons, property, and traffic on land or sea.

dangerous material. Any material that because of its properties is flammable, corrosive, an oxidizing agent, explosive, toxic, radioactive, or unduly magnetic. (Unduly magnetic is construed to mean that sufficient magnetic field strength is present to cause significant navigational deviations to the compass sensing devices of an aircraft.)

dark adaptation. The process by which the iris and retina of the eye adjust to allow maximum vision in dim illumination, following exposure of the eye to a relatively brighter illumination.

dart configuration. A **configuration** of an **aerodynamic vehicle** in which the control surfaces are at the tail of the vehicle. Contrast *canard.*

data-acquisition station. A ground station at which various functions to **control** satellite operations and to obtain data from the **satellite** are performed.

data automation. The use of electronic, electromechanical, or mechanical equipment and associated techniques to automatically record, communicate, and process data and to present the resultant information.

data display. The visual presentation of processed data by specially designed electronic or electromechanical devices through interconnection (either on or off-line) with digital computers or component equipments. Although line printers and punch cards may display data, they are not usually categorized as displays but as output equipments.

data link. Any communications **channel** or circuit used to transmit data from a **sensor** to a **computer**, a **readout** device, or a **storage** device.

data point. A unit of fundamental information obtained through the processing of raw data.

data processing. Application of procedures, mechanical, electrical, computational, or other, whereby data are changed from one form into another.

data processing center. A computer installation providing data processing service for others, sometimes called customers, on a reimbursable or nonreimbursable basis.

data processing installation. The organizational facility where electronic data processing equipment, punch card accounting machines and/or other data processing equipment is located and operated.

data processor. A machine for handling information in a sequence of reasonable operations.

data reduction. The action or process of reducing data to usable form, usually by means of electronic computers and other electronic equipment.

data smoothing. The mathematical process of fitting a smooth curve to dispersed data points.

data system. The means, either manual or automatic, of converting data into action and/or decision information, including the forms, procedures, and processes which together provide an organized and interrelated means of recording, communicating, processing, and presenting information relative to a definable function or activity.

data transcription equipment. Those devices or equipment designed to convey data from its original state to a data processing media. The operation of this equipment is usually characterized by one of the following methods:

(a) from manual key strokes to data processing media;

(b) from manual key strokes to data processing media with simultaneous production of documents;

(c) character reading from documents and translation to data processing media through optical or magnetic scanning techniques.

datum. Any numerical or geometrical quantity or set of such quantities which can serve as a reference or a base for measurement of other quantities.

datum line. A base line or reference line from which calculations or measurements are taken.

day. 1. The duration of one **rotation** of the earth, or another celestial body, on its axis.

A day is measured by successive transits of a reference point on the celestial sphere over the meridian, and each type takes its name from the reference used. Thus, for a *solar day* the reference is the sun; a mean solar day if the mean sun; and an *apparent solar day* if the apparent sun. For a *lunar day* the reference is the moon; for a *sidereal day* the vernal equinox; for a *constituent day* an astre fictif or fictitious star. The expression *lunar day* refers also to the duration of one rotation of the moon with respect to the sun. A *Julian day* is the consecutive number of each day, beginning with January 1, 4317 BC.

2. A period of 24 hours beginning at a specified time, as the civil day beginning at midnight, or the astronomical day beginning at noon.

daylight saving time. See time.

D-day. The day on which an operation commences or is due to commence. This may be the commencement of hostilities or any other operation.

dead band. An arrangement incorporated in a control system which prevents an error from being corrected until that error exceeds a specified magnitude.

dead reckoning. Finding one's position by means of a compass and calculations based on speed, time elapsed, effect of wind, and direction from a known position.

dead spot. In a **control** system, a region centered about the neutral control position where small movements of the actuator do not produce any response in the system.

dead stick landing. A landing (of an airplane) without engine power.

debarkation. The unloading of troops, equipment, or supplies from a ship or aircraft.

debug. 1. To isolate and remove malfunctions from a device, or mistakes from a **routine** or **program. 2.** Specifically, in electronic manufacturing, to operate equipment under specified environmental and test conditions in order to eliminate early failures and to stabilize equipment prior to actual use. Also called *burn-in.*

debug (data automation). 1. To locate and correct any errors in a computer program. **2.** To detect and correct malfunctions in the computer itself. **3.** To test-run and check out a program of machine instructions for a digital computer for the purpose of eliminating mistakes.

decade. 1. The interval between any two quantities having the ratio of 10:1. **2.** A group or series of.10.

decayed object. An object once, but no longer, in orbit.

decay, radioactive. The spontaneous transformation of one nuclide into a different nuclide or into a different energy state of the same nuclide. The process results in a decrease, with time, of the number of the original radioactive atoms in a sample. It involves the emission from the *nucleus* of *alpha particles, beta particles* (or electrons), *or gamma rays;* or the nuclear capture or ejection of orbital electrons; or fission. Also called *radioactive disintegration.* (See **half-life, nuclear reaction, radioactive series.**)

decay time. 1. In computer operations, the time required for a pulse to fall to one-tenth of its peak value. See **rise time. 2.** In charge-storage tubes, the time interval during which the magnitude of the stored charge decreases to a stated fraction of its initial value.

The fraction is usually 1/*e* where *e* is the base of natural logarithms.

3. Approximately the lifetime of an orbiting object in a nonstable orbit.

Decay time is usually applied only to objects with short orbit lifetimes caused by atmospheric drag.

Decca. A long-range, ambiguous, two-dimensional navigation system using continuous-wave transmission to provide **hyperbolic lines of position** through the radio frequency phase comparison techniques from four transmitters.

Frequency band, 68 to 150 kilocycles.

deceleration. The retarding or slowing down of an object, i.e., it is the decrease in the rate of change of velocity. Negative **acceleration.**

deceleration parachute. A parachute attached to a craft and deployed to slow the craft, especially during landing. Also called a *brake parachute, drogue parachute, parabrake.*

decentralized control. In air defense, the normal mode whereby a higher echelon monitors unit actions, making direct target assignments to units only when necessary to insure proper fire distribution or to prevent engagement of friendly aircraft.

deception. Those measures designated to mislead the enemy by manipulation, distortion, or falsification of evidence to induce him to react in a manner prejudicial to his interests.

deci *(abbr* **d***).* A prefix meaning multiplied by 10−¹; one-

tenth.

decibel *(abbr* **db***).* **1.** A dimensionless measure of the ratio of two **powers,** equal to 10 times the logarithm to the base 10 of the ratio of two powers P_1/P_2 **2.** One-tenth of a **bel.**

The power P_2 may be some reference power; in electricity, the reference power is sometimes taken as 1 milliwatt *(abbr* **dbm***);* in acoustics, the decibel is often taken as 20 times the common logarithm of the sound pressure ratio, with the reference pressure as 0.0002 dyne per square centimeter.

decibel per second. A unit used to measure the **rate of decay** of a sound.

deck landing mirror sight. An landing aid on an aircraft carrier incorporating a system of lights and a mirror, so stabilized as to indicate to the pilot the correct descent path on to the flight deck.

decimal digit. 1. One of the digits used in decimal notation, i.e., 1, 2, 3, 4, 5, 6, 7, 8, 9, or 0. **2.** One of 10 possible conditions.

decimal notation. A mathematical system in which each **digit** is the coefficient of some power of 10.

decimal point. The **base point** in decimal notation.

decimal-to-binary converstion. The mathematical process of converting a quantity from **decimal notation** to the equivalent **binary notation.** For example: 1 = 1; 7 = 111; 23 = 10111, etc. See **binary notation.**

decision element. In computer operations, any device which as the result of the **input** of data issues one or two or more possible **instructions.**

decision height (DH). With respect to the operation of aircraft means the height at which a decision must be made, using an ILS or PAR instrument approach, to either continue the approach or to execute a missed approach.

declassification. The determination that classified information no longer requires, in the interests of national defense, any degree of protection against unauthorized disclosure, coupled with a removal or cancellation of the classification designation.

declination. 1. Angular distance north or south of the **celestial equator;** the arc of an **hour circle** between the celestial equator and a point on the **celestial sphere,** measured northward or southward from the celestial equator through 90°, and labeled N or S to indicate the direction of measurement.

2. Magnetic declination. See **equatorial system.**

decoder. 1. A device for translating electrical **signals** into predetermined functions. **2.** In computer operations, a **network** or device in which one of two or more possible outputs results from a prescribed combination of inputs. Also called *many-to-few matrix.*

decompression sickness. A disorder experienced by deep sea divers, pilots and astronauts caused by reduced barometric pressure and evolved gas bubbles in the body, marked by pain in the extremities (shoulders, arms, legs), pain in the chest (chokes), occasionally leading to sever central nervous symptoms and neurocirculatory collapse. See **bends, dysbarism.**

decommutator. Equipment for separation, **demodulation,** or demultiplexing commutated signals. See **commutator.**

decontamination. The act of removing chemical, biological, or **radiological** contamination from, or neutralizing it on, a person, item, or area.

decoupled. Of circuits or devices, interconnected through

*A **deceleration parachute** is used to reduce the landing roll of high speed aircraft. The landing roll of this LTV Corsair II is reduced by one-third when using the parabrake.*

any means which passes only the static characteristics of a **signal.**

decoy. A device or devices used to divert or mislead enemy defensive systems so as to increase the probability of penetration and weapon delivery.

decrement. A decrease in the value of a variable. See **increment.**

decrypt. To convert a cryptogram into plain text by a reversal of the encryption process. This does not include solution by cryptanalysis.

defense classification. A category or grade assigned to defense information or material which denotes the degree of danger to national security that would result from its unauthorized disclosure and for which standards in handling, storage, and dissemination have been established. These categories are defined as follows:

    a. confidential—Defense information or material, the unauthorized disclosure of which could be prejudicial to the defense interests of the Nation.

    b. secret—Defense information or material, the unauthorized disclosure of which could result in serious damage to the Nation, such as jeopardizing the international relations of the United States, endangering the effectiveness of a program or policy of vital importance to the national defense, or compromising important military or defense plans, scientific or technological developments important to national defense, or information revealing important intelligence operations.

    c. top secret—Defense information or material which requires the highest degree of protection. The top secret classification shall be applied only to that information or material, the defense aspect of which is paramount, and the unauthorized disclosure of which could result in exceptionally grave damage to the Nation, such as leading to a definite break in diplomatic relations affecting the defense of the United States, an armed attack against the United States or its allies, a war, or the compromise of military or defense plans, or intelligence operations, or scientific or technological developments vital to the national defense.

defense emergency. An emergency condition which exists when: **a.** a major attack is made upon United States forces overseas, or on allied forces in any theater and is confirmed either by the commander of a command established by the Secretary of Defense, or higher authority; or **b.** an overt attack of any type is made upon the United States and is confirmed either by the commander of a command established by the Secretary of Defense, or higher authority.

definition. In imagery interpretation, the degree of clarity and sharpness of an image.

deflagration. A sudden or rapid burning, as opposed to a **detonation** or **explosion.**

deflection of the vertical. The angular difference, at any place, between the direction of a plumb line (the **vertical**) and the perpendicular (the normal) to the **reference spheroid.** This difference seldom exceeds 30 seconds of arc. Also called *station error.*

    When measured at the earth's surface the deflection of the vertical is equal to the angle between the **geoid** and the reference spheroid.

deflector. A plate, baffle, or the like that diverts something in its movement or flow, as: **(a)** a plate that projects into the airstream on the underside of an airfoil to divert the airflow, as into a slot—sometimes distinguished from a **spoiler; (b)** a conelike device placed or fastened beneath a rocket launched from the vertical position, to deflect the exhaust gases to the sides; **(c)** any of several different devices used on jet engines to reverse or divert the exhaust gases; **(d)** a **baffle** or the like to deflect and mingle fluids prior to combustion, as in certain jet engines.

defoliant operations. The employment of defoliating agents on vegetated areas in support of military operations.

defoliating agent. A chemical which causes trees, shrubs, and other plants to shed their leaves prematurely.

degas. To remove gas from a material, usually by heating under **high vacuum.** Compare **get.**

degradation. Gradual deterioration in performance.

degree of freedom. 1. A mode of motion, either angular or linear, with respect to a **coordinate system,** independent of any other mode.

    A body in motion has six possible degrees of freedom, three linear and three angular.

2. Specifically, of a gyro the number of **orthogonal axes** about which the spin axis is free to rotate. **3.** In an unconstrained dynamic or other system, the number of independent variables required to specify completely the state of the system at a given moment.

    If the system has constraints, i.e., kinematic or geometric relations between the variables, each such relation reduces by one the number of degrees of freedom of the system. In a continuous medium with given boundary conditions, the number of degrees of freedom is the number of normal modes of **oscillation. 4.** Of a mechanical system, the minimum number of independent generalized **coordinates** required to define completely the positions of all parts of the system at any instant of time.

    In general, the number of degrees of freedom equals the number of independent generalized displacements that are possible.

de-icing. The protection of aircraft against icing by allowing ice to build up and causing it to be removed by mechanical, chemical or thermal means. Compare **anti-icing.**

deka (*abbr* **da**). A prefix meaning multiplied by 10. Sometimes spelled deca.

de Laval nozzle. (After Dr. Carl Gustaf Patrik de Laval (1845—1913), Swedish engineer.) A converging-diverging **nozzle** used in certain rockets. Also called *Laval nozzle.*

delay. The time (or equivalent distance) displacement of some characteristic of a **wave** relative to the same characteristic of a reference wave; that is, the difference in **phase** between the **two waves.** Compare **lag.**

    In one-way radio propagation, for instance, the phase delay of the reflected wave over the direct wave is a measure of the extra distance traveled by the reflected wave in reaching the same receiver.

delayed drop. A live parachute descent begun by a free fall, for a distance greater than that normally allowed for the opening parachute to clear the aircraft.

delayed neutrons. Neutrons emitted by excited nuclei in a radioactive process, so called because they are emitted an appreciable time after the **fission.**

delay line. In electronic computers, any device for producing a time delay of a **signal.**

delay-line storage. A **storage** or memory device consisting of a **delay line** and means for regenerating and reinserting information into the delay line.

delta wing. A triangularly shaped wing of an aircraft.

demand oxygen system. An **oxygen system** in which oxygen flows to the user during inspiration only.

demilitarize. To mutilate, disarm or accomplish any other action required to render explosives unusable for military use.

demilitarized zone. A defined area in which the stationing or concentrating of military forces, or the retention or establishment of military installations of any description, is prohibited.

demodulation. The process of recovering the **modulating wave** from a modulated carrier.

demonstration and shakedown operation. (military) Tests conducted by operational commands, assisted by the service technical agency, in an operational environment using operational procedures. During these tests, operational and logistical procedures are refined; basic system capability and limitations are demonstrated and the determination is made that the system is sufficiently stabilized to perform its intended mission.

denitrogenation. The removal of nitrogen dissolved in the blood and body tissues, usually by breathing of pure oxygen for an extended period of time in order to prevent **aeroembolism** at high altitudes.

densitometer. An instrument for the measurement of **optical density** (photographic transmission, photographic reflection, visual transmission, etc.) of a material, generally of a photographic image.

density. Weight per unit volume expressed in pounds per cubic ft., grams per cubic centimeter, etc.

density altitude. Density altitude is pressure altitude corrected for non-standard temperature. Density altitude must be computed for high altitude and/or high temperature conditions to determine take-off run, for example. Navigation problems normally do not require computation of density altitude. Also see **altitude.**

density specific impulse. The product of the **specific impulse** of a **propellant** combination and the average **specific gravity** of the propellants.

Department of the Air Force. The executive part of the Department of the Air Force at the seat of government, and all field headquarters, forces, reserve components, installations, activities, and functions under the control or supervision of the Secretary of the Air Force. See also **Military Department.**

Department of the Army. The executive part of the Department of the Army at the seat of government, and all field headquarters, forces, reserve components, installations, activities, and functions under the control or supervision of the Secretary of the Army. See also **Military Department.**

Department of the Navy. The executive part of the Department of the Navy at the seat of government; the headquarters, United States Marine Corps; the entire operating forces of the United States Navy, including naval aviation, and of the United States Marine Corps, including the reserve components of such forces; all field activities, headquarters, forces, bases, installations, activities, and functions under the control or supervision of the Secretary of the Navy; and the United States Coast Guard when operating as a part of the Navy pursuant to law. See also **Military Department.**

departure. Any aircraft taking off from an airport is referred to as a departure.

departure control. A function of approach control providing service for departing IFR aircraft and on occasion, VFR aircraft.

departure point. A navigational check point used by aircraft as a marker for setting course.

dependent variable. Any **variable** considered as a function of other variables, the latter being called *independent.* Compare **parameter.**

Whether a given quantity is best treated as a dependent or independent variable depends upon the particular problem.

depleted uranium. Uranium having a smaller percentage of uranium-235 than the 0.7% found in natural uranium. It is obtained from the spent (used) fuel elements or as by-product *tails,* or residues, of uranium *isotope separation.* (Compare **natural uranium, spent fuel.**)

deploy. Of a parachute, to release so as to let it fill out or to unfold and fill out.

deploy (military). **1.** In a strategic sense, to relocate forces to desired areas of operations. **2.** To extend or widen the front of a military unit, extending from a close order to a battle formation.

deployment. The series of functions which transpire from the time a packed parachute is placed in operation until it is fully opened and is supporting its load.

descending node. That point at which a planet, planetoid, or comet crosses to the south side of the **ecliptic;** that point at which a **satellite** crosses to the south side of the equatorial plane of its primary. Also called *southbound node. The opposite is ascending node or northbound node.*

design gross weight. The **gross weight** at take-off that an aircraft, rocket, etc., is expected to have, used in design calculations.

design studies. Studies conducted to determine the characteristics of a system needed to satisfy a particular requirement.

desorption. The process of removing **sorbed** gas.

destroyed. A condition of a target so damaged that it cannot function as intended nor be restored to a usable condition. In the case of a building, all vertical supports and spanning members are damaged to such an extent that nothing is salvageable. In the case of bridges, all spans must be dropped and all piers must require replacement.

destruct. The deliberate action of destroying a **rocket** vehicle after it has been launched, but before it has completed its course.

Destructs are executed when the rocket gets off its plotted course or functions in a way so as to become a hazard.

destruction area. An area in which it is planned to destroy or defeat the enemy airborne threat. The area may be further subdivided into air intercept, missile (long, medium, and short-range) or antiaircraft gun zones.

destruct line. On a rocket test range, a boundary line on each side of the downrange course beyond which a rocket cannot fly without being destroyed under destruct procedures, or a line beyond which the **impact point** cannot pass. See **command destruct.**

detached shock wave. A **shock wave** not in contact with the body which originates it. See **bow wave.** Also called *detached shock.*

detachment. A particular state of isolation in which man is separated or detached from his accustomed behavioral **environment** by inordinate physical and psychological distances. This condition may compromise his performance.

detector (nucleonics). Material or a device that is sensitive to radiation and can produce a response signal suitable for measurement or analysis. **A radiation detection instrument.**

detector. 1. same as **sensor,** sense **1. 2.** An instrument employing a sensor, sense **1,** to detect the presence of something in the surrounding environment.

deterrence. The prevention from action by fear of the consequences. Deterrence is a state of mind brought about by the existence of a credible threat of unacceptable counter action.

detonation. A rapid chemical reaction which propagates at a **supersonic** velocity.

detonation wave. A **shock wave** in a combustible mixture, which originates as a **combustion wave.**

deuterium. An *isotope* of hydrogen whose nucleus contains one neutron and one proton and is therefore about twice as heavy as the nucleus of normal hydrogen, which is only a single proton. Deuterium is often referred to as *heavy hydrogen;* it occurs in nature as 1 atom to 6500 atoms of normal hydrogen. It is nonradioactive. (See **heavy water, hydrogen.**)

development. The process of working out and extending the theoretical, practical, and useful applications of a basic design, idea, or scientific discovery. The design, building, modification, or improvement of the prototype of a vehicle, engine, instrument or the like as determined by the basic idea or concept.

development project. An undertaking:

(a) to develop a subsystem, assembly, accessory, attachment, end item (or principal component thereof) or materiel;

(b) to develop a related family of components; or

(c) to explore a field of knowledge in search of scientific information.

development testing and evaluation. That testing and evaluation used to measure progress, verify accomplishment of development objectives, and to determine: if theories, techniques, and material are practicable; and if systems or items under development are technically sound, reliable, safe, and satisfy specifications.

deviation. 1. In statistics, the difference between two numbers. Also called *departure.*

It is commonly applied to the difference of a variable from its mean, or to the difference of an observed value from a theoretical value.

2. magnetic deviation. 3. In radio transmission, the apparent variation of frequency above and below the unmodulated or center frequency.

deviation card. A card recording the deviations of a particular compass and indicating the compass direction corresponding to any desired magnetic direction. Also see **compass compensation.**

deviation correction. The correction to be applied to a compass reading to correct for deviation error.

device, nuclear. A *nuclear explosive* used for peaceful purposes, tests or experiments. The term is used to distinguish these explosives from nuclear weapons, which are packaged units ready for transportation or use by military forces. (Compare **nuclear weapons.**)

dew. Atmospheric moisture condensed, in liquid form, upon objects cooler than the air, especially at night.

dew point. The temperature to which air must be cooled, at constant pressure and moisture content, in order for

saturation to occur.

DF (*abbr*)–**direction finder.** See **radio direction finder.**

diabatic process. A process in a **thermodynamic** system in which there is a transfer of heat across the boundaries of the **system.**

Diabatic process is preferred to *nonadiabatic process.*

diamonds. The pattern of **shock waves** often visible in a **rocket** exhaust which resembles a series of diamond shapes placed end to end.

dichotomy. In astronomy, a **configuration** of three bodies so that they form a right triangle; specifically, such a configuration in the **solar system** with the sun at the apex of the 90° angle.

dielectric. A substance capable of supporting an electric stress, sustaining an electric field and undergoing electric polariation. All insulators are dielectrics and a vacuum is also a dielectric.

dielectric constant. Ratio of the **capacitance** of a **condenser** with a **dielectric** between the plates to the capacitance of the same condenser with a **vacuum** between the **plates.**

dielectric strength. A measure of the resistance of a **dielectric** to electrical breakdown under the influence of strong electric fields; usually expressed in volts per centimeter. Sometimes called *breakdown potential.*

difference of latitude. The shorter arc of any **meridian** between the **parallels** of two places, expressed in angular measure.

difference of longitude. The smaller angle at the **pole** or the shorter arc of a **parallel** between the **meridians** of two places, expressed in angular measure.

differential analyzer. An **analog computer** designed and used primarily for solving differential equations.

differential pressure. The pressure difference between two systems or volumes.

differential thermal analysis. The technique of detecting **endothermic** and **exothermic** phase changes and other processes within a heated material by the corresponding temperature changes.

differentiator. 1. In computer operations, a device whose output is proportional to the derivative of an input signal. **2.** In electronics, a **transducer** whose output **waveform** is the time derivative of its input waveform.

diffraction. The bending of waves while they pass an angle or travel through a slit.

diffraction propagation. **Wave** propagation around objects, or over the horizon, by **diffraction.**

diffuser. A specially designed **duct, chamber,** or section, sometimes equipped with guide vanes, that decreases the volocity of a fluid, as air, and increases its pressure, as in a jet engine, a wind tunnel, etc. See **supersonic diffuser.**

diffuse radiation. Radiant energy propagating in many different directions through a given small volume of space; to be contrasted with parallel radiation.

The ideal form of diffuse radiation is **isotropic radiation.** Careful distinction should be made between this concept and that of a **perfectly diffuse radiator.**

diffuse reflection. Any reflection process in which the reflected radiation is sent out in many directions usually bearing no simple relationship to the angle of incidence; the opposite of **specular reflection.** See **diffuse reflector, perfectly diffuse reflector.**

A term frequently applied to the process by which solar radiation is scattered by dust and other suspensoids in the atmosphere. See **diffuse sky radiation.**

diffuse reflector. Any surface which reflects incident rays in a multiplicity of directions, either because of irregularities in the surface or because the material is optically inhomogeneous, as a paint, although optically smooth; the opposite of a **specular reflector.** See **perfectly diffuse reflector.**

Ordinary writing papers are good examples of diffuse reflectors, whereas mirrors or highly polished metal plates are examples of specular reflectors. Almost all terrestrial surfaces (except calm water) act as diffuse reflectors of incident solar radiation.

diffuse sky radiation. **Solar radiation** reaching the earth's surface after having been **scattered** from the direct solar beam by molecules or suspensoids in the atmosphere. Also called *skylight, diffuse skylight, sky radiation.*

Of the total light removed from the direct solar beam by scattering in the atmosphere (approximately 25 percent of the incident radiation), about two-thirds ultimately reaches the earth as diffuse sky radiation.

diffusion. In an atmosphere, or in any gaseous system, the exchange of **fluid parcels** between regions, in apparently random motions of a scale too small to be treated by the **equations of motion. 2.** In materials, the movement of atoms of one material into the crystal lattice of an adjoining material, e.g., penetration of the atoms in a ceramic coating into the lattice of the protected metal. **3.** In ion engines, the migration of neutral atoms through a porous structure incident to ionization at the emitting surface.

difluence. The **divergence** vector of adjacent **streamlines.** The opposite of **confluence.**

digit. 1. A single symbol or **character** representing an integral quantity. **2.** Any one of the symbols used in **positional notation** as coefficients of each power, or order, of the base. See **binary digit, decimal digit.**

digital. Using **discrete** expressions to represent **variables.**

digital autopilot. An **autopilot** utilizing a **digital computer** to provide control inputs in response to **sensors.**

digital computer. A device which processes information represented by combinations of discrete or discontinuous data as compared with an analogue computer for continuous data. More specifically, it performs sequences of arithmetic and logical operations, not only on data but its own program. Still more specifically, it is a stored program digital computer capable of performing sequences of internally stored instructions, as opposed to calculators, on which the sequence is impressed manually.

digital output. Transducer output that represents the magnitude of the stimulus in the form of a series of discrete quantities coded to represent **digits** in a system of notation. Compare **analog output.**

digitize. To express an **analog** measurement of a variable in **discrete** units.

digitizer. A device which converts **analog data** into numbers expressed in **digits** in a system of notation. Also called *analog-to-digital converter.*

dihedral angle. The acute angle between two intersecting planes or between lines representative of planes. An airplane wing design incorporating dihedral is used to improve **lateral stability.**

dioptric light. A light concentrated into a collimated beam by means of **refracting** lenses or prisms.

One collimated by means of a reflector is a catoptric light.

dip, of a compass needle. The angle included between a perfectly poised magnetic needle and its horizontal plane. The "dip" is called, by scientists, the "magnetic inclination." Magnetic needles are weighted so that they normally rotate in a horizontal plane, thus overcoming the "dip".

diplexer. A device permitting an **antenna** system to be used simultaneously or separately by two transmitters. Compare **duplexer.**

diplex transmission. The simultaneous transmission of two signals using a common **carrier wave.** Compare **duplex operation, multiplexing.**

dipole. 1. A system composed of two, separated, equal electric or magnetic charges of opposite sign. **2. dipole antenna.**

dipole antenna. An **antenna** composed of two **conducters** in line and **fed** at the middle. The total length of the antenna is equal to one-half of the wave length to be transmitted or received. Dipoles are short (for ulta high **frequency**), self supporting and close together at the middle.

direct air cycle. A **thermodynamic** propulsion cycle involving a nuclear **reactor** and gas turbine or ramjet engine, in which air is the **working fluid.** Also called *direct cycle.*

Air is successively compressed in the compressor section, heated in the nuclear reactor, and expelled through the turbine-tailpipe section to obtain thrust.

direct air support center. A subordinate operational component of a tactical air control system designed for control and direction of close air support and other tactical air support operations and is normally collocated with fire support coordination elements. See also **direct air support center (airborne.)**

direct air support center (airborne). An airborne aircraft equipped with the necessary staff personnel, communications, and operations facilities to function as a direct air support center. See also **direct air support center.**

direct-current discharge. The conduction of direct current through two **electrodes** immersed in a gas. See **Townsend discharge, glow discharge, arc discharge.**

direct-cycle reactor system. A nuclear power plant system in which the *coolant* or heat transfer fluid circulates first through the reactor and then directly to a turbine. (Compare **indirect-cycle reactor system.**)

directed energy weapon. A generic term encompassing both **particle beam** and **high-energy** weapons.

direct fire. Gunfire delivered on a target, using the target itself as a point of aim for either the gun or the director.

direct indicating compass. A magnetic compass in which the dial, scale or index is carried on the detecting element.

direct lift control system (DLC). A system utilizing **spoilers** to vary wing lift for flight path control such as precision glide-slope tracking. The result is translational movement without angular (pitch) motion.

direction (navigation).

Azimuth. The initial direction of the arc of a great circle; the angle between the plane of the great circle and the meridian of the place. As used in air navigation, it is measured from the north, in a clockwise direction, from 0" to 360".

Bearing. In air navigation, the same as azimuth.

Course. The direction of the rhumb line, or the line of constant direction. As used in air navigation with the Lambert projection, it is measured at the meridian nearest halfway between the starting point and the destination.

Heading. The direction in which the airplane is pointed, in contradistinction to its path over the ground.

Track. All rhumb-line direction of the actual flight path of an aircraft over the ground.

(All the above directions may be true, magnetic, or compass, according to whether they are referred to as "true north," "magnetic north," or "compass north.")

Wind. Always the true direction *from* which the wind blows.

direction angle. In tracking, the angle between the antenna **baseline** and an imaginary line connecting the center of the baseline with the target.

direction center. A major facility of the North American Air Defense Command from which the Air Defense Commander and his staff exercise operational control over assigned forces. This is the highest command level from which active control of the air battle is accomplished. A direction center may be either semi-automated or manual.

direction cosine. 1. The cosine of the angle formed by the intersection of a line, as a line of sight to an orbiting body, with an axis of a rectangular coordinate system with the origin on the line. **2.** Specifically, in **tracking,** the cosine of the angle between a **baseline** and the line connecting the center of the baseline with the target.

direction finder. See **automatic direction finder.**

direction finding (DF). See **automatic direction finder.**

direction finding penetration and approach procedures (military). A procedure whereby a direction finding controller furnishes the pilot with directional information during the penetration and approach to the airport. Radar may be used to complete the final phase of the approach.

directional antenna. An **antenna** that radiates or receives radio signals more efficiently in some directions than in others. See **Adcock antenna, loop antenna, sense antenna.**

A group of antennas arranged for this purpose is called an *antenna array.*

directional gyro. 1. A two-degree-of-freedom **gyro** with a provision for maintaining its spin axis approximately horizontal. **2.** A flight instrument incorporating a gyro that holds its position in **azimuth** and thus can be used as a directional reference.

directional stability. The property of an aircraft, rocket, etc., enabling it to restore itself from a **yawing** or sideslipping condition. Also called *weathercock stability.*

directive. 1. A military communication in which policy is established or a specific action is ordered. **2.** A plan issued with a view to placing it in effect when so directed, or in the event that a stated contingency arises. **3.** Broadly speaking, any communication which initiates or governs action, conduct, or procedure.

directivity. The ability of an **antenna** to radiate or receive more energy in some directions than in others. See **beam.**

The directivity of an antenna implies a maximum value, and it is equal to the ratio of the maximum **field intensity** to the average field intensity at a given dis-

tance.

director (D), (military designation). Aircraft capable of controlling a drone aircraft or a missile. See **military aircraft types.**

direct solar radiation. In **actinometry,** that portion of the radiant energy received at the instrument direct from the sun, as distinguished from **diffuse sky radiation, effective terrestrial radiation,** or radiation from any other source. See **global radiation.**

Direct solar radiation is measured by **pyrheliometers.**

dirty bomb. A *fission bomb* or any other weapon which would distribute relatively large amounts of radioactivity upon explosion, as distinguished from a fusion weapon. (Compare **clean bomb.**)

disarmament. The reduction of a military establishment to some level set by international agreement. See also **arms control; arms control agreement.**

discone antenna. An **antenna** formed of a disk and a cone whose apex approaches and becomes common with the outer conductor of the coaxial feed at its extremity.

The center conductor terminates at the center of the disk which is perpendicular to the axis of the cone.

discontinuity. A break in sequence or continuity of anything.

discrete. Composed of distinct or discontinuous elements.

discrete frequency. A frequency assigned a particular function.

discrete spectrum. A **spectrum** in which the component wavelengths (and wave numbers and frequencies) constitute a **discrete** sequence of values (finite or infinite in number) rather than a **continuum** of values. See **continuous spectrum.**

discrete variable. A quantity that may assume any one of a number of individually distinct or separate values.

discriminator. In general, a **circuit** in which output depends upon the difference between an input signal and a reference signal.

dish, radar. A parabolic reflector used as a portion of a radar antenna.

disk area (rotor). The area of the circle described by the tips of the blades.

disk loading (rotor). The thrust of the rotor devided by the rotor disk area.

dispersal airfield. An airfield, military or civil, to which aircraft might move before H-hour on either a temporary duty or permanent change of station basis and be able to conduct operations. See also **airfield.**

dispersed operating base. A United States military air base that is equipped, manned, and maintained in a reduced operational status, with a capability to expand operations on short notice for accommodation of dispersed theater forces and/or Continental United States augmentative forces during periodic exercises, contingency, or general war operations.

dispersion (military). 1. A scattered pattern of hits, by bombs dropped under identical conditions or by projectiles fired from the same weapon or group of weapons with the same firing data. **2.** In antiaircraft gunnery, the scattering of shots in range and deflection about the mean point of impact. As used in flak analysis, the term includes scattering due to all causes, and the mean point of impact is assumed to be the target. **3.** The spreading or separating of troops, materiel, establishments, or activities which are usually concentrated in limited areas, to reduce vulnerability to enemy action. **4.** In chemical operations, the dissemination of agents in liquid or aerosol form from bombs and spray tanks.

dispersion. 1. In rocketry, (**a**) deviation from a prescribed **flight path,** (**b**) specifically, **circular dispersion. 2.** A measure of the scatter of data points around a mean value or around a regression curve.

Usually expressed as a standard-deviation estimate, or as a standard error of estimate. Note that the scatter is not centered around the true value unless systematic errors are zero.

3. The process in which **radiation** is separated into its component **wavelengths.**

Dispersion results when an optical process, such as **diffraction, refraction,** or **scattering,** varies according to wavelength.

4. In spectroscopy, a measure of the resolving power of a **spectroscope** or spectrograph, usually expressed in angstroms per millimeter. **5.** As applied to materials, a scattering of very fine particles (e.g., ceramics) within the body of a metallic material usually resulting in over-all strengthening of the composite material.

dispersive medium. A medium in which the **phase velocity** of a wave, either **electromagnetic** or **hydromagnetic,** is a function of the **frequency.**

A **plasma** is a dispersive medium whereas **free space** is not, since waves of all frequencies travel in free space with the velocity of light.

displaced threshold. A **threshold** that is located at a point on the runway other than the beginning.

displacement (military). In air interception, separation between target and interceptor tracks established to position the interceptor in such a manner as to provide sufficient maneuvering and acquisition space.

displacement. Distance from a standard point (which is taken as the origin) measured in a given direction. The origin is the center of the path in describing simple harmonic motion.

display. The graphic presentation of the **output** data of any device or system.

disposable load. For a military aircraft: the fuel, oil and armament stores. For a civil aircraft: the crew, fuel, oil and pay load.

disposition (military). 1. Distribution of the elements of a command within an area, usually the exact location of each unit headquarters and the deployment of the forces subordinate to it. **2.** A prescribed arrangement of all the tactical units composing a flight or group of aircraft.

dissociation. The separation of a complex **molecule** into constituents by **collision** with a second body, or by absorption of a photon.

The product of dissociation of a molecule is two **ions,** one positively charged and one negatively charged.

distance. One-dimensional space which is measured or can be measured.

distance marker. A reference marker indicating distance, particularly such a marker on a radar indicator, to indicate distance of a target from the radar antenna. On a plan position indicator it is usually one of a series of concentric circles. Also called *range marker.* See **range ring.**

distance measuring equipment (DME). Electronic equipment used to measure in nautical miles, the distance of the aircraft from a navigation aid.

In the operation of DME, paired pulses at a specific spacing are sent out from the aircraft (this is the in-

terrogation) and are received at the ground station. The ground station (transponder) then transmits paired pulses back to the aircraft at the same pulse spacing but on a different frequency. The time required for the round trip of this signal exchange is measured in the airborne DME unit and is translated into distance (Nautical Miles) from the aircraft to the ground station.

DME operates on frequencies in the UHF spectrum between 962 MHz and 1213 MHz. Aircraft equipped with TACAN equipment will receive distance information from a VORTAC automatically, while aircraft equipped with VOR must have a separate DME airborne unit.

distortion. 1. An undesired change in **waveform.**

Noise and certain desired changes in waveform, such as those resulting from modulation or detection, are not usually classed as distortion.

2. In a system used for transmission or reproduction of sound, a failure by the system to transmit or reproduce a received wave form with exactness. **3.** An undesired change in the dimensions or shape of a structure as, *distortion of a fuel tank due to abnormal stresses or extreme temperature gradients.*

ditching. Controlled landing of a distressed aircraft on water.

diurnal. Having a period of, occurring in, or related to a **day.**

diurnal aberration. Aberration caused by the rotation of the earth. The value of diurnal aberration varies with the latitude of the observer and ranges from zero at the poles to 0.31 second of arc.

dive. A steep descent, with or without power in which the air speed is greater than the maximum speed in horizontal flight.

dive brake. See **air brake.**

dive flap. A flap-type air brake used to reduce the limiting velocity of an aircraft.

dive toss. A weapon delivery maneuver in which the aircraft is dived to a predetermined altitude and point in space, pulled up and the weapon released in such a way that it is tossed onto the target.

divergence. 1. The expansion or spreading out of a **vector** field; also a precise measure thereof.

In mathematical discussion *divergence* is considered to include *convergence,* i.e., negative divergence.

2. A static instability of a lifting surface or of a body on a vehicle wherein the aerodynamic loads tending to deform the surface or body are greater than the elastic restoring forces.

diversionary missile. A missile decoy.

divert. 1. Proceed to divert field or carrier as specified. **2.** To change the target, mission, or destination of an airborne flight.

division. 1. A tactical unit/formation as follows: **a.** a major administrative and tactical unit/formation which combines in itself the necessary arms and services required for sustained combat, larger than a regiment/brigade and smaller than a corps; **b.** a number of naval ships of similar type grouped together for operational and administrative command, or a tactical unit of a naval aircraft squadron, consisting of two or more sections; and **c.** an air division is an air combat organization normally consisting of two or more wings with appropriate service units. The combat wings of an air division will normally contain similar type units. **2.**

An organizational part of a headquarters that handles military matters of a particular nature, such as personnel, intelligence, plans and training, or supply and evacuation. **3.** A number of personnel of a ship's complement grouped together for operational and administrative command.

DLC. Direct lift control system.

DME (*abbr*) *distance measuring equipment.*

DME fix. A geographical postion determined by reference to a navaid which provides distance and azimuth information and defined by a specified distance in nautical miles and a radial in degrees magnetic from that aid.

DME separation. Spacing of aircraft in terms of distance determined by reference to distance measuring equipment (DME).

Dobson spectrophotometer. A photoelectric **spectrophotometer** which is used in the determination of the ozone content of the atmosphere. The instrument compares the solar energy at two wavelengths in the **absorption band** of ozone by permitting the two radiations to fall alternately upon a photocell. The stronger radiation is then attenuated by an optical wedge until the photoelectric system of the photometer indicates equality of incident radiation. The ratio of radiation intensity is obtained by this process and the ozone content of the atmosphere is computed from the ratio.

docking. The act of coupling two or more orbiting objects; the operation of mechanically connecting together, or in some manner bringing together, orbital payloads.

document. Any recorded information regardless of its physical form or characteristics, including, without limitation, written or printed material; data processing cards and tapes; maps, charts, photographs, negatives; moving or still films; film strips; paintings; drawings; engravings; sketches; reproduction of such things by any means or process; and sound, voice, or electronic recordings in any form.

documentation (general). Any tangible media created or acquired to record and communicate temporarily or permanently, knowledge, acts, interest, or events.

dog leg. A route containing a major alteration of course (as opposed to a straight line course).

dolly (military). Airborne data link equipment.

domestic air service. A scheduled air service which passes through the air space over only one country.

domestic air traffic. Air traffic within continental United States.

doping. Treatment of a fabric surface to tauten, strengthen or render it air-tight.

doping. Addition of impurities to a **semiconductor** or production of a deviation from **stoichiometric** composition to achieve a desired characteristic.

Doppler effect. The change in **frequency** with which energy reaches a receiver when the receiver and the energy source are in motion relative to each other. Also called *Doppler shift.*

Doppler error. In using **Doppler radar,** the error in the measurement of target **radial velocities** due to atmospheric refraction. Compare **range error, azimuth error.**

Doppler navigation. Dead reckoning performed automatically be a device which gives a continuous indication of position by integrating the speed derived from measurement of the **Doppler effect** of echoes from

directed beams of radiant energy transmitted from the craft. See **Doppler radar.**

Doppler radar. A **radar** which detects and interprets the **Doppler effect** in terms of the **radial velocity** of a target.

Doppler ranging. A continuous wave **trajectory-measuring system** which utilizes the **Doppler effect** to measure the distances between a transmitter, a rocket transponder, and several receiving stations.

From these measurements trajectory data are computed. In contrast to less sophisticated systems, Doran obviates the necessity of continuously recording the Doppler signal by making simultaneous distance measurements with four different frequencies.

Doppler shift. 1. Same as **Doppler effect. 2.** The magnitude of the Doppler effect, measured in cycles per second.

Doppler system. In radar, any system utilizing the **Doppler effect** for obtaining information.

Doppler, velocity and position. A continuous-wave **trajectory-measuring system** using the **Doppler effect** caused by a target moving relative to a ground transmitter and receiving stations.

The transmitter interrogates a frequency doubling transponder and the output is received at three or more receiver sites for comparison with the interrogation frequency. The intersection of ellipsoids formed by the transmitter and each receiver site provides the spatial postion of the target.

dorsal. Pertaining to the back.

dorsal fin. A verticle surface or **fin** located on the **aft** and upper side of a **fuselage.** The fin extends aft from a point well forward of the tail section and gradually slopes upward to the larger part of the fin. It is somewhat triangular in shape.

dose (nucleonics). See **absorbed dose, biological dose, maximum permissible dose, threshold dose.**

dose rate (nucleonics). The radiation dose delivered per unit time and measured, for instance, in rems per hour. (See **absorbed dose, rem.**)

dosimeter. 1. An instrument for measuring the **ultraviolet** in solar and sky radiation. Compare **actinometer. 2.** A device, worn by persons working around radioactive material which indicates the *dose* of radiation to which they have been exposed.

double agent. Agent in contact with two opposing intelligence services only one of which is aware of the double contact or quasi-intelligence services.

double-base propellant. A **solid rocket** propellant using two unstable compounds, such as nitrocellulose and nitroglycerin.

The unstable compounds used in a double-base propellant do not require a separate oxidizer.

double channel simplex (radio communications). Simplex using two frequency channels, one in each direction.

double drift. A method of determining the wind by observing drift on an initial true heading and two other true headings flown in specific pattern.

double-entry compressor. A **centrifugal compressor** that takes in air or fluid on both sides of the impeller, with vanes on each side to accelerate the fluid into the diffuser. The double-entry compressor is not a **multistage compressor.**

double-integrating gyro. A single-degree-of-freedom **gyro** having essentially no restraint of its spin axis about the output axis. In this gyro an **output** signal is produced by gimbal angular displacement, relative to the base, which is proportional to the double integral of the angular rate of the base about the input axis.

double local oscillator. An **oscillator** mixing system which generates two radio-frequency signals accurately spaced a few hundred cycles apart and mixes these signals to give the difference **frequency** which is used as the reference.

This equipment is used in an **interferometer** system to obtain a detectable signal containing the **phase** information of an antenna pair and the reference signal to allow removal of the phase data for use.

double stars. Stars which appear as single points of light to the eye but which can be resolved into two points by a telescope.

A *double star* is not necessarily a *binary*, a two-star system revolving about a common center, but may be an *optical double*, two unconnected stars in the same line of sight.

doubling time (nucleonics). The time required for a breeder reactor to produce as much fissionable material as the amount usually contained in its core plus the amount tied up in its fuel cycle (fabrication, reprocessing, etc.). It is estimated as 10 to 20 years in typical reactors. (See **breeder reactor, fuel cycle.**)

down current. A downward movement of air. May be due to sinking air on the leeward side of large objects or to a descending body of cool air.

downgrade. To determine that classified information requires, in the interests of national defense, a lower degree of protection against unauthorized disclosure than currently provided, coupled with a changing of the classification designation to reflect such lower degree.

down range. The airspace extending down stream on a given rocket test range.

downtime. A period during which equipment is not operating correctly because of machine failure.

downwash. The downward component of power plant(s) derived windblast directly under the aircraft.

downwind leg. A flight path in the traffic pattern parallel to the landing runway in the direction opposite to landing. It extends to the intersection of the base leg.

DR (*abbr*) **dead reckoning.**

dracontic month. The average period of **revolution** of the moon about the earth with respect to the moon's **ascending node,** a period of 27 days 5 hours 5 minutes 35.8 seconds, or approximately 27 1/4 days. Also called *nodical month.*

drag (*symbol* D). A retarding force acting upon a body in motion through a **fluid,** parallel to the direction of motion of the body. It is a component of the total fluid forces acting on the body. See **aerodynamic force.**

drag coefficient (*symbol* C_D). A coefficient representing the **drag** on a given airfoil or other body, or a coefficient representing a particular element of drag. See **Rayleigh formula.**

drag parachute. 1. drogue parachute. 2. Any of various types of parachutes attached to high-performance aircraft that can be deployed, usually during landings, to decrease speed and also, under certain flight conditions, to control and stabilize the aircraft.

drag polar. The curve of the lift coefficient verses the drag coefficient.

drift. 1. The lateral divergence from the prescribed **flight path** of an aircraft, a rocket, or the like, due primarily to the effect of a crosswind. **2.** A slow movement in one direction of an instrument pointer or other marker. **3.** A

slow change in **frequency** of a radio transmitter. **4.** The angular deviation of the spin axis of a **gyro** from a fixed reference in space. **5.** In semiconductors, the movement of **carriers** in an electric field.

drift angle. The angle measured in degrees between the heading of an aircraft and the track made good.

drift correction. Correction for drift, expressed as degrees plus or minus and applied to true course to obtain true heading.

drift rate. The amount of **drift,** in any of its several senses, per unit time.

Drift rate has many specific meanings in different fields. The type of drift rate should always be specified.

drizzle. Precipitation from stratus clouds consisting of numerous tiny droplets.

drogue. 1. A device, usually shaped like a funnel or cone, dragged or towed behind something and used, e.g., as a sea anchor. **2.** A funnel-shaped part at the end of the hose of a tanker aircraft, used in air refueling to drag the hose out and stabilize it and to receive the probe of the receiving aircraft. **3. drogue parachute.**

drogue chute. Same as **drogue parachute.**

drogue parachute. 1. A type of parachute attached to a body, used to slow it down; also called *deceleration parachute* or *drag parachute.* **2.** A parachute used specifically to pull something, usually a larger parachute, out of stowage, as, a *drogue parachute deploys a drag parachute.*

drogue recovery. A type of recovery system for space vehicles or space capsules after initial **reentry** into the atmosphere using deployment of one or more small parachutes to diminish speed, to reduce aerodynamic heating, and to stabilize the vehicle so that larger recovery parachutes can be safely deployed at lower aititudes without too great an opening shock.

drone. A remotely controlled **aircraft.**

drone (Q), (military designation). Aircraft capable of being controlled from a point outside the aircraft. See **military aircraft types.**

Drone Antisubmarine Helicopter. Small, lightweight, remotely controlled helicopter capable of operating from a destroyer and delivering an antisubmarine warfare weapon to any enemy submarine. It provides destroyers with a stand-off weapon. Popular name is Dash.

drop or dropping zone. A specified area upon which airborne troops, equipment, or supplies are dropped.

dropsonde. A radiosonde dropped, or designed to be dropped, as by parachute from a high-flying aircraft to measure weather conditions and report them back to the aircraft. Used over water or other areas in which no ground station can be maintained.

drop tank. An external tank designed to be dropped in flight.

dry adiabatic lapse rate (meteorology). A rate of decrease of temperature with height approximately equal to 1° C per 100 meters (1.8°F. per 328 feet). This is close to the rate at which an ascending body of unsaturated air (clean air), will cool due to adiabatic expansion.

dry bulb thermometer. A name given to an ordinary thermometer used to determine the temperature of the air, in order to distinguish it from the wet bulb.

dry emplacement. A launch emplacement that has no provision for water cooling during launch. See **wet emplacement.**

dry weight. The weight of a rocket vehicle without its fuel.

(This term, appropriate especially for liquid rockets, is sometimes considered to include the payload.)

DSIF (*abbr*). *Deep Space Instrumentation Facility*. A worldwide network of tracking stations operated for the NASA by the Jet Propulsion Laboratory.

dual control. A double set of cockpit controls permitting either pilot or co-pilot, the instructor or student, to fly the aircraft.

dual-cycle reactor system. A reactor-turbine system in which part of the steam fed to the turbine is generated directly in the reactor and part in a separate heat exchanger. A combination of **direct-cycle** and **indirect-cycle reactor systems.**

dual instruction time. Flight time during which a person is receiving flight instruction from an instructor-pilot on board the aircraft.

dual-purpose reactor. A reactor designed to achieve two purposes, for example, to produce both electricity and new fissionable material.

dual thrust. A rocket thrust derived from two propellant grains using the same propulsion section of a missile. The dual-thrust technique is considered to provide what is in effect a two-stage propulsion system without the disadvantages of jettisoning the booster unit or of sequential ignition of the sustainer grain, and with the advantages of lower weight, shorter length, and lower cost.

duck (military). Trouble headed your way (usually followed by "bogey, salvos," etc.)

duct. Specifically, a tube or passage that confines and conducts a fluid, as a passage for the flow of air to the compressor of a gas-turbine engine, a pipe leading air to a supercharger, etc.

ducted fan. 1. A **fan** enclosed in a **duct. 2. ducted-fan engine.**

ducted-fan engine. An aircraft engine incorporating a **fan** or propeller enclosed in a **duct**; especially, a **jet engine** in which an enclosed fan or propeller is used to ingest ambient air to augment the gases of combustion in the jetstream.

The air may be taken in at the front of the engine and passed around the combustion section, or it may be taken in aft of the combustion chamber. In the former case the ducted fan may be considered a type of bypass engine.

ducted rocket. Same as **rocket ramjet.**

duct propulsion. A means of propelling a vehicle by ducting a surrounding **fluid** through an engine, adding **momentum** by mechanical or thermal means, and ejecting the fluid to obtain a reactive force. Compare **rocket propulsion.**

dud. Explosive munition which has not been armed as intended or which has failed to explode after being armed.

dummy. 1. In computer operations, an artificial and intrinsically useless unit of information inserted solely to fulfill certain prescribed conditions such as word length or block length. **2.** In rocketry, an inert **stage,** i.e., no propellant.

dummy antenna. A device which has the necessary impedance characteristics of an **antenna** and the necessary power-handling capabilities, but which does not **radiate** or receive radio waves. Also called *artificial antenna.*

In receiver practice, that portion of the impedance not included in the signal generator is often called *dummy antenna.*

dummy run. Any simulated firing practice, particularly a dive bombing approach made without release of a bomb. Same as dry run.

dump. In computer operation, (**a**) to destroy intentionally or accidentally stored information, (**b**) to transfer all or part of the contents of one section of **storage** into another section.

duplexer. A device which permits a single **antenna** system to be used for both transmitting and receiving.

Duplexer should not be confused with *diplexer,* a device permitting an antenna system to be used simultaneously or separately by two transmitters.

duplex operation. The operation of associated transmitting and receiving apparatus in which the processes of transmission and reception are concurrent. Compare **diplex transmission.**

Duster (antiaircraft weapon). A self-propelled, twin 40-mm antiaircraft weapon for use against low-flying aircraft.

dutch roll (aircraft). A lateral oscillation with a pronounced rolling component.

duty factor. 1. In computer operations, the ratio of active time to total time. **2.** In a **pulse carrier** composed of pulses that recur at regular intervals, the product of the **pulse duration** and the pulse repetition **frequency.**

duty ratio. In a **pulse radar** or similar system the ratio of average to peak pulse power.

dye marker. A substance which, when placed in water, spreads out and colors the water immediately surrounding so as to make a spot readily visible from the air.

dynamical mean sun. A fictitious sun conceived to move eastward along the ecliptic at the average rate of the **apparent sun.** See **mean sun.**

The dynamical mean sun and the apparent sun occupy the same position in January, when the earth is **at perihelion.**

dynamic balance. (Running dynamic balance). A rotating body in a condition wherein all rotating forces are balanced within themselves so that no vibration is produced while the body is in motion.

dynamic load. A load imposed by dynamic action, as distinguished from a static load. Specifically, with respect to aircraft, rockets, or spacecraft, a load due to an acceleration of craft, as imposed by gusts, by maneuvering, by landing, by firing rockets, etc.

dynamic meteorology. The branch of meteorology that treats of the motions of the atmosphere and their relations to other meteorological phenomena.

dynamic model. A model of an aircraft or other object having its linear dimensions and its weight and **moments of inertia** reproduced in scale in proportion to the original.

dynamic pressure (*symbol q*). The **pressure** of a fluid resulting from its motion, equal to one-half the fluid density times the fluid velocity squared $(1/2\rho V^2)$. In incompressible flow, dynamic pressure is the difference between **total pressure** and **static pressure.** Also called *kinetic pressure.* Compare **impact pressure.**

dynamic scale. The scale of the flow about a model relative to a flow about its prototype.

If two such flows have the same **Reynolds number,** both flows are said to be at the same dynamic scale.

dynamic similarity. The relationship existing between a model and its prototype when, by virtue of similarity between their geometric dimensions and mass distributions or elastic characteristics, the motion of the model in some respect (such as linear velocity, acceleration, vibration, flutter, etc.) is similar to the motion of the prototype; also, the similarity between the fluid flows about a scale model and its prototype when the flows have the same **Reynolds number.**

dynamic stability. The characteristics of a body, such as an aircraft or rocket, that causes it, when disturbed from an original state of steady flight or motion, to **damp** the oscillations set up by restoring moments and gradually return to its original state; specifically, the **aerodynamic characteristics.** See **stability.**

dynamic storage. In computer operations, information **storage** in which the information is continuously changing position, as, for example, *delay-line storage,* or *magnetic-drum storage.*

dynamic viscosity. Of a fluid, the ratio of the **shearing stress** to the **shear** of the motion. It is independent of the velocity distribution, the dimensions of the system, etc., and for a gas it is independent of pressure except at very low pressures. Also called *coefficient of molecular viscosity, coefficient of viscosity.*

dynamometer. An instrument for measuring **power** or force; specifically, an instrument for measuring the power, torque, or **thrust** of an aircraft engine or rocket. See **thrust meter.**

dyne (dine). A unit of force in the metric system of measurement. It is numerically equal to 1/981 gram. It is that force, under whose influence, one gram would be accelerated one centimeter per sec. during each second. Pertains to **c.g.s. system of measurements.**

dysbarism. A condition of the body resulting from the existence of a pressure differential between the total **ambient** pressure and the total pressure of dissolved and free gases within the body tissues, fluids, and cavities.

Characteristic symptoms, other than **hypoxia,** caused by decreased barometric pressure are **bends** and abdominal gas pains at altitudes above 25,000 to 30,000 feet. Increased barometric pressure, as in descent from high altitude, is characterized by painful distention of the ear drums.

*The camera peers into the **electron gun** of the Experimental Testing Accelerator (ETA), a $5 million accelerator, funded by the U.S. Department of Defense at Lawrence Livermore Laboratory (University of California). The ETA is designed to study the physics and technology of the production and propagation of electron pulses.*

The ETA produces a 10,000-amp electron pulse 40 trillionths of a second long and accelerates it to an energy of 5 million electron volts. Since each electron has a negative charge and repels other electrons, the pulse must be focused by strong magnets so it does not disperse as it travels down the accelerator toward a target 30 feet away.

E

E. Military mission designation for **special electronic installation** aircraft.

early warning. See **air defense early warning.**

earth. See **planet.**

earth axis. Any one of a set of mutually perpendicular reference **axes** established with the upright axis (the Z-axis) pointing to the center of the earth, used in describing the position or performance of an aircraft or other body in flight.

The earth axes may remain fixed or may move with the aircraft or other object.

earth current. A large-scale surge of electric charge within the earth's crust, associated with a disturbance of the **ionosphere.**

Current patterns of quasi-circular form and extending over areas the size of whole continents have been identified and are known to be closely related to solar-induced variations in the extreme upper atmosphere.

earthlight. The illumination of the dark part of the moon's disk produced by sunlight reflected onto the moon from the earth's surface and atmosphere. Also called *earthshine.*

earth point. The point where the forward straight-line projection of a **meteor** trajectory intersects the surface of the earth.

earth radiation. Same as **terrestrial radiation.**

earth-rate unit. A unit of angular **drift,** as of a **gyro,** equal to the rate of angular movement of the earth with respect to the stars, 15° per hour.

earth's rate correction. A command rate applied to a **gyro** to compensate for the apparent **precession** of the gyro spin axis with respect to its base caused by the rotation of the earth.

earth satellite. A body that orbits about the Earth; specifically such body placed in orbit by man, otherwise called "artificial Earth satellite".

earth tide. A periodic movement of the earth's crust caused by the tide-producing forces of the moon and sun.

EAS. Equivalent Air Speed.

ebullism. The formation of bubbles, with particular reference to water vapor bubbles and the boiling effect in biological fluids caused by reduced barometric pressure.

eccentric. 1. Of an orbit: Deviating from the line of a circle so as to form an ellipse. **2.** The state or degree of being eccentric, expressed by the difference of greatest and least distances between the two centers of mass, divided by the sum of the greatest and least distances. (The eccentricity of Earth orbit is 0.016751.)

eccentricity. Deviation from a common center.

echelon (military). 1. A subdivision of a headquarters, i.e., forward echelon, rear echelon. **2.** Separate level of command. As compared to a regiment, a division is a higher echelon; a battalion is a lower echelon. **3.** A fraction of a command in the direction of depth, to which a principal combat mission is assigned, i.e., attack echelon, support echelon, reserve echelon. **4.** A formation in which the subdivisions are placed one behind another, extending beyond and unmasking one another wholly or in part.

echo. 1. A **wave** that has been reflected or otherwise returned with sufficient magnitude and delay to be detected as a wave distinct from that directly transmitted. **2.** In radar, a pulse of reflected radiofrequency energy; the appearance on a radar **indicator** of the energy returned from a target. Also called *blip.*

echo intensity. The brightness or brilliance of a radar **echo** as displayed on an **intensity-modulated indicator.** Echo intensity is, within certain limits, proportional to the voltage of the target signal or to the square root of its power. Compare **echo power.**

echo power. The electrical strength, or **power,** of a radar **target signal.** Echo power is normally measured in watts or dbm (decibels referred to a milliwatt).

echo pulse. A **pulse** of radio energy received at the radar after reflection from a target; that is, the **target signal** of a **pulse radar.**

echo signal. Same as **target signal.**

eclipse. 1. The reduction in visibility or disappearance of a nonluminous body by passing into the shadow cast by another nonluminous body. **2.** The apparent cutting off, wholly or partially, of the light from a luminous body by a dark body coming between it and the observer.

eclipse year. The interval between two successive **conjunctions** of the sun with the same **node** of the moon's orbit, averaging 346 days 14 hours 52 minutes 52.42 seconds in 1962, and increasing at the rate of 0.0276 second annually. See **year.**

ecliptic. The apparent annual path of the sun among the stars; the intersection of the plane of the earth's **orbit** with the **celestial sphere.**

The ecliptic is a **great circle** of the celestial sphere inclined at an angle of about 23°27' to the celestial equator.

ecliptic longitude. Same as **celestial longitude.**

ecliptic pole. On the **celestial sphere,** either of the two points 90° from the **ecliptic.**

ecliptic system of coordinates. A set of **celestial coordinates** based on the **ecliptic** as the **primary great circle.**

The points 90° from the ecliptic are the north and south ecliptic poles. Angular distance north or south of the ecliptic, analogous to latitude, is celestial latitude. Celestial longitude is measured eastward along the ecliptic from the vernal equinox through 360°.

ecological system. A habitable **environment,** either created artifically, as in a manned space vehicle, or occurring naturally, such as the environment on the surface of the earth, in which man, animals, or other organisms can live in mutual relationship with one another and the environment.

Ideally the environment furnishes the sustenance for life, and the resulting waste products revert or cycle back into the environment to be used again for the continuous support of life.

ecology. The study of the environmental relations of organisms. See **environment.**

ecosphere. 1. biosphere. 2. A volume of space surrounding the Sun, extending from the orbit of Venus past the orbit of Mars, in which some biologists believe conditions are favorable for the development and maintenance of life.

eddy. In a **fluid,** any circulation drawing its energy from a **flow** of much larger scale and brought about by pressure irregularities.

eddy. (**meteorology**) A more or less fully developed **vortex** in the atmosphere, constituting a local irregularity in a wind. All winds near the earth's surface contain eddies, which at any given place produces "gusts" and "lulls". Air containing numerous eddies is said to be "turbulent". Small eddies are caused by a non-streamlined body traveling through a fluid.

EDP (*abbr*). **electronic data processing.**

effective atmosphere. That part of the atmosphere which effectively influences a particular process of motion, its outer limits varying according to the terms of the process of the motion considered. (For example, an Earth satellite orbiting at 250 miles altitude remains within the ionosphere, but because the air particles are so rare at this altitude as to cause no appreciable friction or deflection, the satellite may be considered to be outside the effective atmosphere. For movement of space vehicles the effective atmosphere ends at about 120 miles altitude.)

Effective Perceived Noise Level (EPNL). Perceived Noisiness Level (PNL) is a noisiness rating with units of PNdb (perceived noise **decibels**), which is calculated from weighted sound pressure levels in octave or one-third octave bands. It is commonly used in the United States aircraft industry for exterior noise evaluation. When empirical tone and duration adjustments are made to PNL, the result is called Effective Perceived Noise Level (EPNL) with units of EPNdb. The maximum allowable exterior noise levels for certification of new aircraft established by the FAA in FAR Part 36, are in terms of EPNL.

effective range. The maximum distance at which a weapon may be expected to fire accurately to inflict casualties or damage.

effective temperature. 1. In astrophysics, a measure of the temperature of a star deduced by means of the **Stefan-Boltzmann law,** from the total energy emitted per unit area. Compare **brightness temperature, color temperature.**

Effective temperature is always less than actual temperature.

2. In physiology, the temperature at which motionless, saturated air would induce, in a sedentary worker wearing ordinary indoor clothing, the same sensation of comfort as that induced by the actual conditions of temperature, humidity, and air movement. Compare **sensible temperature, operative temperature.**

Effective temperature is used as a guide in air-conditioning practice, and, on the comfort chart (American Society of Heating and Air Conditioning Engineers) it appears as a family of curves which serves as one coordinate in defining comfort zones.

effective terrestrial radiation. The amount by which outgoing infrared **terrestrial radiation** of the earth's surface exceeds downcoming infrared **counter-radiation** from the atmosphere. Also called *nocturnal radiation, effective radiation.* See **actinometer.**

It is to be emphasized that this amount is a positive quantity, of the order of several tenths of a langley per minute, at all times of day (except under conditions of low overcast clouds). It typically attains its diurnal maximum during the midday hours when high soil temperatures create high rates of outgoing terrestrial

radiation. (For this reason the synonym *nocturnal radiation* is apt to lead to slight confusion.) However, in daylight hours the effective terrestrial radiation is generally much smaller than the **insolation,** while at night it typically dominates the energy budget of the earth's surface.

effector. Any device used to maneuver a **rocket** in flight, such as an aerodynamic surface, a gimbaled motor, or a jet.

efficiency. The ratio of output to input, usually expressed in percentage. The ratio of the quantity of power or work obtained from a machine to the amount of power or work used to operate it.

egads (*abbr*). **electronic ground automatic destruct sequencer.**

egads button. A button used by the range safety officer to initiate destruction of a rocket vehicle in flight if its course, as plotted during flight, is predicted to go beyond the destruct line. See **egads, impact predictor system.**

EHF (*abbr*). **Extremely High Frequency.** See **frequency band.**

eigenvalue. See **characteristic-value problem.**

eight ball. Common name given to a flight **attitude indicator.**

Einstein equation. (See **mass-energy equation.**)

ejection (**aircrew**). Escape from an aircraft by means of explosively propelled seats.

ejection capsule. 1. In an aircraft or manned spacecraft, a detachable compartment serving as a cockpit or cabin, which may be ejected as a unit and parachuted to the ground. **2.** A satellite, probe, or unmanned spacecraft, a box-like unit, usually containing recording instruments or records of observed data, which may be ejected and returned to earth by a parachute or other deceleration device.

ejection seat. A seat capable of being ejected in any emergency to carry the occupant and his equipment clear of the aircraft.

ejector. A device consisting of a **nozzle,** mixing tube, and **diffuser** utilizing the kinetic energy of a **fluid** stream to pump another fluid from a low pressure region by direct mixing and ejecting both streams.

Ekman layer. The layer of transition between the **surface boundary layer,** where shearing stress is constant, and the **free atmosphere,** where the atmosphere is treated as an ideal fluid in approximate **geostrophic** equilibrium. Also called *spiral layer.*

In Ekman's analysis (see **Ekman spiral**), the coefficient of eddy viscosity is assumed constant within this layer; subsequent calculations have relaxed this assumption.

Ekman spiral. As used in meteorology, an idealized mathematical description of the wind distribution in the **planetary boundary layer** of the atmosphere, within which the earth's surface has an appreciable effect on the air motion. The model is simplified by assuming that within this layer eddy viscosity and density are constant, the motion is horizontal and steady, the isobars are straight and parallel, and the **geostrophic wind** is constant with height.

elasticity of a material. Is that property which enables a body deformed by stress to regain its original dimensions when the stress is removed.

elasticizer. An elastic substance or fuel used in a solid rocket propellant, especially to prevent cracking of the propellant grain and to bind it to the combustion chamber case.

elastic model. A model in which the distribution of stiffness as well as the linear dimensions are so represented as to make the aeroelastic behaviour of the model correspond to that of the full-scale aircraft.

elastomers. Rubber-like compounds.

Elastomers are used as pliable components, as in tires, seals, or gaskets.

E-layer. A dimision of the **ionosphere,** usually found at an altitude between 60 and 75 miles in the **E-region.** It exhibits one or more distinct maximums and sharp gradients of **free electron** density. It is most pronounced in the daytime but does not entirely disappear at night. Also called E_1-layer, Kennelly-Heaviside layer, Heaviside layer. See **atmospheric shell, ionosphere.**

There is some evidence to indicate a second layer above the normal E-layer located at about 150 kilometers, and called the E_2-layer.

electrical. Involving the flow of electricity in a conductor. Compare **electronic.**

electrical element. See **element,** sense 2.

electrical engine. A **rocket engine** in which the **propellant** is accelerated by some electrical device. Also called electric propulsion system, electric rocket.

Electrical engines can be classified as electrothermal, electrostatic, or electromagnetic, depending on the nature of the accelerating device.

electrical equipment. Individual items which together make up an electrical system in an aircraft, missile or spacecraft.

electrical system. A group of electrical equipment with all items correlated for a specific purpose.

electric dipole. A pair of equal and opposite charges an infinitesimal distance apart.

In electromagnetics, the term dipole is often applied to two equal and opposite oscillating charges an infinitesimal distance apart; in this sense, it is synonymous with an electric-current element.

electric discharge. The flow of electricity through a gas, resulting in the **emission** of radiation that is characteristic of the gas and of the intensity of the current. Also called discharge, gaseous electric discharge, gaseous discharge. See **corona discharge, point discharge, spark discharge, lightning discharge.**

electric field. 1. A region in which a charged particle would experience an electrical force; the geometric array of the imaginary **electric lines of force** that exist in relation to points of opposite charge.

An electric field is a vector field in which magnitude of the vector is the electric-field strength and the vector is parallel to the lines of force.

2. **electric-field strength.**

electric-field strength. The electrical force exerted on a unit positive charge at a given point in space. Electric-field strength is expressed, in the practical system of electrical units, in terms of volts/centimeter. It is a vector quantity, being the magnitude of the electric-field vector. Also called electric field, electric intensity, electric field intensity, electric potential gradient, field strength.

The electric-field strength of the atmosphere is commonly referred to as the atmospheric electric field.

electric intensity. electric-field strength.

electric lines of force. Imaginary lines defined by the paths traced by unit charges placed in an electric field. Lines of force are everywhere parallel to the **electric field strength** vector. Their principal use is as a con-venient means of picturing the geometry of an electric field. See **magnetic lines of force.**

electric potential. In electrostatics, the work done in moving unit positive charge from infinity to the point whose potential is being specified. Sometimes shortened to potential.

electric propulsion. A general term encompassing all the various types of propulsion in which the **propellant** consists of charged electrical particles which are accelerated by electrical or magnetic fields, or both; for example, electrostatic propulsion, electro-magnetic propulsion, electrothermal propulsion.

electric starter. An electric motor used to rotate the engine for starting.

electroacoustic transducer. A **transducer** for receiving **waves** from an electric system and delivering waves to an **acoustic** system, or vice versa.

Microphones and earphones are electroacoustic transducers.

electrochemical. See **chemical energy.**

electrochemical transducer. A **transducer** which uses a chemical change to indicate the input parameter.

electrode. 1. A terminal at which electricity passes from one medium into another. The positive electrode is called anode; the negative electrode is called cathode. 2. In a semi-conductor device, an **element** that performs one or more of the functions of emitting or collecting **electrons** or **holes,** or of controlling their movements by an electric field. 3. In electron tubes, a conducting **element** that performs one or more of the functions of emitting, collecting or controlling, by an electro-magnetic field, the movements of electrons or ions. See **anode** (electron tubes) and **cathode.**

electrodynamics. The science of electric currents and their interaction and action.

electrojet. Current sheet or stream moving in an ionized layer in the upper atmosphere of a planet. On Earth, electrojets move around the equator following the subsolar point and also in polar regions where they give rise to auroral phenomena; generally caused by solar activity.

electroluminescence. Emission of light caused by an application of electric fields to solids or gases.

In gas electroluminescence, light is emitted when the kinetic energy of electron or ions accelerated in an electric field is transferred to the atoms or molecules of the gas in which the discharge takes place.

electromagnetic. An adjective referring to the magnetic field created by an electric current; also referring to the combined magnetic and electric fields accompanying the movements of electrons through conductors of any type.

electromagnetic energy. Same as **electromagnetic radiation.**

electromagnetic radiation. Energy propagated through space or through material media in the form of an advancing disturbance in electric and magnetic fields existing in space or in the media. The term radiation, alone, is used commonly for this type of energy, although it actually has a broader meaning. Also called electromagnetic energy or simply radiation. See **electromagnetic spectrum.**

electromagnetic rockets. plasma rockets.

See **electric propulsion.**

electromagnetic spectrum. The ordered array of known **electromagnetic radiations,** extending from the shortest cosmic rays, through gamma rays, X-rays, ultraviolet radiation, visible radiation, infrared radiation, and including microwave and all other wavelengths of radio energy. See **absorption spectrum.**

The division of this continuum of wavelengths (or frequencies) into a number of named subportions is rather arbitrary and, with one or two exceptions, the boundaries of the several subportions are only vaguely defined. Nevertheless, to each of the commonly identified subportions there correspond characteristic types of physical systems capable of emitting radiation of those wavelengths. Thus, gamma rays are emitted from the nuclei of atoms as they undergo any of several types of nuclear rearrangements; visible light is emitted, for the most part, by atoms whose planetary electrons are undergoing transitions to lower energy states; infrared radiations are associated with characteristic molecular vibrations and rotations; and radio waves, broadly speaking, are emitted by virtue of the accelerations of free electrons as, for example, the moving electrons in a radio antenna wire.

electromagnetic theory. See **electromagnetic radiation.**

electromagnetic wave. The wave associated with an electromagnetic field moving through material substances or moving through space. It has both electric and magnetic fields perpendicular to each other. The electromagnetic waves are also known as electic waves, radio waves, heat rays, light, X-rays and by other names.

electromagnetism. 1. Magnetism produced by an electric current. **2.** The science dealing with the physical relations between electricity and magnetism.

electromechanical transducer. A **transducer** for receiving **waves** from an electric system and delivering waves to a mechanical system, or vice versa.

electrometeor. A visible or audible manifestation of **atmospheric electricity.** This includes, therefore, not only visible electric discharges (igneous meteors) but also the sounds produced by them, principally thunder.

electrometer. An instrument for measuring differences of **electric potential.**

electron. The **subatomic particle** that possesses the smallest possible negative electric charge (4.80298 X 10^{-10} electrostatic units).

The mass of the electron is approximately equal to 1/1836 that of a hydrogen atom; its theoretical **rest mass** is equal to (9.1091) 10^{-28} grams and its rest energy is equal to 0.511006 million electronvolt.

The term *electron* is usually reserved for the orbital or extranuclear particle, whereas the term *beta particle* refers to a nuclear electron.

electron beam. Specifically, a focused stream of **electrons** used for neutralization of the positively charged ion beam in an **ion engine.** Also used to melt or weld materials with externally high melting points.

electron device. A device in which electricity is conducted principally by **electrons** moving through a vacuum, gas, or semiconductor.

electron gun. An **electrode** structure which produces and may control, focus, deflect, and converge one or more **electron beams.**

electronic. 1. Involving the flow of **electrons** in a vacuum or through **semiconductors. 2.** Of or pertaining to electronics, i.e., to that branch of physics that treats of the emission, transmission, behavior, and effects of electrons, especially as applied by means of vacuum tubes,

cathode-ray tubes, photoelectric cells, and the like, together with the associated electrical devices.

electronic counter-countermeasures. That division of electronic warfare involving actions taken to insure friendly effective use of the electromagnetic spectrum despite the enemy's use of electronic warfare.

electronic countermeasures. That division of electronic warfare involving actions taken to prevent or reduce an enemy's effective use of the electromagnetic spectrum. See also **electronic counter-countermeasures; barrage** (part **3**); **electronic deception; electronic jamming; jamming; electronic warfare support measures.**

electronic data processing. The use of **electronic** devices and systems in the processing of data so as to interpret the data and put them into usable form.

electronic deception. The deliberate radiation, reradiation, alternation, absorption, or reflection of electromagnetic energy in a manner intended to mislead an enemy in the interpretation or use of information received by his electronic systems. See also **electronic countermeasures; radio deception.**

electronic defense evaluation. A mutual evaluation of radar(s) and aircraft, with the aircraft trying to penetrate the radars, area of coverage in an electronic countermeasure environment.

electronic intelligence. The intelligence information product of activities engaged in the collection and processing, for subsequent intelligence purposes, of foreign, noncommunications, electromagnetic radiations emanating from other than nuclear detonations and radioactive sources. Also called ELINT.

electronic jamming. The deliberate radiation, reradiation, or reflection of electromagnetic energy with the object of impairing the use of electronic devices, equipment or systems being used by an enemy.

electronic line of sight. The path traversed by electromagnetic waves which is not subject to reflection or refraction by the atmosphere.

electronic reconnaissance. The detection, identification, evaluation, and location of foreign, electromagnetic radiations emanating from other than nuclear detonations or radioactive sources.

electronics. That branch of physics that treats of the emission, transmission, behavior, and effects of **electrons.** See **electronic.**

electronic warfare. Military action involving the use of electromagnetic energy to determine, exploit, reduce or prevent hostile use of the electromagnetic spectrum and action which retains friendly use of the electromagnetic spectrum.

electronic warfare support measures. That division of electronic warfare involving actions taken to search for, intercept, locate, record, and analyze radiated electromagnetic energy for the purpose of exploiting such radiations in support of military operations. Thus, electronic warfare support measures provide a source of electronic warfare information required to conduct electronic countermeasures, electronic counter-countermeasures, threat detection, warning, avoidance, target acquisition and homing.

electron tube. A device in which conduction by **electrons** takes place through a vacuum or gaseous medium within a gastight envelope.

electron-volt. A unit of **energy** equal to the energy required to move an electron through a potential difference of 1 volt. Often shorten to *volt*.

One electron volt equals 1.6020 X 10^{-9} joule.

electrostatic rocket. Same as **ion rocket.** See **electric propulsion, ion engine.**

electrostatic storage. In a computer, **storage** of information in the form of **electrostatic** charges.

electrostatic-storage tube. A **cathode-ray tube** in which inoformation is stored as positive or negative charges on a **dielectric** surface.

element. 1. One of the simple parts of which a complex entity is composed. **2.** In chemistry, a substance which cannot be broken down by ordinary chemical means into simpler components. **3.** In an **electron tube,** a constituent part of the tube that contributes directly to the electrical operation of the tube. **4.** In a **circuit,** any electrical device (such as inductor, resistor, capacitor, generator, line, electron tube) with terminals at which it may be directly connected to other electrical devices. **5.** In a **semiconductor device,** any integral part of the semiconductor device that contributes to its operation. **6. orbital element.**

elementary particles (nucleonics). The particles of which all *matter* and *radiation* are composed. All are short-lived, do not exist independently under normal conditions (except electrons, protons and *neutrinos* in the form of *cosmic rays*), and are of less than atomic size. Originally this term was applied to any particles which could not be further subdivided; now it is applied to *nucleons (protons* and *neutrons), electrons, mesons, antiparticles* and *strange particles,* but not to alpha particles or deuterons. Also called *fundamental particles.*

elevation. Same as **angle of elevation.**

elevator. A hinged, horizontal control surface used to raise or lower the tail in flight thereby impressing a pitching moment to the airplane. The elevator is usually hinged to the horizontal stabilizers and connected to the pilot's control wheel or stick.

elevons. Control surfaces combining the functions of ailerons and elevators. When placed on the tail they are sometimes called **tailerons.**

ELF (*abbr*) **e***xtremely* **l***ow* **f***requency.* See **frequency band.**

ellipse. A plane curve constituting the locus of all points the sum of whose distances from two fixed points called *focuses* or *foci* is constant; an elongated circle.

The **orbits** of planets, satellites, planetoids, and comets are ellipses, the primary being at one focus.

ellipsoid. A surface whose plane sections (cross sections) are all **ellipses** or circles, or the solid enclosed by such a surface. Also called *ellipsoid of revolution, spheroid.*

elliptic. Pertaining to an **ellipse,** or in the form of an ellipse.

elliptical system. A tracking or navigation system where **ellipsoids** of position are determined from time or **phase** summation relative to two or more fixed stations which are the focuses for the ellipsoids.

ellipticity. The amount by which a **spheroid** differs from a sphere or an **ellipse** differs from a circle, calculated by dividing the difference in the length of the axes by the length of the major axis. Also called *compression.* See **flattening.**

ellipticity ratio. 1. The ratio of the **major axis** to the **minor axis** of an **ellipse.**

2. As a measure of **elliptical polarization,** the power ratio of the maximum to the minimum electric vectors of an elliptically polarized antenna.

elongation. A change in length or dimension.

elsse (*abbr*) **E***lectronic* **s***ky* **s***creen* **e***quipment.* An electronic device which indicates the departure of a rocket from a predetermined **trajectory.**

EMA (*abbr*) **e***lectronic* **m***issile* **a***cquisition.*

embarkation. The loading of troops with their supplies and equipment into ships and/or aircraft.

embolism. Large amounts of air in the blood stream which, reaching the heart, cause it to fail; small amounts are resorbed arid cause no sysmptoms.

emergency locator transmitter (ELT). A radio transmitter, attached to the aircraft structure, which operates from its own power source on 121.5 MHz and 243 MHz, transmitting a distinctive downward swept audio tone for homing purposes, and is designed to function without human action after an accident.

emergency scramble. Carrier(s) addressed immediately launch all available fighter aircraft as combat air patrol. If all available are not required, numerals and/or type may be added.

emergency war order. The order issued by competent authority to launch combat-ready weapon systems maintained in readiness for or generated for first strike wartime operations.

emission. 1. With respect to **electromagnetic radiation,** the process by which a body emits electromagnetic radiation as a consequence of its temperature only. Compare **reflection, transmission.** See **emittance, emissivity. 2.** With respect to **electric propulsion** and energy conversion, the sending out of charged particles from a surface causing the generation of these particles; e.g., emission of ions from an ionizing surface in ion engines.

emission line. A minute range of **wavelength** (or **frequency**) in the **electromagnetic spectrum** within which radiant energy is being emitted by a radiating substance. See **spectral line, emission spectrum.**

emission spectrum. The array of **wavelengths** and relative intensities of **electromagnetic radiation** emitted by a given radiator.

Each radiating substance has a unique, characteristic emission spectrum, just as every medium of transmission has its individual absorption spectrum.

emissivity. A property of a material, measured as the emittance of a specimen of the material that is thick enough to be completely opaque and has an optically smooth surface.

emittance. 1. The radiant flux per unit area emitted by a body. **2.** The ratio of the emitted radiant flux per unit area of a sample to that of a **black body** radiator at the same temperature and under the same conditions.

Spectral emittance refers to emittance measured at a specified wavelength.

Because of the two common meanings of *emittance,* it should be defined when used unless the context allows no misinterpretation.

empennage. The rear part of an airplane, usually consisting of a group of stabilizing planes (horizontal stabilizers and vertical fin.) to which are attached the control surfaces (elevators and rudders). Sometimes this portion of the airplane is referred to as: airplane tail assembly, tail group, or **tail unit.**

emplacement. 1. A prepared position for one or more weapons or pieces of equipment, for protection against hostile fire or bombardment, and from which they can execute their tasks. **2.** The act of fixing a gun in a prepared position from which it may be fired.

employment. The tactical usage of aircraft in a desired area of operation. In airlift operations a movement of forces into a combat zone, or objective area, usually in the assault phase.

emulsion. In photography, a light-sensitive coating on a film, plate, or paper.

enamel. A thin **ceramic** coating, usually of high glass content, applied to a **substrate,** generally a metal.

encipher. To convert a plain-text message into unintelligible language by means of a cipher system.

encoder. Analog to digital converter.

encoding altimeter. An aneroid barometer type altimeter which presents the usual altitude display to the pilot on the instrument face but, in addition, incorporates a digitized output to a transponder for transmission of altitude information to a ground based radar scope. Thus the aircraft's altitude is automatically reported to the ground based Air Traffic Control (ATC) system.

encrypt. To convert a plain-text message into unintelligible form (cryptogram) by means of a cryptosystem.

end-fire array. A linear **antenna array** whose direction of maximum radiation is along the axis of the array.

end item. A final combination of end products, component parts, and/or materials which is ready for its intended use, e.g., ship, tank, mobile machine shop, aircraft.

endurance. The time an aircraft can continue flying under given conditions without refueling.

energy. Any quantity with dimensions mass X length squared divided by time squared. Compare **entropy.**

energy conversion efficiency. The efficiency with which a **nozzle** converts the energy of the working substance into **kinetic energy,** expressed as the ratio of the kinetic energy of the jet leaving the nozzle to the kinetic energy of a hypothetical ideal jet leaving an ideal nozzle using the same working substance at the same initial state and under the same conditions of velocity and expansion.

energy equation. See **thermodynamic energy equation, total energy equation.**

energy level. Any one of different values of **energy** which a **particle, atom,** or **molecule** may adopt under conditions where the possible values are restricted by quantizing conditions.

During transitions from one energy level to another, quanta of radiant energy are emitted or absorbed, their frequency depending on the difference between the energy levels.

engage. Attack designated contact.

engagement control. In air defense, that degree of control exercised over the operational functions of an air defense fire unit that are related to detection, identification, engagement, and destruction of hostile targets.

engine. A machine or apparatus that converts energy, especially heat energy, into work. Also called *motor.*

engine log book. The same as an aircraft log book except that engine log book entries are kept for each individual engine and the entries consist of current reports of line and periodic inspections, the duration of the running time of each engine both on the ground and in the air, changes in the engine installation, and of tye overhaul and any alteration of, and damage to, the engine.

engine mount. A structure used for attaching an engine to a **vehicle.**

engine pod. The housing for each externally mounted engine on a multiengine aircraft.

engine pressure ratio (EPR). For a **turbojet engine,** the ratio between the engine air inlet pressure and the exhaust gas pressure measured downstream of the turbine. EPR as displayed to the flight crew is an indication of engine **thrust.**

enplaned passengers. The total number of revenue passengers boarding aircraft, including originating, stopover, and transfer passengers, in scheduled and nonscheduled services.

enriched material (nucleonics). Material in which the percentage of a given *isotope* present in a material has been artificially increased, so that it is higher than the percentage of that isotope naturally found in the material. Enriched *uranium* contains more of the fissionable isotope uranium -235 than the naturally occurring percentage (0.7%). (See **isotopic enrichment.**)

enrichment (nucleonics). *isotopic enrichment.*

en route air traffic control service. Air traffic control service provided aircraft on an IFR flight plan, generally by centers, when these aircraft are operating between departure and destination terminal areas.

en route support team (military). A functional package of personnel and material consisting of selected personnel, skills, equipment and supplies necessary to service and perform limited specialized maintenance on tactical aircraft at enroute bases so that the aircraft can proceed to their destination base with a minimum of delay.

enthalpy. A mathematically defined thermodynamic function of state.

entropy. 1. A measure of the extent to which the energy of a system is unavailable. A mathematically defined thermodynamic function of state, the increase in which gives a measure of the energy of a system which has ceased to be available for work during a certain process.

See **third law of thermodynamics.**

In an adiabatic process, the entropy increases if the process is irreversible and remains unchanged if the process is reversible. Thus, since all natural processes are irreversible, it is said that in an isolated system the entropy is always increasing as the system tends toward equilibrium, a statement which may be considered a form of the second law of thermodynamics.

2. In communication theory, **average information content.**

entry corridor. Depth of the region between two **trajectories** which define the design limits of a vehicle which will enter a planetary **atmosphere.**

envelope. 1. Of a variable, a curve which bounds the values which the **variable** can assume, but does not consider possible simultaneous occurences or correlations between different values. **2.** The bounds within which a certain system can operate as a **flight envelope,** especially a graphic representation of these bounds showing interrelationships of operational parameters.

environics. The study of the interrelation of man and his total environment, physical, biological, and social. Application of systems approach to the total man/technology interaction.

environment. An external condition or the sum of such conditions, in which a piece of equipment, a living organism, or a system operates as in *temperature environment, vibration environment,* or *space environment.*

Environments are usually specified by a range of values, and may be either natural or artificial.

environmental chamber. A chamber in which humidity, temperature, pressure, fluid contents, noise, and movement may be controlled so as to simulate different environments.

environmental services. The various combinations of scientific, technical and advisory activities (including modification processes, i.e., the influence of man-made and natural factors) required to acquire, produce, and supply information on the past, present, and future states of spaces, atmospheric, oceanographic, and terrestrial surroundings for use in flight planning and decision-making processes, or to modify those surrounding to enchance flight operations.

ephemeris (*plural*, **ephemerides**). A periodical publication tabulating the predicted positions of **celestial bodies** at regular intervals, such as daily, and containing other data of interest to astronomers.

A publication giving similar information useful to a navigator is called an *almanac*.

ephemeris day. 86,400 **ephemeris seconds.** See **ephemeris time.**

ephemeris second. The fundamental unit of time of the International System of Units of 1960: 1/31556925.9747 of the tropical year defined by the mean motion of the sun in longitude at the epoch 1900 January 0 day 12 hours. See **ephemeris time.**

ephemeris time (*abbr* **E.T.**). The uniform measure of time defined by the laws of dynamics and determined in principle from the orbital motions of the planets, specifically the orbital motion of the earth as represented by Newcomb's Tables of the Sun. Compare **universal time.**

Beginning with the volume for 1960 the American Ephemeris and Nautical Almanac uses ephemeris time as the tabular argument in the fundamental ephemerides of the sun, moon, and planets.

A gravitational ephemeris expresses the position of a celestial body as a function of ephemeris time; and, at any instant, the measure of ephemeris time is the value of the argument at which the ephemeris position is the same as the actual position at the instant. The ephemeris time at any instant is obtained from observation by directly comparing observed position of the sun, moon, and planets with gravitational ephemerides of their coordinates; observations of the moon are the most effective and expeditious for this purpose. An accurate determination, however, requires observations over a more or less extended period; in practice, it takes the form of determining the time correction ΔT that must be applied to universal time (U.T.) to obtain ephemeris time:

$$E.T. = U.T. + \Delta T$$

The universal time at any instant may be obtained with little delay from observations of the diurnal motions.

The fundamental epoch from which ephemeris time is reckoned is the epoch that Newcomb designated as 1900 January 0, Greenwich mean noon, but which actually is 1900 January 0 day 12 hours E.T.; the instant to which this designation is assigned is the instant near the beginning of the calendar year A.D. 1900 when the geometric mean longitude of the Sun referred to the mean equinox of date was 279 degrees 41 minutes 48.04 seconds. Ephemeris time is the measure of time in which Newcomb's Tables of the Sun agree with obser-

vation.

The primary unit of ephemeris time is the tropical year, defined by the mean motion of the sun in longitude at the epoch 1900 January 0 day 12 hours E.T.; its length in ephemeris days is determined by the coefficient of T in Newcomb's expression for the geometric mean longitude of the sun L referred to the mean equinox of date, given among the elements of the sun.

epoch. A particular instant for which certain data are valid, as the data for which an astronomical catalogue is computed.

Eppley pyrheliometer. A pyrheliometer of the thermoelectric type. Radiation is allowed to fall on two concentric silver rings, the outer covered with magnesium oxide and the inner covered with lampblack. A system of **thermocouples** (thermopile) is used to measure the temperature difference between the rings. Attachments are provided so that measurements of direct and diffuse solar radiation may be obtained.

This instrument has been adopted by the U.S. Weather Bureau for station use.

EPR. engine pressure ratio.

equation of state. An equation relating temperature, pressure, and volume of a system in thermodynamic equilibrium.

A large number of such equations have been devised to apply equally to gaseous and liquid phases throughout a wide range of temperatures and pressures. Of these, the simplest are the **perfect gas law** and **Van der Waal** equation.

equation of time. Prior to 1965, the difference between **mean time** and **apparent time,** usually labed + or - as it is to be applied to mean time to obtain apparent time. After 1965, the correction to be applied to 12 hours + local mean time (LMT) to obtain the local hour angle (LHA) of the sun.

equations of motion. A set of equations which give information regarding the motion of a body or of a point in space as a function of time when initial position and initial velocity are known. See **Newton laws of motion, Eulerian coordinates.**

equator. The **primary great circle** of a sphere or spheroid, such as the earth, perpendicular to the **polar axis**; or a line resembling or approximating such a circle.

equatorial bulge. The excess of the earth's equatorial diameter over the polar diameter.

equatorial satellite. A **satellite** whose **orbit** plane coincides, or almost coincides, with the earth's equatorial plane.

equilibrium flow. Gas **flow** in which energy is constant along **streamlines** and composition of the gas at any point is not time dependent.

equilibrium glide. Gliding flight in which the sum of the vertical components of the aerodynamic **lift** and **centrifugal force** is equal to the force of **gravity.**

equilibrium vapor pressure. The **vapor pressure** of a system in which two or more phases of a substance coexist in equilibrium. See **vapor tension.**

In meteorology the reference is to water substance unless otherwise specified.

equinox. 1. One of the two points of intersection of the **ecliptic** and the **celestial equator,** occupied by the sun when its declination is 0°. Also called *equinoctial point*.

That point occupied on or about March 21, when the

sun's declination changes from south to north, is called *vernal equinox, March equinox,* or *first point of Aries;* that point occupied on or about September 23, when the declination changes from north to south, is called *autumnal equinox, September equinox,* or *first point of Libra.*

    Equinox is often used to mean *vernal equinox,* when referring to the origin of measurement of **right ascension** and **celestial longitude.**

2. That instant the sun occupies one of the equinoctial points.

equipment operationally ready (military). A condition status of a major item of equipment weapon system which indicates that it is capable of safe use and that all subsystems necessary for the performance of its primary mission are ready.

equisignal zone. A region in which equal signal strength is received from two intersecting lobes of radiation.

equivalent airspeed. See **airspeed.**

equivalent pendulum. A device, usually incorporating **accelerometers** and **gyros,** which has the same response to acceleration as a pendulum with a specific period.

equivalent potential temperature (meteorology). The temperature that a given sample of air would have if it were brought adiabatically to the top of the atmosphere (i.e., to zero pressure) so that along its route all the water vapor present were condensed and precipitated, the latent heat of condensation being given to the sample, and then the remaining dry air compressed adiabatically to a pressure of 1,000 millibars. The equivalent potential temperature at any point is therefore determined by the values of absolute temperature, pressure, and humidity. It is one of the most conservative of **air mass** properties.

equivalent temperature (meteorology). The temperature a particle of air would have if it were made to rise **adiabatically** to the top of the **atmosphere,** that is to zero pressure, in such a manner that all the heat of condensation of the water vapor were added to the air and the sample of dry air were then brought back adiabatically to its original pressure.

eradiation. 1. Same as **radiation,** with respect to **emission. 2. terrestrial radiation.**

erase. In computer terminology, to expunge, wipe out, or destroy stored information, usually without destroying the **storage** media, as in demagnetizing a magnetic tape.

erector. A vehicle used to support a **rocket** for transportation and for placing the rocket in an upright position within a **gantry.**

E-region. The region of the **ionosphere** in which the E-layer tends to form. See **atmospheric shell.**

    The E-layer has been observed to be subdivided into two or more *layers,* and these are then assigned the designation, E_1, E_2, etc. Patchy and intermittent clouds of fairly high ionization, known as *sporadic E-layers,* also form in the same general region.

erg. The unit of energy or work in the **centimeter-gram-second system;** the work performed by a force of 1 dyne acting through a distance of 1 centimeter.

ergometer. An instrument for measuring muscular **work.**

erosion gage. An instrument for measuring the effect of **dust** and **micrometeors** on materials exposed to space environment.

erratic error. An error caused by an incomplete element in an instrument.

    An example of an erratic error is backlash in a gear train.

error. 1. In mathematics, the difference between the true value and a calculated or observed value.

    A quantity (equal in magnitude to the error) added to a calculated or observed value to obtain the true value is called a *correction.*

    2. In a computer or data-processing system, any incorrect step, process, or result.

error band. An error value, usually expressed in percent of full scale, which defines the maximum allowable error permitted for a specified combination of **transducer parameters.**

error signal. A voltage the magnitude of which is proportional to the difference between an actual and a desired position.

    In radar use, error signals are obtained from **selsyns** and from automatic gain control circuits and are used to control a **servo** system so that the resultant motions tend to correct the error.

escalation (military). An increase in scope or violence of a conflict, deliberate or unpremediated.

escape. Of a particle or larger body; to achieve an escape velocity and a flight path outward from a primary body so as neither to fall back to the body nor to orbit it.

escape rocket. A small rocket engine attached to the leading end of an escape tower, which may be used to provide additional thrust to the capsule to obtain separation of the capsule from the booster vehicle in an emergency.

escape tower. A trestle tower placed on top of a space capsule, which during liftoff connects the capsule to the escape rocket.

    The escape tower is of such length as to protect the capsule from the heat of the escape rocket in case the rocket is used to separate the capsule from the booster vehicle during ascent. The tower is ultimately separated from the capsule if ascent is normal.

escape velocity. The **radial speed** which a particle or larger body must attain in order to escape from the gravitational field of a planet or star. Also called *escape speed.*

    Escape velocity from Earth is 7 miles/sec; from Mars it is 3.2 miles/sec; and from the Sun it is 390 miles/sec. In order for a celestial body to retain an atmosphere for long periods of time, the mean velocity of the atmospheric molecules must be considerably below the escape velocity.

escort. 1. To convoy. **2.** A combatant unit or units assigned to accompany and protect another force. **3.** Aircraft assigned to protect other aircraft during a mission.

espionage. Actions directed toward the acquisition of information through clandestine operations.

essential industry. Any industry necessary to the needs of a civilian or war economy. The term includes the basic industries as well as the necessary portions of those other industries which transform the crude basic raw materials into useful intermediate or end products, e.g., the iron and steel industry, the food industry, and the chemical industry.

establishment. An installation, together with its personnel and equipment, organized as an operating entity.

estimated time of arrival (ETA). In **navigation,** the expected time of arrival of an aircraft at a checkpoint and/or destination based on pre-flight calculations.

ETA. Estimated Time of Arrival.

Eulerian coordinates. Any system of **coordinates** in which properties of a **fluid** are assigned to points in space at each given time, without attempt to identify individual **fluid parcels** from one time to the next. See **equations of motion.** Compare **Lagrangian coordinates.**

Eulerian coordinates are to be distinguished from **Lagrangian coordinates.** The particular coordinate system used to identify points in space (Cartesian, cylindrical, spherical, etc.) is quite independent of whether the representation is Eulerian or Lagrangian.

Eulerian equations. Any of the fundamental equations of hydrodynamics expressed in **Eulerian coordinates.** These are so commonly used that the designation *Eulerian* is often omitted.

EVA. Extra Vehicular Activity.

evaporation. That process by which a liquid under the influence of heat becomes gas. (Also influenced by pressure, humidity, and agitation of the vapor.)

evection. A **perturbation** of the moon in its orbit due to the attraction of the sun. This results in an increase in the **eccentricity** of the moon's orbit when the sun passes the moon's **line of apsides** and a decrease when perpendicular to it. See **lunar inequality.**

Evection amounts to 1 degree 15 minutes in the moon's longitude at maximum.

excess reactivity (nucleonics). More reactivity than that needed to achieve criticality. Excess reactivity is built into a reactor (by using extra fuel) in order to compensate for fuel *burnup* and the accumulation of fission-product *poisons* during operation. (See **criticality, reactivity.**)

excitation. 1. An external force, or other input, applied to a system that causes the system to respond in some way. Also called *stimulus*. **2.** The increase in the internal energy of an atomic or molecular system caused by a collision with another particle of greater energy.

For atoms in a discharge, *excitation* usually refers to increasing the energy level of a bound electron.

excited state (nucleonics). The state of a molecule, atom, electron or nucleus when it possesses more than its normal *energy*. Excess nuclear energy is often released as a gamma ray. Excess molecular energy may appear as fluorescence or heat.

exclusion area (nucleonics). An area immediately surrounding a nuclear reactor where human habitation is prohibited to assure safety in the event of accident.

excursion (nucleonics). A sudden, very rapid rise in the power level of a reactor caused by *supercriticality*. Excursions are usually quickly suppressed by the negative temperature coefficient of the reactor and/or by automatic control rods.

exercise. A military maneuver or simulated wartime operation involving planning, preparation, and execution. It is carried out for the purpose of training and evaluation. It may be a combined, unified, joint, or single Service exercise, depending on participating organizations.

exhaust deflecting ring. A type of **jetavator** consisting of a ring so mounted at the end of a **nozzle** as to permit it to be rotated into the exhaust stream.

exhaust gas temperature. The average temperature of the exhaust gas stream.

exhaust manifold. A pipe or chamber into which exhaust gases are led from a number of cylinders.

exhaust trail. A **condensation trail** that forms when the water vapor of an aircraft exhaust is mixed with and saturates (or slightly supersaturates) the air in the wake of the aircraft. Exhaust trails are of more common occurrence and of longer duration than **aerodynamic trails.** Also called *engine-exhaust trail.*

exhaust velocity. The velocity of gaseous or other particles (exhaust stream) that exhaust through the **nozzle** or a **reaction engine,** relative to the nozzle.

exobiology. That field of biology which deals with the effects of extraterrestrial environments on living organisms and with the search for extraterrestrial life.

exosphere. The outermost, or topmost, portion of the **atmosphere.** Its lower boundary is the **critical level of escape,** variously estimated at 500 to 1000 kilometers (30 to 60 miles) above the earth's surface. Also called *region of escape*. See **atmospheric shell.**

In the exosphere, the air density is so low that the **mean free path** of individual particles depends upon their direction with respect to the local vertical, being greatest for upward moving particles (see **cone of escape**). It is only from the exosphere that atmospheric gases can, to any appreciable extent, escape into outer space.

exotic fuel. Unusual fuel combinations for aircraft and rocket use with the purpose of attaining far greater thrust.

exotic material. Any structural material which is not presently used in great quantities in conventional applications. Usually, materials with melting points above 3000°F.

expandable space structure. A structure which can be packaged in a small volume for **launch** and then erected to its full size and shape outside the earth's atmosphere.

expansion wave. A simple **wave** or progressive distrubance in the **isentropic flow** of a compressible fluid, such that the pressure and density of a fluid particle decrease on crossing the wave in the direction of its motion. Also called *rarefaction wave.*

expected approach clearance time (EAC). The time at which it is expected that an arriving aircraft will be cleared to begin approach for a landing.

expected further clearance time (EFC). (air traffic control). the time at which it is expected that additional clearance will be issued to an aircraft.

experimental model. A model of the complete equipment to demonstrate the technical soundness of the basic idea. This model need not have the required final form or necessarily contain parts of final design.

explosion. 1. The sudden production of a large quantity of gas, usually hot, from a much smaller amount of a gas, liquid, or solid. **2.** Specifically, an explosion, sense **1,** produced by combustion of a **fuel** and an **oxidizer.**

The distinction between an explosion, sense **2,** and a detonation is that in an explosion the heat release rate and the number of molecules per unit volume increase with time more or less uniformly, whereas a detonation is propagated by an advancing **shock front** behind which exothermic reactions take place and thus is (spatially) nonuniform.

explosion turbine. A **turbine** rotated by gases from an intermittent combustion process taking place in a constant-volume chamber.

explosive bolt. A bolt incorporating an explosive which can be detonated on command, thus destroying the bolt. Explosive bolts are used, for example, in separating a **satellite** from a **rocket.**

explosive decompression. A very rapid reduction of air pressure inside a cabin, coming to a new static condi-

Astronaut Edwin E. Aldrin Jr. walks on the surface of the moon near the Lunar Module during the Apollo 11 **Extra Vehicular Activity (EVA).** *Astronaut Neil A. Armstrong took this photo during man's first lunar landing mission.*

Excellent view of the docked Apollo 9 Command and Service Modules, with the earth in background during Astronaut David R. Scott's standup **Extra Vehicular Activity (EVA).** *Astronaut Russell L. Schweickart, lunar module pilot, took this photograph of Scott from the porch of the Lunar Module.*

tion of balance with the external pressure.

exposure suit. A suit designed to protect a person from a harmful natural **environment,** such as cold water.

extended over-water operation. An operation over water at a horizontal distance of more than 50 nautical miles from the nearest shore line. This is an FAA definition.

extent of damage. The visible plan area of damage to a target element, usually expressed in units of 1,000 square feet in detailed damage analysis and in approximate percentages in immediate-type damage assessment reports (e.g., 50 percent structural damage.)

exterior ballistics. That branch of **ballistics** that deals with the motion of projectiles in flight.

external load. A load that is carried, or extends, outside of the aircraft fuselage. Examples of external loads are fuel tanks, armament, missles and cargo containers.

external-load attaching means. The structural components used to attach an external load to an aircraft, including external-load containers, the backup structure at the attachment points, and any quick-release device used to jettison the external load.

external storage. In computer terminology, **storage** media separate from the machine but capable of retaining information in a form acceptable to the machine, as decks of **punched cards** or removable reels of magnetic tape.

extinction. The attenuation of light; that is, the reduction in **illuminance** of a collimated beam of light as the light passes through a medium wherein **absorption** and **scattering** occur.

extinction coefficient. In meteorology, a measure of the space rate of diminution, or extinction, of any transmitted light; thus, it is the **attenuation coefficient** applied to **visible** radiation.

When so used, the extinction coefficient equals the sum of the medium's absorption coefficient and scattering coefficient, each computed as a weighted average over all wavelengths in the visible spectrum. As long as scattering effects are primary, as in the lower atmosphere, the value of the extinction coefficient is a function of the particle size of atmospheric suspensoids. It varies in order of magnitude from 10 per kilometer with very low visibility to 0.01 per kilometer in very clear air.

extraction parachute. An auxiliary parachute designed to release and extract cargo from aircraft in flight and deploy cargo parachutes.

extragalactic. Outside our galaxy, which is the **Milky Way.**

extraterrestrial life. Life forms evolved and existing outside the terrestrial **biosphere.**

extraterrestrial radiation. In general, **solar radiation** received just outside the earth's atmosphere.

Extra Vehicular Activity (EVA). A term used in spaceflight operations for functions performed by an **astronaut** outside of a spacecraft. Appropriate protective garments and life support requirements are provided.

extremely high frequency (*abbr* EHF). See **frequency bands.**

extremely low frequency (*abbr* ELF). See **frequency bands.**

eyeballs in, eyeballs out, eyeballs down, eyeballs up, eyeballs left, eyeballs right. See **physiological acceleration.**

*Present and future **fighter** aircraft are the U.S. Air Force McDonnell Douglas F-15 (above) and mockup of an advanced technology, forward swept wing aircraft by Rockwell International, (lower photo).*

F

F. Military mission designation for **fighter** aircraft.

FAA. Federal Aviation Administration.

FAR. Federal Aviation Regulations.

facility. **1.** A physical plant, such as real estate and improvements thereto, including buildings and equipment, which provides the means for assisting or making easier the performance of a function, e.g., base arsenal, factory. **2.** Any part or adjunct of a physical plant, or any item of equipment which is an operating entity and which contributes or can contribute to the execution of a function by providing some specific type of physical assistance.

facsimile. In electrical communications, the process, or the result of the process, by which fixed graphic material including pictures or images is scanned and the information converted into signals which are used either locally or remotely to produce in record form a likeness (facsimile) of the subject copy.

factor of safety. A design factor used to provide for the possibility of loads greater than those assumed, and for uncertainties in design and fabrication.

faculae. Large patches of bright material forming a veined network in the vicinity of **sunspots** and are probably due to elevated clouds of luminous gas.

fade. Of a radiant energy signal, to decrease, often temporarily, in received **signal strength** without a change of receiver controls.

The opposite is *build*.

faded. Contact has disappeared from reporting station's scope, and any position information given is estimated.

fadeout. A type of **fading** in which the received signal strength is reduced to a value below the **noise level** of the receiver. The most common cause of fadeout is a disturbed **ionosphere.** Also called *radio fadeout, Dellinger effect, Mogel-Dellinger effect.* See **blackout.**

fading. The variation of radio **field strength** caused by changes in the transmission medium with time.

Fahrenheit temperature scale. (*abbr* F). A temperature scale with the **ice point** at 32° and the **boiling point** of water at 212°.

Conversion with the Celsius (**centigrade**) temperature scale (*abbr* C) is by the formula

$$F = 9/5 \, C + 32$$

fail safe system. A system used to minimize risk in case of a malfunction.

fairing. An auxiliary member or structure whose primary function is to reduce **drag** of the part to which it is fitted. Fairings are not strength members and therefore not intended to carry any of the principal loads placed upon the airplane structure.

faker. A known strike aircraft engaged in an air defense exercise.

fall. Of a spacecraft or spatial body, to drop toward another spatial body under the influence of the latter's **gravity.**

fallaway section. Any section of a rocket vehicle that is cast off and falls away from the vehicle during flight, especially such a section that falls back to earth. See **blowoff, companion body.**

fallout. Air-borne particles containing radioactive material which fall to the ground following a nuclear explosion. "Local fallout" from nuclear detonations falls to the earth's surface within 24 hours after the detonation. "Tropospheric fallout" consists of material injected into the troposphere but not into the higher altitudes of the stratosphere. It does not fall out locally, but usually is deposited in relatively narrow bands around the earth at about the latitude of injection. "Stratospheric fallout" or "worldwide fallout" is that which is injected into the stratosphere and which then falls out relatively slowly over much of the earth's surface. (Compare **background radiation.**)

familiarization training. Individual training for personnel having a fundamental technical knowledge to acquaint them with a specific system.

fan. **1.** (**a**) Any vaned rotary device for producing a current or stream of air. (**b**) Specifically, a multivaned wheel or rotor used to take in air in a bypass engine or **ducted-fan engine.** It may be either a mere blower or a low-pressure compressor. See **ducted fan. 2.** A propeller, especially when the emphasis is upon its function of moving air rather than propelling.

fan jet. See **turbofan**

fan marker beacon. A type of radio beacon, the emissions of which radiate in a vertical, fan-shaped pattern. The signal can be keyed for identification purposes. See also **beacon, marker beam.**

fan noise. General term for the noise generated within the fan stage of a turbofan engine; includes both discrete frequencies and random noise.

fanned-beam antenna. A unidirectional antenna so designed that transverse cross sections of the major **lobe** are approximately elliptical.

farad. The capacity of a condenser when it will hold one coulomb of electricity under a pressure of one volt, i.e., it is a unit of electrical capacity.

farm gate type operations. Operational assistance and specialized tactical training provided a friendly foreign air force by the United States Armed Forces to include, under certain specified conditions, the flying of operational missions in combat by combined United States/foreign aircrews as a part of the training being given when such missions are beyond the capability of the foreign air force.

fast breeder reactor. A reactor that operates with fast neutrons and produces more *fissionable material* than it consumes. (See **breeder reactor, fast reactor.**).

fatigue. **1.** A weakening or deterioration of metal or other material occurring under load, especially under repeated cyclic, or continued loading.

Self-explanatory compounds include: *fatigue crack, fatigue failure, fatigue load, fatigue resistance, fatigue test.*

2. State of the human organism after exposure to any type of physical or psychological stress (e.g., pilot fatigue).

fatigue strength. The maximum **stress** that can be sustained for a specified number of cycles without failure, the stress being completely reversed within each cycle unless otherwise stated. Also called *fatigue limit*.

FBO. See **fixed-base operator.**

feasibility test. A test to determine whether or not a plan is within the capacity of the resources which can be made available.

feathered pitch (propeller). The pitch setting which, in flight with the engine stopped, gives approximately the minimum drag, and corresponds with a windmilling torque of approximately zero.

feathering propeller. A propeller with blades that can be rotated in the hub so that its leading and trailing edges are parallel, or nearly so, with the line of flight of the airplane. A full-feathering propeller is one that can be rotated a full 180° angle of **pitch,** the purpose of which is to decrease air resistance in case of engine failure, so that the propeller will not be rotated by the air. A disabled engine that has stopped completely during flight will eliminate the possibility of structural damage to the engine.

Federal Airways System. Each Federal Airway is based on a centerline that extends from one navigation aid or intersection to another navigation aid (or through several navigation aids or intersections) specified for that airway. The VICTOR airways are based upon the VHF omni-Directional Range (VOR).

The enroute airspace structure consists of three strata: Airways (generally 700 feet or 1200 feet above the surface up to but not including 18,000 MSL), the Jet Route structure (18,000 feet to flight level 450—45,-000 feet), and the airspace above FL 450, which is for point-to-point operation.

Like highways, Victor Airways are designated by number—generally north/south airways are *odd;* east/west airways are *even.*

Also see **controlled airspace.**

Federal Aviation Administration (FAA). The Federal Aviation Administration is the arm of the Department of Transportation responsible for the promotion, regulation and safety of civil aviation, and for safe and efficient use of airspace which is shared by both civil and military aircraft.

Federal Aviation Regulations (FAR). Mandatory requirements and standards issued by the FAA to govern all civil aviation activities.

feed. 1. To provide a **signal. 2.** The point at which a signal enters a circuit or device, as *antenna feed.* **3.** The signal entering a circuit or device; **input.**

feedback. 1. The return of a portion of the **output** of a device to the **input;** positive feedback adds to the input, negative feedback subtracts from the input. **2.** Information, as to progress, results, etc., returned to an originating source. **3.** In aeronautics, the transmittal of forces initiated by aerodynamic action on control surfaces or rotor blades to the cockpit **controls;** the forces so transmitted.

feedback control loop. A closed transmission path (loop), which includes an active **transducer** and which consists of a forward path, a feedback path, and one or more mixing points arranged to maintain a prescribed relationship between the loop **input** signal and the loop **output** signal.

feedback control system. A **control** system, comprising one or more **feedback control loops,** which combines functions of the controlled signals with functions of the **commands** to tend to maintain prescribed relationships between the commands and the controlled signals.

feedback path. In a **feedback control loop** the transmis-

sion path from the loop **output** signal to the loop **feedback** signal.

feel. The sensation or impression that a pilot has or receives as to his, or his craft's attitude, orientation, speed, direction of movement or acceleration, or proximity to nearby objects, or, as most often used, as to the aircraft's stability and responsiveness to control. See **control feel.**

fence. 1. A line of **readout** or **tracking** stations for pickup of signals from an orbiting **satellite. 2.** A line or network of radar or radio stations for detection of a satellite in orbit. **3.** A stationary plate or vane projecting from the upper surface of an airfoil, substantially parallel to the airflow, used to prevent spanwise flow.

ferry flight. A flight for the purpose of returning an aircraft to base or moving an aircraft to and from a maintenance base. Ferry flights, under certain conditions, are conducted under the terms of a special flight permit.

fertile material (nucleonics). A material, not itself fissionable by thermal neutrons, which can be converted into a fissionable material by irradiation in a reactor. There are two basic fertile materials, uranium-238 and thorium-232. When these fertile materials capture neutrons, they·are partially converted into fissionable plutonium-239 and uranium-233, respectively. (Compare **fissionable material.**)

fibre optics system. A bundle or system of very thin, transparent homogeneous fiber of glass or plastic that is enclosed by material of lower index of refraction and transmits light throughout its length by internal reflections. See **fly-by-light.**

fidelity. The accuracy to which an electrical system, such as a radio, reproduces at its **output** the essential characteristics of its **input signal.**

fiducial mark. An internally generated identification mark on a film; two or more of these are generally used for orienting a film for reading, and for determining the geometric center of the film.

The L-shaped corner marks and the + mark near the picture center, which are on the focal plane of the Tiros vidicon camera are fiducial marks. Their appearance on the image permits various calibrations such as determination of the degree of enlargement needed to fit the picture to the retification grids, etc.

field. A region of space within which each point has a definite value of a given physical or mathematical quantity has some definite value.

One may speak of a *gravitation field, magnetic field, electric field, pressure field, temperature field,* etc. If the quantity specified at each point is a vector quantity, the field is said to be a *vector field.*

field maintenance. That maintenance authorized and performed by designated Army, Air Force, and Marine Corps maintenance activities in direct support of using organizations. It is normally limited to replacement of unserviceable parts, subassemblies, or assemblies.

field strength. 1. For any physical **field,** the **flux density, intensity,** or **gradient** of the field at the point in question. Also called *field intensity.*

Although field *intensity* is commonly used, it should be noted that this does not follow the strict radiometric definition of intensity, i.e., flux per unit solid angle. **2. signal strength,** in radar. **3. electric field strength.**

fighter (F), (military designation). Aircraft designed to intercept and destroy other aircraft and/or missiles. (Includes multi-purpose aircraft designed for ground support and interdiction.) See **military aircraft types.**

fighter controller. The officer on the staff of a tactical air controller charged with coordination and evaluation of air warning reports and operational control of aircraft allocated to him. See also **air controller; fighter director; tactical air controller; tactical air director.**

fighter cover. The maintenance of a number of fighter aircraft over a specified area or force for the purpose of repelling hostile air activities. See also **airborne alert.**

fighter direction ship or aircraft. A ship or aircraft properly equipped and manned for effectively directing fighter aircraft operations.

fighter director. The officer on the staff of a tactical air director responsible for direction of such air warning facilities and aircraft as may be allocated to him for the defense of his area. See also **air controlled; fighter controller; tactical air controller; tactical air director.**

fighter interceptor. A fighter aircraft designed to intercept its target. It may or may not carry devices to assist in interception and in aiming its weapons.

fighter sweep. An offensive mission by fighter aircraft to seek out and destroy enemy aircraft or targets of opportunity in an allotted area of operations.

filamentary structure. A shell or membrane structure constructed of woven or layered filaments embedded in a suitable matrix.

fillet. A fairing at the junction of two surfaces to improve the air-flow.

film badge. A light-tight package of photographic film worn like a badge by workers in nuclear industry or research, used to measure possible exposure to **ionizing radiation.** The *absorbed dose* can be calculated by the degree of film darkening caused by the irradiation. (See **dosimeter.**)

film cooling. The cooling of a body or surface, such as the inner surface of a rocket **combustion chamber,** by maintaining a thin fluid layer over the affected area. Compare **transpiration cooling.**

filter center (military). The location in an aircraft control and warning system at which information from observation posts is filtered for further dissemination to air defense control centers and air defense direction centers.

filtering. 1. The decomposition of a **signal** into its **harmonic** components. **2.** The separation of a wanted component of a **time series** from any unwanted residue (noise).

filtering (military). The process of interpreting reported information on movements of aircraft, ships, and submarines in order to determine their probable true tracks and, where applicable, heights or depths.

fin. 1. A fixed or adjustable **airfoil** or **vane** attached longitudinally to an aircraft, rocket, or similar body to provide a stabilizing effect. **2.** A projecting flat plate or structure, as a *cooling fin.*

final approach. The part of an approach from the time the aircraft has completed the last turn into landing or has crossed a specified position to the point where a landing can be made.

final approach fix. The fix from or over which final approach (IFR) to an airport is executed.

final approach (IFR). The flight path of an aircraft which is inbound to the airport on an approved final instrument approach course, beginning at the final approach fix and extending to the airport or the point where circling for landing or missed approach is executed.

final approach (VFR) A flight path of a landing aircraft in the direction of landing along the extended runway centerline from the base leg to the runway.

final mass. The mass of a **rocket** after its **propellants** are consumed.

fineness ratio. The ratio of the length to the maximum diameter of a streamline body, e.g., an airship hull, a streamline strut, or a fuselage, etc.

fire. 1. To ignite a **rocket engine.**

Usage is sometimes restricted to period of main chamber burning when small igniter chambers are used, especially with igniter idle provisions where the igniter may burn for some significant period prior to main chamber fire.

2. To **launch** a rocket.

fire (military). The command given to discharge a weapon(s).

fireball. A bright **meteor** with **luminosity** which equals or exceeds that of the brightest planets.

fireball (nucleonics). The luminous ball of hot gases that forms a few millionths of a second after a nuclear explosion. (See **atomic cloud.**)

fire control radar. Radar used to provide target information inputs to a weapon fire control system.

fire point. The temperature at which a substance, as lubricating oil, will give off a vapor that will burn continuously after ignition. Compare **flashpoint.**

firepower. 1. The amount of fire which may be delivered by a position, unit, or weapon system. **2.** Ability to deliver fire.

firepower umbrella. An area of specified dimensions defining the boundaries of the airspace over a naval force at sea within which the fire of ships' antiaircraft weapons can endanger aircraft, and within which special procedures have been established for the identification and operation of friendly aircraft.

fireproof. 1. With respect to materials and parts used to confine fire in a designated fire zone, means the capacity to withstand heat at least as well as steel, in dimensions appropriate for the purpose for which they are used, under the most severe conditions of fire and duration likely to occur in that zone; and **2.** With respect to other materials and parts, means the capacity to withstand heat at least as well as steel in dimensions appropriate for the purpose for which they are used.

fire resistant. 1. With respect to sheet or structural members, means the capacity to withstand heat at least as well as aluminum alloy in dimensions appropriate for the purpose for which they are used; and **2.** With respect to fluid-carrying lines, other flammable fluid system parts, wiring, air ducts, fittings, and powerplant controls, means the capacity to perform the intended functions under the heat and other conditions likely to occur at the place concerned.

firing. 1. The action or event of igniting a **rocket engine.** **2.** The action or event of launching a rocket.

first law of thermodynamics. A statement of the conservation of energy for **thermodynamic** systems (not necessarily in equilibrium). The fundamental form requires that the heat absorbed by the system serve either to raise the internal energy of the system or to do work on the environment.

first officer (F/O). Second in command of an aircraft. In airline service the first officer occupies the right or **copilot's** seat. See **captain, second officer** and **flight engineer.**

first pilot. A pilot, fully qualified on type, responsible for the flying of an aircraft.

first quarter. The **phase** of the moon when it is near east **quadrature,** when the western half of it is visible to an observer on the earth. See **phases of the moon.**

fishbone antenna. An **antenna** consisting of a series of coplanar elements arranged in colinear pairs, loosely coupled to a balanced transmission line.

fission. The splitting of a heavy *nucleus* into two approximately equal parts (which are nuclei of lighter elements), accompanied by the release of a relatively large amount of energy and generally one or more neutrons. Fission can occur spontaneously, but usually is caused by nuclear absorption of gamma rays, neutrons or other particles. (Compare *fusion;* see **chain reaction, nuclear reaction.**)

fission weapon. An **atomic bomb.**

fissionable material. Any material fissionable by *slow neutrons.* The three basic ones are uranium-235, plutonium-239, and uranium-233. (Compare **fertile material.**)

fission yield. The amount of energy released by fission in a thermonuclear (fusion) explosion as distinct from that released by fusion. Also the amount (percentage) of a given nuclide produced by fission. (Compare **yield;** see **thermonuclear reaction, TNT equivalent.**)

fix. A geographical position determined by visual reference to the surface by reference to one or more radio navaids, by celestial plotting, or by another navigational device.

fixed-area exhaust nozzle. On a jet engine, an exhaust **nozzle** exit opening which remains constant in area. Compare **variable-area exhaust nozzle.**

fixed-base operator (FBO). An individual or firm operating at an airport and providing general aircraft services such as maintenance, storage, ground and flight instructions, etc.

fixed-pitch propeller. A propeller having no provision for changing the pitch setting.

fixed point. 1. Positional notation in which corresponding places in different quantities are occupied by coefficients of the same power of the base. Contrast to **floating point. 2.** A notation in which the **base point** is assumed to remain fixed with respect to one end of the numeric expressions.

fixed price incentive contract. A fixed price type of contract with provision for the adjustment of profit and price by a formula based on the relationship which final negotiated total cost bears to negotiated target cost as adjusted by approved changes.

fixed price type contract. A type of contract which generally provides for a firm price, or under appropriate circumstances may provide for an adjustable price, for the supplies or services which are being procured. Fixed price contracts are of several types so designed as to facilitate proper pricing under varying circumstances.

fixed satellite. A **satellite** that orbits the earth from west to east at such a speed as to remain fixed over a given place on the earth's equator at approximately 35,900 kilometers altitude. See **stationary orbit, 24-hour satellite, synchronous satellite.**

fixer network or system. A combination of radio or radar direction-finding installations which, operating in con-junction, are capable of plotting the position relative to the ground of an aircraft in flight.

flag stop. A special stop by a scheduled airlift mission aircraft at a station at which a stop is not normally made, to load or to unload traffic as required.

flak suppression fire. Fire used to suppress antiaircraft fire immediately prior to and during an air attack on enemy positions.

flame attenuation. Attenuation of a radio **signal** by the **ionization** produced in the rocket exhaust.

flame bucket. An opening built into the pads of some rockets into which the hot gases of the rocket pour as thrust is built up. The flame bucket is directly under the rocket positioned for launch. One of its sides turns inward to form the flame deflector; the opposite side is open.

flame deflector. 1. In a vertical **launch,** any of variously designed obstructions that intercept the hot gases of the **rocket engine** so as to deflect them away from the ground or from a structure. **2.** In a captive test, an elbow in the exhaust conduit or **flame bucket** that deflects the flame into the open.

flame out. Of a jet engine or gas-turbine engine: To cease burning in the combustion section from cause other than deliberate shutoff.

flame resistant. Not susceptible to combustion to the point of propagating a flame, beyond safe limits, after the ignition source is removed.

flame tube. Same as **inner liner.**

flammable. With respect to a fluid or gas, means susceptible to igniting readily or to exploding.

Flamsteed number. A number sometimes used with the possessive form of the Latin name of the **constellation** to identify a star, as *72 Ophiuchi.*

The Flamsteed number is used for stars numbered in Flamsteed's British Catalogue of 1725. For stars which do not appear in Flamsteed's catalog, numbers from other catalogs are used.

flap. An appendage to an airfoil, usually the wing, for changing its lift characteristics to permit slower landings. See **high lift devices.**

flap extended speed. The highest speed permissible with wing flaps in a prescribed extended position.

flare. 1. A bright eruption from the sun's chromosphere. Compare **prominence.**

Flares may appear within minutes and fade within an hour. They cover a wide range of intensity and size, and they tend to occur between **sunspots** or over their penumbrae.

Flares are related to radio **fadeouts** and terrestrial magnetic disturbances.

Flares eject high energy **protons** which present a serious hazard to men in unshielded spacecraft. **2.** Pyrotechnic devices used for signalling or to provide illumination. **3.** An expansion at the end of a cylindrical body as at the base of a rocket.

are. A flight maneuver involving the transition from a glide to a landing.

flare out. To round out a landing by decreasing the rate of descent and airspeed by slowly raising the nose.

flash burn (nucleonics). A skin burn due to flash of *thermal radiation.* It can be distinguished from a flame burn by the fact it occurs on unshielded parts of the body that are in a direct line with the origin of the thermal radiation. (See **ionizing radiation, thermal burn.**)

flash point. The temperature at which the vapor of a fuel or oil will flash or ignite momentarily.

flash resistant. Not susceptible to burning violently when ignited.

flat spin. A spin at a large mean angle of attack with the longitudinal axis more nearly horizontal than vertical.

flattening. Of the earth, the ratio of the difference between the equatorial radius (major semiaxis) and the polar radius (minor semiaxis) of the earth to the equatorial radius. Also called *compression.* See **astronomical constants.**

fleet. An organization of ships, aircraft, marine forces and shore-based fleet activities, all under the command of a commander or commander in chief who may exercise operational as well as administrative control.

fleet ballistic missile submarine. A nuclear-powered submarine designed to deliver ballistic missile attacks against assigned targets from either a submerged or surfaced condition.

fleet in being. A fleet (force) which avoids decisive action but which, because of its strength and location, causes or necessitates counter-concentrations and so reduces the number of opposing units available for operations elsewhere.

flicker control. Control of an aircraft, rocket, etc. in which the control surfaces are deflected to their fullest degree with any motion of the remote control. Compare **proportional control.** See **bang-bang control.**

flight. **1.** The movement of an object through the atmosphere or through space, sustained by aerodynamic, aerostatic, or reaction forces, or by orbital speed; especially, the movement of a man-operated or man-controlled device, such as a rocket, a space probe, a space vehicle, or aircraft. **2.** An instance of such a movement.

flight. A flight begins when the aircraft begins to move forward on the takeoff run or takes off vertically from rest at any point of support, as applicable, and ends after airborne flight when the aircraft is on the surface and either: **1.** The engines are stopped, or **2.** The aircraft has been on the surface for 5 minutes, whichever occurs first between (1) and (2), or **3.** A change is made in the crew. A series of practice landings is considered part of one flight, and the provisions of (2) above do not apply. **4.** In Navy and Marine Corps usage, a specified group of aircraft usually engaged in a common mission. **5.** The basic tactical unit in the Air Force, consisting of four or more aircraft in two or more elements. **6.** A single aircraft airborne on a nonoperational mission.

flight advisory. A message dispatched to aircraft in flight, or to interested stations to advise of any deviation or irregularity.

flight attendant. Same as **steward/stewardess.**

flight attitude. The attitude of an aircraft, rocket, etc., in flight; specifically, the attitude of an aircraft with respect to the **relative wind.**

flight characteristic. A **characteristic** exhibited by an aircraft, rocket, or the like in flight, such as a tendency to stall or to yaw, an ability to remain stable at certain speeds, etc.

flight check. A general term usually applicable to a proficiency check of a pilot in actual flight.

flight controls. The controls used by the pilot and/or automatic pilot to control the **attitude** and **direction** of flight of an aircraft.

flight control system. The arrangement of all control elements which enable control forces and torques to be brought into play by the human pilot or otherwise. Also see **vehicle control system.**

flight crew. Those members of the aircrew whose primary concern is the operation and navigation of the aircraft and its safety in flight.

flight crewmember. A **pilot, flight engineer,** or **flight navigator** assigned to duty in an aircraft during flight time.

flight deck. **1.** In certain airplanes, an elevated compartment occupied by the crew for operating the airplane in flight. **2.** The upper deck of an aircraft carrier that serves as a runway.

flight director. An airborne device which indicates to the pilot, by visual means, the correct control application for the operation of an aircraft in accordance with a preselected flight plan; normal instrument data is fed to a computer for modification, combination, and calculation leading to an output to a simple visual display.

flight engineer. A **flight crew member** on a large, multi-engine aircraft who performs non-flying duties with reference to the aircraft powerplants, systems, fuel management and ground servicing. See **flight engineer certificate, airman certificates.**

flight engineer certificate. A certificate of competency issued by the **FAA** to a person meeting the requirements of the applicable **Federal Aviation Regulations.** Also see **flight engineer, airman certificates.**

flight instructor. A **pilot** authorized to instruct other persons to pilot an aircraft. See **flight instructor certificate.**

flight instructor certificate. A certificate of competency issued by the **FAA** to a **pilot** meeting the requirements of the applicable Federal Aviation Regulations. Flight instructor **category ratings** are: **airplane, rotorcraft** and/or **gliders and instrument.** Also see **airman certificates.**

flight level (FL). A level of constant atmospheric pressure related to a reference datum of 29.92 inches of mercury. Each is stated in three digits that represent hundreds of feet. For example, FL 250 represents a barometric altimeter indication of 25,000 feet. FL 255 indicates 25,500 feet.

flight line. A general term indicating the ramp area of an airport, exclusive of runways and taxiways, where aircraft are parked and serviced for flight.

flight Mach number. A free-stream **Mach number** measured in flight as distinguished from one measured in a **wind tunnel.**

flight management system (FMS). The flight management system is an extension of the **automatic flight control system** and provides fully automatic navigation and **performance management** under control of a **digital computer.** Aircraft flight path is accurately maintained along the selected route. Airplane speed and engine thrust are accurately controlled to selected schedules during climb, cruise and descent.

flight navigator. A **flight crew member** who conducts **navigation** duties usually with regard to long range over water flight operations. See **airmen certificates.**

flightpath. The path made or followed in the air or in space by an aircraft, rocket, etc.; the continuous series of positions occupied by a flying body; more strictly, the path of the center of gravity of the flying body, referred to the earth or other fixed reference.

flightpath angle. The angle between the **horizontal** and a tangent to the **flightpath** at a point.

flight path deviation indicator (FPDI). An instrument that indicates the deviation of the actual flight path from the planned flight path, expressed in angular or linear measurement.

flight plan. Specified information relating to the intended flight of an aircraft that is filed orally or in writing with an air traffic control facility.

flight profile. A graphic portrayal or plot of the flight path of an aeronautical vehicle in the vertical plane.

flight readiness firing. A missile system test of short duration conducted with the propulsion system operating while the missile is secured to the launcher. Such a test is performed to determine the readiness of the missile system and launch facilities prior to flight test.

flight recorder. An instrument for automatically recording certain elements of the **performance** of an aircraft.

flight service station (FSS). A facility operated by the FAA to provide flight assistance service.

flight simulator. A training device or apparatus that simulates certain conditions of actual flight or of flight operations.

flight space. The space above and beyond the earth's surface now used, or potentially to be used, for flight of aircraft, spacecraft, or rockets.

flight station. Same as **cockpit.**

flight surgeon (military). A physician specially trained in aviation medical practice whose primary duty is the medical examination and medical care of aircrew.

flight test. 1. A test by means of actual or attempted flight to see how an aircraft, spacecraft, space-air vehicle, or missile flies. **2.** A test of a component part of a flying vehicle, or of an object carried in such a vehicle, to determine its suitability or reliability in terms of its intended function by making it endure actual **flight.**

flight test. A flight in which a pilot applicant demonstrates prescribed maneuvers to an accompanying inspector pilot or examiner for the purpose of meeting the requirements of an additional **rating** or certificate. See **Federal Aviation Regulations, airman certificates.**

flight test vehicle. A test vehicle for the conduct of flight tests, either to test its own capabilities or to carry equipment requiring flight test.

flight time. The time from the moment the aircraft first moves under its own power for the purpose of flight until the moment it comes to rest at the next point of landing. ("Block-to-block" time.)

flight visibility. The average forward horizontal distance from the cockpit of an aircraft in flight at which prominent unlighted objects may be seen and identified by day and prominent lighted objects may be seen and identified by night. See **visibility, prevailing.**

flightworthy. Of an aircraft, missile, or spacecraft: Ready and sufficiently sound in all respects to meet and endure the stresses and strains of flight. See **airworthy.**

flip-flop. 1. A device having two stable states and two input terminals (or types of input signals) each of which corresponds with one of the two states. The **circuit** remains in either state until caused to change to the other state by application of the corresponding signal. **2.** A similar bistable device with an input which allows it to act as a single-stage **binary counter.**

float. A completely enclosed water-tight structure attached to an aircraft in order to give it buoyancy and stability when in contact with the surface of the water. In float seaplanes the crew is carried in a fuselage or nacelle separate from the float. Sometimes called pontoon.

float gear. A term sometimes used as descriptive of floats which have replaced the landing gear on a land plane.

float plane. Same as **seaplane.**

floating point. In computer operations, a **positional notation** in which corresponding places in different quantities are not necessarily occupied by coefficients of the same power of the base. Compare **fixed point.**

Floating point corresponds to multiplication using powers of 10; for example, 186,000 can be represented as 1.86×10^5. By shifting the point so that the number of significant digits in any quantity does not exceed machine capacity, widely varying quantities can be handled. The scale factor may be fixed for each problem, or indicated along with the digits and sign for each quantity.

flocculi. Patches of relatively dense, dark or bright clouds in the sun's atmosphere. They appear in photographs taken with the **spectroheliograph.**

flotation gear. Gear or apparatus, commonly inflatable bags, vests, rafts, and the like, carried aboard a vehicle to support the vehicle or persons if downed in water.

flow. A stream or movement of air or other **fluid,** or the rate of fluid movement, in the open or in a duct, pipe, or passage; specifically, an **airflow.**

flow chart. A graphical representation of a sequence of operations using symbols to represent the operations.

A flow chart is a more detailed representation than a diagram.

flow, laminar. See **laminar flow.**

flow, turbulent. See **turbulent flow.**

fluid. A substance which, when in static equilibrium, cannot sustain a **shear** stress; a liquid or a gas.

This concept is only approximated by actual liquids and gases.

fluid fuel reactor (nucleonics). A type of reactor (for example, a *fused-salt reactor*) whose fuel is a fluid form.

fluidity. Reciprocal of **viscosity.**

fluidized bed reactor (nucleonics). A reactor design in which the *fuel* ranges in size from small particles to pellets. Although the fuel particles are solid, their entire mass behaves like a fluid because a stream of liquid or gas *coolant* keeps them moving.

fluid parcel. In any **fluid,** an imaginary portion of that fluid which for theoretical studies may be considered to have all the basic dynamic and thermodynamic properties of the fluid but which is small enough so that its motion with respect to the surrounding fluid does not induce marked compensatory movements. Also called *parcel.*

The size of the fluid parcel cannot be given precise numerical definition but it must be large enough to contain a great number of molecules and small enough so that the properties assigned to it are approximately uniform within it.

fluoroscope. An instrument with a fluorescent screen suitably mounted with respect to an X-ray tube, used for immediate indirect viewing of internal organs of the body, internal structures in apparatus or masses of metals, by means of X rays. A fluorescent image, really a kind of X-ray shadow picture, is produced. (See **X**

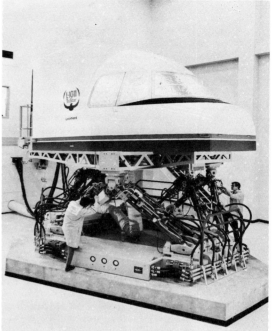

A **flight simulator** *realistically duplicates the flight characteristics of an aircraft. It is an effective and economical device for pilot checkout and proficiency training.*

ray.)

fluorescence. Emission of light or other radiant energy as a result of and only during **absorption** of **radiation** of a different wavelength from some other source. Also called *photoluminescence.* See **luminescence.** Compare **phosphorescence.**

fluorescence. Many substances can absorb energy (as from X rays, ultraviolet light, or radioactive particles), and immediately emit this energy as an electromagnetic *photon,* often of visible light. This emission is fluorescence. The emitting substances are said to be fluorescent. (Compare **luminescence, scintillation;** see **excited state.**)

flutter. An **aeroelastic** self-excited vibration in which the external source of energy is the **airstream** and which depends on the elastic, inertial and dissipative forces of the system in addition to the **aerodynamic forces.**

flutter speed. The lowest equivalent airspeed at which flutter occurs.

flux. 1. The rate of **flow** of some quantity, often used in reference to the flow of some form of **energy.** Also called *transport.* See **power. 2.** In nuclear physics generally, the number of **radioactive particles** per unit volume times their mean velocity.

flux density. The flux (rate of flow) of any quantity, usually a form of **energy,** through a unit area of specified surface. (Note that this is not a volumetric density like **radiant density.**) Compare **luminous density.**

   In radar, flux density commonly is referred to as *power density.* It is essential to understand that the flux density of radiation is in no sense a vector quantity, because it is the sum of the flux corresponding to all ray directions incident upon one *side* of the unit area.

fluxgate. The magnetic direction sensing element of most remote indicating compasses. The detector is used to give an electrical signal proportional to the intensity of the external magnetic field acting along its axis.

fluxgate compass. An instrument employing the fluxgate principle to indicate, subject to certain corrections, the direction of the observer's magnetic meridian.

fly. 1. To operate an aircraft in flight. **2.** To ride as a passenger in an aircraft.

flyby. An interplanetary mission in which the vehicle passes close to the target planet but does not impact it or go into orbit around it.

fly-by-light. An aircraft flight control system whereby pilot inputs to the controls are transmitted, utilizing **fiber optic** light transmissions, to control operation of the primary flight control system **servos.**

fly-by-wire. An aircraft flight control system whereby pilot inputs to the controls are transmitted electronically to a computer which provides signals to a hydraulic control system to operate the control surfaces.

flying. The movement of an aircraft or a person in an aircraft, through the air; the act of such movement. The piloting of aircraft; the art or skill of piloting an aircraft or a spacecraft.

flying boat. A seaplane supported, when resting on the surface of the water, by a hull or hulls providing floatation in addition to serving as fuselages. For the central hull type, lateral stability is usually provided by wingtip floats.

flying stabilizer. Same as **stabilator.**

flying test bed. An aircraft, rocket, or other flying **vehicle** used to carry objects or devices being flight tested.

FM. See **frequency modulation.**

FM/AM (abbr). **1. Amplitude modulation** of a **carrier** by subcarrier(s) which is (are) frequency modulated by information. **2.** Alternate FM or AM operation.

FMS. Flight management system.

focal length. The distance between the optical center of a lens, or the surface of a mirror, and its **focus.**

focal plane. A plane parallel to the plane of a lens or mirror and passing through the **focus.**

focal point. Same as **focus,** in optics.

focus (*plural* **focuses**). **1.** That point at which parallel rays of light meet after being refracted by a lens or reflected by a mirror. Also called *focal point.* **2.** A point having specific significance relative to a geometrical figure. See **ellipse, hyperbola, parabola.**

foehn. A dry wind with a strong downward component, warm for the season, characteristic of many mountainous regions. The air is cooled dynamically in ascending the mountains, but this leads to condensation, which checks the fall in temperature through the liberation of latent heat. The wind deposits its moisture as rain or snow. In descending the opposite slope it is strongly heated dynamically and arrives in the valleys beyond as a warm and very dry wind. Some writers apply this term to any wind that is dynamically heated by descent; e.g., the sinking air of an anticyclone.

fog. A cloud at the earth's surface. Fog consists of numerous droplets of water, which are so small that they cannot readily be distinguished by the naked eye. In ordinary speech the term "fog" generally implies an obscurity of the atmosphere sufficiently great to interfere with marine and air navigation. Also see **advection fog, radiation fog.**

folded dipole antenna. An **antenna** composed of two parallel, closely spaced **dipole antennas** connected at their ends with one of the dipole antennas fed at its center.

folding fin. A **fin** hinged at its base to lie flat, especially a fin on a **rocket** that lies flat until the rocket is in flight.

follow-on. Any object, group of objects, technique, or procedure considered to be a second or subsequent generation in the development of the object, group of objects, technique, or procedure. See **generation.**

follow-on developmental tests. Tests during the acquisition phase of the system life cycle which occur after completion of the formal category II tests. Test responsibility is normally retained by Air Force research and development agencies. Test consists of developmental testing and updating changes, or additions to system, subsystems, and components not normally resulting in a series change and which were not available in usable form during the normal test cycle. It is conducted on a somewhat reduced scale, concentrating on the specific changes or additions to the development program.

foot (*abbr* **ft**). The foot (international) is exactly 0.3048 meter.

   The American Survey foot is 0.3048006 meter.

   The old U. S. foot, used prior to July 1, 1959, was 0.3048006 meter.

foot-candle. A unit of **illuminance,** incident light, or **illumination** equal to 1 lumen per square foot. This is the illuminance provided by a light source of one candle at a distance of 1 foot, hence the name. Compare **lux.**

   Full sunlight with zenith sun produces an illuminance of the order of 10,000 foot-candles on a horizontal surface at the earth's surface. Full moonlight provides an illuminance of only about 0.02 foot-candle also at earth's surface. Adequate illumination for steady reading is taken to be about 10 foot-candles; that for close machine work is about 30 to 40 foot-candles.

foot-to-head acceleration. See **physiological acceleration.**

footward acceleration. See **physiological acceleration.**

force. The cause of the **acceleration** of material bodies measured by the rate of change of **momentum** produced on a free body.

force (military). 1. A body of troops, ships, or aircraft, or combination thereof. **2.** A major subdivision of a fleet.

force balance transducer. A **transducer** in which the output from the sensing member is amplified and fed back to an element which causes the force-summing member to return to its rest position.

force combat air patrol. A patrol of fighters maintained over the task force to destroy enemy aircraft which threaten the force. See also **combat air patrol.**

forced landing. A landing, either on land or water, made when it is impossible for the aircraft to remain airborne as a result of mechanical failure such as the propulsion system.

force rendezvous (air). A navigational checkpoint over which formations of aircraft join and become part of the main force.

foreign air carrier. Any person other than a citizen of the United States, who undertakes directly, by lease or other arrangement, to engage in air transportation.

foreign air commerce. The carriage by aircraft of persons or property for compensation or hire, or the carriage of mail by aircraft, or the operation or navigation of aircraft in the conduct or furtherance of a business or vocation, in commerce between a place in the United States and any place outside thereof; whether such commerce moves wholly by aircraft or partly by aircraft and partly by other forms of transportation.

foreign air transportation. The carriage by aircraft of persons or property as a common carrier for compensation or hire, or the carriage of mail by aircraft, in commerce between a place in the United States and any place outside of the United States, whether that commerce moves wholly by aircraft or partly by aircraft and partly by other forms of transportation.

foreign military sales. That portion of United States miliary assistance authorized by the Foreign Assistance Act of 1961, as amended. This assistance is for both defense articles and services (including training). This assistance differs from Military Assistance Program Grant Aid in that it is purchased by the recipient country.

formability. The relative ease with which a metal can be shaped through plastic deformation.

formation. 1. An ordered arrangement of troops and/or vehicles for a specific purpose. **2.** An ordered arrangement of two or more ships, units, or aircraft proceeding together.

forward acceleration. See **physiological acceleration.**

forward aeromedical evacuation. That phase of evacuation which provides airlift for patients between points within the battlefield, from the battlefield to the initial point of treatment, and to subsequent points of treatment within the combat zone.

forward air controller. An officer (aviator/pilot) member of the tactical air control party who, from a forward ground or airborne position, controls aircraft engaged in close air support of ground troops.

forward air control post. A highly mobile United States Air Force tactical air control system radar facility subordinate to the control and reporting center and/or post used to extend radar coverage and control in the forward combat area.

forward area. An area in proximity to combat.

forward oblique air photograph. Oblique photography of the terrain ahead of the aircraft.

forward observer. An observer operating with front line troops and trained to adjust ground or naval gunfire and pass back battlefield information. In the absence of a forward air controller he may control close air support strikes. See also **spotter.**

forward operating base. An airfield used to support tactical operations without establishing full support facilities. The base may be used for an extended time period. Support by a main operating base will be required to provide backup support for a forward operating base.

forward scatter. The **scattering** of **radiant energy** into the hemisphere of space bounded by a plane normal to the direction of the incident radiation and lying on the side toward which the incident radiation was advancing; the opposite of **backward scatter.**

forward slip (flight). A slight side slip sometimes made during an approach to a landing. The chief objective is to lose altitude faster than a normal glide permits but with no increase in ground speed; the flight path being maintained in the same straight line (**track**) over the ground. Also see **slip.**

forward supply point (military). An enroute or turnaround station at which selected aircraft spares, peculiar to the mission, design and series of aircraft are prepositioned for the support of the assigned mission(s).

four minute turn. See **standard rate turn.**

fox away. Missile has fired or been released from aircraft.

fractocumulus. Ragged cumulus clouds in which the different parts show continual change. Also see **cloud, cumuliform clouds.**

fractostratus. A stratus cloud broken up into irregular, ragged fragments. Also see **cloud, stratiform clouds.**

frame. In photography, any single exposure contained within a continuous sequence of photographs.

Fraunhofer lines. Dark lines in the **absorption spectrum** of **solar radiation** due to absorption by gases in the outer portions of the sun and in the earth's atmosphere.

Fraunhofter spectrum. The visible solar spectrum.

freak. Frequency in megacycles.

free atmosphere. That portion of the earth's **atmosphere,** above the **planetary boundary layer,** in which the effect of the earth's surface friction on the air motion is negligible, and in which the air is usually treated (dynamically) as an ideal fluid. The base of the free atmosphere is usually taken as the **geostrophic wind** level. Also called *free air.*

free drop. The dropping of equipment or supplies from an aircraft without the use of parachutes. See also **airdrop; air movement; free fall.**

free electron. An **electron** which is not bound to an atom.

free fall. A parachute maneuver in which the parachute is manually activated at the discretion of the jumper or automatically at a pre-set altitude. See also **airdrop; air movement; free drop.**

free flight. Unconstrained or unassisted **flight,** as: (**a**) the flight of a rocket after consumption of its propellant or after motor shutoff; (**b**) the flight of an unguided projectile; (**c**) the flight in certain kinds of wind tunnel of an unmounted model.

free gyro. 1. A two-degree-of-freedom **gyro** whose spin axis may be oriented in any specified attitude.

In a free gyro, output signals are produced by gimbal angular displacements which correspond to components of the angular displacement of the base. **2.** A **gyro** not provided with an erection system, i.e., a gyro free to move about its axes.

free molecule flow. 1. A **flow** regime in aerodyamics in which molecules emitted from an object, as it passes through a resistive medium, do not affect the flow of oncoming molecules by scattering interactions, i.e., the **mean free path** of the emitted molecules is much longer than a characteristic linear dimension of an object. **2.** **Flow** about a body in which the number of **collisions** between the molecules of the fluid is negligible compared with the collisions between these molecules and the body. Also called *free molecular flow*. See **rarefied gas dynamics**, note.

free-radical. An atom or a group of atoms broken away from a stable compound by application of external energy. They may remain in the free state for extended periods.

free rocket. A rocket not subject to guidance or control in flight.

free space. An ideal, perfectly homogeneous medium possessing a **dielectric constant** of unity and in which there is nothing to reflect, refract or absorb energy. A perfect vacuum possesses these qualities. Compare **homogeneous atmosphere.**

Radio signal strength measurements are often expressed in terms of decibels above or below free-space values at a given distance from the transmitter. A free-space radiation pattern would show only the minor and major lobes of the antenna and not the interference pattern normally produced by reflection from the earth's surface.

free stream. 1. The stream of **fluid** outside the region affected by a body in the fluid. **2.** Pertaining to the free stream, sense **1,** as in *free-stream dynamic pressure, free-stream flow, free-stream Mach number, free-stream static pressure, free-stream temperature, free-stream turbulence, free-stream velocity.*

free turbine. In a turbine engine, a **turbine wheel** that drives the output shaft and is not connected to the shaft driving the **compressor.**

free-vortex compressor. An **axial-flow compressor** designed so as to impart to the fluid tangential velocities that are inversely proportional to the distance from the axis of rotation, as in a **vortex.**

freezeout method. A method for controlling **humidity** by passing moist air over a cold surface, thus condensing and freezing out water vapor and possibly carbon dioxide.

freight. Cargo transported, including mail and unaccompanied baggage.

frequency. Of a function periodic in time, the reciprocal of the primitive **period.** The unit is the **cycle** per unit time and must be specified.

In the International System the cycle per second is called the hertz (Hz).

frequency assignment. The specific **frequency** or frequencies authorized by competent authority; expressed for each radio channel by: (**a**) the authorized **carrier frequency,** the **frequency tolerance,** and the authorized

emission-bandwidth, (**b**) the authorized emission-bandwidth in reference to a specific assigned frequency (when a carrier does exist), or (**c**) the authorized **frequency band** (when a carrier does not exist).

frequency band. A continuous range of frequencies extending between two limiting frequencies.

Specific frequency bands used in radio and radar are often designated by names, numbers, or letters. The band designations as decided upon by the Atlantic City Radio Convention of 1947 and later modified by Comite Consultatif International Radio (CCIR) Recommendation No. 142 in 1953 are:

Band number	Frequency range		Metric sub-division waves	Atlantic City frequency subdivision	
	kc				
4	3-	30	Myriametric	Very-low	VLF
5	30-	300	Kilometric	Low	LF
6	300-	3,000	Hectometric	Medium	MF
7	3,000-	30,000	Decametric	High	HF
	mc				
8	30-	300	Metric	Very-high	VHF
9	300-	3,000	Decimetric	Ultra-high	UHF
10	3,000-	30,000	Centimetric	Super-high	SHF
11	30,000-	300,000	Millimetric	Extremely high	EHF
12	300,000-	3,000,000	Decimillimetric		

Note that band N extends from 0.3×10^{N} to 3×10^{N} cycles; thus band 4 designates the frequency range 0.3×10^{4} to 3×10^{4} cycles. The upper limit is included in each band; the lower limit is excluded.

Description of bands by means of adjectives is arbitrary and the CCIR recommends that it be discontinued.

The designation ELF, extremely low frequency, has recently been proposed for the band extending from 3 kilocycles down to 1 cycle per second. These frequencies have been used for years in the study of lightning and associated phenomena and may be useful in communicating with spacecraft.

The frequency bands used by **radar** (radar frequency bands) were first designated by letters for military secrecy. Those designations were:

Frequency band	Approximate frequency range, gigacycles	Approximate wavelength range, centimeters
P-band	0.225 to 0.39	140 to 76.9
L-band	0.39 to 1.55	76.9 to 19.3
S-band	1.55 to 5.20	19.3 to 5.77
X-band	5.20 to 10.90	5.77 to 2.75
K-band	10.90 to 36.00	2.75 to 0.834
Q-band	36.00 to 46.00	0.834 to 0.652
V-band	46.00 to 56.00	0.652 to .536

The C-band, 3.9 to 6.2 gigacycles, overlaps the S-and X-bands. These letter designations have no official sanction.

Also see **frequency, Hertz.**

frequency bias. A constant **frequency** purposely added to the frequency of a signal to prevent the signal frequency from going to zero.

frequency channel. 1. The band of **frequencies** which must be handled by a **carrier** system to transmit a specific quantity of information. **2.** A band of radio frequencies within which a station must maintain its modulated **carrier frequency** to prevent interference with stations on adjacent channels. **3.** Any **circuit** over which telephone, telegraph, or other signals may be sent by an electric current.

frequency departure. The amount of variation of a **carrier**

frequency or center frequency from its assigned value.

frequency modulation (*abbr* **FM**). **Angle modulation** of a sine-wave **carrier** in which the instantaneous frequency of the modulated wave differs from the carrier frequency by an amount proportional to the instantaneous value of the modulating wave. Compare **pulse modulation, amplitude modulation, phase modulation, intensity modulation.**

Combinations of phase and frequency modulation are commonly referred to as *frequency modulation*.

frequency response. 1. The portion of the **frequency spectrum** which can be sensed by a device within specified limits of amplitude error. **2.** Response of a system as a function of the frequency of **excitation.**

frequency tolerance. The extent to which a **carrier frequency** (or when a carrier is not present, a frequency coinciding with the center of an emission **bandwidth**) is permitted to depart, solely because of frequency instability, from the authorized carrier frequency (or when a carrier is not present from the assigned frequency).

friction layer. Same as **planetary boundary layer.**

friendly. (military). A contact positively identified as friendly.

frigate. A warship designed to operate independently, or with strike, antisubmarine warfare, or amphibious forces against submarine, air, and surface threats. (Normal armament consists of 3" and 5" dual-purpose guns and advanced antisubmarine warfare weapons.)

front (military). 1. The lateral space occupied by an element, measured from the extremity of one flank to the extremity of the other flank. **2.** The direction of the enemy. **3.** The line of contact of two opposing forces. **4.** When a combat situation does not exist or is not assumed, the direction toward which the command is faced.

front (meteorology). A line or narrow belt marking the boundary at the Earth's surface between two air masses of different characteristics. A front is usually associated with a belt of cloud and precipitation and a more or less sharp change in wind. Also see **cold front, warm front, occluded front, stationary front.**

frontal attack. In air interception, an attack by an interceptor aircraft which terminates with a heading crossing angle greater than 135 degrees.

frontogenesis. The term used to describe the process which creates a **front**, i.e., produces a discontinuity in a continuous field of the meteorological elements; also applied to the process which increases the intensity of a pre-existing front. Frontogenesis is generally set up by the horizontal convergence of air currents possessing widely different properties.

frontolysis (meeorology). The disappearance or marked weakening of a front.

frost. Small frozen drops of dew which freeze into crystalline ice structure with the temperature below 32 deg. F. It is water vapor, which upon contact with the surface of an object which is colder than 32 deg. F., will form into frozen drops. It sometimes forms on an airplane when the airplane passes from a cold layer of air into a strata of air which is higher in temperature and of a higher moisture content.

FSS. Flight service station.

fuel. Any substance used to produce heat, either by chemical or **nuclear** reaction, as used, e.g., in a **heat engine.** See **rocket propellant.**

With a liquid propellant rocket engine, *fuel* is or-

dinarily distinguished from *oxidizer* where these are separate.

fuel (nucleonics). Fissionable material used or usable to produce energy in a *reactor*. Also applied to a mixture, such as *natural uranium,* in which only part of the atoms are readily fissionable, if the mixture can be made to sustain a chain reaction. (See **fissionable material.**)

fuel cell. 1. A fuel tank, especially one of a number of fuel tanks, as in an airplane's wing; also, a compartment within a fuel tank. **2.** A device which converts chemical energy directly into electrical energy but differing from a storage battery in that the reacting chemicals are supplied continuously as needed to meet output requirements.

fuel consumption. The using of **fuel** by an engine or power plant; the rate of this consumption, measured, e.g., in gallons or pounds per minute.

fuel control unit (turbine engine). A device for controlling the fuel supply to an engine in accordance with pilot demand, ambient conditions and engine limitations.

fuel cooled. Cooled by **fuel.** Said of a rocket engine, an oil cooler, etc. See **regenerative cooling.**

fuel cycle (nucleonics). The series of steps involved in supplying fuel for nuclear power reactors. It includes mining, refining, the original fabrication of fuel elements, their use in a reactor, chemical processing to recover the fissionable material remaining in the spent fuel, re-enrichment of the fuel material, and refabrication into new fuel elements.

fuel element (nucleonics). A rod, tube, plate, or other mechanical shape or form into which nuclear fuel is fabricated for use in a reactor. (Not to be confused with *element.*) (See **nuclear reactor.**)

fuel grade. The quality of a fuel as defined by its knock rating.

full command. The military authority and responsibility of a superior officer to issue orders to subordinates. It covers every aspect of military operations and administration and exists only within national Services. See also **command.**

full moon. The moon at **opposition,** with a **phase angle** of 0°, when it appears as a round disk to an observer on the earth because the illuminated side is toward him. See **phases of the moon.**

full pressure suit. A suit which completely encloses the body and in which a gas pressure sufficiently above **ambient pressure** for maintenance of function, may be sustained.

function. A **magnitude** so related to another magnitude that for any value of one there is a corresponding value of the other.

For instance, the area of a circle is a function of its radius. The radius is also a function of the area.

functional reserves. The ability of the body to accomplish additional muscular or other activity and useful work beyond the normal level of activity of an individual.

fundamental frequency (rdo). The lowest component **frequency** of a periodic **wave** or quantity.

fundamental mode of vibration. Of a mechanical system, the **mode** having the lowest **natural frequency.**

fused ceramic. A **ceramic** body or coating prepared by heating ceramic powders above the melting point, then cooling to form a coherent mass or film.

fused-salt reactor (nucleonics). A type of reactor that uses molten salts of uranium for both *fuel* and *coolant.*

fusee. A special pyrotechnic-type squib that is installed in a solid-propellant rocket chamber for the purpose of igniting the propellant charge.

fuselage. The structure of an aircraft of approximate **streamline** form, which houses the crew, passengers, or cargo, and to which is attached the wings, and **tail units,** and the **engine mount** in most single-engined airplanes. In some designs the tail unit is attached to **outriggers** or **tail booms** and the fuselage is simply a streamlined structure for housing the crew, etc. In the latter case it is generally called a **nacelle.**

fusion (nucleonics). The formation of a heavier *nucleus* from two lighter ones (such as hydrogen isotopes), with the attendant release of energy (as in a *hydrogen bomb*). (Compare *fission;* see **nuclear reaction, thermonuclear reaction.**)

fusion weapon. An atomic weapon using the energy of nuclear *fusion,* such as a **hydrogen bomb.**

G

G or g. A symbol used to denote gravity. See **acceleration of gravity, gravity, G-force.**

gadget (military) Radar equipment. (Type of equipment may be indicated by a letter as listed in operation orders.) May be followed by a color to indicate state of jamming. Colors will be used as follows: **a.** green—Clear of jamming. **b.** amber—Sector partially jammed. **c.** red—Sector completely jammed. **d.** blue—Completely jammed.

gage pressure. In engineering literature, a term used to indicate the difference between **atmospheric pressure** and **absolute pressure,** as read from a **differential manometer.**

gaging (nucleonics). See **gauging.**

gain. 1. A general term used to denote an increase in **signal power** in transmission from one point to another. Gain is usually expressed in decibels and is widely used to denote **transducer gain. 2.** An increase or amplification. In radar there are two general usages of the term: (a) antenna gain, or gain factor, is the ratio of the power transmitted along the beam axis to that of an **isotropic radiator** transmitting the same total power; (b) receiver gain, or video gain, is the amplification given a signal by the receiver.

galactic. 1. Pertaining to our **galaxy,** the Milky Way. **2.** Pertaining to the **galactic system of coordinates,** as *galactic latitude.*

galactic radio waves. Radio waves emanating from our **galaxy.** See **cosmic radio waves.**

galactic system of coordinates. An astronomical **coordinate system** using latitude measured north and south from the galactic equator and longitude measured in the sense of increasing **right ascension** from 0° to 360° See **coordinate.**

galaxy. A vast assemblage of stars, nebulae, etc., composing an island universe separated from other such assemblages by great distances.

The sun and its family of planets is part of a galaxy commonly called the *Milky Way.* The nearest galaxy to the Milky Way is the spiral galaxy *Andromeda* at a distance of approximately 800,000 light years.

gale. (meteorology). Beaufort number 8: Breaks twigs off trees, generally impedes progress (wind speed: 34—40 knots).

gamma rays. (Symbol γ (gamma)) High-energy, short-wavelength *electromagnetic radiation.* Gamma radiation frequently accompanies alpha and beta emissions and always accompanies *fission.* Gamma rays are very penetrating and are best stopped or shielded against by dense materials, such as lead or depleted uranium. Gamma rays are essentially similar to X rays, but are usually more energetic, and are nuclear in orgin. (Compare *X ray;* see **decay radioactive, excited state, photon.**)

gantry. A frame structure that spans over something, as an elevated platform that runs astride a work area, supported by wheels on each side; short for *gantry crane* or *gantry scaffold.*

gantry scaffold. A massive scaffolding structure mounted on a bridge or platform supported by a pair of towers or trestles that normally run back and forth on parallel tracks, used to assemble and service a large **rocket** as the rocket rests on its launching pad. Often shortened to *gantry.*

This structure is a latticed arrangement of girders, tubing, platforms, cranes, elevators, instruments, wiring, floodlights, cables, and ladders—all used to attend the rocket.

garbage. Miscellaneous objects in **orbit,** usually material ejected or broken away from a **launch vehicle** or **satellite.**

garble. An error in transmission, reception, encryption, or decryption which renders a message or a portion thereof incorrect or undercryptable.

gas. The state of matter in which the **molecules** are practically unrestricted by intermolecular forces so that the molecules are free to occupy any space within an enclosure.

In vacuum technology the word *gas* has been loosely applied to the noncondensable gas and vapor within a **vacuum system.**

gas cap. The gas immediately in front of a body as it travels through the atmosphere.

This gas is compressed and heated. If the speed is sufficiently high, the gas becomes incandescent; it is to this condition that the term is usually applied, as in the *gas cap* of a **meteor.**

gas constant (*symbol R*). The constant factor in the **equation of state** for perfect gases. The universal gas constant is

$$R = 8.3143 \text{ joules}/°\text{K-mol}$$

The gas constant for a particular gas, specific gas constant,

$$r = R/m$$

where m is the molecular weight of the gas. See **Boltzmann constant.**

gas-cooled reactor. A nuclear reactor in which a gas is the coolant.

gaseous diffusion (plant). (nucleonics.) A method of isotopic separation based on the fact that gas atoms or molecules with different masses will diffuse through a porous barrier (or membrane) at different rates. The method is used by the AEC to separate uranium-235 from uranium-238; it requires large gaseous-diffusion plants and enormous amounts of electric power.

gaseous electronics. The study of the conduction of electricity through gases, involving study of the **Townsend, glow,** and **arc discharges,** and all the collision phenomena on an atomic scale. Formerly called *gaseous discharges.*

gas laws. The thermodynamic laws applying to perfect gases: **Boyle-Mariotte law, Charles-Gay-Lussac law, Dalton law, equation of state.** Also called *perfect-gas laws, ideal-gas laws.*

gas scrubbing. The contacting of a gaseous mixture with a liquid for the purpose of removing gaseous contaminants or entrained liquids or solids.

gas turbine. 1. A **turbine** rotated by expanding gases, as in a **turbojet engine** or in a turbo-supercharger. **2.** A **gas-turbine engine.**

gas-turbine engine. An engine incorporating as its chief element a **turbine** rotated by expanding gases. In its most usual form, it consists essentially of a rotary air **compressor** with an air intake, one or more **combustion chambers,** a turbine, and an exhaust outlet.

gate. 1. To control passage of a **signal** as in the circuits of a **computer. 1.** A **circuit** having an output and inputs so designed that the output is energized only when a definite set of input conditions are met. In computers, called *AND-gate.*

gate (military). Fly at maximum possible speed (or power). (To be maintained for a limited time only, depending on type of aircraft.) Use of afterburners. rockets, etc., in accordance with local doctrine.

gauging (nucleonics). The measurement of the thickness, density or quantity of material by the amount of *radiation* it absorbs. This is the most common use of radioactive isotopes in industry. Also spelled *gaging.*

gauss. A unit of **magnetic induction** (or **magnetic flux density**) equal to 1 dyne per unit cgs magnetic pole.

Prior to 1932, the gauss was used both as a unit of magnetic induction and as a unit of **magnetic field intensity,** but the latter quantity is now measured in *oersteds.*

GCA (*abbr*) **ground-controlled approach.**

GCI (abbr) **ground-controlled intercept.**

geared tab. A balance tab mechanically linked to a control surface so that its angular movement is determined by that of the control surface.

gee. A suffix meaning earth, as in *perigee, apogee.* See **perigee,** note.

gegenschein. Faint light area of the sky always opposite the position of the Sun on the celestial sphere. Believed to be the reflection of sunlight from particles moving beyond the Earth's orbit.

Geiger-Müller counter (Geiger-Müller tube). A radiation detection and measuring instrument. It consists of a gasfilled (Geiger-Müller) tube containing electrodes, between which there is an electrical voltage but no current flowing. When *ionizing radiation* passes through the tube, a short, intense *pulse* of current passes from the negative electrode to the positive electrode and is measured or counted. The number of pulses per second measures the *intensity* of radiation. It is also often known as *Geiger counter;* it was named for Hans Geiger and W. Müller who invented it in the 1920s. (See **counter**).

general aviation. That portion of civil aviation which encompasses all facets of aviation except air carriers holding a certificate of convenience and necessity from the Civil Aeronautics Board, and large aircraft commercial operators.

General Aviation District Office (GADO) An FAA field office serving a designated geographical area, staffed with Flight Standards personnel who have responsibility for serving the aviation industry and the general public on all matters relating to the certification and operation of general aviation aircraft.

general staff. A group of officers in the headquarters of Army or Marine divisions, Marine brigades and aircraft wings, or similar or larger units which assist their commanders in planning, coordinating, and supervising operations. A general staff may consist of four or more principal functional sections: personnel (G-1), military intelligence (G-2), operations and training (G-3), logistics (G-4), and (in Army organizations) civil affairs/military government (G-5). (A particular sec-

tion may be added or eliminated by the commander, dependent upon the need which has been demonstrated.) The comparable Air Force staff is found in the wing and larger units, with sections designated Personnel, Operations, etc. G-2 Air and G-3 Air are Army officers assigned to G-2 or G-3 at division, corps, and Army headquarters level, who assist in planning and coordinating joint operations of ground and air units. Naval staffs ordinarily are not organized on these lines, but, when they are, they are designated N-1, N-2, etc. Similarly, a joint staff may be designated J-1, J-2, etc. In Army brigades and smaller units and in Marine Corps units smaller than a brigade or aircraft wing, staff sections are designated S-1, S-2, etc., with corresponding duties; referred to as a unit staff in the Army and as an executive staff in the Marine Corps. See also **staff.**

general war. Armed conflict between major powers in which the total resources of the belligerents are employed, and the national survival of a major belligerent is in jeopardy.

generation. In any technical or technological development, as of missile, jet engine, or the like, a stage or period that is marked by features of performances considered to be primitive, sophisticated, maturing, or matured, as in "the first generation of rockets used liquid propellants."

genetic effect of radiation. Inheritable changes, chiefly mutations, produced in living organisms by the absorption of **ionizing radiations.** On the basis of present knowledge these effects are purely additive, and there is no recovery.

geo. A prefix meaning earth, as in *geology, geophysics.*

Some writers use the established terms such as *geology* to refer to the same concept on other bodies of the solar system, as the *geology of Mars,* rather than *areology* or *marsology, geology of the Moon,* rather than *selenology.*

geocentric. Relating to or measured from the center of the Earth; having, or relating to, the Earth as a center.

geocorona. 1. The shell of hydrogen surrounding the earth at the limit of the atmosphere.

In Shlovsky's system of nomenclature, the geocorona includes the metasphere, the outer, fully ionized zone, and the protosphere, the inner zone.

2. Van Allen radiation belts.

The use of *geocorona* in sense **2** should be discouraged, since it conflicts with the relatively well-established usage in sense **1.**

geodesic line. The shortest line on the curved surface of the Earth between two points.

geodetic. Pertaining to, or determined by the branch of mathematics which determines the exact position of points and the figures and areas of large portions of the Earth's surface, or the shape and size of the Earth, and the variations of terrestrial gravity.

geodetic coordinates. Quantities which define the position of a point on the **spheroid** of reference with respect to the planes of the geodetic equator and of a reference meridian. Compare **geographic coordinates.**

geodetic equator. That **great circle** midway between the poles of revolution of the earth, connecting points of 0° geodetic latitude. See **astronomical equator.**

geodetic latitude. Angular distance between the plane of the **equator** and a normal to the **spheroid.** It is the **astronomical latitude** corrected for the meridional component of the **deflection of the vertical.** Also called *geographic latitude, topographical latitude.*

This is the latitude used for charts.

geodetic longitude. The angle between the plane of the reference **meridian** and the plane through the polar axis and the normal to the **spheroid.** It is the **astronomical longitude** corrected for the prime vertical component of the **deflection of the vertical** divided by the cosine of the **latitude.** Also called *geographic longitude.*

This is the longitude used for charts.

geodetic lines. Lines upon a curved surface which follows the path of the shortest distance between any two points on the surface.

geodetic meridian. A line connecting points of equal **geodetic longitude.** Also called *geographic meridian.* See **astronomical meridian.**

geodetic parallel. A line connecting points of equal **geodetic latitude.** Also called *geographic parallel.* See **astronomical parallel.**

geodetic position. A position of a point on the surface of the earth expressed in terms of **geodetic latitude** and **geodetic longitude.**

A geodetic position implies an adopted geodetic **datum,** which must be stated for a complete record of the position.

geodetic survey. 1. A survey which takes into account the size and shape of the earth. **2.** An organization engaged in making geodetic surveys, sense **1.**

geographical meridian. The meridians of the earth, which pass through the earth's **geographic poles.** Sometimes called **true meridian.**

geographical mile. The length of 1 minute of arc of the equator, or 6087.08 feet.

geographical poles. The points where the axis about which the earth revolves meets the surface of the earth.

geographical position. 1. That point on the earth at which a given **celestial body** is in the **zenith** of a specified time.

The geographical position of the sun is also called the *subsolar point,* of the moon the *sublunar point,* and of a star the *substellar* or *subastral point.*

2. Any position on the earth defined by means of its **geographic coordinates,** either astronomical or geodetic.

geographic coordinates. Coordinates defining a point on the surface of the earth, usually **latitude** and **longitude.** Also called *terrestrial coordinates, geographical coordinates. See* **coordinate,** *for relationship between* **geographic coordinates** *and* **celestial coordinates.**

Geographic coordinates can refer to either astronomical or geodetic coordinates.

geographic latitude. Same as **geodetic latitude.**

geographic longitude. Same as **geodetic longitude.**

geoid. The figure of the earth as defined by the **geopotential surface** which most nearly coincides with mean sea level over the entire surface of the earth.

Because of variations in the direction of gravity, to which it is everywhere perpendicular, the geoid is not quite an **ellipsoid of revolution,** the sea-level surface being higher under mountainous areas. Compare **equilibrium spheroid, geosphere.**

geoidal horizon. That circle of the **celestial sphere** formed by the intersection of the celestial sphere and a plane through a point on the **geoid** perpendicular to the zenith-nadir line. See **horizon.**

geomagnetic. Of or pertaining to **geomagnetism.**

geomagnetic coordinates. A **system of coordinates** based on the best fit of a centered **dipole** to the actual **magnetic field** of the earth.

geomagnetism. 1. The **magnetic** phenomena, collectively considered, exhibited by the earth and its atmosphere and by extension of the magnetic phenomena in interplanetary **space. 2.** The study of the **magnetic field** of the earth. Also called *terrestrial magnetism.*

geomagnetic pole. Either of two antipodal points marking the intersection of the earth's surface with the extended axis of a **dipole** assumed to be located at the center of the earth and approximating the source of the actual **magnetic field** of the earth.

That pole in the Northern Hemisphere (latitude, 78 1/2° N; longitude, 69° W) is designated *north geomagnetic pole,* and that pole in the Southern Hemisphere (latitude, 78 1/2° S, longitude, 111° E) is designated *south geomagnetic pole.* The great circle midway between these poles is called *geomagnetic equator.* The expression geomagnetic pole should not be confused with magnetic pole, which relates to the actual magnetic field of the earth. See **geomagnetic latitude.**

geophysics. The physics of the earth and its environment, i.e., earth, air, and (by extension) space.

Classically, geophysics is concerned with the nature of and physical occurrences at and below the surface of the earth including, therefore, geology, oceanography, geodesy, seismology, hydrology, etc. The trend is to extend the scope of geophysics to include meteorology, geomagnetism, astrophysics, and other sciences concerned with the physical nature of the universe.

geopotential. The **potential energy** of a unit mass relative to sea level, numerically equal to the work that would be done in lifting the unit mass from sea level to the height at which the mass is located; commonly expressed in terms of **geopotential height.**

geopotential height. The height of a given point in the atmosphere in units proportional to the potential energy of unit mass **(geopotential)** at this height, relative to sea level.

geopotential surface. A surface of constant geopotential, i.e., a surface along which a particle of matter could move without undergoing any changes in its **potential energy.** Also called *equigeopotential surface, level surface.*

Geopotential surfaces almost coincide with surfaces of constant geometric height. Because of the poleward increase of the acceleration of gravity along a constant geometric-height surface, a given geopotential surface has a smaller geometric-height over the poles than over the equator. See **potential, geopotential height.**

georef (*abbr*) **World** *Geo***graphic** *Ref***erence System.** Pronounced as a word.

geosphere. The solid and liquid portions of the earth; the **lithosphere** plus the **hydrosphere.** Compare **geoid.**

Above the geosphere lies the atmosphere and at the interface between these two regions is found almost all of the biosphere, or zone of life.

geostropic wind. A wind, the direction of which is determined by the deflective force due to the rotation of the earth.

geostrophic wind level. The lowest level at which the wind becomes **geostrophic** in the theory of the **Ekman spiral.**

In practice it is observed that the geostrophic wind level is between 3000 and 5000 feet and it is assumed that this marks the upper limit of frictional influence of the earth's surface.

The geostrophic wind level may be considered to be the top of the **Ekman layer** and **planetary boundary layer,** i.e., the base of the **free atmosphere.**

get. To remove gas from a vacuum system by **sorption.**

getter. 1. A material which is included in a vacuum system or device for removing as by **sorption. 2.** To remove gas by **sorption.** Also called *get.*

G-force. An **inertial force** usually expressed in multiples of terrestrial **gravity.**

giant planets. The **planets** Jupiter, Saturn, Uranus, and Neptune. Also called *Jovian planets.*

giga (*abbr* G). A prefix meaning multiplied by 10^9.

gimbal. 1. A device with two mutually perpendicular and intersecting **axes of rotation,** thus giving free angular movement in two directions, on which an engine or other object may be mounted. **2.** In a **gyro,** a support which provides the **spin axis** with a **degree of freedom. 3.** To move a **reaction engine** about on a gimbal so as to obtain pitching and yawing **correction moments. 4.** To mount something on a gimbal.

gimbal freedom. In a **gyro,** the maximum angular displacement about the output axis of a gimbal.

It is expressed in degrees of output angle or in equivalent angular input.

gimbal lock. A condition of a two-degree-of-freedom **gyro** wherein the alinement of the spin axis with an **axis of freedom** deprives the gyro of a **degree of freedom,** and therefore of its useful properties.

gimballed motor. A rocket motor mounted on a gimbal, i.e., on a contrivance having two mutually perpendicular axes of rotation, so as to obtain pitching and yawing correction moments.

G-indicator. 1. G-display. **2.** A display that shows the amount of **inertial force** acting on a body.

glaze. A coating of ice, generally clear and smooth, but containing some air pockets. Synonymous with clear ice, it forms on exposed objects by the freezing of super-cooled water deposited by rain, drizzle, fog, or possibly condensed from supercooled water vapor.

glide. 1. A controlled descent by a heavier-than-air aeronautical vehicle under little or no engine thrust in which forward motion is maintained by **gravity** and vertical descent is controlled by **lift** forces. **2.** A descending flight path of a glide, sense **1,** as, a shallow *glide.* **3.** To descend in a glide, sense **1.**

glide bomb. A bomb fitted with airfoils to provide lift, carried and released in the direction of a target by an airplane.

glide path. 1. The **flight path** of an aeronautical vehicle in a **glide,** seen from the side. **2.** The path used by an aircraft or spacecraft in approach procedure and which is generated by an instrument-landing facility, such the electronic glide path componet of the **instrument landing system (ILS).**

The UHF glide path transmitter, operating on one of the twenty ILS channels within the frequency range 329.3 MHz to 335.0 MHz radiates its signals principally in the direction of the final approach.

The glide path transmitter is located between 750' and 1250' from the approach end of the runway (down the runway) and offset 400—600' from the runway centerline. It transmits a glide path beam 1.4° wide.

The glide path projection angle is normally adjusted to 2.5 degrees to 3 degrees above horizontal so that it intersects the middle marker at about 200 feet and the outer marker at about 1400 feet above the runway elevation.

glider. A heavier-than-air aircraft, that is supported in flight by the dynamic reaction of the air against its lifting surfaces and whose free flight does not depend principally on an engine.

glide ratio. The ratio of the horizontal distance traveled to the vertical distance descended in a **glide.** Also called *gliding ratio.*

glide slope. 1. An inclined surface which includes a **glide path** and which is generated by an instrument-landing facility. **2. slope angle. 3. gliding angle.**

gliding angle. The angle between the horizontal and the **glide path** of an aircraft. Also called *glide angle* or *glide slope.*

global commands (U. S. Air Force). Those major commands with units in both the Continental United States and in oversea areas, i.e., Aerospace Defense Command. Air Force Communications Service, Air Force Systems Command, Military Airlift Command, Strategic Air Command, United States Air Force Security Service.

global radiation. The total of **direct solar radiation** and **diffuse sky radiation** received by a unit horizontal surface.

Global radiation is measured by **pyranometers.**

global velocities. The range of velocities, slightly less than **circular velocity,** that permit sustained flight once around the earth in equilibrium glide. Compare **orbital velocity.**

glove box (nucleonics). A sealed box in which workers, using gloves attached to and passing through openings in the box, can handle radioactive materials safely from the outside.

glow discharge. An **electrical discharge** which produces **luminosity.**

Thus **corona discharge** is a glow discharge, but **point discharge** is not. Relatively high electric field strengths are required for glow discharges, for the density of radiatively recombining gas atoms and molecules must be high. See **gaseous electronics.**

G-meter. A meter that indicates **acceleration.**

GMT (abbr) *Greenwich mean time.*

goldie (military). The term, peculiar to air support radar team operations, indicating aircraft automatic flight control system and ground control bombing system are engaged and awaiting electronic ground control commands.

goldie lock (military). The term, peculiar to air support radar team operations, indicating ground controller has electronic control of the aircraft.

goniometer. An instrument for measuring **angles.**

go-no-go. The condition or state of operability of a component or system: "go," functioning properly; or "no-go," not functioning properly.

gox. Gaseous oxygen.

gradient. l. The space rate of decrease of a function. Of a function in three space dimensions, the **vector** normal to surfaces of constant value of the function and directed toward decreasing values, with magnitude equal to the rate of decrease of the function in this direction. **2.** Often loosely used to denote the **magnitude** of the gradient or ascendant. **3.** Either the rate of change of a quantity (as temperature, pressure, etc.) or a diagram or curve representing this.

gradient wind. A wind of the velocity which is necessary to balance the **pressure gradient.** The direction of the "gradient wind" is along the isobars, and the velocity is so adjusted that there is equilibrium between the force pressing the air toward the region of low pressure, and the centrifugal action to which the moving air is subject in consequence of its motion.

grain (unit of weight). 1/7000 of a pound avoirdupois, also equal to .0648 grams.

grain. 1. An elongated molding or extrusion of **solid propellant** for a **rocket,** regardless of size. 2. In photography, a small particle of metallic silver remaining in a photographic emulsion after development and fixing. In the agglomerate, these grains form the dark area of a photographic image. 3. An individual crystal in a polycrystalline metal or alloy.

gram. The principal unit of measure for weight in the metric system of measurements.

gram-centimeter. The CGS gravitation unit of work.

gram-molecule. The mass in grams of a substance numerically equal to its **molecular weight.**

grand slam. All enemy aircraft originally sighted are shot down.

graph. A diagram indicating the relationship between two or more **variables.**

Grashof number. A nondimensional parameter used in the theory of **heat transfer.**

The Grashof number is associated with the **Reynolds number** and the **Prandtl number** in the study of **convection.**

grass. 1. Sharp, closely spaced discontinuities in the trace of a **cathode-ray tube,** produced by random interference; so named because of their resemblance to blades of lawn grass. 2. In radar, a descriptive colloquialism used to refer to the indication of noise on an 'A' or similar type of **display.** See **noise.**

graticule. 1. The network of lines representing **parallels** and **meridians** on a map, chart, or plotting sheet. See **grid.** 2. A scale at the focal plane of an optical instrument to aid in the measurement of objects. See **reticle.**

gravireceptors. Highly specialized nerve endings and receptor organs located in skeletal muscles, tendons, joints, and in the inner ear which furnish information to the brain with respect to body position, equilibrium, and the direction of gravitational forces. See **gravitation.**

gravitation. The **acceleration** produced by the mutual attraction of two masses, directed along the line joining their centers of masses, and of magnitude inversely proportional to the square of the distance between the two centers of mass.

gravity. 1. Viewed from a frame of reference fixed in the earth, force imparted by the earth to a mass which is **at** rest relative to the earth. Since the earth is rotating, the force observed as gravity is the resultant of the force of **gravitation** and the **centrigugal force** arising from this rotation and the use of an earthbound rotating frame of reference. It is directed normal to sea level and to its geopotential surfaces.

The magnitude of the force of gravity at sea level decreases from the poles, where the centrifugal force is zero, to the equator, where the centrifugal force is a maximum but directed opposite to the force of gravitation. This difference is accentuated by the shape of the earth, which is nearly that of an oblate spheroid of revolution slightly depressed at the poles. Also, because of the asymmetric distribution of the mass of the earth,

the force of gravity is not directed precisely toward the earth's center.

The magnitude of the force of gravity is usually called either *gravity, acceleration of gravity, or apparent gravity.*

2. acceleration of gravity. 3. By extension, the attraction of any heavenly body for any mass; as *Martian gravity.*

gravity well. Analogy in which the gravitational field is considered as a deep pit out of which a space vehicle has to climb to escape from a planetary body.

gray body. A hypothetical body which absorbs some constant fraction, between zero and one, of all **electromagnetic radiation** incident upon it, which fraction is the **absorptivity** and is independent of **wavelength.** As such, a gray body represents a surface of absorptive characteristics intermediate between those of a **white body** and a **black body.** No such substances are known in nature. Also called *grey body.*

grayout. A temporary condition in which vision is hazy, restricted, or otherwise impaired, owing to insufficient oxygen. Compare **blackout.**

great circle. Any circle indicated on the surface of the earth by a plane surface which passes through the center of the sphere.

great circle course. The route between two points on the earth's surface that is measured along the shorter segment of the circumference of the great circle between the two points. A great circle course establishes the shortest distance over the surface of the earth between any two terrestrial points. Also see **course.**

great circle route. The route which follows the shortest arc of a great circle between two points.

Great Red Spot. An oval feature in the visible cloud surface of Jupiter, at latitude 20° to 25° S. It is about 25,000 miles long in the planet's east-west direction, and about 7000 miles wide in the north-south direction. It is often reddish in color, but may be white or grey, or nearly invisible compared to its surroundings.

great year. The period of one complete cycle of the **equinoxes** around the **ecliptic,** about 25,800 years. Also called *platonic year.*

green flash. A brilliant green coloring of the upper edge of the sun as it appears at sunrise or disappears at sunset when there is a clear, distinct horizon.

The green flash is due to **refraction** by the atmosphere, which disperses the first (or last) spot of light into a spectrum. The green is bent more than red or yellow and hence is visible sooner at sunrise and later at sunset.

greenhouse effect. The heating effect exerted by the atmosphere upon the earth by virtue of the fact that the atmosphere (mainly, its water vapor) absorbs and reemits **infrared radiation.** In detail: the shorter wavelengths of **isolation** are transmitted rather freely through the atmosphere to be absorbed at the earth's surface. The earth then reemits this as long-wave (infrared) **terrestrial radiation,** a portion of which is absorbed by the atmosphere and again emitted (see **atmospheric radiation**). Some of this is emitted downward back to the earth's surface (**counter-radiation**).

The mean surface temperature for the entire world, 14° C, is almost 40° C higher than the mean temperature required for radiative equilibrium of a black body at the earth's mean distance from the sun. It is essential, in understanding the concept of the greenhouse effect, to note that the important additional warming is due to the counterradiation from the atmosphere. The glass panes of a greenhouse function in this manner, hence the name.

Greenwich apparent time (*abbr* GAT). **Local apparent time** at the **Greenwich meridian;** the arc of the celestial equator, or the angle at the celestial pole, between the lower branch of the Greenwich celestial meridian and the hour circle of the apparent or true sun, measured westward from the lower branch of the Greenwich celestial meridian through 24 hours; **Greenwich hour angle** of the apparent or **true sun,** expressed in time units, plus 12 hours.

Greenwich civil time (*abbr* GCT) **Greenwich mean time.** (United States terminology from 1925 through 1952).

Greenwich hour angle (*abbr* GHA). Angular distance west of the Greenwich **celestial meridian;** the arc of the celestial equator, or the angle at the celestial pole, between the upper branch of the Greenwich celestial meridian and the hour circle of a point on the **celestial sphere,** measured westward from the Greenwich celestial meridian through 360°; local hour angle at the Greenwich meridian.

Greenwich mean time (*abbr* GMT). **Local mean time** at the **Greenwich meridian;** the arc of the celestial equator, or the angle at the celestial pole, between the lower branch of the Greenwich celestial meridian and the hour circle of the mean sun, measured westward from the lower branch of the Greenwich celestial meridian through 24 hours; **Greenwich hour angle** of the **mean sun,** expressed in time units, plus 12 hours. Called *Greenwich civil time* in U. S. terminology from 1925 through 1952. Also called *universal time, Z-time.*

Mean time reckoned from the upper branch of the Greenwich meridian is called *Greenwich astronomical time.*

Greenwich meridian. The meridian through Greenwich, England, serving as the reference for Greenwich time.

The Greenwich meridian is accepted almost universally as the prime meridian, or the origin of measurement of **longitude.**

Greenwich sidereal time (*abbr* GST). **Local sideral time** at the **Greenwich meridian;** the arc of the celestial equator, or the angle at the celestial pole, between the upper branch of the Greenwich celestial meridian and the hour circle of the vernal equinox, measured westward from the upper branch of the Greenwich celestial meridian through 24 hours; **Greenwich hour angle** of the **vernal equinox,** expressed in time units.

Gregorian calendar. The calendar now in common use, in which each year has 365 days except leap years. See **calendar year.**

grid. 1. A series of lines, usually straight and parallel, superimposed on a chart or plotting sheet to serve as a directional reference for navigation. See **graticule.** 2. Two sets of mutually perpendicular lines dividing a map or chart into squares or rectangles to permit location of any point by a system of rectangular coordinates. See **military grid, world geographic referencing system.** 3. An **electrode** with one or more openings to permit passage of **electrons** or **ions.** It usually consists of a wire meshelectrode placed between the anode and cathode of an electron tube to serve as a control of the current flowing between them. 4. Pertaining to or measured from a reference grid, as *grid azimuth, grid latitude, grid meridian.*

gross thrust. The total **thrust** of a **jet engine** without deduction of the **drag** due to the momentum of the incoming air (ram drag).

The gross thrust is equal to the product of the mass rate of fluid flow and the velocity of the fluid relative to the nozzle, plus the product of the nozzle exit area and the difference between the exhaust pressure and ambient pressure.

gross weight. The total weight of an aircraft, rocket, etc., as loaded; specifically, the total weight with full crew, full tanks, payload, etc. Also called *take-off weight.* See **design gross weight.**

ground. 1. The earth's surface, especially the earth's land surface. Used in combination to form adjectives, as in *ground-to-air, ground-to-ground,* and *air-to-ground.* See **surface.** 2. The domain of nonflight operations that normally take place on the earth's surface or in a vehicle or on a platform that rests upon the surface, as in *ground support.* 3. electrical ground.

ground alert. The status in which aircraft on the ground/deck are fully serviced and armed, with combat crews in readiness to take off within a specified short period of time (usually 15 minutes) after receipt of a mission order.

ground avoidance radar. Same as **radar altimeter** and **radio altimeter.**

ground clutter. ground return.

ground control. A control tower position established to control taxiing aircraft and vehicular traffic operating on the airport movement area.

ground-controlled approach (*abbr* GCA). A ground **radar** system providing information by which aircraft approaches may be directed via radio communications.

ground-controlled intercept. (*abbr* GCI). A **radar** system by means of which a controller may direct an aircraft to make interception of another aircraft.

ground cushion. The phenomenon of powered lift experienced when proximity to another surface interferes with free flow of air from the vehicle so as to generate sufficient pressure under the vehicle to counterbalance its weight.

ground effect. The temporary gain in lift during flight at very low altitudes due to the compression of the air between the wings of an airplane and the ground.

ground effect machine. A machine which normally flies within the zone of the ground effect or ground cushion. See also **hovercraft.**

ground environment. 1. The environment that surrounds and affects a system or a piece of equipment that operates on the ground. 2. That system or part of a system, as of a guidance system, that functions on the ground, the aggregate of equipment, conditions, facilities, and personnel that go to make up a system, or part of a system, functioning on the ground.

ground fire. Small arms ground-to-air fire directed against aircraft.

ground fog. Shallow fog rarely extending more than a few feet above the surface of the ground and caused by cooling of the earth through **radiation.**

groundhandling equipment. Equipment on the ground used to move, lift, or transport a space vehicle, a rocket, or component parts.

Such equipment includes the gantry, the transporter, and the forklift.

ground instructor. A general term applicable to persons engaged in non-flight aviation education. See **ground in-**

structor certificate.

ground instructor certificate. A certificate of competency issued by the **FAA** to a person meeting the requirements of the applicable **Federal Aviation Regulations.**

ground liaison officer. An officer trained in offensive air support activities. Ground liaison officers are normally organized into parties under the control of the appropriate army commander to provide liaison to air force and naval units engaged in training and combat operations.

ground loop. An uncontrollable violent turn of an airplane while taxiing, or during the landing or take-off run.

ground observer team. Small units or detachments deployed to provide information of aircraft movements over a defended area, obtained either by aural or visual means.

ground proximity extraction system (military). A system of extracting cargo from an aircraft in flight. A hook is extended from the cargo pallet to engage a preplaced energy absorber which extracts the cargo while the aircraft is about five feet above ground level.

ground radar aerial delivery system (military). This airdrop is essentially a high altitude container delivery system airdrop. A special device delays full parachute deployment which reduces wind effect and increases accuracy.

ground readiness (military). That status wherein aircraft can be armed and serviced and personnel alerted to take off within a specified length of time after receiving orders.

ground return. Radar **echoes** reflected from the terrain. Also called *ground clutter, land return.*

Echoes from the sea are called *sea clutter* or *sea return.*

ground school. A school in which instruction is given to a student undergoing flight training.

ground speed (GS). Ground speed (GS) is the rate of motion of the airplane over the ground. It is the result of inter-action between the airplane's speed through the *air* (TAS) and the *wind* speed in their relative directions of motion. Ground speed must be computed for navigation purposes.

ground start. An *igition sequence* of a rocket's main stage, initiated and cycled through on the ground. Compare **air start, inflight start.**

In large rockets, the ground start may be fueled from pressurized tanks external to the rocket, permitting takeoff with the rocket's own internal propellant load intact.

ground-support equipment (*abbr* GSE). That equipment on the ground, including all implements, tools, and devices (mobile or fixed), required to inspect, test, adjust, calibrate, appraise, gage, measure, repair, overhaul, assemble, disassemble, transport, safeguard, record, store, or otherwise function in support of a rocket, space vehicle, or the like, either in the research and development phase or in an operational phase, or in support of the guidance system used with the missile, vehicle, or the like.

The GSE is not considered to include land or buildings; nor does it include the guidance-station equipment itself, but it does include the test and checkout equipment required for operation of the guidance-station equipment.

ground visibility. Prevailing horizontal visibility near the earth's surface as reported by an accredited observer.

groundwash. The outward flow of the power plant(s) derived windblast over the ground.

ground wave. A **radio wave** that is propagated over the earth and is ordinarily affected by the presence of the earth's surface and the **troposphere.** The ground wave includes all components of a radio wave over the earth except **ionospheric** and **tropospheric** waves. Compare **sky wave.**

The ground wave is refracted because of variations in the dielectric constant of the troposphere including the condition known as a **surface duct.**

ground zero. The point on the surface of land or water vertically below or above the center of a burst of a nuclear explosion. For a burst over or under water, the term *surface zero* is preferable.

group. 1. A flexible administrative and tactical unit composed of either two or more battalions or two or more squadrons. The term also applies to combat support and service support units. **2.** A number of ships and/or aircraft, normally a subdivision of a force, assigned for a specific purpose.

GSE (*abbr*) **ground-support equipment.**

g-suit or **G-suit.** A suit that exerts pressure on the abdomen and lower parts of the body to prevent or retard the collection of blood below the chest under positive **acceleration.** Compare **pressure suit.**

g-tolerance. A tolerance in a person or other animal, or in a piece of equipment, to an **acceleration** of a particular value.

guerrilla warfare. Military and paramilitary operations conducted in enemy-held or hostile territory by irregular, predominantly indigenous forces. See also **unconventional warfare.**

guidance. The process of directing the movements of an aeronautical vehicle or space vehicle, with particular reference to the selection of a flight path. See **control.**

In preset guidance a predetermined path is set into the guidance mechanism and not altered, in inertial guidance accelerations are measured and integrated within the craft, in command guidance the craft responds to information received from an outside source. Beam-rider guidane utilizes a beam; terrestrial-reference guidance, some influence of the earth; celestial guidance, the celestial bodies and particularly the stars; and homing guidance, information from the destination. In active homing guidance the information is in response to transmissions from the craft, in semiactive homing guidance the transmissions are from a source other than the craft, and in passive homing guidance natural radiations from the destination are utilized. Midcourse guidance extends from the end of the launching phase to an arbitrary point enroute and terminal guidance extends from this point to the destination.

guided air-to-surface missile. An air launched guided missile for use against surface targets.

guided missile. 1. Broadly, any **missile** that is subject to, or capable of, some degree of guidance or direction after having been launched, fired, or otherwise set in motion. **2.** Specifically, an unmanned, self-propelled flying vehicle (such as a pilotless aircraft or **rocket**) carrying a destructive load and capable of being directed or of directing itself after launching or take-off, responding either to external direction or to direction originating from devices within the missile itself. **3.** Loosely, by extension, any steerable projectile. See ballistic missile.

guided missile (air-to-air). An air-launched guided missile for use against air targets.

guided missile cruiser. These ships are full conversion of heavy cruisers. All guns are removed and replaced with missile launchers. The CGN is a nuclear-powered, long-range ship equipped with missile launchers.

guided missile destroyer. This destroyer type is equipped with guided missile launchers, improved naval gun battery, long-range sonar, and antisubmarine warfare weapons.

guided missile submarine. A submarine designed to have an additional capability to launch guided missile attacks from surfaced condition.

guided missile (surface-to-air). A surface-launched guided missile for use against air targets.

gun. A cannon with relatively long barrel, operating with relatively low angle of fire, and having a high muzzle velocity.

guns/weapons free. Fire may be opened on all aircraft not recognized as friendly.

guns/weapons tight. Do not open fire, or cease firing on any aircraft (or on bogey specified, or in section indicated) unless target(s) known to be hostile.

gust. A sudden brief increase in the force of the wind. Most winds near the earth's surface display alternate gusts and lulls. Also see **down current, up current.**

gust tunnel. A **wind tunnel** in which gusts are simulated.

Specifically, a tunnel in which models are passed over a vertical jet or jets simulating gusts.

gyro. 1. A device which utilizes the angular momentum of a spinning mass (rotor) to sense angular motion of its base about one or two axes orthogonal to the spin axis. Also called *gyroscope.*

This definition does not include more complex systems such as **stable platforms** using gyros as components.

2. Short for *directional gyro, gyrocompass,* etc.

gyro horizon. An **artificial horizon** or an **attitude gyro.**

gyro pickoff. A device which produces a **signal,** generally a voltage, as a function of the angle between two **gyro** gimbals or between a **gimbal** and the base.

gyrodyne. A rotorcraft whose rotors are normally engine-driven for takeoff, hovering, and landing, and for forward flight through part of its speed range, and whose means of propulsion, consisting usually of conventional propellers, is independent of the rotor system.

gyroplane. A rotorcraft whose rotors are not engine-driven except for initial starting, but are made to rotate by action of the air when the rotorcraft is moving; and whose means of propulsion, consisting usually of conventional propellers, is independent of the rotor system.

gyroscope. gyro.

gyroscopic inertia. The property of a rotor of resisting any force which tends to change its axis of rotation. See **gyro.**

H

H. Military mission designation for **search/rescue** aircraft. Also designates **helicopters.**

hail. Precipitation consisting of balls of irregular lumps of ice often of considerable size; a single unit of hail is called a hailstone. Large hailstones usually have a center surrounded by alternating layers of clear and cloudy ice. Hail falls almost exclusively in connection with thunderstorms. The largest hailstone observed in the United States was 17 inches in circumference and weighed 1 1/2 pounds.

hail storms. Storms occurring only with the large Cumulo Nimbus clouds and are due to rapid rising and falling air currents which quite often reach velocities of 30 to 40 miles per hour.

half-life (nucleonics). The time in which half the atoms of a particular radioactive substance disintegrate to another nuclear form. Measured half-lives vary from millionths of a second to billions of years. (**See decay, radioactive.**)

half-life, effective (nucleonics). The time required for a radionuclide contained in a biological system, such as a man or an animal, to reduce its activity by half as a combined result of radioactive decay and biological elimination. (Compare **biological half-life;** see **half-life.**)

hand and foot counter (nucleonics). A monitoring device arranged to give a rapid radiation survey of hands and feet of persons working with radioactive materials, to detect radioactive contamination. (**See counter, monitor, personnel monitoring, radioactive contamination.**)

handoff. The passing of control authority of an aircraft from one control agency to another control agency. Handoff action is complete when the receiving controller acknowledges assumption of control authority.

handover (tracks, aircraft). The passing of control authority of an interceptor from one control agency to another control agency. Handover action may be accomplished between control agencies of separate Services when conducting join operations or between control agencies within a single command and control system. Handover action is complete when the receiving air weapons controller acknowledges assumption of control authority.

hangar. A building or other suitable shelter for housing aircraft.

hangfire. A faulty condition in the **ignition system** of a **rocket engine.**

harassing (air). The attack of any target within the area of land battle not connected with interdiction or close air support. It is designed to reduce the enemy's combat effectiveness.

harassment. An incident in which the primary objective is to disrupt the activities of a unit, installation, or ship, rather than to inflict serious casualties or damage.

hardened site. A site constructed to withstand the blast and associated effects of a nuclear attack and likely to be protected against a chemical, biological, or radiological attack.

hard landing. An impact landing of a spacecraft on the surface of a **planet** or natural **satellite** destroying all equipment except possibly a very rugged package.

hard missile base. A launching base that is protected against a nuclear explosion.

hardness. Resistance of metal to plastic deformation usually by indentation. However, the term may also refer to stiffness or temper, or to resistance to scratching, abrasion, or cutting. Indentation hardness may be measured by various hardness tests, such as Brinnell, Rockwell, and Vickers.

hard radiation. Radiation of high penetrating power; that is, **radiation** of high **frequency** and short **wavelength.** A 10-centimeter thickness of lead is usually used as the criterion upon which the relative penetrating power of various types of radiation is based. Hard radiation will penetrate such a shield; soft radiation will not.

hardstand. 1. A paved or stabilized area where vehicles are parked. **2.** Open ground area having a prepared surface and used for storage of materiel.

hard vacuum. A very **high vacuum,** usually considered to be a pressure less than about 10^{-7} torr.

hardware. Physical equipment as contrasted to ideas or design that may exist only on paper.

hard wire telemetry. wire link telemetry.

harmonic. 1. An integral multiple or submultiple of a given **frequency;** a **sinusoidal** component of a periodic **wave. 2.** A **signal** having a frequency which is a harmonic (sense **1**) of the fundamental frequency.

harmonic motion. The projection of circular motion on a diameter of the circle of such motion. Simple harmonic motion is produced if the circular motion is of constant speed. The combination of two or more simple harmonic motions results in compound harmonic motion.

harness. An assembly of straps or cords worn by a parachutist or employed to suspend an inanimate load to which the parachute is attached.

haze. Fine dust or salt particles scattered through a portion of the atmosphere. Particles are so small that they cannot be seen individually, but they diminish horizontal visibility.

H-BOMB. A *hydrogen bomb.*

heading. The horizontal direction in which a craft is pointed, expressed as angular distance from a reference direction, usually from 0° at the reference direction clockwise through 360°. Heading is often designated as *true, magnetic, compass,* or *grid* as the reference direction is true, magnetic, compass, or grid north, respectively. Also see **direction.**

heading crossing angle. In air interception, the angular difference between interceptor heading and target heading at the time of intercept.

head pressure. Same as **inlet pressure.**

Headquarters Air Force Reserve. A separate operating agency under Headquarters United States Air Force with the procedural functions and responsibilities of a Major Command, to which all Air Force Reserve units organized to train and function as units, and certain individual Air Force Reservists, are assigned. AFRES serves as the field Command Headquarters in the management structure of the Air Force Reserve receiving technical guidance and direction from the Office of Air Force Reserve (AF/RE) Headquarters United States Air Force.

heads up. Enemy got through (part or all) or I am not in position to engage target.

head-to-foot acceleration. See **physiological acceleration.**

head up display. A device by means of which flight and weapon delivery information (air speed, altitude, heading, angle of attack, and target location) is projected onto a transparent screen in the pilot's line of sight through the aircraft's wind screen.

heat. **Energy** transferred by a **thermal** process. Heat can be measured in terms of the dynamical units of energy, as the erg, joule, etc., or in terms of the amount of energy required to produce a definite thermal change in some substance, as, for example, the energy required per degree to raise the temperature of a unit mass of water at some temperature (**calorie, Btu**).

heat balance. **1.** The equilibrium which exists on the average between the **radiation** received by **a planet** and its atmosphere from the sun and that emitted by the planet and atmosphere. That the equilibrium does exist in the mean is demonstrated by the observed long-term constancy of the earth's surface temperature. On the average, regions of the earth nearer the equator than about 35° latitude receive more energy from the sun than they are able to radiate, whereas latitudes higher than 35° received less. The excess of heat is carried from low latitudes to higher latitudes by atmospheric and oceanic circulations and is reradiated there. **2.** The equilibrium which is known to exist when all sources of heat gain and loss for a given region or body are accounted for. In general this balance includes advective, evaporative (etc.) terms as well as a radiation term.

heat engine. A system which receives **energy** in the form of heat and which, in the performance of an energy transformation, does work. See **thermodynamic efficiency, Carnot engine.** The atmosphere itself is a heat engine.

heat exchanger. Any device that transfers heat from one fluid (liquid or gas) to another or to the environment.

heating muff. A chamber, surrounding an exhaust pipe or manifold, to provide hot air.

heat of ablation. A measure of the effective heat capacity of an **ablating material,** numerically the heating rate input divided by the mass loss rate which results from **ablation.**

heat of fusion. See **latent heat.**

heat of sublimation. See **latent heat.**

heat of vaporization. See **latent heat.**

heat pulse. Specifically, the sudden rise and subsequent fall in the temperature of a vehicle on reentry.

heat shield. **1.** Any device that protects something from heat. **2.** Specifically, the protective structure necessary to protect a **reentry** body from **aerodynamic heating.** See **heat sink.**

heat sink. **1.** In thermodynamic theory, a means by which **heat** is stored, or is dissipated or transferred from the **system** under consideration. **2.** A place toward which the heat moves in a system. **3.** A material capable of absorbing heat; a device utilizing such a material and used as a thermal protection device on a **spacecraft** or **reentry** vehicle. **4.** In nuclear propulsion, any thermodynamic device, such as a **radiator** or **condenser,** that is designed to absorb the excess heat energy of the **working fluid.** Also called *heat dump.*

heat transfer. The transfer or exchange of **heat** by radiation, conduction, or convection within a substance and between the substance and its surroundings. Radiation represents the transfer of radiant energy from one region to another by electromagnetic waves, with or without an intervening medium. Conduction, or diffusuion of heat, implies the elastic impact of fluid molecules, without any net transfer of matter. Convection arises from the mixing of relatively large volumes of fluid because of the fluid motion and may be due either to local temperature inequalities (free convection) or to an applied pressure gradient (forced convection).

heat-transfer coefficient. **1.** The rate of **heat transfer** per unit area per unit temperature difference, a quantity having the dimensions of reciprocal length. **2.** A misnomer for **Nusselt number.**

heat treatment. Heating and cooling a solid metal or alloy in such a way as to obtain desired conditions or properties. Heating for the sole purpose of hot-working is excluded from the meaning of this definition.

heavenly body. Same as **celestial body.**

heavier-than-air aircraft. Any aircraft deriving its lift in flight chiefly from aerodynamic forces.

heaviside layer. Same as **E-layer.**

heavy hydrogen. See **deuterium.**

heavy water. (Symbol D_2O) Water containing significantly more than the natural proportion (one in 6500) or heavy hydrogen (*deuterium*) atoms to ordinary *hydrogen* atoms. Heavy water is used as a moderator in some reactors because it slows down neutrons effectively and also has a low cross section for absorption of neutrons.

height *(symbol h).* **1.** Vertical distance; the distance above some reference point or plane, as, height above sea level. See **altitude. 2.** The vertical dimension of anything the distance which something extends above its foot or root, as *blade height.*

height. (Military) The vertical distance of an object, point, or level above the ground or other established reference plane. Height may be indicated as follows:

very low—Below 500 feet (above ground level)
low—500 to 2,000 feet (above ground level)
medium—2,000 to 25,000 feet
high—25,000 to 50,000 feet
very high—Above 50,000 feet.

height above airport (HAA). Indicates the height of the **MDA** above the published airport elevation. This is published in conjunction with circling minimums during an **instrument approach.**

Height above touchdown (HAT). Indicates the height of the **DH** or **MDA** above the highest elevation in the touchdown zone. This is published in conjunction with straight-in minimums during an **instrument approach.**

helicopter. A rotorcraft that, for its horizontal motion, depends principally on its engine-driven rotors.

helicopter (H), (Military designation). A rotary-wing aircraft designed with the capability of flight in any plane, e.g., horizontal, vertical, or diagonal. See **military, aircraft types.**

helicopter assault force. A task organization combining helicopters, supporting units, and helicopter-borne troop units for use in helicopter-borne assault operations.

helicopter break-up point. A control point at which helicopters returning from a landing zone break forma-

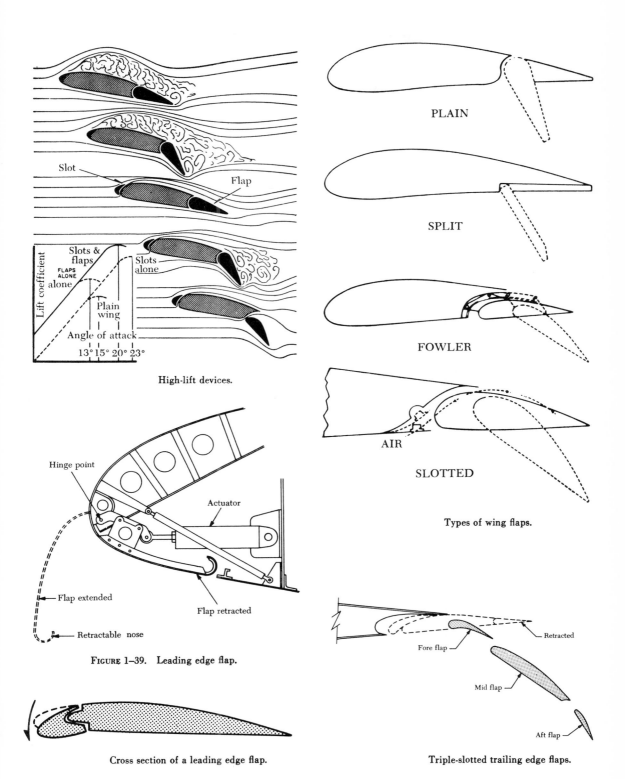

Slot

Flap

Slots & flaps

FLAPS ALONE

Slots alone

Lift coefficient

Plain wing

alone

Angle of attack

13° 15° 20° 23°

High-lift devices.

PLAIN

SPLIT

FOWLER

AIR

SLOTTED

Types of wing flaps.

Hinge point

Actuator

Flap extended

Flap retracted

Retractable nose

FIGURE 1–39. Leading edge flap.

Retracted

Fore flap

Mid flap

Aft flap

Cross section of a leading edge flap.

Triple-slotted trailing edge flaps.

High lift devices

tion and are released to return to base or are dispatched for other employment.

heliocentric. Relative to the sun as a center, as a *heliocentric orbit.*

heliocentric parallax. The difference in the **apparent positions** of a **celestial body** outside the solar system, as observed from the earth and sun. Also called *stellar parallax.* See **parallax.**

heliographic. Referring to positions on the sun measured in **latitude** from the sun's equator and in **longitude** from a reference meridian.

helipad. A prepared area designated and used for the takeoff and landing of helicopters. (Includes touchdown or hover point).

heliport. A facility designated for operating, basing, servicing, and maintaining helicopters.

Helmholtz function. A mathematically defined **thermodynamic function of state,** the decrease in which during a reversible **isothermal process** is equal to the work done by the system.

hemispherical. Referring to thermal radiation properties, in all possible directions from a flat surface.

henry. Unit of **inductance,** the inductance present which will cause one **volt** to be induced if the **current** changes at the rate of one **ampere** per second.

hertz *(abbr* **Hz**). The unit of **frequency,** cycles per second.

Hertzian waves. Electromagnetic waves of any **frequency** between 10 kilocycles per second and 300,000 megacycles per second. Now generally called *radio waves.* See **frequency bands.**

heterodyne. To mix two radio **signals** of different **frequencies** to produce a third signal which is of lower frequency; i.e., to produce **beating.** Radar receivers are of the heterodyne type (as contrasted to the superregenerative type) because the very high radio frequencies used in radar are difficult to amplify. A target signal is heterodyned with a current of lower frequency produced by a klystron oscillator and the resulting intermediate-frequency signal can them be highly amplified for subsequent presentation or analysis.

heterosphere. The upper portion of a two-part division of the **atmosphere** according to the general hemogeniety of atmospheric composition; the layer above the **homosphere.** The heterosphere is characterized by variation in composition and mean molecular weight of constituent gases. See **atmospheric shell.** This region starts at 80 to 100 kilometers above the earth, and therefore closely coincides with the ionosphere and the thermosphere.

H-hour. (Military) **1.** The specific hour on D-day on which hostilities commence. **2.** When used in connection with planned operations, it is the specific hour on which the operation commences.

hibernating spacecraft. A **spacecraft** maintaining an **orbit** without using **propellant** power and without maintaining orientation within the orbit, but with inherent power capability. A hibernating spacecraft could be in an orbit around the sun for months or years before power is triggered from a station on earth at an opportune time.

high. (Military) A height between twenty-five thousand and fifty thousand feet.

high. metgy. Often used when meaning a high pressure area.

high altitude. (Military) Conventionally, an altitude above 10,000 meters (33,000 feet). See also **altitude.**

high altitude bombing. Horizontal bombing with the height of release over 15,000 feet.

high altitude burst. The explosion of a nuclear weapon which takes place at a height in excess of 100,000 feet. See also **types of burst.**

high energy laser weapon. A **laser** beam provided with a large amount of energy. A laser beam delivers its **photon** "bullets" at the speed of light thus the lethal **flux** arrives at the target almost instantaneously, eliminating the need to lead the target.

high frequency *(abbr* **HF**). See **frequency bands.**

high frequency communications/HF communications. High radio frequencies (HF) between 3 and 30 MHz used for air-to-ground voice communication in overseas operations.

high or cirro clouds. High or cirro clouds include **cirrostratus** and **cirro-cumulus.** These clouds occur from 20,000 feet upwards. Also see **cloud, cummuliform clouds, stratiform clouds.**

high-lift devices. Methods employed to increase the lift of a basic wing. The increase in lift is utilized to secure a steeper glide path and slower landing speed and sometimes used to enable the plane to take off with a greater load. The most common forms are **wing flaps** and **wing slots.**

high-pass filter. A **wave filter** having a single transmission **band** extending from some critical or cutoff frequency, not zero, up to infinite frequency.

high pressure area. (**Meteorology**) A high pressure area is an area of high barometric pressure and caused by the downward movement of cool air. Such an atmospheric condition usually results in clear and good weather.

high vacuum. The condition in a gas-filled space at pressures less than 10^{-3} **torr.** The term *high vacuum* has frequently been defined as a pressure less than some upper limit. *High vacuum* (and similar vacuum terms) should not be defined as a pressure but rather as the condition or state in a gas-filled space at pressures less than some upper limit or within specified limits. The following classification of degrees of *high vacuum* has been proposed:

Condition	Pressure Range
high vacuum........10^{-3} to 10^{-6} torr (microtorr range)
very high vacuum...10^{-5} to 10^{-9} torr (nanotorr range)
ultrahigh vacuum...10^{-9} torr and below

high-wing monoplane. A monoplane in which the wings are located at or near the top of the fuselage.

hinge moment. The moment about the hinge axis of a control or other hinged surface due to aerodynamic forces.

Hohmann orbit. A minimum energy **transfer orbit.**

hold. 1. During a **countdown** to stop counting and to wait until an impediment has been removed so that the countdown can be resumed, as in *T minus 40 and holding.* Compare **count, recycle. 2.** In computer terminology, to retain information in one **storage** device after copying it into another storage device. **3.** As applied to air traffic, to keep an aircraft within a specified space or location which is identified by visual or other means in accordance with air traffic control instructions.

holddown test. The testing of some system or subsystem in a **rocket** while the rocket is firing but restrained in a **test stand.**

holding. A predetermined maneuver which keeps an aircraft within a specified airspace while awaiting further clearance. See **hold, sense 3.**

holding pattern. The flight track an aircraft is directed to follow at a **holding point.** See **holding.**

holding point. A specified location, identified by visual or other means, in the vicinity of which the position of an aircraft is maintained in accordance with air traffic control instructions. Also called holding fix.

home. To follow a path of energy **waves,** especially **radio** or **radar** waves, by means of a **directional antenna,** radar equipment, or other sensing devices, to or toward the point of transmission or reflection of the waves.

homer. Same as **homing beacon.**

homing. The following of a path of energy **waves** to or toward their source or point of reflection. See **home, active homing, passive homing, semiactive homing guidance.**

homing beacon. A beacon providing **homing guidance.** Also called *homer.* See **beacon, radio beacon (RBN).**

homing guidance. Guidance in which a craft or missile is directed toward a destination by means of information received from the destination. It is *active* homing guidance if the information received is in response to transmission from the craft, *semiactive* homing guidance if in response to transmissions from a source other than the craft, and *passive* homing guidance if natural radiations from the destination are utilized.

homing phase. The period of flight of a missile between the end of midcourse guidance and arrival in the vicinity of the target. See also **terminal guidance.**

homogeneous atmosphere. 1. A hypothetical **atmosphere** in which the density is constant with height. **2.** With respect to radio propagation, an atmosphere which has a constant **index of refraction,** or one in which radio waves travel in straight lines at constant speed. Free space is the ideal *homogeneous atmosphere* in this sense.

homopause. The top of the **homosphere,** or the level of transition between it and the **heterosphere.** See **atmospheric shell.** The homopause probably lies between 80 and 90 kilometers, where molecular oxygen begins to dissociate into atomic oxygen. The homopause is somewhat lower in the daytime than at night.

homosphere. The lower portion of a two-part division of the **atmosphere** according to the general homogeniety of atmospheric composition; opposed to the **heterosphere.** The region in which there is no gross change in atmospheric composition, that is, all the atmosphere from the earth's surface to about 90 kilometers. See **atmospheric shell.** The homosphere is about equivalent to the **neutrosphere,** and includes the troposphere, stratosphere, and mesosphere; it also includes the ozonosphere and at least part of the chemosphere.

honeycomb core. A lightweight strengthening material of a structure resembling a honeycomb mesh. See **sandwich construction.**

hood. a collapsible hood or opaque shield used in instrument flight instruction to prevent the person practicing instrument flight from seeing outside the cabin or cockpit.

hop. Travel of a **radio wave** to the **ionosphere** and back to earth. The number of hops a radio signal has experienc-

ed is usually designated by the expressions *one hop, two hop, multihop,* etc. The number of hops is called the *order of reflection.*

horizon. That **great circle** of the **celestial sphere** midway between the **zenith** and **nadir,** or a line resembling or approximating such a circle. That line where earth and sky appear to meet, and the projection of this line upon the celestial sphere, is called *visible* or *apparent horizon.* A line resembling the visible horizon but above or below it is called a *false horizon.* An artifical horizon is a device for indicating the horizontal. A gyro horizon is a gyroscopic instrument for indicating the attitude of an aircraft with respect to the horizontal. A radio horizon is the line at which direct rays from a transmitting antenna become tangent to the earth's surface. A radar horizon is the radio horizon of a radar antenna.

horizontal scanning. In **radar scanning,** rotating the antenna in **azimuth** around the horizon or in a sector. Also called **search-lighting.**

horizontal stabilizer. See **stabilizer.**

horizontal stratification. Uniform meteorological conditions at a given altitude, over the area under consideration.

horn antenna. A flared radiator used with a radar waveguide, which matches the impedance of the wave-guide to free air and also produces a directive pattern.

horn balance. A localized balance area at the tip of a control surface. This may be shielded, by a surface in front.

horsepower. A unit for measurement of power output of an engine. It is the power required to raise 550 pounds one foot in one second.

hostile. A contact postively identified as enemy.

hostile track. The classification assigned to a track which, based upon established criteria, is determined to be an enemy airborne, ballistic and/or orbiting threat.

hot cell. (Nucleonics.) A heavily shielded enclosure in which radioactive materials can be handled by persons using remote *manipulators* and viewing the materials through shielded windows or periscopes. (See **shield**).

hot test. A propulsion system test conducted by actually firing the **propellants.** Compare **cold-flow test.**

hour angle. Angular distance west of a **celestial meridian** or **hour circle;** the arc of the celestial equator, or the angle at the celestial pole, between the upper branch of a celestial meridian or hour circle and the hour circle of a **celestial body** or the **vernal equinox,** measured westward through 360°. Hour angle is usually further designated as *local, Greenwich,* or *sidereal* as the origin of measurements is the local or Greenwich celestial meridian or the hour circle of the vernal equinox. See **meridian angle.**

hour circles. Great circles of the **celestial sphere** formed by projecting the earth's meridians of longitude to the "celestial sphere". The "hour circle" which contains the zenith of an observer is called the Celestial meridian of the observer.

hourly report. See **sequence report.**

hover. To remain stationary relative to the air mass.

hovercraft. A vehicle that hovers or moves just above the surface by varying the direction and volume of a cushion of supporting air in conjunction with conventional propellor or jet thrust. See also **ground effect machine.**

hovering. A self-sustaining maneuver whereby a fixed, or nearly fixed position is maintained relative to a spot on the earth's surface.

hovering ceiling. The highest altitude at which the helicopter is capable of hovering in standard atmosphere. It is usually stated in two figures; hovering in ground effect and hovering out of ground effect.

hovering flight. Flight primarily supported by power plant(s) derived lift.

hull. The main structural and flotation body of a flying boat or amphibian.

human engineering. The application of knowledge of man's capabilities and limitations to the planning, design, development, and testing of aerospace systems, equipment, and facilities to achieve optimum personnel safety, comfort, and effectiveness compatible with systems requirements.

human factors. The study of psychophysical, psychological, and physiological variables which affect man's performance in an operational system. See **human engineering.**

humidity. That state of the atmosphere in reference to water vapor. The degree of humidity at any time in any one of the several ways following: **Relative humidity** is the ratio of the actual amount of vapor present to the saturation amount at the same temperature. It is expressed in per cent. **Absolute humidity** is the mass of water vapor present per unit volume. It may be expressed as so many grains per cubic goot, or its partial pressure may be given. **Specific humidity** expresses the mass of water vapor per unit mass of air. Specific humidity is considered as one of the least variable of air mass qualitites as it only changes when moisture is taken out of the air sample by precipitation.

hunt. 1. Of an aircraft, rocket, etc.,: to weave about its flightpath, as if seeking a new direction or another angle of attack, specifically, to **yaw** back and forth. **2.** Of a control surface: to rotate up and down or back and forth without being deflected by the pilot. **3.** Of a control system to oscillate about a selected value. **4.** Of an indicator on a display: to swing back and forth or to **oscillate,** especially rather slowly.

hunter-killer force. A naval force consisting of an antisubmarine warfare carrier, associated aircraft and escorts combining specialized searching, tracking, and attacking capabilities of air and surface antisubmarine warfare units operated as a coordinated group for the conduct of offensive antisubmarine operations in an area of submarine probability.

hunting. Fluctuation about a midpoint due to **instability,** as **oscillations** of the needle of an instrument about the zero point, or alternate lead and lag of a synchronous motor with respect to the alternating current.

hurricane. A tropical cyclone with wind speeds of 64 knots or greater.

hush house. A small hangar attached to a jet engine test cell which will house an entire fighter aircraft. This facility will control the noise level and permit the run-up testing of the jet engines without removal from the aircraft. It is used principally for fighter aircraft.

Huygens principle. A very general principle applying to all forms of **wave motion** which states that every point on the instantaneous position of an advancing **phase front** (wave front) may be regarded as a source of secondary spherical wavelets. The position of the phase front a moment later is then determined as the envelope of all the secondary wavelets (ad infinitum). This principle, stated by the Dutch physicist Christian Huygens (1629-

95), is extremely useful in understanding effects due to refraction, reflection, diffraction, and scattering of all types of radiation, including sonic radiation as well as electromagnetic radiation and applying even to ocean-wave propagation.

hydraulic system. 1. A particular method of transmitting power by means of liquid under pressure. **2.** Of an aircraft: the complete hydraulic installation.

hydrodynamics. The science or study of the laws of motion and action of fluids, especially water.

hydrofoil. A surface, similar in form to an aerofoil, on a seaplane or amphibian hull or float to facilitate take-off by providing hyrdodynamic lift.

hydrogen bomb. A nuclear weapon that derives its energy largely from *fusion.* (See **thermonuclear reaction.**)

hydrographic chart. A nautical chart showing depths of water, nature of bottom, contours of bottom and coastline, and tides and currents in a given sea or sea and land area.

hydroplaning. The condition in which moving aircraft tires are separated from a pavement surface by a water or liquid rubber film, or by steam resulting in a derogation of mechanical braking effectiveness.

hydrosphere. That part of the earth that consists of the oceans, seas, lakes, and rivers; a similar part of any other spatial body if such a body exists.

hydrostatic equilibrium. 1. The state of a **fluid** whose surfaces of constant pressure and constant mass (or density) coincide and are horizontal throughout. Complete balance exists between the force of **gravity** and the pressure force. **2.** Of a rotating body, a state in which the body maintains, or returns to, the figure generated by this rotation in spite of small disturbances.

hydyne. A hydrazine-base **liquid rocket fuel.** Also called *hidyne.*

hygrometer. An instrument used to determine the amount of **humidity** of the air.

hyperbaric. Pertaining to breathing atmosphere pressures above sea level normal.

hyperbarism. Disturbances in the body resulting from an excess of the **ambient pressure** over that within the body fluids, tissues, and cavities.

hyperbola. A conic section made by a plane intersecting a cone of revolution at an angle smaller than that of a parabola. The value of its eccentricity is greater than one.

hyperbolic. Of or pertaining to a **hyperbola.**

hyperbolic fix. A fix established by means of **hyperbolic lines of position.**

hyperbolic guidance. The guidance of a missile or the like in which radio signals, transmitted simultaneously from two ground stations, arrive at the guided object with a constant time difference, thereby establishing a hyperbolic path which the object follows.

hyperbolic line of position. A line of position in the shape of a **hyperbola,** determined by measuring the difference in distance to two fixed points. **Loran** lines of position are an example.

hyperbolic navigation. Radio navigation in which a **hyperbolic line of position** is established by signals received from two stations at a constant time difference.

hyperbolic system. A system where **lines of position** are determined from time or **phase** differences relative to two or more fixed stations which are the focuses of **hyperbolas.** In a three-dimensional system, the lines of position become hyperbolic surfaces of position.

hyperbolic velocity. A velocity sufficient to allow escape from the solar system. Comets unless captured by the sun have hyperbolic velocities and their trajectories are hyperbolas.

hypergolic propellants. Rocket **propellants** that ignite spontaneously when mixed with each other.

hyperoxia. A condition in which the total oxygen content of the body is increased above that normally existing at sea level.

hypersonic. 1. Pertaining to **hypersonic flow. 2.** Pertaining to speeds of Mach 5 or greater.

hypersonic flow. In aerodynamics, **flow** of a fluid over a body at speeds much greater than the speed of sound and in which the shock waves start at a finite distance from the surface of the body. Compare **supersonic flow.**

hypersonic glider. An unpowered vehicle, specifically a **reentry vehicle,** designed to fly at **hypersonic** speeds.

hypersonics. That branch of **aerodynamics** that deals with **hypersonic flow.**

hypervelocity. Extremely high velocity. Applied by physicists to speeds approaching the speed of light, but generally implies speeds of the order of **satellite** speed and greater.

hyperventilation. Overbreathing. A respiratory-minute volume, or pulmonary ventilation, that is greater than normal. Hyperventilation often results in an abnormal loss of carbon dioxide from the lungs and blood, which may lead to dizziness, confusion, and muscular cramps.

hyperventilation syndrome. The syndrome of blurring of vision, (feeling of) tingling of the extremities, faintness, and dizziness, which may progress to unconsciousness, and convulsions, caused by reduction of the normal car-

bon dioxide tension of the human body, due to increased **pulmonary ventilation.**

hypobaric. Pertaining to low atmospheric pressure, particularly the low atmospheric pressure of high altitudes.

hypobarism. Disturbances resulting from a decrease of **ambient pressure** to less than that within the body fluids, tissues, and cavities.

hypocapnia. Deficiency of carbon dioxide in the blood and body tissues, which may result in dizziness, confusion, and muscular cramps.

hypoventilation. A respiratory-minute volume, or **pulmonary ventilation,** that is less than normal. Also called *underbreathing.*

hypoxaemia. The condition of reduction of the normal oxygen tension in the blood. Also called *anoxaemia.*

hypoxia. Oxygen want or deficiency; any state wherein a physiologically inadequate amount of oxygen is available to, or utilized by, tissue without respect to cause or degree. Compare **anoxia.**

hypsometric tinting. A method of showing relief on maps and charts by coloring in different shades those parts which lie between different levels. Sometimes referred to as elevation tints, altitude tints, and layer tints.

hysteresis. 1. Any of several effects resembling a kind of internal friction, accompanied by the generation of heat within the substance affected. Magnetic hysteresis occurs when a ferromagnetic substance is subjected to a varying magnetic intensity; electric hysteresis occurs when a dielectric is subjected to a varying electric intensity. Elastic hysteresis is the internal friction in an elastic solid subjected to varying stress. **2.** The delay of an indicator in registering a change in a parameter being measured.

I

IAS. Indicated Airspeed. See **airspeed.**

ICAO *(abbr). International Civil Aviation Organization.* Usually pronounced as a word.

ICAO Standard Atmosphere. See **standard atmosphere.**

ice. See **clear ice, rime ice.**

ice frost. A thickness of ice that gathers on the outside of a rocket vehicle over surfaces supercooled by liquid oxygen inside the vehicle. This ice frost is quickly shaken loose and falls to the ground once the vehicle begins its ascent.

ice point. The temperature at which a mixture of air-saturated pure water and pure ice may exist in equilibrium at a pressure of one standard atmosphere. By decision of the Tenth General Conference on Weights and Measures, Paris, October 1954 the ice point was established as 273.15° K.

ice-up. The formation of ice on the leading edges of wings or any external surfaces of an airplane or the carburetor venturi, when passing through air of relative high humidity at freezing temperature.

ICSU *(abbr). International Council of Scientific Unions.* Usually pronounced as a word.

ideal gas. A gas which conforms to **Boyle law** and has zero heat of free expansion (or also obeys **Charles law**). Also called *perfect gas.*

ideal rocket. A theoretical rocket postulated for parameters that may be corrected in practice. An ideal rocket assumes a homogeneous and invariant propellant, an observance of perfect gas laws, a nofriction condition, a no-heat-transfer condition, a steady and constant propellant flow, an axially directed velocity of all exhaust gases, a uniform gas velocity across any section, and a chemical equilibrium established in the combustion chamber.

"ident" feature. The special feature in ATCRBS equipment and the "I/P" feature in certain military equipment used to distinguish one displayed select code from other codes.

identification. 1. In air defense and antisubmarine warfare, the process of determining the friendly or hostile character of a detected contact. **2.** In arms control, the process of determining which nation is responsible for the detected violations of an arms control measure.

Identification Friend or Foe. (IFF). A system using electronic transmissions to which equipment carried by friendly forces automatically responds, for example, by emitting pulses, thereby distinguishing themselves from enemy forces. It is a method of determining the friendly or unfriendly character of aircraft and ships by other aircraft or ships and by ground forces using electronic detection equipment and associated Identification Friend or Foe units. See also **selective identification feature.**

Identification Friend or Foe personal identifier. The discreet Identification Friend or Foe code assigned to a particular aircraft, ship, or other vehicle for identification by electronic means.

identification maneuver. A maneuver performed for identification purposes.

IFF. Identification Friend or Foe.

IFR. Instrument Flight Rules. The symbol used to designate Instrument Flight Rules as prescribed by Federal Aviation Regulations. See **IFR conditions.**

IFR conditions. Bad weather conditions which prevail generally when visibility or ceiling falls below those prescribed for **Visual Flight Rules.** Aircraft must be operated according to **Instrument Flight Rules** under these conditions. See **instrument flight.**

IFR over-the-top. With respect to the operation of aircraft, means the operation of an aircraft over-the-top of clouds on an IFR flight plan when cleared by air traffic control to maintain "VFR conditions" or "VFR conditions on top".

igneous meteor. In U.S. weather observing practice, a visible electrical discharge in the atmosphere. Compare **electrometeor.** Lightning is the most common and important type, but types of **corona discharge** are also included.

igniter. A device used to begin combustion, such as a spark plug in the combustion chamber of a jet engine, or a squib used to ignite the fuel in a rocket.

ignition delay. The time lapse occurring between the instance of an igniting action of a fuel and the onset of a specified burning reaction. Also called *ignition lag.*

ignore. In computer terminology, a **code** group or **character** which indicates that the associated information is to be disregarded.

Igor *(abbr). Intercept ground optical recorder.* A long-focal-length telescopic camera used to observe **attitude** and other details of a **rocket** in flight.

IGY *(abbr). International Geophysical Year.*

illuminance. The total **luminous flux** received on a unit area of a given real or imaginary surface, expressed in such units as the **footcandle, lux,** or **phot.** Illuminance is analogous to **irradiance,** but is to be distinguished from the latter in that *illuminance* refers only to light and contains the luminous efficiency weighting factor necessitated by the nonlinear wavelength-response of the human eye. Compare **luminous intensity.** The only difference between *illuminance* and *illumination* is that the latter always refers to light incident upon a material surface. A distinction should be drawn, as well, between *illuminance* and *luminance.* The latter is a measure of the light coming from a surface; thus, for a surface which is not self-luminous, luminance is entirely dependent upon the illuminance upon that surface and its reflection properties.

illumination. See **illuminance,** note.

ILS. Abbreviation for **instrument landing system.**

imagery. Collectively, the representations of objects reproduced electronically or by optical means on film, electronic display devices or other media.

imagery interpretation. The process of location, recognition, identification, and description of objects, activities and terrain represented on imagery.

imagery sortie. 1. One flight by one aircraft for the purpose of recording air imagery. **2.** All the imagery obtained by one air vehicle on one imagery sortie.

immediate mission request. A request for an air strike on a target which by its nature could not be identified sufficiently in advance to permit detailed mission coordination and planning.

immediate operational readiness. Those operations directly related to the assumption of an alert or quick-reaction posture. Typical operations include strip alert, airborne alert/indoctrination, no-notice launch of an alert force, and the maintenance of missiles in an alert configuration.

Immelmann turn. An airplane flight maneuver in which the airplane changes its direction 180 degrees and at the same time gains altitude by means of a half-loop followed by a half-roll to normal flight on top of the loop. Named after Max Immelmann, German World War I fighter pilot.

impact. 1. A single **collision** of one mass in motion with a second mass which may be either in motion or at rest. **2.** Specifically, the action or event of an object, such as a **rocket,** striking the surface of a planet or natural satellite, or striking another object; the time of this event, as in *from launch to impact.* **3.** To strike a surface or an object. **4.** Of a rocket or fallaway section: To collide with a surface or object, as in *the rocket impacted 10 minutes after launch.*

impact acceleration. The **acceleration** generated by very sudden starts or stops of a vehicle. The term is usually applied in the context of **physiological acceleration.**

impact area. The area in which a rocket strikes the surface of the earth or other celestial body. Used specifically in reference to the *impact area* of a rocket range.

impact line. One of two lines each on the outside of a destruct line and running parallel to it, which marks the outer limits of impact for a missile destroyed under destruct procedures.

impact microphone. An instrument that picks up the vibration of an object impinging upon another, used especially on space probes to record the impact of small meteoroids.

impact point. The point on the drop zone where the first parachutist or air-dropped cargo item should land.

impact predictor. A device which takes information from a **trajectory measuring system** and continuously computes the point (in **real time**) at which the rocket will strike the earth; based on the assumption that the rocket power is shut off at that instant and the remaining trajectory is **ballistic** in nature.

impact pressure. 1. That **pressure** of a moving **fluid** brought to rest which is in excess of the pressure the fluid has when it does not flow, i.e., **total pressure** less **static pressure.** Impact pressure is equal to dynamic pressure in incompressible flow, but in compressible flow impact pressure includes the pressure change owing to the compressibility effect. **2.** A measured quantity obtained by placing an open-ended tube, known as an impact tube or **pitot tube,** in a gas stream and noting the pressure in the tube on a suitable **manometer.** Since the pressure is exerted at a **stagnation point,** the impact pressure is sometimes referred to as the *stagnation pressure* or *total pressure.*

impeller. 1. A device that imparts motion to a fluid; specifically, in a **centrifugal compressor,** a rotary disk which, faced on one or both sides with radial vanes, accelerates the incoming fluid outward into a **diffuser.** Also called *impeller wheel.* **2.** That part of a centrifugal compressor comprising this disk and its housing.

impeller vane. Any one of the vanes on the **impeller** of a

centrifugal compressor, serving to take in air and accelerate it radially outward. Also called *impeller blade.* Compare **compressor blade.**

impeller wheel. The vaned rotary disk in a centrifugal compressor. Usually called the *impeller.*

implosion. The rapid inward collapsing of the walls of a vacuum system or device as the result of failure of the walls to sustain the **ambient pressure.**

implosion weapon. A weapon in which a quantity of fissionable material, less than a *critical mass* at ordinary pressure, has its volume suddenly reduced by compression (a step accomplished by using chemical explosives) so that it becomes supercritical, producing a nuclear explosion. (See **supercritical mass.**)

impulse. Mathematically, it is an unbalanced force acting on an object multiplied by the brief time during which the force acts.

impulse total. The product obtained by multiplying the thrust from the rocket motor or engine by the burning time in seconds.

impulse turbine. A type of turbine having rotor blades shaped such that the wheel is turned from the impact of the fluid against the blades, no pressure drop occurring across the blades. Compare **reaction turbine.**

inactive aircraft. Aircraft placed in storage, bailment, on loan outside the Defense establishment, or otherwise not available to the military Services.

inactive duty training. That training performed by Reservists while not on active duty for which point credit is authorized. It includes unit training assemblies, training periods, instruction, preparation of instruction, appropriate duties, equivalent training, military flying duty, and completion of correspondence courses through the United States Air Force Extension Course Institute or other approved programs.

inadvertant release. The unintentional release of external stores.

inbound traffic. (Military) Traffic originating in an area outside Continental United States destined for or moving in the general direction of Continental United States.

incandescence. Emission of light due to high temperature of the emitting material. Any other emission of light is called *luminescence.*

incentive type contract. A contract which may be of either a fixed price, or cost reimbursement nature, with a special provision for adjustment of the fixed price or fee. It provides for a tentative target price and a maximum price or maximum fee, with price or fee adjustment after completion of the contract for the purpose of establishing a final price or fee based on the contractor's actual costs plus a sliding scale of profit or fee which varies inversely with the cost but which in no event shall permit the final price or fee to exceed the maximum price or fee stated in the contract.

inch *(abbr* **in.)** Exactly 2.540 centimeters. Prior to July 1, 1959, the inch was 2.54005 centimeters although the conversion factor 2.540 has actually been in use in industry in the United States since 1933.

incidence. 1. Partial coincidence, as a circle and a tangent line. **2.** The impingement of a **ray** on a surface. See **angle of incidence.**

incident ray. A **ray** impinging on a surface.

incidents. Brief clashes or other military disturbances generally of a transitory nature and not involving protracted hostilities.

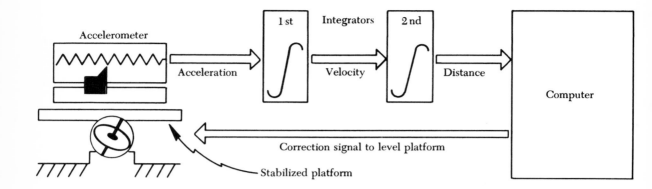

*A basic **inertial navigation** system.*

*An **inertial navigation** system is self-contained, requiring no outside visual, radio or radar contact. The lower photo shows the Carousel IV, manufactured by the Delco Electronics division of General Motors Corporation as installed in the Boeing 747 jet liner. This is similar to the guidance and navigation systems used for space navigation of the Apollo spacecraft on their lunar missions.*

inclination. 1. magnetic dip. 2. *(symbol i).* The angle between the plane of an **orbit** and a reference plane. The **equator** is the reference plane for geocentric orbits and the **ecliptic** is the reference plane for heliocentric orbits.

included angle. In aerodynamics, the angle between **free-stream** velocity and the **longitudinal axis** of the vehicle.

increment. Any arbitrary change (either plus or minus) in the value of an independent variable, i.e., any stepping up or down of quantities or magnitudes, etc.

independent variable. Any of those **variables** of a problem, chosen according to convenience, which may arbitrarily be specified, and which then determine the other or dependent variables of the problem. The independent variables are often called the *coordinates,* particularly in problems involving motion in space. Dependent and independent variables can be interchanged, e.g., height and pressure.

index of refraction. 1. A measure of the amount of **refraction** (a property of a **dielectric** substance.) It is the ratio of the wavelength or phase velocity of an **electromagnetic wave** in a vacuum to that in the substance. Also called *refractive index, absolute index of refraction, absolute refractive index, refractivity.* **2.** A measure of the amount of **refraction** experienced by a ray as it passes through a refractive interface, i.e., a surface separating two media of different densities. It is the ratio of the absolute indices of refraction of the two media (see sense **1** above). Also called *refractive index, relative index of refraction.*

indicated airspeed. See **airspeed.**

indicated altitude. Indicated altitude is the altitude read on the altimeter assuming that the altimeter is correctly adjusted to show the approximate height of the aircraft above mean sea level (MSL). This is accomplished by setting in the current barometric pressure. Altitudes assigned to aircraft in controlled airspace under Instrument Flight Rules (IFR) are *indicated altitudes,* except for flights operating in the high altitude route structure. Also see **altitude.**

indicator. A device which makes information available but in which there is no provision for storage of such information, as a *radar indicator.*

indirect air support. All forms of air support provided to land or naval forces which do not immediately assist those forces in the tactical battle.

indirect-cycle reactor system. A reactor system in which a *heat exchanger* transfers heat from the reactor *coolant* to a second fluid which then drives a turbine. (Compare **closed-cycle reactor system, direct-cycle reactor system.**)

induced drag. That part of the drag **induced** by the lift, i.e., it is the resistance which would be encountered if the air had no **viscosity.**

induced magnetism. Magnetism acquired by a piece of magnetic material while it is in a **magnetic field.** See **permanent magnetism.**

induced radioactivity. Nucleonics. Radioactivity that is created when substances are bombarded with neutrons, as from a nuclear explosion or in a reactor, or with charged particles produced by accelerators.

industrial mobilization. The transformation of industry from its peacetime activity to the industrial program necessary to support the national military objectives. It includes the mobilization of materials, labor, capital, production facilities, and contributory items and services essential to the industrial program.

industrial property. As distinguished from military property, means any contractor-acquired or Government-furnished property, including materials, special tooling and industrial facilities furnished or acquired in the performance of a contract or subcontract.

inelastic collision. A collision between two particles in which changes occur both in the internal energy of one or both of the particles and in the sums, before and after collision, of their **kinetic energies.**

inert atmosphere. A gaseous medium that because of its lack of chemical reaction is used to enclose tests or equipment.

inert gas. Any one of six gases, helium, neon, argon, kyrpton, xenon, and radon, all of whose shells of planetary electrons contain stable numbers of electrons so that the atoms are almost completely chemically inactive. Also called *rare gas.* All these gases are found in the earth's atmosphere but, with the exception of argon, are found only in very small amounts. Fluorine compounds of the rare gases have only recently been discovered.

inertia. Resistance to **acceleration.**

inertial axes. Axes that are not undergoing **acceleration** or **rotation.**

inertial coordinate system. A system in which the (vector) **momentum** of a particle is conserved in the absence of external forces. Thus, only in an inertial system can **Newton laws of motion** be appropriately applied. When **relative coordinate systems** are used, moving with respect to the inertial system, apparent forces arise in Newton laws, such as the **coriolis force.**

inertial force. A **force** in a given coordinate system arising from the **inertia** of a parcel moving with respect to another coordinate system. The inertial force is proportional and directionally opposite to the accelerating force. Also called *inertia force.* For example, the **coriolis acceleration** on a parcel moving with respect to a coordinate system fixed in space becomes an inertial force, the **coriolis force,** in a coordinate system rotating with the earth.

inertial guidance. Guidance by means of the measurement and integration of **acceleration** from within the craft.

inertial navigation. Dead reckoning performed automatically by a device which gives a continuous indication of position by integration of accelerations since leaving a starting point.

inertial navigation system. A self-contained navigational system capable of automatically determining dead-reckoning position through double integration of the outputs of accelerometers having pre-aligned gyrostabilized axes.

inertial orbit. The type of orbit described by all celestial bodies, according to **Kepler laws** of celestial motion. This applies to all satellites and spacecraft provided they are not under any type of propulsive power, their driving force being imparted by the momentum at the instant propulsive power ceases.

inertial space. An assumed stationary frame of reference. A nonrotating set of coordinates in space-relative to which the trajectory of a space vehicle or longrange missile is calculated.

inertial velocity. Velocity with respect to a fixed system of **coordinates.**

inertia starter. A device by which energy is stored in a small high-speed flywheel and, for starting, transmitted to the engine through a slipping clutch, the flywheel being energized either by hand or otherwise.

inferior planets. The planets with **orbits** smaller than that of the earth: Mercury and Venus.

infinity. 1. A point, line, or region, beyond measurable limits. A source of light is regarded as at infinity if it is at such a great distance that rays from it can be considered parallel. See **parallax. 2.** Any quantity larger than the largest quantity which can be stored in a register of a specific **computer.**

inflection. Reversal of direction of curvature. A point at which reversal takes place is called *point of inflection* or *inflection point.*

in-flight start. An engine **ignition sequence** after take-off and during flight. Compare **air start, ground start.** This term includes starts both within and above the sensible atmosphere.

information. An facts or data which can be used, transferred, or communicated.

infrahuman. A live animal other than man used as a substitute for a human in life-science experiments.

infrared *(abbr* **IR). 1. infrared radiation. 2.** Pertaining to **infrared radiation,** as an *infrared absorber.*

infrared countermeasures. Countermeasures used specifically against enemy threats operating in the infrared spectrum.

infrared detector. A thermal device for observing and measuring infrared radiation.

infrared guidance. A system for reconnaissance of targets and navigation using infrared heat sources.

infrared film. Film carrying an emulsion especially sensitive to "near-infrared". Used to photograph through haze, because of the penetrating power of infrared light, and in camouflage detection to distinguish between living vegetation and dead vegetation or artificial green pigment.

infrared imagery. That imagery produced as a result of sensing electromagnetic radiations emitted or reflected from a given target surface in the infrared portion of the electromagnetic spectrum (approximately 0.72 to 1,000 microns).

infrared photography. Photography employing an optical system and direct image recording on film sensitive to near-infrared wave length (infrared film). (*Note:* Not to be confused with infrared imagery.)

infrared radiation *(abbr* **IR). Electromagnetic radiation** lying in the wavelength interval from about 75 microns to an indefinite upper boundary sometimes arbitrarily set at 1000 microns (0.01 centimeter). Also called *long-wave radiation.* At the lower limit of this interval, the infrared radiation spectrum is bounded by visible radiation, whereas on its upper limit it is bounded by microwave radiation of the type important in radar technology. See **electromagnetic spectrum.** Whereas visible radiation is generated primarily by intra-atomic processes, infrared radiation is generated almost wholly by larger scale intramolecular processes, chiefly molecular rotations and internal vibrations of many types. Electrically symmertric molecules, such as the nitrogen and oxygen molecules which comprise most of the earth's atmosphere, are not capable of absorbing or emitting infrared radiation, but several of the triatomic gases, such as water vapor, carbon dioxide, and ozone, are infrared active and play important roles in the propagation of infrared radiation in the atmosphere. Since a black body at terrestrial temperature radiates with maximum intensity in the infrared spectrum (near

10 microns), there exists a complex system of infrared radiation currents within the earth's atmosphere.

infrared reconnaissance. Reconnaissance by use of the infrared dection principle.

infrared sensors. An electro-optical sensor which is capable of producing either photographic or TV type images from the thermal radiation viewing since no background illumination is required. In the Forward Looking Mode the scene is rapidly scanned in two dimensions to provide a visual image. In the Downward Looking Mode the scene is scanned only in the direction perpendicular to the line of flight. The forward motion of the aircraft is employed to provide scanning in the second dimension to produce a complete image.

infrasonic frequency. A frequency below the **audiofrequency** range. The word *infrasonic* may be used as a modifier to indicate a device or system intended to operate at an infrasonic frequency. The term *subsonic* was once used in acoustics synonymously with *infrasonic;* such usage is now discouraged.

infrasonic sound. Sound whose **frequency** is below the lower pitch limit, below about 15 cycles per second.

inherent stability. Stability of an aircraft due solely to the disposition and arrangement of its fixed parts, i.e., that property which causes it, when disturbed, to return to its normal attitude of flight without the use of controls or the interposition of any mechanical devices.

inhibitor. A substance bonded, taped, or dip-dried onto a solid propellant to restrict the burning surface and to give direction to the burning process.

initial approach. That part of the instrument approach procedure consisting of the first approach to the first navigational facility associated with the procedure, or to a predetermined fix. When not associated with an instrument approach procedure, that portion of the flight of an aircraft immediately prior to arrival over the airfield of destination or over the reporting point from which the final approach to the airfield is commenced.

initial mass. The take-off mass of a rocket.

initial nuclear radiation. Radiation emitted from the *fireball* of a nuclear explosive during the first minute (an arbitrary time interval) after detonation. (Compare **residual nuclear radiation.**)

initial operational capability. The first attainment of the capability to employ effectively a weapon, item of equipment, or system of approved specific characteristics, and which is manned or operated by an adequately trained, equipped, and supported military unit or force.

initial vector. The initial command heading to be assumed by an interceptor after it has been committed to intercept an airborne object.

injection. 1. The introduction of fuel, fuel and air, fuel and oxidizer, water, or other substance into an engine induction system or **combustion chamber. 2.** The time following launching when nongravitational forces (thrust, lift, and drag) become negligible in their effect on the **trajectory** of a rocket or spacecraft. **3.** The process of putting a spacecraft up to **escape velocity.**

injector. A device designed to introduce propellants into the combustion chamber of a rocket or aircraft engine.

inland search and rescue region. The inland areas of continental United States, except waters under the jurisdiction of the United States. See also **search and rescue region.**

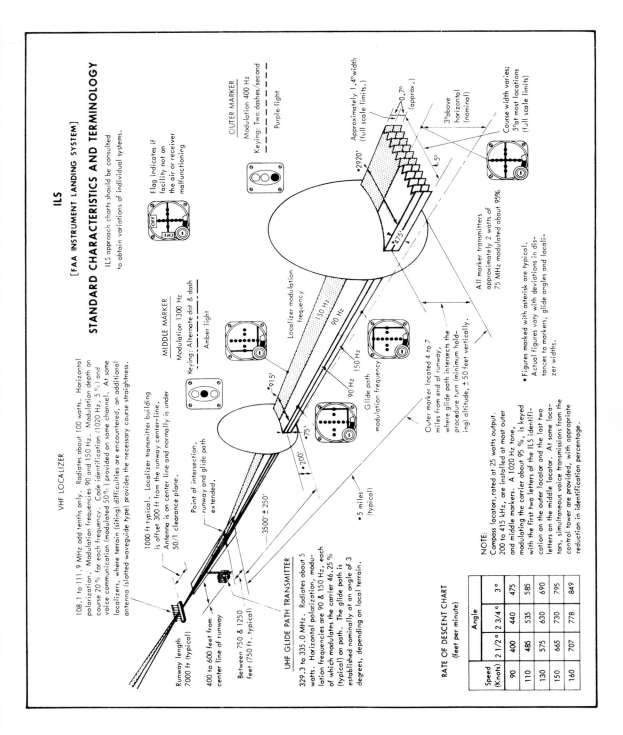

inlet. An entrance or orifice for the admission of fluid. Frequently used in compounds, such as *inlet air, inlet air temperature, inlet casing, inlet duct, inlet guide vane, inlet port, inlet valve,* etc.

inlet noise. The fan noise which propagates backward through the turbofan engine inlet.

inlet pressure. In connection with performance data on pumps, when not otherwise specified, the total **static pressure** measured in a standard testing chamber by a **vacuum gage** located near the inlet port. Also called *intake pressure, fine pressure, head pressure.*

inner liner. Specifically, a tube mounted coaxially inside the outer cover or shell of a **combustion chamber.** Also called a *flame tube* or a *combustion-chamber liner.*

inner planets. The four **planets** nearest the sun: Mercury, Venus, Earth, and Mars.

in phase. The condition of two or more **cyclic** motions which are at the same part of their cycles at the same instant. Also called *in step.* Two or more cyclic motions which are not at the same part of their cycles at the instant are said to be *out of phase* or *out of step.*

input axis. In a **gyro,** an **axis** normal to the **spin axis** about which a rotation of the base causes a maximum output as a function of this rotation.

insertion. The process of putting an **artificial satellite** or **spacecraft** into **orbit.**

insolation. (Contracted from *incoming solar radiation.*) **1.** In general, **solar radiation** received at the earth's surface. See **terrestrial radiation, extraterrestrial radiation, direct solar radiation, global radiation, effective terrestrial radiation, diffuse sky radiation, atmospheric radiation. 2.** The rate at which **direct solar radiation** is incident upon a unit horizontal surface at any point on or above the surface of the earth. Compare **solar constant.**

inspect and repair as necessary. (IRAN). A method of maintenance for aircraft.

inspection. In arms control, physical process of determining compliance with arms control measures.

instability. 1. The condition of a body if, when displaced from a state of equilibrium, it continues, or tends to continue, to depart from the original condition. Compare **stability. 2. Combustion instability.**

instability (Meteorology). See **unstable air.**

installation. A grouping of facilities, located in the same vicinity, which support particular functions. Installation may be elements of a base. See also **Base, Base Complex.**

instantaneous readout. Transmission of data by a radio transmitter instantaneous with the computation of data to be transmitted. See **readout station, real time.**

instruction. 1. Information which tells a **computer** where to obtain the **operands,** what operations to perform, what to obtain the next instruction. **2. command.**

instruction code. An artificial language for describing or expressing the instructions which can be carried out by a **digital computer.**

instrument. A device using an internal mechanism to show visually or aurally the attitude, altitude, or operation of an aerospace vehicle or part. It includes electronic devices for automatically controlling a vehicle in flight.

instrument. To provide a **vehicle** or **component** with **instrumentation.**

instrument approach. An approach to an airport of an aircraft by instruments and radio guidance, designed to break through an overcast within the required weather minimums and to permit a landing after break-through by visual contact.

instrument approach light systems. A group of high intensity and/or condenser-discharge flashing lights extending from the approach end of an instrument runway to provide a means for transistion from instrument flight using electronic approach aids to visual flight and landing. Also see **instrument landing system.**

instrument approach procedure. A series of predetermined maneuvers for the orderly transfer of an aircraft under instrument flight conditions from the beginning of the initial approach to a landing, or to a point from which a landing may be made visually.

instrumentation. 1. The installation and use of electronic, gyroscopic, and other instruments for the purpose of detecting, measuring, recording, telemetering, processing, or analyzing different values or quantities as encountered in the flight of an aircraft, a missile or spacecraft. Instrumentation applies to both flightborne and ground-based equipment. **2.** The assemblage of such instrument in an aerospace vehicle, each instrument designed and located so as to occupy minimum space, achieve minimum weight, yet function effectively. **3.** A special field of engineering concerned with the design, composition, and arrangement of such instruments.

instrument flight. Flight in which the path and attitude of the aircraft are controlled solely by reference to instruments. See **airman certificates. IFR (Instrument Flight Rules)**

Instrument Flight Rules (IFR). See **IFR (Instrument Flight Rules).**

instrument flight time. Time during which a pilot is piloting an aircraft solely by reference to instruments and without external reference points.

instrument flying hood. a collapsible hood or opaque shield used in instrument flight instruction to prevent the person practicing instrument flight from seeing outside the cabin or cockpit.

instrument landing system (ILS). The instrument landing system is designed to provide an approach path for exact alignment and descent of an aircraft on final approach to a runway. The ground equipment consists of two highly directional transmitting systems and, along the approach, three (or fewer) **marker beacons.** The directional transmitters are known as the **localizer** and **glide path** transmitters. The system may be divided functionally into three parts:

Guidance information—localizer, glide path
Range information—marker beacons
Visual information—approach lights, touchdown and centerline lights, runway lights

Compass locators located at the outer marker or middle marker may be substituted for these marker beacons. **DME** when located at the glide slope antenna may be substituted for the outer marker.

instrument landing system (ILS) categories.

ILS Category I—An ILS approach procedure which provides for approach to a height above touchdown of not less than 200 feet and with runway visual range of not less than 1800 feet.

ILS Category II—An ILS approach procedure which provides for approach to a height above touchdown of not less than 100 feet and with runway visual range of not less than 1200 feet.

ILS Category III.

IIIA—An ILS approach procedure which provides for approach without a decision height minimum and with runway visual range of not less than 700 feet.

IIIB—An ILS approach procedure which provides for approach without a decision height minimum and with runway visual range of not less than 150 feet.

IIIC—An ILS approach procedure which provides for approach without a decision height minimum and without runway visual range minimum.

instrument rating. A rating issued to a **private** or **commercial pilot** by the **FAA** upon meeting the requirements of the applicable **Federal Aviation Regulations.** An instrument pilot can exercise the privilege of piloting an aircraft under **IFR conditions. See instrument flight, airman certificates.**

instrument runway. A runway specially marked and lighted and served by a nonvisual navigation aid and intended for landings under instrument weather conditions.

integer. A whole number; a number that is not a fraction.

integral. 1. Of or pertaining to an **integer. 2.** Serving to form a whole or a part of a whole, as an *integral tank.*

integrally stiffened. Of structures, referring to thin-walled components in which increased section wall stiffeners and wall are formed as a single structural member rather than as two separate pieces.

integral tank. A fuel or oxidizer tank built within the normal contours of an aircraft of missile and using the skin of the vehicle as part of the walls of the tank.

integrated fire control system. (Military) A system which performs the functions of target acquisition, tracking, data computation, and engagement control, primarily using electronic means assisted by electromechanical devices.

integrated trajectory system *(abbr* **ITS).** A multiple **trajectory measuring system** composed of several angle-measuring-equipment and distance-measuring-equipment sites whereby in-flight selection of station combination can be made to provide the best geometrical solution to space position at any given time of rocket flight.

integrating accelerometer. A mechanical and electrical device which measures the forces of acceleration along the longitudinal axis, records the velocity, and measures the distance traveled.

integrator. 1. In **digital computers,** a device for accomplishing a **numeric** approximation of the mathematical process of integration. **2.** A device whose output is proportional to the integral of an input signal.

intelligence. (Military) The product resulting from the collection, evaluation, analysis, integration, and interpretation of all information concerning one or more aspects of foreign countries or areas, which is immediately or potentially significant to the development and execution of plans, policies, and operations.

input. 1. The path through which **information** is applied to any device. **2.** The means for supplying information to a machine. See **input equipment. 3.** Information or **energy** entering into a system. Compare **output. 4.** The quantity to be measured, or otherwise operated upon, which is received by an instrument. Also called *input signal.* For a thermometer, temperature is the input.

intensity. 1. In general, the degree or amount, usually expressed by the elemental time rate or spatial distribution of some condition or physical quantity, such as electric field, sound, magnetism, etc. **2.** With respect to

electromagnetic radiation, a measure of the **radiant flux** per unit solid angle emanating from some source. Frequently, it is desirable to specify this as *radiant intensity* in order to distinguish is clearly from *luminous intensity.* Compare **emittance.** Occasionally, *intensity* is used as synonymous to *flux density.* This usage does not coincide with accepted photometric and radiometric usage, but is of long standing in meteorology. See **sound intensity.**

intensity-modulated indicator. One of two general classes of radar **indicators,** in which **echoes** from **targets** are presented as spots or areas of light whose intensity or brilliance is normally a function of the power of the echo signal. Compare **amplitude-modulated indicator.** See **radarscope.**

ntensity modulation. The change of the brilliance (or intensity) of the **trace** on the screen of a **cathode-ray tube** in accordance with the strength of the applied signal.

intensive student jet training area. Airspace which contains the intensive training activities of military student jet pilots and in which restrictions are imposed on IFR flight.

intercardinal points. The directions, north-east, south-east, south-west, north-west.

intercepting search. A type of search designed to intercept an enemy whose previous position is known and the limits of whose subsequent course and speed can be assumed.

interceptor. A manned **fighter** aircraft utilized for identification and/or engagement of airborne objects.

interceptor controller. An officer who controls fighter aircraft allotted to him for interception purposes.

interceptor/fighter. See **fighter interceptor.**

interceptor missile. A missile designed to counter enemy offensive forces. See also **surface-to-air missile.**

intercept point. A computed point in space toward which an interceptor is vectored to complete an interception.

interchangeability. A condition which exists when two or more items possess such functional and physical characteristics as to be equivalent in performance and durability, and are capable of being exchanged one for the other without alteration of the items themselves or of adjoining items, except for adjustment, and without selection for fit and performance.

intercom. See **interphone.**

intercontinental ballistic missile. (**ICBM**) A ballistic missile with a range capability from about 3,000 to 8,000 nautical miles.

intercooler. A component installed on the delivery side of a supercharger or compressor to cool either the compressed air or the mixture.

interdict. To prevent or hinder, by any means, enemy use of an area or route.

interdiction fire. Fire placed on an area or point to prevent the enemy from using the area or point. See also **fire.**

interface. 1. A common boundary between two parts of a **system,** whether material or nonmaterial. **2.** Specifically, in a **rocket** vehicle or other mechanical assembly, a common boundary between two components. See **mating. 3.** Specifically, in fluid dynamics, a surface separating two fluids across which there is a discontinuity of some fluid property such as density or velocity or of some derivative of these properties in a direction normal to the interface. The equations of motion do not apply to the interface but are replaced by the **boundary**

conditions.

interference. **1.** Extraneous signals, noises, etc. that hinder proper reception of the desired **signal** in electronic equipment. See **babble, clutter, cosmic noise crosstalk, jitter, static. 2.** The mutual effect of two or more meeting waves or vibrations of any kind. Sometimes called *wave interference*. **3.** The aerodynamic influence of bodies on one another.

interference guard bands. The two **frequency bands** additional to and on either side of the authorized frequency band, which may be provided to minimize the possibility of **interference** between different radio channels.

interference region. That region in space in which **interference** between **wave trains** occurs. In microwave propagation, it refers to the region bounded by the ray path and the surface of the earth which is above the radio horizon. Interference lobes and height-gain patterns are formed in this region by the addition of the direct and the surface-reflected wave. In contrast is the diffraction zone which lies below the radio horizon.

interferometer. An apparatus used to produce and measure **interference** from two or more **coherent** wave trains from the same source. See **radio interferometer.** Interferometers are used to measure wavelengths, to measure angular width of sources, to determine the angular position of sources (as in satellite tracking), and for many other purposes.

interior ballistics. That branch of **ballistics** that deals with the propulsion of **projectiles,** i.e., the motion and behavior of projectiles in a gun barrel, the temperatures and pressures developed inside a gun barrel or rocket, etc. Sometimes called *internal ballistics.*

intermediate frequency. The beat frequency used in heterodyne receivers, usually the difference between the received radio-frequency signal and a locally generated signal.

intermediate orbit. An orbit tangent to an actual orbit and having the same coordinates but not the same velocity at the point of tangency.

intermediate-range ballistic missile. (IRBM) A ballistic missile with a range capability from about 1,500 to 3,-000 nautical miles.

intermediate staging base. A strategically located base, with support facilities where final preparations are made for airborne assault forces in support of an exercise or contingency.

intermittent pressure breathing. Pressure breathing in which different pressures are used at different points in the respiratory cycle, usually with a high pressure during inspiration and lower pressure during expiration.

intermodulation. The **modulation** of the **components** of a complex wave by each other in a nonlinear system.

internal efficiency. The **efficiency** with which a **reaction engine,** such as a rocket, converts the available thermal energy of its combustion gases into **kinetic energy** in the exhaust jet, expressed as a ratio.

International airport. Any airport designated by the Contracting State (Country) in whose territory it is situated as an airport of entry and departure for international air traffic, where the formalities incident to customs, immigration, public health, agricultural quarantine and similar procedures are carried out.

International air service. An air service which passes through the air space over the territory of more than one State (Country).

international arms control organization. An appropriately constituted organization established to supervise, and verify the implementation of arms control measures.

international call sign. A call sign assigned in accordance with the provisions of the International Telecommunications Union to identify a radio station. The nationality of the radio station is identified by the first or the first two characters. (When used in visual signaling, international call signs are referred to as signal letters.) See also **call sign.**

International Civil Aviation Organization (ICAO). An international organization set up to provide uniform standards for the regulation and safety of civil aviation and the efficient use of international airspace.

International Geophysical Year *(abbr* **IGY).** By international agreement, a period during which greatly increased observation of world-wide geophysical phenomena is undertaken through the cooperative effort of participating nations. July 1957 to December 1958 was the first such *year;* however, precedent was set by the International Polar Years of 1882 and 1932.

international standard atmosphere. ICAO. Standard Atmosphere; see **standard atmosphere.**

International System of Units *(abbr* **SI).** The metric system of units based on the **meter, kilogram, second, ampere, Kelvin degree,** and **candela.** Also called *MSKA system.* Other SI units are **hertz, radian, newton, joule watt, coulomb, volt, ohm, fared, weber,** and **tesla.**

interphone. (intercommunication) An electrically operated intercommunication system between various members of the crew of an aircraft or spacecraft.

interpreter. In computer terminology, a **circuit** or device which translates an instruction from **pseudocode** into an **instruction** or series of instructions which the **computer** can understand and obey.

interrogation. Transmission of a radio **signal** or combination of signals intended to trigger a **transponder** or group of transponders.

interrogator. 1. A radar set or other electronic device that transmits an **interrogation. 2.** An **interrogator-responsor** or the transmitting component of an interrogator-responsor. See **transponder, IFF, air traffic control radar beacon system.**

interrogator-responsor. A radio transmitter and receiver combined to **interrogate** a **transponder** and display the resulting replies. Often shortened to *interrogator* and sometimes called *challenger.*

intersection. In Bollean algebra, the operation in which concepts are described by stating that they have all the characteristics of the classes involved. *Intersection* is expressed as **AND.**

intersite communications. The communications systems and facilities employed by dispersed missile units to provide communications between the support base and the missile launch complexes, and between the dispersed launch complexes.

interstate air commerce. The carriage by aircraft of persons or property for compensation or hire, or the carriage of mail by aircraft, or the operation or navigation of aircraft in the conduct or furtherance of a business or vocation in commerce between a place in any State of the United States, or the District of Columbia, and a place in any other State of the United States, or the District of Columbia; or between places in the same State of the United States through the airspace over any place outside thereof; or between places in the same territory or possession of the United States, or the District of Columbia.

interstate air transportation. The carriage by aircraft of persons or property as a common carrier for compensation or hire, or the carriage of mail by aircraft, in commerce—(1) Between a place in a State or the District of Columbia and another place in another State or the District of Columbia; (2) Between places in the same State through the airspace of any place outside that State; or (3) Between places in the same possession of the United States; whether that commerce moves wholly by aircraft or partly by aircraft and partly by other forms of transportation.

intertropical front; metgy. The boundary between the trade wind systems of the northern and southern hemispheres. It manifests itself as a fairly broad zone of transition commonly known as the doldrums.

interval. 1. The space between adjacent aircraft measured from front to rear in units of time or distance.
2. The time lapse between photographic exposures.

intervalometer. Any device that may be set so as to accomplish automatically a series of like actions, such as the taking of photographs, or the closure of electrical circuits, at constant predetermined intervals.

in the dark. Not visible on my scope.

intruder operations. Offensive operations by day or night over enemy territory with the primary object of destroying enemy aircraft in the vicinity of their bases.

inversion (Meteorology). An abbreviation for "inversion of the vertical gradient of temperature." The temperature of the air is ordinarily observed to become lower with increasing height, but occasionally the reverse is the case, and when the temperature increases with height there is said to be an "inversion."

inverter. 1. A device for changing direct current to alternating current. **2.** In computers, a device or circuit which inverts the **polarity** of a **pulse.** Also called *NOT circuit.*

inviscid. Not viscous, not clinging or sticky; frictionless, as in *inviscid flow.*

ion. An atom or molecule that has lost or gained one or more *electrons.* By this *ionization* it becomes electrically charged. Examples: an alpha particle, which is a helium atom minus two electrons; a proton, which is a hydrogen atom minus its electron. (Compare **atom, elementary particles, molecule.**)

ion column. The trail of ionized gases in the trajectory of a **meteoroid** entering the upper atmosphere; a part of the composite phenomenon known as a **meteor.** A type of **meteor train.** See **meteor.** Compare **gas cap.**

ion engine. An engine which provides thrust by expelling accelerated or high velocity ions. Ion engines using energy provided by nuclear reactors are proposed for space vehicles.

ionic conduction. Any electrical **conduction** where the current is sustained by the motion of **ions** (as opposed to **electrons**) within the conductor. All electrical conduction in the atmosphere is of this type.

ionization. The process of adding one or more *electrons* to, or removing one or more electrons from atoms or molecules, thereby creating *ions.* High temperatures, electrical discharges, or nuclear radiations can cause ionization.

ionization gage. A **vacuum gage** with a means of ionizing the gas molecules and a means of correlating the number and type of **ions** produced with the pressure of the *gas.* Various types of ionization gage are distinguished according to the method of producing the ionization. Some common types are: **hot-cathode ionization gage; cold-cathode ionization gage; radioactive ionization gage.**

ionizing radiation. Any radiation displacing electrons from atoms or molecules, thereby producing *ions.* Examples: alpha, beta, gamma radiation, shortwave ultraviolet light. Ionizing radiation may produce severe skin or tissue damage. (See **radiation, radiation burn, radiation illness.**)

ionosphere. The region of the atmosphere, extending from roughly 40 to 250 miles altitude, in which there is appreciable ionization. The presence of charged particles in this region profoundly affects the propagation of electromagnetic radiations of long wavelengths (radio and radar waves). See also **atmospheric shell.**

ionospheric storm. Disturbance of the **ionosphere,** resulting in anomalous variations in its characteristics and effects on radio communication. See **sudden ionospheric disturbance.**

IR *(abbr).* **infrared** or **infrared radiation.**

irradiation. Exposure to radiation, as in a nuclear reactor. (See *spent fuel.*)

irreversible control system. A flight control in which the control surface can be moved freely by the pilot but cannot be moved by aerodynamic forces alone.

isentropic. Of equal or constant **entropy** with respect to either space or time.

isobar. A line on a chart or diagram drawn through places or points having the same barometric pressure. (Isobars are customarily drawn on weather charts to show the horizontal distribution of atmospheric pressure reduced to sea level or the pressure at some specified altitude).

isobaric charts. 1. Charts which show the distribution of pressure at any particular time, as those made daily to forecast weather. **2.** Those showing the average pressure for any particular period of time as for a month or a year. **3.** Charts showing the normal pressure for any period of time as a month, year, or season. On these charts the isobars are lines joining points of equal pressure.

isoclinic lines. Lines on a map or chart connecting points of equal magnetic dip, i.e., the **dip of a compass needle.**

isogonic lines. Lines on a map or chart which join locations of equal magnetic delclination.

isogram. See **isopleth.**

isolation. In **vibration** studies, a reduction in the capacity of a **system** to respond to an **excitation,** attained by the use of a resilient support.

isolator. Same as **vibration isolator.**

isopleth. On a chart or graph, a line of constant value of a given quantity with respect to either space or time. Also called *isogram.*

isotherm. A line connecting points of equal temperature on a weather map.

isothermal process. Any **thermodynamic change of state** of a system that takes place at constant temperature.

isotope. One of two or more atoms with the same *atomic number* (the same chemical element) but with different *atomic weights.* An equivalent statement is that the nuclei of isotopes have the same number of protons but different numbers of neutrons. Isotopes usually have very nearly the same chemical properties, but somewhat different physical properties. See **radioisotope.**

J

jacket. 1. A covering or casing of some kind. 2. Specifically, a shell around the **combustion chamber** of a **liquid-fuel rocket,** through which the propellant is circulated in **regenerative cooling.** 3. A coating of one material over another to prevent oxidation, micrometeoroid penetration, etc.

jamming. Intentional **transmission** or reradiation of radio **signals** in such a way as to interfere with reception of desired signals by the intended receiver.

JATO, Jato, or **jato.** (From *jet-assisted take-off*). 1. A take-off utilizing an auxiliary jet-producing unit or units, usually rockets, for additional.thrust. Hence *JATO bottle, Jato unit,* etc; a rocket or unit so used. Where rockets are the auxiliary units, RATO (which see) is the more specific term. 2. A JATO bottle or unit; the complete auxiliary power system used for assisted take-off.

jet. 1. A strong well-defined stream of **fluid** either issuing from an orifice or moving in a contracted duct, such as the jet of combusion gases issuing from a **reaction engine,** or the jet in the test section of a **wind tunnel.** See **free jet.** 2. A tube, nozzle, or the like through which fluid passes, or from which it issues, in a jet, such as a jet in a carburetor. See **metering jet.** 3. A jet engine, as, *an airplane with jets slung in pods.*

jet advisory service. The service provided certain civil aircraft while operating within radar and non-radar jet advisory areas. Within radar jet advisory areas, civil aircraft receiving this service are provided radar flight following, radar traffic information, and vectors around observed traffic. In non-radar jet advisory areas, civil aircraft receiving this service are afforded standard instrument flight rules separation from all other aircraft known to Air Traffic Control to be operating within these areas.

jetavator. A control surface that may be moved into or against a rocket's **jetstream,** used to change the direction of the jet flow for thrust vector control. Compare **jet vane.**

jet conventional low altitude bombing system. A maneuver used by jet aircraft to loft conventional ordnance by means of a low altitude bombing system.

jet engine. 1. Broadly, any **engine** that ejects a **jet** or stream of gas or fluid, obtaining all or most of its thrust by reaction to the ejection. See **reaction engine. 1.** Specifically, an aircraft engine that derives all or most of its thrust by reaction to its ejection of combustion products (or heated air) in a jet and that obtains oxygen from the atmosphere for the combustion of its fuel (or outside air for heating, as in the case of the nuclear jet engine), distinguished in this sense from a **rocket engine.** A jet engine of this kind may have a **compressor,** commonly turbine-driven, to take in and compress air (turbojet), or it may be compressorless, taking in and compressing air by other means (**pulsejet, ramjet**).

jettison. The selective release of stores from an aircraft other than normal attack.

jet nozzle. A **nozzle,** usually specially shaped, for producing a **jet,** such as the exhaust nozzle on a jet or rocket engine. See **rocket nozzle.**

jet propulsion. Reaction propulsion in which the propulsion unit obtains oxygen from the air as distinguished from rocket propulsion in which the unit carries its own oxygen-producing material. In connection with aircraft propulsion, the term refers to a gasoline or other fuel turbine jet unit which discharges hot gas through a tail pipe and a nozzle, affording a thrust which propels the aircraft. See also **rocket propulsion.**

jet routes. A high altitude route system at 18,000 feet MSL to Flight Level 450 inclusive. Jet routes are predicated on high altitude navigational aids.

jetstream. A jet issuing from an orifice into a medium with much lower velocity, such as the stream of combustion products ejected from a **reaction engine.** In the meteorological sense *jet stream* is two words, see following definition, but in the sense defined above, one word.

jet stream. A strong band of wind or winds in the upper **troposphere** or in the **stratosphere,** moving in a general direction from west to east and often reaching velocities of hundreds of miles an hour. See **jetstream,** note.

jet thrust. The **thrust** of a **fluid,** especially as distinguished from the thrust of a propeller. The thrust of a rocket engine is calculated in the same manner as gross thrust of a jet engine. See *gross thrust.*

jettisonable tank. A tank which can be jettisoned in emergency. Also see **drop tank.**

jet vane. A vane, either fixed or movable, used in a **jetstream,** especially in the jetstream of a **rocket,** for purposes of stability or control under conditions where external **aerodynamic** controls are ineffective. Also called *blast vane.* Compare **air vane.**

jitter. 1. Instability of the signal or **trace** of a **cathode-ray tube.** 2. Small rapid variations in a **waveform** due to deliberate or accidental electrical or mechanical disturbances or to changes in the supply voltages, in the characteristic of components, etc.

Jodrell Bank. The site of a large **radio telescope,** located near Manchester, England; by extension, the radio telescope itself. The radio telescope has a paraboloidal receiver 250 feet in diameter, 60 feet deep.

joint. (military) 1. Connotes activities, operations, organizations, etc., in which elements of more than one Service of the same nation participate. See also **combined.** 2. When prefixed to any of the materiel terms applicable to joint usage, connotes that the definition of the designated term is enlarged to embrace the sum of the Army, Navy, Air Force, and Marine Corps quantities.

joint airborne training. Training operations or exercises involving airborne and appropriate troop carrier units. This training includes: **a.** air delivery of personnel and equipment; **b.** assault operations by airborne troops and/or air transportable units; **c.** loading exercises and local orientation flights of short duration; and **d.** maneuvers/exercises as agreed upon by the Services concerned and/or as authorized by the Joint Chiefs of Staff.

Joint Air Training Command Facility. An Air Force Air Training Command facility where in the division of operational responsibility is clearly defined for the military and other agency, usually Federal Aviation Administration.

joint Army-Navy-Air Force publications. A series of publications produced by supporting agencies of the Joint Chiefs of Staff and intended for distribution through the approved offices of distribution within the Army, Navy, and Air Force.

joint force. A general term applied to a force which is composed of significant elements of the Army, the Navy or the Marine Corps, and the Air Force, or two or more of these Services, operating under a single commander authorized to exercise unified command or operational control over such joint forces.

Joint Long Range Proving Ground. The earliest predecessor organization and facility of the Atlantic Missile Range, activated at Cape Canaveral 1 October 1949 as a joint undertaking of the Air Force, Army, and Navy under the executive control of the Chief of Staff, USAF. This facility became the sole responsibility of the Air Force on 16 May 1950, and became known as the Florida Missle Test Range.

joint task force. A force composed of assigned or attached elements of the Army, the Navy or the Marine Corps, and the Air Force, or two or more of these Services, which is constituted and so designated by the Secretary of Defense or by the commander of a unified command, a specified command, or an existing joint task force.

joint use restricted area. A restricted area within which IFR and/or VFR flight operations may be authorized by the controlling agency (a FAA facility) when not in use by the using agency.

joule *(abbr* **j**). A unit of energy or work in the **MKS system;** the work done when the point of application of 1 **newton** is displaced a distance of 1 meter in the direction of the force. 1 joule = 10^7 ergs = 1 watt second.

Joule constant. The ratio between heat and work units from experiments based on the **first law of thermodynanics;** 4.1858 x 10^7 ergs per 15° calorie. Also called *mechanical equivalent of heat.*

Joule cycle. (After James Prescott Joule, 1818-89, English physicist.) An ideal **cycle** for engines consisting of **isentropic** compression of the working substance, addition of heat at constant pressure, isentropic expansion to **ambient pressure,** and exhaust at constant pressure. Also called *Brayton cycle.*

Joule-Thomson effect. The decrease in temperature which takes place when a gas expands through a throttling device as a nozzle. Also called *Joule-Kelvin effect.*

Jovian planet. Any one of the giant planets: Jupiter, Saturn, Uranus, or Neptune. Usually in plural *Jovian planets.*

JP *(abbr). jet propellant.* Compare **RP.**

JP-4. A liquid fuel for jet and rocket engines, the chief ingredient of which is kerosene.

Julian day. The number of each day, as reckoned consecutively since the beginning of the present Julian period on January 1, 4713 B.C. The Julian day is used primarily by astronomers to avoid confusion due to the use of different calendars at different times and places. The Julian day begins at noon, 12 hours later than the corresponding **civil day.** The day beginning at noon January 1, 1965, is Julian day 2,438,395.

jump. In computer programming, to cause the next **instruction** to be selected conditionally or unconditionally from a specified **storage** location.

jumper. short length of conductor used to complete an electrical **circuit,** usually temporary, between terminals, or to bypass an existing circuit.

jumpmaster. The assigned airborne qualified individual who controls parachutists from the time they enter the aircraft until they exit.

Jupiter. See **planet,** table.

K

K. Military mission designation for **tanker** aircraft.

Karman vortex street. (After Theodore von Karman. 1881-1963, Hungarian-born American scientist.) A double trail of **vortices** formed alternately on both sides of a cylinder or similar body moving at right angles to its axis through a **fluid,** the vortices in one row rotating in a direction opposite to that of the other row.

K-band. A frequency band used in **radar** extending approximately from 10.9 gigacycles per second to 36 gigacycles per second.

Kelvin temperature scale (*abbr* **K**). An **absolute temperature scale** independent of the thermometric properties of the working substance. On this scale, the difference between two temperatures T_1 and T_2 is proportional to the heat converted into mechanical work by a **Carnot engine** operating between the **isotherms** and **adiabats** through T_1 and T_2. Also called *absolute temperature scale, thermodynamic temperature scale.* For convenience the Kelvin degree is identified with the Celsius degree. The ice point in the Kelvin scale is 273.15° K. The triple point of water, the fundamental reference point, is 273.16° K. See **absolute, zero, approximate absolute temperature scale, Rankine temperature scale.**

Kennelly-Heaviside layer. Same as **E-layer.**

Kepler equation. In celestial mechanics $M = E - e \sin E$ where M is mean anomaly; E is eccentric anomaly; and e is eccentricity of the orbital ellipse. See **anomaly, note.**

Keplerian. Pertaining to motion in conformance with **Kepler laws,** as *Keplerian trajectory, Keplerian ellipse.*

Kepler laws. The three empirical laws governing the motions of planets in their **orbits,** discovered by Johannes Kepler (1571-1630). These are: (a) the orbits of the planets are **ellipses,** with the sun at a common focus; (b) as a planet moves in its orbit, the line joining the planet and sun sweeps over equal areas in equal intervals of time (also called *law of equal areas*); (c) the squares of the periods of revolution of any two planets are proportional to the cubes of their mean distances from the sun.

kill effects. Destructive effects available upon detonation of a weapon. Kill effects are blast, penetration, perforation, fragmentation, cratering, earth shock, fire, nuclear and thermal radiation and combination of these in varying degrees.

kill probability. A measure of the probability of destroying a target.

kilo (*abbr* **k**). Prefix meaning multiplied by 10^3.

kilocycle (*abbr* **kc**). One thousand cycles or 1000 cycles per second.

kilogram (*abbr* **kg**). The unit of mass in the metric system; the mass of the International Prototype Kilogram, a cylinder of platinumiridium alloy, stored at Seures, France, by the International Bureau of Weights and Measures.

kilohertz. One thousand **hertz.**

kilomegacycle. gigacycle.

kilometer (*abbr* **km**). A unit of distance in the **metric system.** 1 kilometer = 3280.8 feet = 1093.6 yards = 1000 meters = 0.62137 statute miles = 0.53996 nautical miles.

kiloton energy. The energy of a nuclear explosion which is equivalent to that produced by the explosion of 1 kiloton (i.e., 1,000 tons) of trinitrotoluene.

kiloton weapon. A nuclear weapon, the yield of which is measured in terms of thousands of tons of trinitrotoluene explosive equivalents, producing yields from 1 to 999 kilotons, See also **nominal weapon; sub-kiloton weapon.**

kinematics. The branch of mechanics dealing with the description of the motion of bodies or fluids without reference to the forces producing the motion.

kinematic viscosity. A **coefficient** defined as the ratio of the **dynamic viscosity** of a **fluid** to its density.

kinesthesia. The sense which detects and estimates motion without reference to vision or bearing.

kinetic energy. The **energy** which a body possesses as a consequence of its motion, defined as one-half the product of its mass m and the square of its speed v, $\frac{1}{2}mv^2$. The kinetic energy per unit volume of a fluid parcel is thus $\frac{1}{2}\rho v^2$, where ρ is the density and v the speed of the parcel. See **potential energy.**

kinetic theorgy. The derivation of the bulk properties of **fluids** from the properties of their constituent **molecules,** their motions, and interactions.

Kirchhoff law. The **radiation law** which states that at a given temperature the ratio of the **emissivity** to the **absorptivity** for a given wavelength is the same for all bodies and is equal to the emissivity of an ideal **black body** at that temperature and wavelength. Loosely put, this important law asserts that good absorbers of a given wavelength are also good emitters of that wavelength. It is essential to note that the Kirchhoff law relates absorption and emission at the same wavelength and at the same temperatures. Also called *Kirchhoff radiation law.*

klystron. An **electron tube** for converting direct-current energy into radio frequency energy by alternately speeding up and slowing down the electrons. See **magnetron.**

knot. A nautical mile per hour, 1.1508 statute miles per hour.

L

lag. 1. The delay between change of conditions and the indication of the change on an instrument. **2.** Delay in human reaction. **3.** The amount one **cyclic** motior is behind another, expressed in degrees. The opposite is *lead*.

Lagrangian coordinates. 1. A system of **coordinates** by which **fluid parcels** are identified for all time by assigning them coordinates which do not vary with time. Examples of such coordinates are: (**a**) the values of any properties of the fluid conserved in the motion; or (**b**) more generally, the positons in space of the parcels at some arbitrarily selected moment. Subsequent positions in space of the parcels are then the dependent variables, functions of time and of the Lagrangian coordinates. Also called *material coordinates.*

Lagrangian point. One of the five solutions by Lagrange to the three-body problem in which three bodies will move as a stable configuration. In three of the solutions the bodies are in line; in the other two the bodies are at the vortices of equilateral triangles. Lagrange predicted in 1772 that if the three bodies form an equilateral triangle revolving about one of the bodies, the system would be stable. This prediction was fulfilled in 1908 with the discovery of the asteroid Achilles approximately 60° ahead of Jupiter in Jupiter's orbit. Since then other asteroids have been discovered 60° ahead and 60° behind Jupiter.

lambert *(abbr* **L** or **l**). A unit of **luminance** (or brightness equal to $1/\pi$ candle per square centimeter. Physically, the lambert is the luminance of a perfectly diffusing white surface receiving an **illuminance** of 1 **lumen** per square centimeter.

Lambert law. A law of physics which states that the **radiant intensity** (flux per unit solid angle) emitted in any direction from a unit radiating surface varies as the cosine of the angle between the normal to the surface and the direction of the radiation. The **radiance** (or luminance) of a radiating surface is, therefore, independent of direction. Also called *Lambert cosine law*. Lambert law is not obeyed exactly by most real surfaces, but an ideal black body radiates according to this law. This law is also satisfied (by definition) by the distribution of radiation from a perfectly diffuse radiator and by the radiation reflected by a perfectly diffuse reflector. In accordance with Lambert law, an incandescent spherical black body when viewed from a distance appears to be simply a uniformly illuminated disk. This law does not take into account any effects that may alter the radiation after it leaves the source.

laminar boundary layer. In **fluid flow,** layer next to a fixed boundary. The fluid velocity is zero at the boundary but the molecular viscous stress is large because the velocity gradient normal to the wall is large. See **turbulent boundary layer.** The equations describing the flow in the laminar boundary layer are the **Navier-Stokes equations** containing only the inertia and molecular viscous terms.

laminar flow. In **fluid flow,** a smooth flow in which no crossflow of fluid particles occur between adjacent **stream lines**; hence, a flow conceived as made up of layers—commonly distinguished from **turbulent flow.**

laminar flow control system. A technology for reducing airplane drag by maintaining **laminar boundary layers.** The laminarization is accomplished by sucking a small amount of the external boundary layer flow through the skin. The system requires a perforated or slotted skin and a compressor to expel the sucked air.

land and sea breezes. The breezes that, on certain coasts and under certain conditions, blow from the land by night and from the water by day. These winds are seldom over 2500 feet deep.

landing. The act of terminating flight and bringing the airplane to rest, used both for land and seaplanes. Also used for missiles, spacecraft and rotorcraft.

landing aids. Any illuminating light, radio beacon, radar device, communicating device, or any system of such devices, for aiding aircraft in an approach and landing.

landing flare. The descent of an aircraft just prior to landing. The transition is a smooth curve from a steep descent to a direction of flight substantially parallel to the landing surface.

landing gear. The apparatus comprising those components of an aircraft or spacecraft that support and provide mobility for the craft on land, water, or other surface. The landing gear consists of wheels, floats, skis, **bogies,** and treads, or other devices, together with all associated struts, bracing, shock absorbers, etc. Landing gear includes all supporting components, such as the tail wheel or tail skid, outrigger wheels or pontoons, etc., but the term is often conceived to apply only to the principal components, i.e., to the main wheels, floats, etc., and the nose gear, if any. See **auxiliary landing gear.**

landing gear extended speed. The maximum speed at which an aircraft can be safely flown with the landing gear extended.

landing gear operating speed. The maximum speed at which the landing gear can be safely extended or retracted.

landing mat. A prefabricated, portable mat so designed that any number of planks (sections) may be rapidly fastened together to form surfacing for emergency runways, landing beaches, etc.

landing minimums/IFR landing minimums. The minimum visibility prescribed for landing a civil aircraft while using an instrument approach procedure. The minimum applies with other limitations set forth in FAR Part 91, with respect to the Minimum Descent Altitude (MDA) or Decision Height (DH) prescribed in the instrument approach procedure as follows:

1. Straight-in landing minimums—A statement of MDA and visibility, or DH and visibility, required for straight-in landing on a specified runway, or
2. Circling minimums—A statement of MDA and visibility required for the circle-to-land maneuver.

Descent below the established MDA or DH is not authorized during an approach unless the aircraft is in a position from which a normal approach to the runway of intended landing can be made, and adequate visual reference to required visual cues is maintained.

landing sequence. The orderly spacing of aircraft landing and approaching to land; accomplished by an airport

Artists conception of a potential future laser weapon in a ship defense role. Such high energy laser weapon systems will be able to be used on board ship to defend high value ships against attack by air-to-surface or surface-to-air missiles.

traffic controller where control tower service is provided.

landing threshold. The beginning of that portion of a runway usable for landing.

land mile. Same as **statute mile** (5280 feet).

landplane. An airplane capable normally of taking off from and landing solely on the ground or a solid platform.

Langmuir probe. A small metallic **conductor** or pair of conductors inserted within a **plasma** in order to sample the plasma current.

language. In electronic **computers: 1.** A system consisting of (**a**) a well-defined, usually finite, set of **characters; (b)** rules for combining characters with one another to form words or other expressions; and (**c**) a specific assignment of meaning to some of the words or expressions, usually for communicating information or data among a group of people, machines, etc. **2.** A system similar to the above but without any specific assignment of meanings. Such systems may be distinguished from sense **1** above, when necessary, by referring to them as *formal* or *uninterpreted* languages See **code, machine language.** Although it is sometimes convenient to study a language independently of any meanings, in all practical cases at least one set of meanings is eventually assigned.

lap belt. A safety belt that fastens across the lap. This is the usual kind of safety belt. Also called a *seat belt*.

Laplacian speed of sound. The **phase speed** of a **sound wave** in a compressible fluid if the expansions and compressions are assumed to be **adiabatic.** The value of this speed under standard conditions in dry air is 740 miles/hr. See **acoustic velocity.**

lapse rate. The decrease of an atmospheric **variable** with height, the variable being temperature, unless otherwise specified.

large aircraft. Aircraft of more than 12,500 pounds, maximum certificated takeoff weight. An FAA definition.

Larmor orbit. The circular motion of a charged **particle** in a uniform **magnetic field.**

laser, (From **light amplification by stimulated emission of radiation.**) A device for producing light by **emission** of **energy** stored in a molecular or atomic system when stimulated by an input **signal.**

laser guided bomb. A general purpose bomb fitted with a special guidance and control kit which will guide the free fall towards a target illuminated by a laser beam.

laser reconnaissance system. A near-covert night reconnaissance system which acquires imagery on standard Air Force film. The laser set is a line scan system that produces a continuous strip. The small, intense laser beam rapidly sweeps the ground in contiguous scans as opposed to the conventional methods of illuminating targets using photoflash cartridges, flares, or condenser discharge strobe lights. The system also increases the coverage obtainable by other systems since it is not restricted by the amount of cartridges or flares that are carried or time required to recharge the strobe light condensers.

last quarter. The **phase of the moon** when it is near west quadrature, when the eastern half of it is visible to an observer on the earth.

latch. A device that fastens one thing to another, as a **rocket** to a **launcher,** but is subject to ready release so that the things may be separated.

latency. Of a computer: the time required to establish communication with a specific **storage** location, not in-

cluding transfer time; equals **access time** less **word time.**

latent heat. The unit quantity of **heat** required for **isothermal** change in state of a unit mass of matter. Latent heat is termed **heat of fusion, heat of sublimation, heat of vaporization,** depending on the change of state involved.

lateral. 1. Of or pertaining to the side; directed or moving toward the side. **2.** Of or pertaining to the **lateral axis;** directed, moving, or located along, or parallel to, the lateral axis.

lateral acceleration. Acceleration substantially along the **lateral axis** of an aircraft, rocket, etc.

lateral control. Control over the rolling movement of an aircraft about the longitudinal axis. With aircraft, this is usually accomplished by ailerons.

lateral separation. In air traffic control, the lateral spacing of aircraft at the same altitude by requiring operations on different routes or in different geographical locations.

lattice (nucleonics). An orderly array or pattern of nuclear *fuel elements* and moderator in a reactor or critical assembly. Also, the arrangement of atoms in a crystal.

latitude. Angular distance from a **primary great circle** or plane. See **coordinate.** Terrestrial latitude is angular distance from the equator, measured northward or southward through 90° and labeled N or S to indicate the direction of measurement; astronomical latitude is angular distance between the direction of gravity and the plane of the equator; geodetic or topographical latitude is angular distance between the plane of the equator and a normal to the spheroid; geocentric latitude is the angle between a line to the center of the earth and the plane of the equator. Geodetic and sometimes astronomical latitude are also called *geographic latitude.* Geodetic latitude is used for charts. Assumed latitude is the latitude at which an observer is assumed to be located for an observation or computation. Fictitious latitude is angular distance from a fictitious equator. Grid latitude is angular distance from a grid equator. Transverse or inverse latitude is angular distance from a transverse equator. Oblique latitude is angular distance from an oblique equator. Difference of latitude is the shorter arc of any meridian between the parallels of two places, expressed in angular measure. Magnetic latitude, magnetic inclination, or magnetic dip is angular distance between the horizontal and the direction of a line of force of the earth's magnetic field at any point. Geomagnetic latitude is angular distance from geomagnetic equator. A parallel of latitude is a circle (or approximation of a circle) of the earth, parallel to the equator, and connecting points of equal latitude; or a circle of the celestial sphere, parallel to the ecliptic. Celestial latitude is angular distance north or south of the ecliptic. Galactic latitude is angular distance north or south of the galactic equator.

launch. 1. To send off a rocket vehicle under its own rocket power, as in the case of guided aircraft rockets, artillery rockets, and space vehicles. **2.** To send off a missile or aircraft by means of a catapult, as in the case of the V-1, or by means of inertial force, as in the release of a bomb from a flying aircraft. **3.** To give a space probe an added boost for flight into space just before separation from its launch vehicle. This term has different connotations than those of *fire* and *shoot.* See

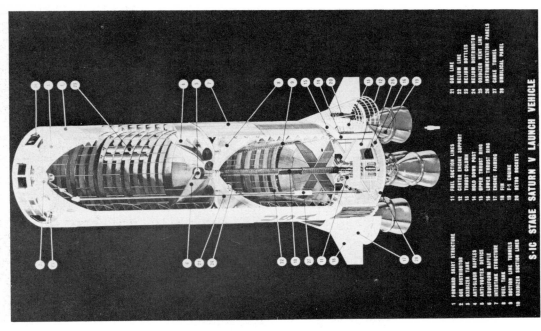

The Saturn V launch vehicle carrying the Apollo 11 spacecraft lifts off from the NASA Kennedy Space Center for the historic first manned lunar landing mission. Details of the first S-1C stage manufactured by the Boeing Company, are shown in the cutaway view at right. The Saturn V was developed under the direction of the NASA-Marshall Space Flight Center.

lift-off.

launch azimuth. The initial heading of a powered vehicle at **launch,** commonly applied to **launch vehicles.**

launch complex. The site, facilities, and equipment used to **launch** a **rocket vehicle.** See **launch site.** The complex differs according to the type rocket or particular rocket, or according to whether land launched or ship launched. The term is sometimes considered to include the launch crew.

launch crew. A group of technicians that prepares and launches a **rocket.**

launch emplacement. A launch pad with associated equipment.

launcher. 1. Specifically, a structure or device, often incorporating a tube, a group of tubes, or a set of tracks, from which self-propelled missiles are sent forth and by means of which the missiles usually are aimed or imparted initial guidance—distinguished in this specific sense from a catapult. **2.** Broadly, a structure, machine, or device, including the catapult, by means of which airplanes, rockets, or the like are directed, hurled, or sent forth.

launching angle. The angle between a horizontal plane and the **longitudinal axis** of a rocket, etc., being launched.

launching base. An area such as Cape Kennedy or Vandenberg Air Force Base that has several **launch sites.**

launching rack. A skeletonlike structure, usually incorporating rails, from which something is launched.

launching rail. A rail that gives initial support and guidance to a **rocket** launched in a nonvertical position.

launch pad. The load-bearing base or platform from which a **rocket vehicle** is launched. Usually called *pad.*

launch point. The **geographic position** from which a **rocket vehicle** is launched.

launch site. 1. A defined area from which a **rocket vehicle** is launched, either operationally or for test purposes; specifically at Cape Kennedy or Vandenberg, any of the several areas equipped to **launch** a rocket. **2.** More broadly, a **launching base.** Also called *launching site.*

launch stand. A facility or station at which a **rocket vehicle** is launched, normally incorporating a **launch pad** with **launcher.** Compare **test stand.**

launch vehicle. A **rocket** or other vehicle used to launch a **probe, satellite,** or the like.

launch window. The postulated opening in the continuum of time or space, through which a **spacecraft** or missile must be launched in order to achieve a desired encounter, rendezvous, impact, or the like. See **window.**

laydown bombing. A very low level bombing technique wherein delay fuzes and/or devices are used to allow the attacker to escape the effects of his bomb.

layer. Of the ionosphere, an apparently stratified distribution of **free electrons.** See **ionosphere,** note.

lazy eight. A turning flight maneuver during which the airplane, if viewed from another flying at the same level, roughly describes a figure 8 lying on its side.

L-band. A frequency band used in **radar** extending approximately from 0.390 gigacycles per second to 1.55 gigacycles per second. See **frequency band.**

L/D ratio. Lift-drag ratio.

lead. The amount one **cyclic** motion is ahead of another, expressed in degrees. The opposite is lag.

lead aircraft. (Military) **1.** The airborne aircraft designated to exercise command of other aircraft within the flight. **2.** An aircraft in the van of two or more aircraft.

leading edge. The forward edge of an airfoil or other body moving through air.

Lead-the-Force Program. (Military) A specified number of aircraft flight scheduled on an accelerated basis. The objective is to accrue flying hours on the specified aircraft ahead of the main body of the fleet in order that weaknesses in structures, systems, subsystems, engines, and components may be identified and corrected in advance of the bulk of the fleet.

leap year. See **calendar year.**

least squares. Any statistical procedure that involves minimizing the sum of squared **differences.**

legend. An explanation of symbols used on a map, chart, sketch, etc., commonly printed in tabular form at the side of the map, etc.

length. Specifically, the dimension of an aircraft, rocket, etc., from nose to tail; the measure of this dimension. Compare **span.**

lethal dose. Nucleonics. A dose of *ionizing radiation* sufficient to cause death. Median lethal dose (MLD or LD-50) is the dose required to kill within a specified period of time (usually 30 days) half of the individuals in a large group of organisms similarly exposed. The LD-50/30 for man is about 400 - 450 roentgens. (See **biological dose, roentgen, threshold dose.**)

lenticular cloud. A cloud having approximately the form of a double-convex lens or almond. Mountain or "standby" waves are large scale disturbances in the horizontal air flow, which under certain conditions develop downstream to the lee of mountain ridges. If the air is humid and the wave is of large amplitude, lenticular (lens-shaped) clouds mark the wave's crests. Also see **cloud.**

level above threshold. In acoustics, the pressure **level** of the **sound** in decibels above its **threshold of audibility** for the individual observer or for a specified group of individuals. Also called *sensation level.*

LF/MF four course radio range. The subject range defined four navigational courses which the pilot of an aircraft can locate and fly using a simple radio receiver operating in the 200-415 kc band. The courses are defined aurally by an interlocking series of dot-dash (A) and dash-dot (N) signals of equal amplitude. These stations defined the Federal (Civil) Airway System of the United States during the 1930's, 1940's and 1950's. They have been completely superceded by **VOR** and **VORTAC.**

LH₂. Liquid hydrogen

library. In computer operations, a collection of **programs, routines,** and **subroutines** by which problems (and parts of problems) of many types can be solved.

libration. A real or apparent oscillatory motion, particularly the apparent **oscillation** of the moon. Because of libration more than half of the moon's surface is revealed to an observer on the earth even though the same side of the moon is always toward the earth, because the moon's periods of rotation and revolution are the same. Other motions regarded as librations are long period orbital motions and periodic perturbations in orbital elements.

licensed material. nucleonics. *Source material, special nuclear material,* or *byproduct material* received, possessed, used or transferred under a general or special license issued by the Atomic Energy Commission or a state.

life sciences. The field of scientific disciplines encompassing biology, physiology, psychology, medicine, sociology, and other related areas.

life support research. Life support research is applied research in the life sciences to protect and sustain functional personnel in flight and ground operation.

lift *(symbol L).* **1.** That component of the total aerodynamic force acting on a body perpendicular to the undisturbed airflow relative to the body. **2.** To lift off, to take off in a vertical ascent. Said of a rocket vehicle. See **lift-off.**

lift coefficient *(symbol C_L).* A **coefficient** representing the **lift** of a given airfoil or other body. The lift coefficient is obtained by dividing the lift by the free-stream dynamic pressure and by the representative area under consideration.

lift-drag ratio. The ratio of **lift** to **drag** obtained by dividing the lift by the drag or the **lift coefficient** by the **drag coefficient.** Also called *L/D ratio.*

lift-off. The action of a **rocket vehicle** as it separates from its **launch pad** in a vertical ascent. Compare **take-off.** *Lift-off* is applicable only to vertical ascent; *take-off* is applicable to ascent at any angle. A *lift-off* is action performed by a rocket; a *launch* is action performed upon a rocket or upon a satellite or spaceship carried by a rocket.

light. Visible radiation (about 0.4 to 0.7 micron in wavelength) considered in terms of its **luminous efficiency,** i.e., evaluated in proportion to its ability to stimulate the sense of sight.

light amplification by stimulated emission of radiation. (LASER) A process of generating coherent light. The process utilizes a natural molecular (and atomic) phenomenon whereby molecules absorb incident electromagnetic energy at specific frequencies, store this energy for short but usable periods, and then release the stored energy in the form of light at particular frequencies in an extremely narrow frequency-band.

light armed reconnaissance aircraft. Any relatively small, light-weight airplane, equipped with offensive armament, capable of performing a reconnaissance, transport/utility and strike mission. Sometimes called LARA.

lighter-than-air aircraft. An aircraft which is supported chiefly by its buoyancy in the air.

light gun. An intense, narrowly focused spotlight with which a green, red, or white signal may be directed at any selected airplane in the traffic on or about an airport. Usually used in control towers.

light hydrogen. Ordinary *hydrogen.*

light intensity. Same as **luminous intensity.**

light intratheater transport. A conceptual vertical short takeoff and landing tilt-wing, turboprop tactical airlift aircraft for providing responsive flow of material, personnel, and supplies to ground and air forces operating in the combat area.

light microsecond. The distance a light **wave** travels in free space in one-millionth of a second.

lightning. A sudden flash of light caused by electrical discharges produced by thunderstorms.

light plane (light airplane). A general term describing small unsophisticated aircraft.

light time. The elapsed time taken by **electromagnetic radiation** to travel from a **celestial body** to the observer at the time of observation. The American Ephemeris and Nautical Almanac uses a light time of 498.8 seconds for 1 astronomical unit.

light water. Ordinary water (H_2O), as distinguished from *heavy water (*D_2O*).*

light-year. A unit of length used in expressing stellar distances equal to the distance **electromagnetic radiation** travels in 1 year. 1 light-year = 9.460 X 10^{12} kilometers = 63,280 astronomical units = 0.3068 parsecs.

limb. The edge of the apparent disk of a **celestial body,** as of the sun.

limb darkening. A condition, sometimes observed on celestial objects, in which the brightness of the object decreases as the edges or **limbs** of the object are approached. The Sun and Jupiter exhibit limb darkening.

limb of the earth. The edge of the earth at the **horizon.**

limited acrobatics. As defined by the **FAA** for aircraft **type certification,** limited aerobatics include the following flight maneuvers: **spins, lazy eights, chandelles,** steep turns if the angle of bank is more than 60 degrees. See **acrobatics.**

limited remote communications outlet LRCO. An unmanned satellite air/ground communications facility which may be associated with a VOR. These outlets effectively extend the service range of the FSS and provide greater communications reliability. LRCOs are depicted on En Route Charts.

limited war. Armed conflict short of general war, exclusive of incidents, involving the overt engagement of the military forces of two or more nations.

limiter. A device whose **output** is constant for all **inputs** above a predetermined value.

limit load. The maximum load anticipated in normal conditions of operation.

linear. 1. Of or pertaining to a line. **2.** Having a relation such that a change in one quantity is accompanied by an exactly proportional change in a related quantity, such as input and output of electronic equipment.

linear acceleration *(symbol a).* The rate of change of linear velocity. See **acceleration.**

linear accelerator. A long straight tube (or series of tubes) in which charged *particles* (ordinarily electrons or protons) gain in energy by the action of oscillating electromagnetic fields. (Compare **cyclotron;** see **accelerator.)**

linear array. An **antenna array** whose elements are equally spaced along a straight line.

linear speed. Rate of motion in a straight line. See **angular speed.**

line of flight. The line in air or space along which an aircraft, spacecraft, etc., flies or travels.

line of force. A line indicating the direction in which a force acts, as in a magnetic field. See **electric lines of force, magnetic lines of force.**

line of nodes. The straight line connecting the two points of intersection of the **orbit** of a planet, planetoid, or comet and the **ecliptic,** or the line of intersection of the planes of the orbits of a **satellite** and its **primary.**

line of position. In navigation, a line representing all possible locations of a craft at a given instant. In space this concept can be extended to *sphere of position, plane of position,* etc.

line of sight. 1. The straight line between the eye of an observer and the observed object or point. Also called *optical path.* **2.** Any straight line between one point and another, or extending out from a particular point. **3.** In radio, a direct **propagation** path that does not go below the radio horizon.

line printer. A printer, often used in conjunction with a **computer,** which is capable of printing an entire line of characters at one time.

liner. (Military) Fly at speed giving maximum cruising range.

lines of communication. All the routes, land, water, and air, which connect an operating military force with a base of operations and along which supplies and military forces move.

line spectra. The spontaneous emission of **electromagnetic radiation** from the bound **electrons** as they jump from high to low energy levels in an atom. This radiation is essentially at a single frequency determined by the jump in energy. Each different jump in energy level, therefore, has its own frequency and the net radiation is referred to as the line spectra. Since these line spectra are characteristic of the atoms, they can be used for identification purposes.

line spectrum. A **spectrum** which contains a finite number of components within a specified **frequency** range.

line squall. See **squall line.**

line width. The finite width, expressed either in **wavelength** units or **frequency** units, of a **spectral line** (e.g., an absorption line).

link (communications). A general term used to indicate the existence of communications facilities between two points.

link trainer. A ground training device used to simulate actual flying conditions.

Linke scale. A type of **cyanometer;** an instrument used to measure the blueness of the sky. The **Linke** scale is simply a set of eight cards of different standardized shades of blue. They are numbered 2 to 16, the odd numbers to be used by the observer if he judges the sky color to lie between any of the given shades. Also called *blue-sky scale.* Sky-blueness study, or cyanometry, is a means of studying atmospheric turbidity.

liquid. A substance in a state in which the individual particles move freely with relation to each other and take the shape of the container, but do not expand to fill the container. Compare **fluid.**

liquid fuel. A rocket fuel which is **liquid** under the conditions in which it is utilized in the rocket. See **liquid propellant.**

liquid hydrogen (LH₂). See **cryogenic liquid.** Liquid hydrogen is being considered as an **alternate fuel** for long-range aircraft.

liquid level manometer. A **displacement manometer** employing a liquid as the movable partition and providing means for observing the change in level of one or both of the free surfaces.

liquid propellant *(abbr LP).* Specifically, a **rocket propellant** in liquid form. Examples of liquid propellants include fuels such as alcohol, gasoline, aniline, liquid ammonia, and liquid hydrogen; oxidants such as liquid oxygen, hydrogen peroxide (also applicable as a monopropellant), and nitric acid; additives such as water; and monopropellants such as nitromethane.

liquid-propellant rocket engine. A rocket engine using a **propellant** or propellants in **liquid** form. Also called *liquid-propellant rocket.* Rocket engines of this kind vary somewhat in complexity, but they consist essentially of one or more combustion chambers together with the necessary pipes, valves, pumps, injectors, etc.

See **liquid propellant, rocket engine.**

listening watch. A continuous receiver watch established for the reception of traffic addressed to, or of interest to, the unit maintaining the watch, with complete log optional.

lithosphere. The solid part of the earth or other spatial body. Distinguished from the **atmosphere** and the **hydrosphere.** See **geosphere, biosphere.**

live testing. The testing of a **rocket** engine, vehicle, or missile by actually launching it. Compare **static testing.**

LMM. Compass locator combined with **middle marker** of **ILS.**

load. 1. The device which receives **signal** power from a **source. 2.** The signal power delivered by a source. The use of **load** in sense **2** is discouraged.

load factor. (Aircraft) The ratio of a specified load to the total weight of the aircraft. The specified load is expressed in terms of any of the following aerodynamic forces, inertia forces, or ground or water reactions. Or, the sum of loads on a structure, including the static and dynamic loads, expressed in "G" units.

loading chart (aircraft). Any one of a series of charts carried in an aircraft which shows the proper location for loads to be transported and which pertains to check lists, balance records, and clearances for weight and balance.

load isolator. A **waveguide** or coaxial device which provides a good energy path from a signal **source** to a load but provides a poor energy path for reflections from a mismatched load back to the signal source.

load manifest (air). A document specifying in detail the payload expressed in terms of passengers and/or freight carried in one aircraft for a specific destination.

lobe. An element of a **beam** of focused radio energy Lobes define surfaces of equal **power density** at varying distances and directions from the radiating **antenna.** Their configuration is governed by two factors: **(a)** the geometrical properties of the antenna reflector and feed system; and **(b)** the mutual interference between the direct and reflected rays for an antenna situated above a flecting surface. In addition to the major lobes of an antenna system, there exist side lobes (or minor lobes) that result from the unavoidable angles from the axis of the beam, and, while objectionable, they normally contain much less energy than that in the major lobe. See **radiation pattern.**

local apparent time. The arc of the **celestial equator,** or the angle at the celestial pole, between the lower branch of the local **celestial meridian** and the **hour circle** of the apparent or **true sun,** measured westward from the lower branch of the local celestial meridian through 24 hours; **local hour angle** of the apparent or true sun, expressed in time units, plus 12 hours.

local astronomical time. Mean time reckoned from the **upper branch** of the **local meridian.**

local civil time *(abbr LCT).* See **local mean time,** note.

local flight. A flight which remains no more more than 250 nautical miles from the departure point, or which terminates at the point of departure, or which does not include a stop of a greater duration than 15 minutes.

local hour angle. See **hour angle.**

local operations. As pertaining to air traffic operations, aircraft operating in the local traffic pattern or within sight of the tower; aircraft known to be departing for, or arriving from, flight in local practice areas located within a 20-mile radius of the control tower; aircraft executing simulated instrument approaches or low passes at the airport.

localizer. An electronic component of the **instrument landing system (ILS).** The localizer transmitter, operating on one of the ILS channels within the frequency range of 108.1 MHz to 111.9 MHz, emits signals which provide the pilot with course guidance to the runway centerline.

local mean time *(abbr* **LMT).** The arc of the **celestial equator,** or the angle at the celestial pole, between the lower branch of the local **celestial meridian** and the **hour circle** of the **mean sun,** measured westward from the lower branch of the local celestial meridian through 24 hours; local **hour angle** of the mean sun, expressed in time units, plus 12 hours. Mean time reckoned from the upper branch of the local meridian is called *local astronomical time.* Local mean time at the Greenwich meridian is called *Greenwich mean time,* or *universal time.* It was called *local civil time* in United States terminology from 1925 through 1952.

local meridian. The meridian through any particular place or observer, serving as the reference for **local time,** in contrast with **Greenwich meridian.**

local sidereal time *(abbr* **LST).** **Local hour angle** of the **vernal equinox,** expressed in time units; the arc of the **celestial equator,** or the angle at the celestial pole, between the upper branch of the local **celestial meridian** and the hour circle of the vernal equinox, measured westward from the upper branch of the local celestial meridian through 24 hours.

local time. Time based upon the **local meridian** as reference, as contrasted with that based upon a zone meridian, or the meridian of **Greenwich.**

locap. Low combat air patrol.

lock, to lock on. 1. Of a **radar** or other sensing and **tracking** device. To acquire a particular object of interest and continue tracking it automatically. **2.** In **phase-lock** radio receivers, to adjust the frequency of the voltage controlled oscillation, to the point where it is controlled by signal power from the detector. **3.** In coded **ranging systems,** to adjust the ground generated code until it exactly matches in time and code the transmitted code.

loft bombing. A method of bombing in which the delivery plane approaches the target at a very low altitude, makes a definite pull-up at a given point, releases bomb at predetermined point during the pull-up and tosses the bomb onto the target. See also **over-the-shoulder bombing; toss bombing.**

log. To make a flight-by-flight record of all operations of an airplane, engine, or pilot, listing flight time, area of operation, and other pertinent information.

logair. Long term contract airlift service within continental United States for the movement of cargo in support of the logistics systems of the military Services (primarily the Army and Air Force) and Department of Defense agencies.

logarithm. The **power** to which a fixed number, called *the base,* usually 10 or *e* (2.7182818), must be raised to produce the value to which the logarithm corresponds. An antilogarithm or inverse logarithm is the value corresponding to a given logarithm. A cologarithm is the logarithm of the reciprocal of a number.

logarithmic. Pertaining to **logarithms;** in a proportion corresponding to the logarithms of numbers, as a *logarithmic scale.*

logarithmic decrement. The **natural logarithm** of the ratio of any two successive amplitudes of like sign in the decay of a single-frequency **oscillation.**

logarithmic scale. A scale graduated in the **logarithms** of uniformly spaced consecutive numbers.

logical design. 1. The planning of a **computer** or **data-processing** system prior to its detailed engineering design. **2.** The synthesizing of a network of **logical elements** to perform a specified function. **3.** The result of **1** and **2** above, frequently called the *logic* of the system, machine, or network.

logical element. In a **computer** or **data-processing** system, the smallest building blocks which can be represented by operators in an appropriate system of symbolic logic. Typical logical elements are the **AND gate** and the **flip-flop,** which can be represented as operators in a suitable symbolic logic.

logical operation. In **computer** operations, (**a**) any non-arithmetical operation (e.g., extract, bitwise multiplication, jump, data transfer, etc.) (**b**) sometimes, only those nonarithmetical operations which are expressible bitwise in terms of the propositional calculus or a two-valued **Boolean algebra.**

logistics. (Military) The science of planning and carrying out the movement and maintenance of forces. In its most comprehensive sense, those aspects of military operations which deal with: **a.** design and development, acquisition, storage, movement, distribution, maintenance, evacuation, and disposition of materiel; **b.** movement, evacuation, and hospitalization of personnel; **c.** acquisition or construction, maintenance, operation, and disposition of facilities; and **d.** acquisition or furnishing of services.

LOM. **Compass locator** combined with **outer marker** of ILS.

longeron. A fore-and-aft member of the framing of an airplane fuselage or nacelle, usually continuous across a number of points of support.

longitude. 1. Angular distance, along a **primary great circle,** from the adopted reference point; the angle between a reference plane through the polar axis and a second plane through that axis. See **coordinate.** Terrestrial longitude is the arc of a parallel, or the angle at the pole, between the prime meridian and the meridian of a point on the earth, measured eastward or westward from the prime meridian through 180°, and labeled E or W to indicate the direction of measurement. Astronomical longitude is the angle between the plane of the reference meridian and the plane of the celestial meridian; geodetic longitude is the angle between the plane of the reference meridian and the plane through the polar axis and the normal to the spheroid. Geodetic and sometimes astronomical longitude are also called *geographic longitude.* Geodetic longitude is used for charts. Assumed longitude is the longitude at which an observer is assumed to be located for an observation or computation. Difference of longitude is the smaller angle at the pole or the shorter arc of a parallel between the meridians of two places, expressed in angular measure. Fictitious longitude is the arc of the fictitious equator between the prime fictitious meridian and any given fictitious meridian. Grid longitude is angular distance between a prime grid meridian and any given grid meridian. Oblique longitude is angular distance between a prime oblique meridian and any given oblique meridian. Transverse or inverse longitude is angular distance between a prime transverse meridian and any given transverse meridian. Celestial longitude is angular distance east of the vernal equinox, along the ecliptic. Galactic longitude is angular distance east of sidereal hour angle 80°, along the galactic equator. **2.** Of a planet in solar system, the sum of two angles: the **celestial longitude** of the **ascending node** of the planetary orbit, and the angle measured eastward from the ascending node along the **orbit** to the position of the planet.

longitudinal axis. The fore-and-aft line through the **center of gravity** of a craft.

longitudinal separation. The longitudinal spacing of aircraft at the same altitude by a minimum distance expressed in units of time or miles.

long-range accuracy *(abbr* **Lorac).** A two-dimensional radio navigation system using continuous-wave transmission to provide **hyperbolic lines of position** through radiofrequency phase comparison techniques from four transmitters. The system is used for surveying or ship-positioning. Frequency band, 1.7 to 2.5 megacycles. Similar to Raydist system in principle.

long-range navigation *(abbr* **loran).** A two-dimensional pulse-synchronized radio navigation system to determine **hyperbolic lines of position** through pulse-time differencing from a **master** compared to two **slave** stations. Loran uses the frequency band 1.7 to 2.0 megacycles; loran C (Cytac) uses transmission at 100 kilocycles and phase compares the continuous wave in the pulse envelopes for greater accuracy using pulse technique for resolving ambiguities.

long-wave radiation. In meteorology, same as **infrared radiation.**

long-wire antenna. A linear **antenna** which, by virtue of its considerable length in comparison with the operating **wavelength,** provides a directional radiation pattern.

look angles. The **elevation** and **azimuth** at which a particular **satellite** is predicted to be found at a specified time. *Look angles* are used in satellite tracking and data acquisition to minimize the amount of searching needed to acquire the satellite in the telescope field of view or the antenna beam.

looming. A **mirage** effect produced by greater-than-normal **refraction** in the lower atmosphere, thus permitting objects to be seen that are usually below the horizon. This occurs when the air density decreases more rapidly with height than in the normal atmosphere. If the rate of decrease of density with height is greater in the region followed by the ray from the top of the object than for the ray from the bottom of the object, the image will be stretched vertically. This stretching is often called *looming* but is more properly termed *towering.* The antonym of *looming* is *sinking* and that of *towering* is *stooping.*

loop. An airplane flight maneuver executed in a vertical plane and during which the airplane passes successively through a climb inverted flight, dive and back to normal flight.

loop antenna. An **antenna** consisting of a conducting coil, of any convenient cross section (generally circular), which emits or receives radio energy. The principal **lobe** of the **radiation pattern** is wide and is in the direction perpendicular to the plane of the coil. Also called *loop.*

Lorac *(abbr). ***long-range accuracy.**

loran *(abbr). ***long-range navigation.**

loss. A decrease in **signal power** in transmission from one point to another. Loss is usually expressed in **decibels.** Also called *transmission loss.*

loudness. The intensive attribute of an auditory sensation, in terms of which **sounds** may be ordered on a scale extending from soft to loud. Loudness is measured in **sones.** Loudness depends primarily upon the **sound pressure** of the stimulus, but it also depends upon the frequency and waveform of the stimulus.

low. mety. An area of low barometric pressure with is attendant system of winds. Also called a barometric depression or cyclone.

low. (Military) A height between five hundred and two thousand feet.

low altitude bombing. Horizontal bombing with the height of release between 900 and 8,000 feet.

low altitude parachute extraction system. (Military) A low level self-contained extraction system capable of delivering heavy loads into an area where airland is not feasible.

low angle loft bombing. Type of loft bombing of free fall bombs wherein weapon release occurs at an angle less than 35 degrees above the horizontal. See also **loft bombing.**

low approach. An approach over an airport or runway following an instrument approach or a VFR approach including the go-around maneuver where the pilot intentionally does not make contact with the runway.

low clouds. Low clouds with bases generally below 6000 feet are **stratus, nimbostratus,** and **stratocumulus.** Also see **cloud.**

lower atmosphere. Generally, and quite loosely, that part of the atmosphere in which most weather phenomena occur (i.e., the **troposphere** and lower **stratosphere**); hence, used in contrast to the common meaning for the **upper atmosphere.**

low frequency *(abbr* **LF).** See **frequency bands.**

low light level television. An electro-optical device with extreme light sensitivity which takes advantage of low background illumination to provide a television picture of objects under nighttime conditions.

low oblique. An oblique photograph in which the apparent horizon is not shown. See also **oblique air photograph.**

low-pass filter. A **wave filter** having a single transmission band extending from zero **frequency** up to some critical or bounding frequency, not infinite.

low pressure area. An area of low barometric pressure, and caused by rising warm air or air being forced aloft by adjacent high pressure areas. A low pressure area may contain bad weather or storms.

low vacuum. The condition in a gas-filled space at pressures less than 760 torr and greater than some lower limit. It is recommended that this lower limit be chosen as 25 torr corresponding approximately to the **vapor pressure** of water at 25° C and to 1 inch of mercury. The following classification scheme has been proposed for the pressure range from 760 to 10^{-3} torr:

Condition	Pressure Range
low vacuum...........	760 to 25 torr
medium vacuum.....	25 to 10^{-3} torr
rough vacuum.........	760 to 1 torr (torr range)
fine vacuum..........	1 to 10-3 torr (millitorr range)

low-wing monoplane. A monoplane in which the wings are located at or near the bottom of the fuselage.

lox. 1. Liquid oxygen. Used attributively as in *lox tank, lox unit.* Also called *loxygen.* **2.** To load the fuel tanks of a rocket vehicle with liquid oxygen. Hence, loxing.

lox-hydrogen engine. An engine using liquid hydrogen as

fuel and liquid oxygen as **oxidizer.**

loz. Liquid ozone.

LP *(abbr). fiquid propellant.*

lubber line. A datum line on an instrument indicating the fore-and-aft axis of the aircraft such as the small reference line used in reading the figures from the card of an aeronautical compass.

lumen. A unit of **luminous flux** equal to the luminous flux radiated into a unit solid angle (steradian) from a point source having a **luminous intensity** of 1 candela. An ideal source possessing an intensity of 1 candela in every direction would radiate a total of 4π lumens.

luminance. In photometry, a measure of the intrinsic **luminous intensity** emitted by a source in a given direction; the **illuminance** produced by light from the source upon a unit surface area oriented normal to the line of sight at any distance from the source, dvided by the solid angle subtended by the source at the receiving surface. Also called *brightness* but *luminance* is preferred. See **Lambert law.** Compare **luminous emittance.** It is assumed that the medium between source and receiver is perfectly transparent; therefore, luminance is independent of **extinction** between source and receiver. The source may or may not be self-luminous. *Luminance* is a measure only of light; the comparable term for electromagnetic radiation in general is *radiance.*

luminescence. Light **emission** by a process in which kinetic heat energy is not essential for the mechanism of **excitation.** *Electroluminescence* is luminescence from electrical discharges—such as sparks or arcs. Excitation in these cases results mostly from electron or ion collision by which the kinetic energy of electrons or ions, accelerated in an electric field, is given up to the atoms or molecules of the gas present and causes light emission. *Chemiluminescence* results when energy, set free in a chemical reaction, is converted to light energy. The light from many chemical reactions and from many flames is of this type. *Photoluminescence,* or *fluorescence,* results from excitation by absorption of light. The term *phosphorescence* is usually applied to luminescence which continues after excitation by one of the above methods has ceased. Compare **incandescence.**

luminosity. Same as **luminous efficiency.**

luminous. 1. In general, pertaining to the **emission of visible radiation. 2.** In photometry, a modifier used to denote that a given physical quantity, such as **luminous emittance,** is weighted according to the manner in which the response of the human eye varies with the wavelength of the light. See **luminous efficiency.**

luminous density. The instantaneous amount of **luminous energy** contained in a unit volume of the propagating medium; to be distinguished from **radiant density** in that it is weighted in accordance with the characteristics of the human eye in its nonuniform response to different wavelengths of light. See **luminous efficiency.** Compare **flux density, illuminance.**

luminous efficiency. For a given wavelength of **visible radiation,** the ratio of the **flux** that is effectively sensed by the human eye to the flux that is intrinsic in the **radiation.** It may be represented as a dimensionless ratio, e.g., lumens per watt. Also called *luminosity.* Thus, luminous efficiency is a weighting factor which is applied to radiation quantities so that they are related physiologically to the response of the human eye, which

varies as a function of wavelength. All quantities which are weighted in this manner should be modified by the term *luminous* (e.g., luminous emittance, luminous flux, etc.)

luminous emittance. The **emittance** of **visible radiation** weighted to take into account the different response of the human eye to different wavelengths of light. See **luminous efficiency.** In photometry, *luminous emittance* is always used as a property of a self-luminous source, and therefore should be distinguished from *luminance.*

luminous energy. The **energy** of **visible radiation,** weighted in accordance with the wavelength dependence of the response of the human eye. See **luminous efficiency.** Also called *light energy.*

luminous flux. Luminous energy per unit time; the **flux** of **visible radiation,** so weighted as to account for the manner in which the response of the human eye varies with the wave length of radiation. See **luminous efficiency.** The basic unit for luminous flux is the **lumen.**

luminous intensity. **Luminous energy** per unit time per unit solid angle; the **intensity** (flux per unit solid angle) of **visible radiation** weighted to take into account the variable response of the human eye as a function of the wavelength of light; usually expressed in **candles.** Also called *candlepower, light intensity.* Compare **luminance, illuminance.** See **luminous efficiency, light intensity.**

lunar. Of or pertaining to the moon.

lunar crater. A depression, usually circular, on the surface of the moon, usually with a raised rim called a **ringwall.** Craters range in size up to 250 kilometers in diameter. The largest craters are sometimes called *walled plains.* The smaller, 15 to 30 kilometers across, are often called *craterlets* and the very smallest, a few hundred meters across, *beads.* Craters are named after people, mainly astonomers.

lunar day. 1. The duration of one **rotation** of the earth on its axis, with respect to the moon. Its average length is about 24 hours 50 minutes of **mean solar time.** Also called *tidal day.* **2.** The duration of one **rotation** of the moon on its axis, with respect to the sun.

lunar distance. The angle, at an observer on the earth, between the moon and another **celestial body.** This was the basis of a method formerly used to determine longitude at sea.

lunar eclipse. The phenomenon observed when the moon enters the shadow of the earth. A lunar eclipse is called **penumbral** if the moon enters only the penumbra of the earth, *partial* if the moon enters the **umbra** without being totally immersed, and *total* if the moon is entirely immersed in the umbra.

lunar gravity. The force imparted by the moon to a mass which is at rest relative to the moon. It is approximately 1/6 of the earth's **gravity.**

lunar inequality. Variation in the moon's motion in its **orbit,** due to attraction by other bodies of the solar system. See **evection, perturbation.**

lunar interval. The difference in time between the **transit** of the moon over the **Greenwich meridian** and a **local meridian.**

lunar month. The period of **revolution** of the moon about the earth, especially a **synodical month.**

lunar noon. The instant at which the sun is over the **upper branch** of any **meridian** of the moon.

lunar orbit. Orbit of a **spacecraft** around the moon.

lunar probe. A **probe** for exploring and reporting on conditions on or about the moon.

lunar satellite. A manmade **satellite** that would make one or more **revolutions** about the moon.

lunar time. 1. Time based upon the **rotation** of the earth relative to the moon. Lunar time may be designated as **local** or **Greenwich** as the local or Greenwich meridian is used as the reference. **2.** Time on the moon.

lunation. Same as **synodical month.**

lune. That part of the surface of a **sphere** bounded by halves of two **great circles.**

lux. A photometric unit of **illuminance** or **illumination** equal to 1 **lumen** per square meter. Compare **foot-candle.**

Lyman-alpha radiation. The radiation emitted by hydrogen at 1216 angstrom, first observed in the solar spectrum by rocket-borne spectrographs. Lyman-alpha radiation is very important in the heating of the upper atmosphere thus affecting other atmospheric phenomena.

M

M. Military mission designation for **Missile Carrier** aircraft.

Mach. Mach number. Some writers use *Mach* as a unit of speed equivalent to a Mach number of 1.00, as a *speed of Mach 3.1.*

Mach angle. The angle between a **Mach line** and the direction of movement of undisturbed flow. See **Mach wave.**

Mach cone. 1. The cone-shaped shock wave theoretically emanating from an infinitesimally small particle moving at supersonic speed through a fluid medium. It is the locus of the **Mach lines. 2.** The cone-shaped shock wave generated by a sharp-pointed body, as at the nose of a high-speed aircraft. See **Mach wave.**

Mach indicator. Machmeter.

machine language. 1. A **language,** occurring within a **computer,** ordinarily not perceptible or intelligible to persons without special equipment or training. **2.** A translation or transliteration of sense **1** into more conventional **characters** but frequently still not intelligible to persons without special training.

machine oriented language. 1. A language designed for interpretation and use by a machine without translation. **2.** A system for expressing information which is intelligible to a specific machine: e.g., a computer or class of computers. Such a language may include instructions which define and direct machine operations, and information to be recorded by or acted upon by these machine operations. **3.** The set of instructions expressed in the number system basic to a computer, together with symbolic operation codes with absolute addresses, relative addresses, or symbolic addresses. Synonymous with machine language.

machine word. For a given **computer,** the number of information **characters** handled in each transfer. This number is usually fixed, but may be variable in some computers.

Mach line. A line representing a **Mach wave;** a Mach wave.

Machmeter. An instrument that measures and indicates speed relative to the **speed of sound,** i.e., that indicates the **Mach number.** Also called *Mach indicator.*

Mach number *(symbol M).* (Pronounced *mock,* after Ernst Mach, 1838-1916, Austrian scientist.) A number expressing the ratio of the speed of a body or of a point on a body with respect to the surrounding air or other **fluid,** or the speed of a **flow,** to the **speed of sound** in the medium; the speed represented by this number. If the Mach number is less than 1, the flow is called *subsonic* and local disturbances can propagate ahead of the flow. If the Mach number is greater than 1, the flow is called *supersonic* and disturbances cannot propagate ahead of the flow with the result that shock waves form. Some authorities use *mach number* but engineering practice is to use a capital *M* in all words and combinations employing *Mach.*

Mach wave. 1. A **shock wave** theoretically occurring along a common line of intersection of all the pressure disturbances emanating from an infinitesimally small particle moving at supersonic speed through a **fluid** medium, with such a wave considered to exert no changes in the condition of the fluid passing through it. The concept of the Mach wave is used in defining and studying the realm of certain disturbances in a supersonic field of flow. **2.** A very weak shock wave appearing, e.g., at the nose of a very sharp body, where the fluid undergoes no substantial change in direction.

macroscopic. Large enough to be visible to the naked eye or under low order of magnification.

magnesyn. (A trade name, from *magne*tic + *syn*chronous; often capitalized.) An electromagnetic device that transmits the direction of a **magnetic field** from one coil to another, used to transmit measurements electrically from a point of measurement to an **indicator** in a remote-indicating system.

magnet. A body which produces a **magnetic field** around itself.

magnetic. 1. Of or pertaining to a magnet. **2.** Of or pertaining to a material which is capable of being magnetized. **3.** Related to or measured from magnetic north.

magnetic bearing (MB). The horizontal angle at a given point, measured from magnetic north, clockwise to the great circle passing through the object or body and the given point. Also see **bearing.**

magnetic compass. An instrument containing a freely suspended magnetic element which indicates the horizontal direction of the earth's magnetic field at the place of observation. Also see **compass magnetic.**

magnetic course. The horizontal angle measured from the direction of magnetic north clockwise to a line representing the course of the aircraft. The aircraft course measured with reference to magnetic north. Magnetic course is obtained by adding Westerly Variation to True Course or by subtracting Easterly Variation from True Course. Also see **course.**

magnetic declination. In terrestrial magnetism; at any given location, the angle between the **geographical meridian** and the **magnetic meridian;** that is, the angle between true north and magnetic north. Also called *declination,* and in navigation, *variation.* Declination is either *east* or *west* according as the compass needle points to the east or west of the geographical meridian. Lines of constant declination are called *isogonic lines* and the one of zero declination is called the *agonic line.*

magnetic deviation. The angle between the **magnetic meridian** and the axis of a compass card, expressed in degrees east or west to indicate the direction in which the northern end of the compass card is offset from **magnetic north.** Also called *deviation.* Compare **variation.** In an aircraft compass, error caused by magnetism within an aircraft. Also see **deviation.**

magnetic dip. The angle between the **horizontal** and the direction of a line of force of the earth's **magnetic field** at any point. Also called *magnetic inclination, magnetic latitude, inclination, dip.*

magnetic disturbance daily variation. A periodic variation of the earth's **magnetic field** that is in phase with solar **(local)** time. It is the difference between the solar daily variation (or the disturbed-day solar daily variation) and the quiet-day solar daily variation. This variation is primarily an effect of enhanced electromagnetic radiation during increased solar activity.

magnetic drum. A memory device used in **computers;** a rotating cylinder on which information may be stored as magnetically polarized areas, usually along several parallel tracks around the periphery.

magnetic equator. That line on the surface of the earth connecting all points at which the **magnetic dip** is zero. Also called *aclinic line.* See **geomagnetic equator.**

magnetic field. 1. A region of space wherein any magnetic **dipole** would experience a magnetic force or torque; often represented as the geometric array of the imaginary **magnetic lines of force** that exist in relation to magnetic poles. **2. magnetic field intensity.**

magnetic field intensity. The magnetic force exerted on an imaginary unit **magnetic pole** placed at any specified point of space. It is a vector quantity. Its direction is taken as the direction toward which a north magnetic pole would tend to move under the influence of the field. If the force is measured in dynes and the unit pole is a cgs unit pole, the field intensity is given in **oersteds.** Also called *magnetic intensity, magnetic field, magnetic field strength.* Prior to 1932 the oersted was called the *gauss;* but the latter term is now used to measure magnetic induction (within magnetic materials), whereas oersted is reserved for magnetic force. By definition, one magnetic line of force per square centimeter (in air) represents the field intensity of 1 oersted.

magnetic field strength. Same as **magnetic field intensity.**

magnetic inclination. Same as **magnetic dip.**

magnetic induction. A measure of the strength of a **magnetic field** existing within a magnetic medium. The relation between the magnetic induction cnd magnetic field intensity is such that the magnetic induction within a small mass of material of magnetic permeability μ is, except for possible hysteresis effects, μ times greater than the external magnetic field intensity. Whereas magnetic field intensity is measured in **oersteds,** magnetic induction is measured in **gausses.**

magnetic lines of force. Imaginary lines so drawn in a region containing a **magnetic field** to be everywhere tangent to the **magnetic field intensity** vector if in vacuum or nonmagnetic material, or parallel to the **magnetic induction** vector if in a magnetic medium. See **electric lines of force.**

magnetic memory. 1. The ability of a material to retain magnetism after the magnetizing force is removed. **2. magnetic storage.**

magnetic north. The direction north at any point as determined by the earth's **magnetic lines of force;** the reference direction for measurement of magnetic directions.

magnetic pole. 1. Either of the two places on the surface of the earth where the **magnetic dip** is 90°, that in the Northern Hemisphere (at, approximately, latitude 73° 8 N, longitude, 101° W in 1955) being designated *north magnetic pole,* and that in the Southern Hemisphere (at, approximately, latitude, 68° S, longitude, 144° E in 1955) being designated *south magnetic pole.* Also called *dip pole.* See **geomagnetic latitude, geomagnetic pole, magnetic latitude. 2.** Either of those two points of a **magnet** where the magnetic force is greatest. **3.** In magnetic theory, a fictitious entity analogous to a unit electric charge of electrostatic theory. In nature only **dipoles,** not isolated magnetic poles, exist.

magnetic storage. In computer terminology, any device which makes use of the **magnetic** properties of materials for the storage of information.

magnetic storm. A worldwide disturbance of the earth's **magnetic field.** See **M-region.** Magnetic storms are frequently characterized by a sudden onset, in which the magnetic field undergoes marked changes in the course of an hour or less, followed by a very gradual return to normality, which may take several days. Magnetic storms are caused by solar disturbances, though the exact nature of the link between the solar and terrestrial disturbances is not understood. They are more frequent during years of high sunspot number. Sometimes a magnetic storm can be linked to a particular solar disturbance. In these cases, the time between solar flare and onset of the magnetic storm is about 1 or 2 days, suggesting that the disturbance is carried to the earth by a cloud of particles thrown out by the sun. When these disturbances are observable only in the auroral zones, they may be termed *polar magnetic storms.*

magnetic tape. A ribbon of paper, metal, or plastic, coated or impregnated with **magnetic** material on which information may be stored in the form of magnetically **polarized** areas.

magnetic variation. 1. Variation, definition **1. 2.** Change in a **magnetic element.**

magnetic wire. Wire made of **magnetic** material on which information may be stored in the form of magnetically **polarized** areas.

magnetometer. An instrument used in the study of **geomagnetism** for measuring a **magnetic element.**

magnetosphere. The region of the earth's atmosphere where ionized gas plays in important part in the dynamics of the atmosphere and where the geomagnetic field, therefore, plays an important role. The magnetosphere begins, by convention, at the maximum of the *T* **layer** at about 350 kilometers and extends to 10 or 15 earth radii to the boundary between the atmosphere and the interplanetary **plasma.**

magnetron. An **electron tube** characterized by the interaction of electrons with the electric field of a circuit **element** in crossed steady electric and magnetic **fields** to produce alternating-current power output.

magnitude. 1. The relative **luminance** of a celestial body. The smaller (algebraically) the number indicating magnitude, the more luminous the body. Also called *stellar magnitude.* See **absolute magnitude.** The ratio of relative luminosity of two celestial bodies differing in magnitude by 1.0 is 2.512, the fifth root of 100. Decrease of light by a factor of 100 increases the stellar magnitude by 5.00; hence, the brightest objects have negative magnitudes (Sun: -26.8; mean full moon: -12.5; Venus at brightest: -4.3; Jupiter at opposition: -2.3; Sirius: -1.6; Vega: 0.2; Polaris: 2.1). The faintest stars visible to the naked eye on a clear dark night are of about the sixth magnitude (though on a perfectly black background the limit for a single luminous point approaches the eighth magnitude). The faintest stars visible with a telescope of aperture a (in inches) is one approximately of magnitude $9 + 5 \log_{10} a$. The magnitude of the faintest stars which can be photographed with the 200-inch telescope is about +22.7. The expression *first magnitude* is often used somewhat loosely to refer to all bodies of magnitude 1.5 or brighter, including negative magnitudes. **2.** Amount; size; greatness. See **order of magnitude.**

main airfield. An airfield planned for permanent occupation in peacetime, at a location suitable for wartime utilization, and with operational facilities of a standard adequate to develop full use of its war combat potential. See also **airfield, alternative airfield; departure airfield; redeployment airfield.**

main bang. The transmitted **pulse,** within a **radar** system.

main rotor. The rotor that supplies the principal lift to a rotocraft.

main stage. 1. In a multistage rocket, the **stage** that develops the greatest amount of **thrust,** with or without **booster engines. 2.** In a **single-stage rocket** vehicle powered by one or more engines, the period when full thrust (at or above 90 percent) is attained. **3.** A **sustainer engine,** considered as a stage after booster engines have fallen away, as in *the main stage of the Atlas.*

maintainability. A characteristic of design and installation which is expressed as the probability that an item will conform to specified conditions within a given period of time when maintenance action is performed in accordance with prescribed procedures and resources.

maintenance. Inspection, overhaul, repair, preservation, and the replacement of parts, but excludes preventive maintenance.

maintenance engineering. The application of techniques, engineering skills and effort, organized to insure that the design and development of aerospace systems and equipment provide adequately for their effective and economical maintenance.

major alteration. An alteration not listed in the aircraft, aircraft engine, or propeller specifications:

(1) That might appreciably affect weight, balance, structural strength, performance, powerplant operation, flight characteristics, or other qualities affecting airworthiness; or

(2) That is not done according to accepted practices or cannot be done by elementary operations.

An **FAA** definition.

major axis. The longest diameter of an **ellipse** or **ellipsoid.**

major command. A major subdivision of the Air Force that is assigned a major segment of the USAF mission. A major command is directly subordinate to HQ USAF. In general, major commands are of two types:

a. operational. Composed in whole or in part of strategic, tactical, or defense forces, or flying forces directly in support of such forces.

b. support. Consists of activities which provide supplies, weapon systems, support systems, operational support equipment, combat materiel, maintenance, surface transportation, administration, personnel training, advanced education, communications, and special services to the Air Force and other supported organizations.

major nuclear power. Any nation that possesses a nuclear striking force capable of posing a serious threat to every other nation.

major planets. The four largest **planets:** Jupiter, Saturn, Uranus, and Neptune.

major repair. A repair

(1) That, if improperly done, might appreciably affect weight, balance, structural strength, performance, powerplant operation, flight characteristics, or other qualities affecting airworthiness; or

(2) That is not done according to accepted practices or cannot be done by elementary operations.

An **FAA** definition.

major subsystem. The major functional part of an aerospace vehicle which is essential to operational completeness. Examples are: airframe, propulsion, armament, guidance and communications.

malfunction. Improper functioning of a **component,** causing improper operation of a **system.**

management. A process of establishing and attaining objectives to carry out responsibilities. Management consists of those continuing actions of planning, organizing, directing, coordinating, controlling, and evaluating the use of men, money, materials, and facilities to accomplish missions and tasks. Management is inherent in command, but it does not include as extensive authority and responsibility as command.

maneuver. 1. A movement to place ships, troops, or fire in a better location with respect to the enemy. **2.** A tactical exercise carried out at sea, in the air, on the ground, or on a map in imitation of war. **3.** The operation of a ship, aircraft, or vehicle to cause it to perform desired movements.

maneuvering ballistic reentry vehicle. A rocket boosted reentry vehicle equipped with navigation and control systems and capable of performing range and cross range maneuvers during or after reentry into the atmosphere.

maneuvering speed (of an aircraft). The maneuvering speed is the maximum speed at which the flight controls can be fully deflected without damage to the aircraft structure. It may be found in the **Airplane Flight Manual,** and is useful for guidance in performing flight maneuvers, or normal operations in severe turbulence.

Manhattan Project. The War Department program during World War II that produced the first atomic bombs. The term originated in the code-name, "Manhattan Engineer District", which was used to conceal the nature of the secret work underway. The *Atomic Energy Commission,* a civilian agency, succeeded the military unit Jan. 1, 1947.

manifold pressure. The pressure existing in the induction system of an internal combustion engine when it is in operation.

maniford pressure gauge. An instrument for indicating the manifold pressure.

manipulators. Nucleonics. Mechanical devices used for safe handling of radioactive materials. Frequently they are remotely operated from behind a protective *shield.*

man-machine integration. The matching of the characteristics and capabilities of man and machine in order to obtain optimum conditions and maximum efficiency of the combined **system.** See **man-machine system.**

man-machine system. A system in which the functions of the man and the machine are interrelated and necessary for the operation of the system.

manned. Of a **vehicle** occupied by one or more persons who normally have control over the movements of the vehicle, as in a manned aircraft or spacecraft, or who perform some useful function while in the vehicle.

Manned Spaceflight Network (MSFN). A multi-station, worldwide radar network used to track in real time, and communicate with manned spacecraft in Earth orbit or in a lunar mission.

manometer. An instrument for measuring pressure of gases and vapors both above and below **atmospheric pressure.** See **vacuum gage.**

man space. (Military) The space and weight factor used to determine the combat capacity of vehicles, craft, and transport aircraft, based on the requirements of one man with his individual equipment. He is assumed to weight between 222-250 pounds and to occupy 13.5 cubic feet of space.

manufacturer's empty weight. Weight of an aircraft, including the structure, power plants, furnishings, systems and other items that are considered an integral part of a particular configuration. Essentially, this is a dry weight, containing only those fluids which are contained in a closed system, such as the hydraulic fluids.

map. A flat surface representation of a portion of the earth's surface, drawn to some convenient scale, and usually dealing with or showing more land than water. See **aeronautical chart.**

map-matching guidance. 1. The **guidance** of a rocket or aerodynamic vehicle by means of a **radarscope** film previously obtained by a reconnaissance flight over the terrain of the route, and used to direct the vehicle by alining itself with radar **echoes** received during flight from the terrain below. **2.** Guidance by **stellar map matching.**

March equinox. Same as **vernal equinox.**

mare. (*pl.* **maria**) Latin for sea. The large, dark, flat areas on the lunar surface, thought by early astronomers to be bodies of water. The term is also applied to less well-defined areas on Mars.

Marine Air Control Squadron. The component of the Marine Air Control Group which provides and operates ground facilities for the detection and interception of hostile aircraft and for the navigational direction of friendly aircraft in the conduct of support missions.

Marine Air Support Squadron. The component of the Marine Air Control Group which provides and operates facilities for the control of support aircraft operating in direct support of ground forces.

Marine division/wing team. A Marine Corps airground team consisting of one division and one aircraft wing, together with their normal reinforcements.

Marine expeditionary force. A Marine air-ground task force built around a Marine division and a Marine aircraft wing. The Marine expeditionary force normally employs the full combat resources of one Marine division/wing team.

maritime search and rescue (SAR) region. The area in which the U.S. Coast Guard exercises the SAR coordinating function. It includes the territories and possessions of the U.S. (except the Canal Zone and the inland region of Alaska) and areas of the high seas designated in the National SAR Plan. The USCG has divided the Maritime Region into subregions and a rescue coordination center in each subregion exercises coordination responsibilities.

marker beacon. Marker beacons serve to identify a particular location in space along an airway or on the approach to an instrument runway. This is done by means of a 75-MHz transmitter which transmits a directional signal to be received by aircraft flying overhead. These markers are generally used in conjunction with Low Frequency Radio Ranges and the **Instrument Landing System** as point designators. Four classes of markers are now in general use: FM, LFM, station location or Z-Markers, and the ILS marker beacons. Also see **beacon.**

mark mark. Command from ground controller for aircraft to release bombs, may indicate electronic ground controlled release or voice command to aircrew.

marking teams. (Military) Personnel landed in the landing area with the task of establishing navigational aids. See also **pathfinder; pathfinders.**

marmon clamp. A ring-shaped clamp, consisting of three equal length segments held together by **explosive bolts,** used to couple the main subsections of a rocket vehicle.

Mars. See **planet.**

maser. An **amplifier** utilizing the principle of *m*icrowave amplification by *s*timulated *e*mission of *r*adiation. **Emission** of energy stored in a molecular or atomic system by a **microwave** power supply is stimulated by the **input signal.**

mass *(symbol **m**)*. A quantity characteristic of a body, which relates the attraction of this body toward another body. Since the mass of a body is not fixed in magnitude, all masses are referred to the standard kilogram, which is a lump of platinum. *Mass* of a body always has the same value; *weight* changes with change in the acceleration of gravity.

mass-balance weight. A mass attached to a control surface of an airplane usually forward of the hinge, for the purpose of reducing or eliminating the inertial coupling between angular movement of the control and some other degree of freedom of the aircraft.

mass density. Mass per unit volume.

mass-energy equation (mass-energy equivalence) (mass-energy relation). The statement developed by Albert Einstein, German-born American physicist, that 'the mass of a body is a measure of its energy content,' as an extension of his 1905 *Special Theory of Relativity.* That statement was subsequently verified experimentally by measurements of mass and energy in nuclear reactions. The equation, usually given as: $E = mc^2$, shows that when the energy of a body changes by an amount, E, (no matter what form the energy takes) the mass, m, of the body will change by an amount equal to E/c^2. (The factor c^2, the square of the speed of light in a vacuum, may be regarded as the conversion factor relating units of mass and energy.) This equation predicted the possibility of releasing enormous amounts of energy (in the *atomic bomb*) by the conversion of mass to energy. It is also called the *Einstein equation.* See **relativity.**

mass flow rate per unit area *(symbol **G**)*. In aerodynamics, the product of **fluid** density ρ and the linear velocity of the fluid v or $G = \rho v$.

mass number. The sum of the neutrons and protons in a *nucleus.* It is the nearest whole number to an atom's *atomic weight.* For instance, the mass number of uranium-235 is 235. (Compare *atomic number.*)

mass ratio. The ratio of the mass of the **propellant** charge of a **rocket** to the total mass of the rocket when charged with the propellant.

mass spectrograph, mass spectrometer. Two related devices for detecting and analyzing *isotopes.* They separate nuclei that have different charge-to-mass ratios by passing the nuclei through electrical and magnetic fields.

master station. In a **hyperbolic navigation system,** such as **loran,** that transmitting station which controls the transmissions of another station or of other stations. See **hyperbolic navigation, slave station.**

mate. To fit together two major **components** of a **system.** Also called *marry.*

materials. In general, the substances of which rockets and space vehicles are composed; specifically, the metals, alloys, ceramics, and plastics used in structural protective, and electronic functions.

mating. 1. The act of fitting together two major **components** of a **system** as *mating of a launch vehicle and a spacecraft.* Also called *marriage.* 2. **interface.**

matrix. 1. Any rectangular array of elements composed of rows and columns; specifically, such an array consisting of numbers or mathematical symbols which can be manipulated according to certain rules. 2. In electronic **computers,** any logical **network** whose configuration is a rectangular array of intersections of its input-output leads, with elements connected at some of these intersections. The network usually functions as an **encoder** or **decoder.** Loosely, any encoder, decoder, or **translator.**

matter. The substance of which a physical object is composed. All materials in the universe have the same inner nature, that is, they are composed of atoms, arranged in different (and often complex) ways; the specific atoms and the specific arrangements identify the various materials. (See **atom, element.**)

maximum continuous rating. A rating which is assigned to engines and propellers and means the applicable power, rpm, or other limits for which the engine or propeller is certificated.

maximum landing weight. The maximum gross weight due to design or operational limitations at which an aircraft is permitted to land.

maximum permissible dose (MPD) (maximum permissible exposure). Nucleonics. That dose of *ionizing radiation* established by competent authorities as an amount below which there is no reasonable expectation of risk to human health, and which at the same time is somewhat below the lowest level at which a definite hazard is believed to exist. An obsolescent term.

maximum take-off power. The horse power developed in standard sea level conditions in the limiting engine operating conditions approved for use in normal take-off, and restricted to a continous period of five minutes.

maximum take-off weight. The maximum gross weight due to design or operational limitations at which an aircraft is permitted to take off.

Maxwellian distribution. The velocity distribution, as computed in the **kinetic theory of gases,** of the molecules of a gas in thermal equilibrium. This distribution is often assumed to hold for **neutrons** in thermal equilibrium with the **moderator** (thermal neutrons).

mayday. Distress call.

McLeod gage. A liquid-level **vacuum gage** in which a known volume of a gas, at the **pressure** to be measured, is compressed by the movement of a liquid column to a much smaller known volume, at which the resulting higher pressure is measured. Particular designs are named after the inventors or by various trade names.

MCW *(abbr).* **modulated continuous wave.**

mean. See **arithmetic mean.**

MCA. Minimum crossing altitude.

MDA. Minimum descent altitude.

MEA. Minimum enroute altitude.

mean aerodynamic chord. The chord of an imaginary airfoil which would have force **vectors** throughout the flight range identical with those of the actual wing or wings. Abbrev., (M.A.C.)

mean center of moon. A central point for a lunar coordinate system; the point on the lunar surface intersected by the lunar radius that is directed toward the earth's center when the moon is at the mean ascending node and when the node coincides with the *mean perigee* or *mean apogee.*

mean free path. Of any **particle,** the average distance that a particle travels between successive collisions with the other particles of an ensemble. In vacuum technology, the ensemble of particles of interest comprises only the molecules in the gas phase.

mean life. Nucleonics. The average time during which an atom, an excited nucleus, a radionuclide or a particle exists in a particular form. (See **scattering**).

mean time between failures. For a particular interval the total measured operating time of the population of a materiel divided by the total number of failures within the population during the measured period.

mean sea level (MSL or msl). The average height of the surface of the sea for all stages of the tide over a 19-year period, usually determined from hourly height readings. Mean sea level is the datum from which heights are measured. In this sense sometimes shortened to *sea level.* See **geoid.**

mean sidereal time. Sidereal time adjusted for **nutation** to eliminate slight irregularities in the rate.

mean solar day. The duration of one **rotation** of the earth on its axis, with respect to the **mean sun.** The length of the mean solar day is 24 hours of mean solar time or 24 hours 3 minutes 56.555 seconds of mean sidereal time. A mean solar day beginning at midnight is called a *civil day;* and one beginning at noon, 12 hours later, is called an *astronomical day.* See **calendar day.**

mean solar second. Prior to 1960 the fundamental unit of time, equal to 1/86,400 of the **mean solar day.** Now replaced by the **ephemeris second.**

mean solar time. See **solar time.**

mean square. Referring to the **arithmetic mean** of the squares of the values under consideration, as *mean-square amplitude, mean-square error.*

mean-square error. The quantity whose square is equal to the sum of the squares of the individual **errors** divided by the number of those errors.

mean sun. A fictitious sun conceived to move eastward along the **celestial equator** at a rate that provides a uniform measure of time equal to the average **apparent time;** the reference for reckoning **mean time, zone time, etc.** See **dynamical mean sun.**

mean time. Time based upon the **rotation** of the earth relative to the **mean sun.** Mean time may be designated as *local* or *Greenwich* as the local or Greenwich meridian is the reference. Greenwich mean time is also called *universal time.* Zone, standard, daylight saving or summer, and war time are also variations of mean time, specified meridians being used as the reference. Mean time reckoned from the upper branch of the meridian is called *astronomical time.* Mean time was called *civil time* in U.S. terminology from 1925 through 1952. See **equation of time, mean sidereal time.**

measurand. A physical quantity, force, property or condition which is to be measured. Also called *stimulus.*

mechanic. See **aviation mechanic.**

mechanical equivalent of heat *(symbol J).* **Joule constant.**

mechanical system. In the study of **vibration,** an aggregate of matter comprising a defined **configuration** of mass, mechanical stiffness, and mechanical resistance.

median. The middle term of a series, or the interpolated value of the two middle terms if the number of terms is even. Compare **mean.**

median line (of an airfoil). A line, each point of which is equidistant from the upper and lower boundaries of the airfoil section, the distances being measured normal to the chord.

medical certificate. Acceptable evidence of physical fitness on a form prescribed by the **FAA.** A medical certificate is required for various **airmen certificates** to be valid.

medium. (Military) A height between two thousand and twenty-five thousand feet.

medium-altitude bombing. Horizontal bombing with the height of release between 8,000 and 15,000 feet.

medium-angle loft bombing. Type of loft bombing wherein weapon release occurs at an angle between 35 degrees and 75 degrees above the horizontal.

medium-range ballistic missile. A ballistic missile with a range capability from about 600 to 1500 nautical miles.

medium frequency (abbr **MF**). See **frequency bands.**

mega (abbr **M**). A prefix meaning multiplied by 10^6.

megacycle. (Abbr **Mc, mc**). One million cycles; one thousand kilocycles. The term is often used as the equivalent of one million cycles per second.

megahertz (MHz). A frequency of one million cycles per second.

megaton energy. The energy of a nuclear explosion which is equivalent to that produced by the explosion of 1,-000,000 tons (or 1,000 kilotons) of trinitrotoluene (TNT).

megaton weapon. A nuclear weapon, the yield of which is measured in terms of millions of tons of trinitrotoluene explosive equivalents.

membrane structure. A shell structure, often **pressurized,** that does not take wall bending or compression loads.

memory. The component of a **computer,** control system, guidance system, instrumented satellite, or the like, designed to provide ready access to data or instructions previously recorded so as to make them bear upon an immediate problem, such as the guidance of a physical object, or the analysis and reduction of data.

Mercury. See **planet.**

mercurial barometer. An instrument used in measuring atmospheric pressure. The instrument employs the principle of balancing the atmospheric pressure with a column of mercury. A cubic inch of mercury weighs 0.49 pounds. 14.7 pounds, standard air pressure at sea level, supports a column of mercury 29.92 inches high.

meridian. A north-south reference line, particularly a **great circle** through the geographical poles of the earth. The term usually refers to the **upper branch,** that half, from pole to pole, which passes through a given place, the other half being called the *lower branch.* See **coordinate.** A terrestrial meridian is a meridian of the earth. Sometimes designated *true meridian* to distinguish it from magnetic meridian, compass meridian, or grid meridian, the north-south lines relative to magnetic, compass, or grid direction, respectively. An astronomical meridian is a line connecting points having the same astronomical longitude. A geodetic meridian is a line connecting points of equal geodetic longitude. Geodetic and sometimes astronomical meridians are also called *geographic* meridians. Geodetic meridians are shown on charts. The prime meridian passes through longitude 0°. A fictitious meridian is one of a series of great circles or lines used in

place of a meridian for certain purposes. A transverse or inverse meridian is a great circle perpendicular to a transverse equator. An oblique meridian is a great circle perpendicular to an oblique equator. Any meridian used as a reference for reckoning time is called a time meridian. The meridian through any particular place or observer, serving as the reference for local time, is called *local meridian,* in contrast with the Greenwich meridian, the reference for Greenwich time. A celestial meridian is a great circle of the celestial sphere, through the celestial poles and the zenith.

meridian angle. Angular distance east or west of the local **celestial meridian;** the arc of the **celestial equator,** or the angle at the celestial pole, between the **upper branch** of the local celestial meridian and the **hour circle** of a **celestial body,** measured eastward or westward from the local celestial meridian through 180°, and labeled E or W to indicate the direction of measurements. See **hour angle.**

mesopause. The base of the inversion at the top of the **mesophere,** usually found at 80 to 85 kilometers (50 to 53 miles). See **atmospheric shell.**

mesosphere. 1. The **atmospheric shell,** in which temperature generally descreases with heights, extending from the **stratopause** at about 50 to 55 kilometers (30 to 35 miles) to the **mesopause** at about 80 to 85 kilometers (50 to 53 miles). **2.** The atmospheric shell between the top of the **ionosphere** (the top of this region has never been clearly defined) and the bottom of the **exosphere.** (This definition has not gained general acceptance.)

message. 1. An ordered selection from an agreed set of symbols, intended to communicate **information. 2.** The original **modulating wave** in a communication system. The term in sense **1** is used in communication theory; the term in sense **2** is often used in engineering practice.

metallic fuels. Of or pertaining to **nuclear fuels** which are a mixture, a pressed powder, or an alloy of a **fissionable** material, such as uranium-235 or plutonium-239, and a metal such as aluminum, zirconium, or stainless steel.

mestastable compound. A chemical compound of comparative stability which, however, becomes unstable under a particular set of conditions.

metastable propellant. A **metastable compound** used as a **propellant.** Nitromethane (CH_3NO_2), for example, may be used as a monopropellant at chamber pressure above 500 pounds per square inch. At lower pressure, it requires an oxidizer for stable combustion.

meteor. In particular, the light phenomenon which results from the entry into the earth's atmosphere of a solid particle from space; more generally, any physical object or phenomenon associated with such an event. See **meteoroid.**

meteoric. Of or pertaining to **meteors** and **meteoroids.**

meteorite. Any **meteoroid** which has reached the surface of the earth without being completely vaporized.

meteoroid. A solid object moving in interplanetary space, of a size considerably smaller than an **asteroid** and considerably larger than an atom or molecule.

meteorological rocket. A **rocket** designed primarily for routine **upper air observation** (as opposed to research) in the lower 250,000 feet of the atmosphere, especially that portion inaccessible to balloons, i.e., above 100,000 feet. Also called *rocketsonde.*

meteorology. The study dealing with the phenomena of

the **atmosphere.** This includes not only the physics, chemistry, and dynamics of the atmosphere, but is extended to include many of the direct effects of the atmosphere upon the earth's surface, the oceans, and life in general. A distinction can be drawn between meteorology and climatology, the latter being primarily concerned with average, not actual, weather conditions. Meteorology may be subdivided, according to the methods of approach and the applications to human activities, into a large number of specialized sciences. The following are of interest to space science: **aerology, aeronomy, dynamic meteorology, physical meteorology.**

meteor path. The projection of the **trajectory** of a **meteor** in the celestial sphere as seen by the observer.

meteor shower. A number of **meteors** with approximately parallel trajectories.

meteor train. Anything, such as light or ionization, left along the trajectory of the meteor after the head of the meteor has passed.

meter. 1. *(abbr* **m)** The basic unit of length of the **metric system.** Effective 1 July 1959 in the U.S. customary system of measures, 1 yard = 0.9144 meter, exactly, or 1 meter = 1.094 yards = 39.37 inches. The standard inch is exactly 25.4 millimeters. **2.** A device for measuring, and usually indicating, some quantity.

metering jet. A jet in a **fuel-injection system.**

method of attributes. In reliability testing, measurement of quality by noting the presence or absence of some characteristic (attribute) in each of the units in the group under consideration and counting how many do or do not possess it. An example of this method is go and no-go gaging of a dimension.

Metonic cycle. A period of 19 years after which the various **phases of the moon** fall on approximately the same days of the year as in the previous cycle. The Metonic cycle is the basis for the golden numbers used to determine the date of Easter. Four such cycles form a Callippic cycle.

metric photography The recording of events by means of photography (either singly or sequentially), together with appropriate **coordinates,** to form the basis for accurate measurements.

metric system. The international decimal system of weights and measures based on the **meter** and the **kilogram.** The use of the metric system in the United States was legalized by Congress in 1866 but was not made obligatory.

metrology. The science of dimensional measurement; sometimes includes the science of weighing.

micro *(abbr).* **1.** A prefix meaning divided by 10^6. **2.** A prefix meaning very small, as in micrometeorite.

microbar. The unit of pressure in the **CGS system** and equal to 1 dyne per square centimeter; the unit of **sound pressure.** In British literature the term *barye* has been used. The term *bar* properly denotes a pressure of 10^6 dynes per square centimeter. Unfortunately, the bar was once used in acoustics to mean 1 dyne per squre centimeter, but this is no longer correct.

microcircuit. A small circuit having a high equivalent circuit element density, which is considered as a single part composed of interconnected elements on or within a single substrate to perform an electronic circuit function. (This excludes printed wiring boards, circuit card assemblies, and modules composed exclusively of discrete electronic parts.)

microelectronics. A branch of electronics that deals with the miniaturization of electronic components and circuits.

micrometeorite. A very small **meteorite** or meteorite particle with a diameter in general less than a millimeter.

micrometeorite penetration. Penetration of the thin outer shell (skin) of space vehicles by small particles travelling in space at high velocities.

micrometer. One of a class of instruments for making precise linear measurements in which the displacements measured correspond to the travel of a screw of accurately known **pitch.**

micron. 1. A unit of length equal to one-thousandth of a millimeter. The micron is a convenient length unit for measuring wavelengths of infrared radiation, diameters of atmospheric particles, etc. **2. micron of mercury.**

micron liter. One liter of gas at a **pressure** of one **micron of mercury.**

micron of mercury. A unit of **pressure** equal to a pressure of 1/1000th of 1 millimeter of mercury pressure at °C and the standard acceleration of gravity; a millitorr 10^{-3} **torr** approximately). See **torr.**

microphone. An **electroacoustic transducer** which receives an **acoustic** signal and delivers a corresponding electric signal.

microsecond. One-millionth of a second.

microwave. Of, or pertaining to, radiation in the **microwave region.**

microwave landing system (MLS). an instrument approach and landing system operating in the microwave frequencies (5.0-5.25 GHz/15.4-15.7 GHz) that provides precision guidance in azimuth, elevation and distance measurement.

microwave refractometer. A device for measuring the **refractive index** of the atmosphere at microwave frequencies—usually in the 3-centimeter region.

microwave region. Commonly, that region of the radio spectrum between approximately 1000 megacycles and 300,000 megacycles. See **frequency band.** Corresponding wavelengths are 30 centimeters to 1 millimeter. The limits of the microwave region are not clearly defined but in general it is considered to be the region in which radar operates.

midcourse guidance. Guidance of a **rocket** from the end of the launching phase to some arbitrary point or at some arbitrary time when **terminal guidance** begins. Also called *incourse guidance.* See **guidance.**

middle or alto clouds. Middle or "alto" clouds are altostratus and altocumulus which usually occur at 6000 to 20,000 feet. Also see **cloud, cumuliform, stratiform.**

middle marker (MM). An **ILS** marker which is located on a **localizer** course line at a distance of about 3500 feet from the approach end of the runway. Also see **marker beacon.**

mid-wing monoplane. A monoplane in which the wings are located approximately midway between the top and bottom of the fuselage.

mil. 1. One-thousandth of an inch. **2.** A unit of angular measurement, 1/6400 of a circle.

mile. A unit of distance. See **statute mile, nautical mile.**

Military Airlift Command. The single manager operating agency for designated airlift service. Also referred to as **MAC.**

military aircraft types. Military aircraft are classified in accordance with their basic mission (example: F-111, fighter) as well as their modified mission (RF-111, reconnaissance version of the F-111) as applicable. See **Attack (A),Cargo/transport (C), Bomber (B), Special Electronic Installation (E), Fighter (F), Helicopter (H), Tanker (K), Observation (O), Patrol (P), Antisubmarine (S), Trainer (T), Utility (U), VTOL and STOL(V), Research (X), Airship (Z), Missile Carrier (M), Drone (Q), Reconnaissance (R), Director (D), Search/Rescue (H), Staff (V).**

Military Department. One of the departments within the Department of Defense created by the National Security Act of 1947, as amended. See also **Department of the Army; Department of the Navy; Department of the Air Force.**

Milky Way. The **galaxy** to which the sun belongs. As seen at night from the earth, the Milky Way is a faintly luminous belt of faint stars.

milli *(abbr m).* A prefix meaning multiplied by $10-^3$.

millibar. A unit of pressure equal to 1000 dynes per square centimeter, or 1/1000 of a bar. The millibar is used as a unit of measure of atmospheric pressure, a standard atmosphere being equal to 1,013.25 millibars or 29.92 inches of mercury.

millimeter *(abbr mm).* One-thousandth of a meter; one-tenth of a centimeter; 0.039370 U.S. inch.

millimeter of mercury *(abbr mm Hg).* A unit of **pressure** corresponding to a column of mercury exactly 1 millimeter high at 0° C under standard acceleration of gravity of 980.665 centimeters per second squared. See **torr.** By *mercury at 0° C* is meant a hypothetical fluid having an invariable density exactly 13.5951 grams per cubic centimeter.

million electron volt *(abbr Mev).* A unit of energy equal to $1.603 \times 10-^8$ ergs.

millisecond *(abbr msec).* One-thousandth of a second.

MIL specification. The Military Specifications and Standards, and indices thereto, prepared and issued by the Department of Defense.

mini. A contraction of *miniature* used in combination, as in *minicomponent, miniradio, minitransistor.*

miniature. Used attributively in reference to equipment, such as gimbals, gyroscopes, computers, etc., made small to fit into confined spaces, as within an earth satellite or rocket vehicle.

miniaturize. To construct a functioning miniature of a part or instrument. Said of telemetering instruments or parts used in an earth satellite or rocket vehicle, where space is at a premium. Hence, *miniaturized, miniaturization.*

minimum-altitude bombing. Horizontal or glide bombing with the height of release under 900 feet. It includes masthead bombing which is sometimes erroneously referred to as "skip bombing." See also **skip bombing.**

minimum attack altitude. The lowest altitude determined by the tactical use of weapons, terrain consideration, and weapons effects which permits the safe conduct of an air attack and/or minimizes effective enemy counteraction.

minimum control speed. The lowest possible speed of a multiengined aircraft at which, at a constant power setting and aircraft configuration, the pilot is able to main-

tain a straight course after failure of one or more engines.

minimum crossing altitudes (MCA). The lowest altitudes at certain radio fixes at which an aircraft must cross when proceeding in the direction of a higher minimum en route IFR altitude.

minimum descent altitude (MDA) Means the lowest altitude, expressed in feet above mean sea level, to which descent is authorized on final approach or during circling-to-land maneuvering in execution of a standard instrument approach procedure where no electronic glide slope is provided.

minimum en route IFR altitude (MEA). The altitude in effect between radio fixes which assures acceptable navigational signal coverage and meets obstruction clearance requirements between those fixes.

minimum fuel. The lowest quantity of usable fuel necessary to assure a safe landing in normal sequence with other aircraft. This flight condition does not warrant priority traffic handling; instead it is an advisory to the traffic controller that any unusual delay will result in an emergency.

minimum holding altitude (MHA). The lowest altitude prescribed for a holding pattern which assures navigational signal coverage, communications, and meets obstruction clearance requirements.

minimum line of detection. An arbitrary line at which aircraft must be detected to provide sufficient time for identification, scramble, interception, and destruction before reaching a vital area. The minimum line of detection normally coincides with the air traffic control line.

minimum line of interception. An arbitrary line at which the enemy forces should be intercepted by aircraft. It is used to provide the surface-to-air missiles and antiaircraft gun artillery adequate time to destroy, prior to bomb release, those enemy weapon systems not intercepted by aircraft.

minimum obstruction clearance altitude (MOCA). The specified altitude in effect between radio fixes on VOR/LF airways, off-airway routes or route segments, which meets obstruction clearance requirements for the entire route segment and which assures acceptable navigational signal coverage only within 22 nautical miles of a VOR.

minimum reception altitude (MRA). The lowest altitude required to receive adequate signals to determine specific VOR/VORTAC/TACAN fixes.

minimum vectoring altitude (MVA). The lowest altitude, expressed in feet above mean sea level, that aircraft will be vectored by a radar controller. This altitude assures communications, radar coverage, and meets obstruction clearance criteria.

minitrack. A satellite tracking system consisting of a field of separate antennas and associated receiving equipment interconnected so as to form interferometers which track a transmitting beacon in the payload itself.

minor alteration. An alteration other than a **major alteration.**

minor axis. The shortest diameter of an **ellipse** or **ellipsoid.**

minor planet. Same as **asteroid.** See **planet.**

minor repair. A repair other than a **major repair.**

minute *(abbr min* or *').* **1.** The sixtieth part of an hour. **2.** The sixtieth part of a degree of arc.

A Poseidon **missile** *is hurled into the sky from a submarine. The Poseidon was developed by Lockheed Missiles & Space Company for the U.S. Navy.*

mirage. A **refraction** phenomenon in the **atmosphere** wherein an image of some object is made to appear displaced from its true position. See **radio duct,** note. Simple mirages may be any one or three types, the inferior mirage, the superior mirage, or the lateral mirage, depending, respectively, on whether the spurious image appears below, above, or to one side of the true position of the object. Of the three, the inferior mirage is the most common, being usually discernible over any heated street in daytime during summer. The abnormal refraction responsible for mirages is invariably associated with abnormal temperature distributions that yield abnormal spatial variations in the refractive index.

missed-approach procedure. The procedure to be followed if, after an instrument approach, a landing is not effected and occurring normally: (**a**) when the aircraft has descended to the decision height and has not established visual contact, or (**b**) when directed by air traffic control to pull up or to go around again.

missile. Any object thrown, dropped, fired, launched, or otherwise projected with the purpose of striking a target. Short for *ballistic missile, guided missile. Missile* should not be used loosely as a synonym for *rocket* or *spacecraft.*

missile assembly-checkout facility. A building, van, or other type structure located near the operational missile launching location designed for the final assembly and checkout of the missile system.

missile carrier (M), (Military designation). Aircraft modified for carrying and launching guided and nonguided missiles as part of the weapon system. See **military aircraft types.**

missile intercept zone. That geographical division of the destruction area where surface-to-air missiles have primary responsibility for destruction of airborne objects. See also **destruction area.**

missile system. A weapon system in which a missile constitutes the aerospace vehicle. See also **weapon system.**

missilry. The art or science of designing, developing, building, launching, directing, and sometimes guiding a rocket **missile;** any phase or aspect of this art or science. This term is sometimes spelled *missilery,* but is then pronounced as a three-syllable word.

mission. 1. The task together with the purpose, which clearly indicates the action to be taken and the reason therefor. **2.** In common usage, especially when applied to lower military units, a duty assigned to an individual or unit; a task. **3.** The dispatching of one or more aircraft to accomplish one particular task. Note: This term is normally not used in civil operations. Mission is used extensively in military and spaceflight operations.

mist. A popular expression for drizzle.

mistake. An **error,** usually large, resulting from a human failing or an equipment malfunction.

mixed-flow compressor. A rotary **compressor** through which the acceleration of fluid is partly radial and partly axial.

mixture control. A device embodied in the fuel-metering system for adjusting the fuel-air ratio of an engine to compensate for the decrease in air density at high altitudes.

mixture ratio. In **liquid-propellant rockets,** the relative **mass flow rates** to the combustion chamber of **oxidizer** and **fuel.**

mix-up, caution. Mixture of friendly and hostile aircraft.

MKSA system. A system of units based on the **meter, kilogram, second,** and **ampere.** Also called **international system.**

MKS system. system of units based on the meter, the kilogram, and the second.

mobilization. 1. The act of preparing for war or other emergencies through assembling and organizing national resources. **2.** The process by which the armed forces or part of them are brought to a state of readiness for war or other national emergency. This includes assembling and organizing personnel, supplies, and material for active military service.

MOCA. Minimum obstruction clearance altitude.

mockup. A full-sized replica or dummy of something, such as a spacecraft, often made of some substitute material such as wood, and sometimes incorporating actual functioning pieces of equipment, such as engines.

mode. A functioning position or arrangement that allows for the performance of a given task. Said of a **spacecraft,** which may move, for example, from a *cruise mode* to an *encounter mode;* or said of controls that permit the selection of a mode, such as a *reentry mode.*

mode. The number or letter referring to the specific pulse spacing of the signal transmitted by an **interrogator.** See air **traffic control radar beacon system.**

mode (identification friend or foe) (Military) The number or letter referring to the specific pulse spacing of the signals transmitted by an **interrogator.**

mode of vibration. In a system undergoing **vibration,** a characteristic pattern assumed by the system in which the motion of every particle is simple **harmonic** with the same **frequency.** Two or more modes of vibration may exist concurrently in a multiple-degree-of-freedom system.

moderator. Nucleonics. A material, such as ordinary water, heavy water or graphite, used in a *reactor* to slow down high-velocity neutrons, thus increasing the likelihood of further *fission.* Compare *reflector;* see absorber.)

modification center. An installation consisting of an airfield and of facilities for modifying standard production aircraft to meet certain requirements which were not anticipated at the time of manufacture.

modified close control. A method of intercept control during which only target position information is transmitted to an interceptor by the semiautomatic ground environment.

modified precision approach radar. A special precision radar approach landing procedure for high performance aircraft. Radar guidance is provided to a landing flare point instead of a runway touchdown point.

modulated continuous wave *(abbr* MCW). A form of **emission** in which the **carrier** is modulated by a constant **audiofrequency** tone.

modulated wave. A **wave** which varies in some characteristics in accordance with the variations of a modulating **signal.** Compare **continuous wave.** See **modulation.**

modulation. 1. The variation in the value of some parameter characterizing a periodic **oscillation. 2.** Specifically, variation of some characteristic of a **radio wave,** called the *carrier wave,* in accordance with instantaneous values of another wave, called the *modulating wave.* Variation of amplitude is *amplitude modulation,* variation of frequency is *frequency modulation,* and

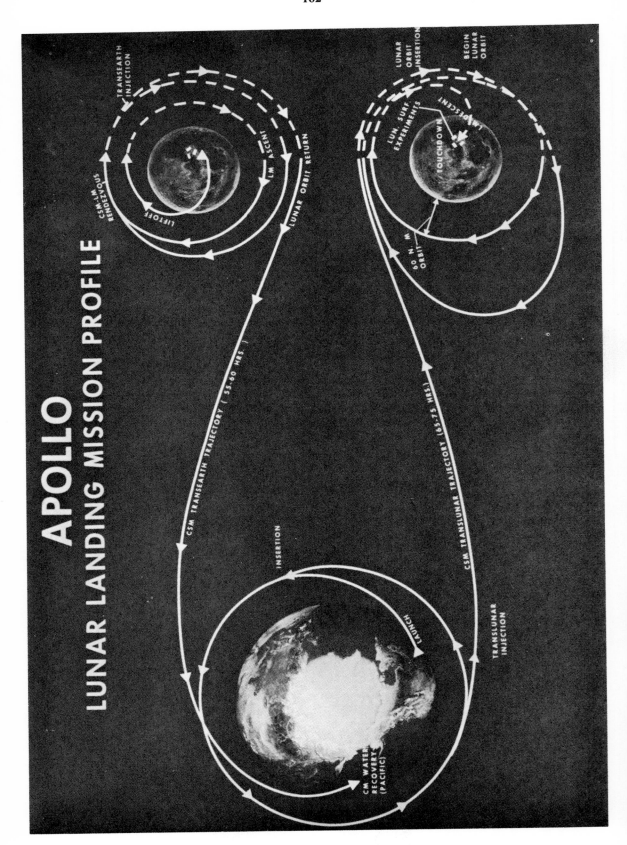

*This outstanding view of a full **moon** was photographed from the Apollo II spacecraft during its transearth journey homeward. When this picture was taken the spacecraft was 10,000 nautical miles from the moon.*

*This picture taken from an orbiting Apollo Command and Service Module shows terrain features typical of the **moon**.*

variation of phase is *phase modulation*. The formation of very short bursts of a carrier wave, separated by relatively long periods during which no carrier wave is transmitted, is *pulse modulation*.

modulator. A device to effect the process of **modulation.**

module. 1. A self-contained unit of a **launch vehicle** or **spacecraft** which serves as a building block for the overall structure. The module is usually designated by its primary function as *command module, lunar landing module,* etc. **2.** A one-package assembly of functionally associated electronic parts, usually a plug-in unit, so arranged as to function as a **system** or **subsystem;** a **black box. 3.** The size of some one part of a **rocket** or other structure, as the semidiameter of a rocket's base, taken as a unit of measure for the proportional design and construction of component parts.

modulus. (*plural* **moduli**) **1.** A real positive quantity which measures the **magnitude** of some number, as *the modulus of a complex number is the square root of the sum of squares of its components.* **2.** A **coefficient** representing some elastic property of a body, such as the *modulus of elasticity* or the *modulus of resilience.*

modulus of elasticity *(symbol E).* Same as **Young modulus.**

moist adiabatic lapse rate. Same as **saturation adiabatic lapse rate.**

molar. Pertaining to a **mole,** or measured in moles.

mole *(abbr* **mol).** The amount of substance containing the same number of atoms as 12 grams of pure carbon. The gram-mole or gram-molecule is the mass in grams numerically equal to the molecular weight.

molecular effusion. The passage of **gas** through a single opening in a plane wall of negligible thickness where the largest dimension of the hole is smaller than the **mean free path.**

molecular flow. The flow of gas through a duct under conditions such that the **mean free path** is greater than the largest dimension of a transverse section of the duct.

molecular weight. The weight of a given **molecule** expressed in **atomic weight units.**

molecule. An aggregate of two or more **atoms** of a substance that exists as a unit.

molten salt reactor **(nucleonics).** A *fused-salt reactor.*

moment *(symbol M).* A tendency to cause **rotation** about a point or axis, as of a control surface about its hinge or of an airplane about its center of gravity; the measure of this tendency, equal to the product of the force and the perpendicular distance between the point of axis of rotation and the line of action of the force.

moment of inertia *(symbol I).* Of a body about an axis, Σmr^2 where *m* is the mass of a particle of the body and *r* is its distance from the axis.

momentum. Quantity of motion.

Linear momentum is the quantity obtained by multiplying the mass of a body by its linear speed. Angular momentum is the quantity obtained by multiplying the moment of inertia of a body by its angular speed.

The momentum of a system of particles is given by the sum of the momentums of the individual particles which make up the system or by the product of the total mass of the system and the velocity of the center of gravity of the system.

The momentum of a continuous medium is given by the integral of the velocity over the mass of the medium or by the product of the total mass of the medium and the velocity of the center of gravity of the medium.

monitor. To observe, listen in on, keep track of, or exercise surveillance over by any appropriate means, as, *to monitor radio signals; to monitor the flight of a rocket by radar; to monitor a landing approach.*

monitor (nucleonics). An instrument that measures the level of **ionizing radiation** in an area. (See **radiation detection instrument, radiation monitoring.**)

monochromatic. Pertaining to a single **wavelength** or, more commonly, to a narrow band of wavelengths.

monocoque. A type of construction, as of a rocket body, in which the most or all of the stresses are carried by the **skin.**

A monocoque may incorporate formers but not longitudinal members such as stringers.

monoplane. An airplane or glider with one pair of wings.

monopropellant. A **rocket propellant** consisting of a single substance, especially a liquid, capable of producing a heated jet without the addition of a second substance. Used attributively in phrases, such as *monopropellant rocket engine or motor, monopropellant rocket fuel, monopropellant system,* etc.

month. 1. The period of the **revolution** of the moon around the earth. The month is designated as *sidereal, tropical, anomalistic, dracontic,* or *synodical,* according to whether the revolution is relative to the stars, the vernal equinox, the perigee, the ascending node, or the sun. **2.** The calendar month, which is a rough approximation to the **synodical month.**

moon. 1. The natural satellite of the earth. **2.** A natural **satellite** of any planet. See **planet.**

moonrise. The crossing of the visible **horizon** by the **upper limb** of the ascending moon.

moonset. The crossing of the visible **horizon** by the **upper limb** of the descending moon.

motion. The act, process, or instance of change of position. Also called *movement,* especially when used in connection with problems involving the motion of one craft relative to another.

motion sickness. The syndrome of pallor, sweating, nausea, and vomiting which is induced by unusual **accelerations.**

motor. See **engine.**

motorboating. Oscillation in a system or component, usually manifested by a succession of **pulses** occurring at a **subaudio** or low-audio repetition **frequency.**

movement area. The runways, taxiways, and other areas of an airport which are used for taxiing, takeoff, and landing of aircraft, exclusive of loading ramps and parking areas.

moving target indicator *(abbr* **MTI).** A device which limits the display of **radar** information primarily to moving **targets.**

MRA. Minimum reception altitude.

M-region magnetic storm. A magnetic storm that is independent of visible solar disk features; it begins gradually and shows a strong tendency to recur within a period of 27 days. The hypothetical region on the solar disk assumed to be the source of the incident corpuscular radiation is called the M-region.

MSFN. Manned Spaceflight Network.

MSL or msl. Mean sea level.

MTI *(abbr).* *moving target indicator.*

multi *(combining form).* More than one. Used in contexts where a category of two or more is distinguished from a category of one, as in *a multipropellent fuel system is more complicated than a monopropellant system.*

multicoupler. A device for connecting several **receivers** to one **antenna** and properly matching the impedances of the receivers and the antenna.

multiple-degree-of-freedom system. A mechanical **system** for which two or more **coordinates** are required to define completely the position of the system at any instant.

multiple ejection rack. A device designed to carry more than one item of ordnance on a single station of an aircraft.

multiple independent re-entry vehicle. The concept whereby one booster carries multiple reentry vehicles which can be aimed at separate targets. The term multiple independently targetable re-entry vehicles is included in this concept.

multiplexer. A mechanical or electrical device for time sharing of a **circuit.**

multiplexing. The simultaneous transmission of two or more **signals** within a single **channel.** The three basic methods of multiplexing involve the separation of signals by time division, frequency division, and phase division.

multipropellant. A **rocket propellant** consisting of two or more substances fed separately to the **combustion chamber.** See **bipropellant.**

multistage compressor. An **axial-flow compressor** having two or more, usually more than two, stages of rotor and stator blades; a radial-flow compressor having two or more **impeller wheels.** Also called a *multiple-stage compressor.*

multistage rocket. A vehicle having two or more **rocket** units, each unit firing after the one in back of it has exhausted its **propellant.** Normally, each unit, or stage, is jettisoned after completing its firing. Also called a *multiplestage rocket* or, infrequently, a *step rocket.*

munitions. A general term applying to all types of nuclear and non-nuclear weapons, general purpose force missiles, and chemical/biological agents designed for use by combat elements, including those elements training to be such, of the armed forces for inflicting or aiding in inflicting damage to or for the neutralization of enemy personnel, equipment or facilities. It includes such items as bombs, rockets, missile warheads, small arms and cartridges, bulk explosives, smoke agents, incendiaries and non-explosive practice and training devices simulating the parent munition. See also **ammunition, air munition, ordance.**

music. Electronic jamming.

N

NACA (*abbr*). **National Advisory Committee** for **Aeronautics.** The predecessor of NASA.

NACA Standard Atmosphere. See **standard atmosphere.**

nacelle. Enclosed shelter for a powerplant or personnel. Usually secondary to the fuselage or cabin.

nadir. That point on the **celestial sphere** vertically below the observer, or 180° from the **zenith.**

napalm. A powder employed to thicken gasoline for use in flame throwers and incendiary bombs.

Napierian base. The logarithmic base, *e.*

NASA (*abbr*). **National Aeronautics and Space Administration.**

NASC (*abbr*). **National Aeronautics and Space Council.**

National Aeronautics and Space Administration (NASA). An agency of the United States Government directed to provide for research into the problems of flight within and outside the Earth's atmosphere. Major NASA facilities are: Headquarters, Ames Research Center, Electronics Research Center, Flight Research Center, Goddard Space Flight Center, Jet Propulsion Laboratory, John F. Kennedy Space Center, Langley Research Center, Lewis Research Center, Manned Spacecraft Center, George C. Marshall Space Flight Center, Nuclear Rocket Development Station, Pacific Launch Operations Office, Wallops Station.

national emergency. A condition declared by the President or the Congress by virtue of powers previously vested in them which authorize certain emergency actions to be undertaken in the national interest. Actions to be taken may include partial or total mobilization of national resources.

nationality mark. The international letter (or letters) used to identify the nationality of the aircraft.

National Military Command Center. The central facility located in the Pentagon to support the national command authorities in their strategic direction of US Armed Forces.

national range. Department of Defense operated flight test network which, because of its size and general purpose facilities, is considered a national asset and is equally available to all United States Government sponsored range users on a common basis. Range capabilities include facilities to support tests of missiles, drones, space launch vehicles, satellites, and space probes. The Air Force Western Test Range, Air Force Eastern Test Range, and White Sands Missile Range are national ranges.

national search and rescue plan. An interagency agreement whose purpose is to provide for the effective utilization of all available facilities in all types of search and rescue missions.

National Transportation Safety Board (NTSB). An independent Federal agency whose function in civil aviation is to investigate aviation accidents; report publicly on their cause or probable cause; and to review on appeal the suspension, revocation, or denial of operating certificates, documents, or licenses issued by the FAA administrator.

National Weather Service. A U.S. Government agency responsible for the nation's weather service. It is the primary source of aviation weather information. Formerly called *Weather Bureau.* **Weather maps, reports** and **forecasts** are prepared specifically for aviation use.

NATO. North Atlantic Treaty Organization.

natural frequency. 1. The frequency of **free oscillation** of a system. For a multiple-degree-of-freedom system, the natural frequencies are the frequencies of the normal **modes** of vibration. **2.**The undamped **resonance frequency** of a physical system. It is expressed in cycles per unit time. The system may be mechanical, pneumatic, or electrical.

natural year. Same as tropical year.

nausea. A feeling of discomfort in the region of the stomach, with aversion to food and a tendency to vomit.

Nautical Almanac. An annual publication of the U.S. Naval Observatory and H.M. Nautical Almanac Office, Royal Greenwich Observatory, listing the Greewich hour angle and declination of various **celestial bodies** to a precision of 0.1 minute of arc at hourly intervals; time of sunrise, sunset, moonrise, moonset; and other astronomical information useful to navigators. Prior to 1960 separate publications were issued by the two observatories entitled the *American Nautical Almanac and the Abridged Nautical Almanac. See **American Ephemeris and Nautical Almanac.***

nautical mile. A unit of distance used principally in navigation. For practical navigation it is usually considered the length of 1 minute of any **great circle** of the earth, the **meridian** being the great circle most commonly used. Also called *sea mile.*

Because of various lengths of the nautical mile in use throughout the world, due to differences in definition and the assumed size and shape of the earth, the International Hydrographic Bureau in 1929 proposed a standard length of 1852 meters, which is known as the international nautical mile. This has been adopted by nearly all maritime nations. The U.S. Departments of Defense and Commerce adopted this value on July 1, 1954. With the yard-meter relationship then in use, the international nautical mile was equivalent to 6076.10333 feet. Using the yard-meter conversion factor effective July 1, 1959, the international nautical mile is equivalent to 6076.11549 international feet.

NAVID. Air Navigation Facility.

naval campaign. An operation or a connected series of operations conducted essentially by naval forces including all surface, subsurface, air, and amphibious troops, for the purpose of gaining, extending, or maintaining control of the sea.

naval or marine (air) base. An air base for support of naval or marine air units, consisting of landing strips, seaplane alighting areas, and all components of related facilities for which the Navy or Marine Corps has operating responsibilities, together with interior lines of communication and the minimum surrounding area necessary for local security. (Normally, not greater than an area of 20 square miles.) See also **base complex.**

NavierStokes Equations. The **equations of motion** for a viscous fluid.

navigation. The practice or art of directing the movement of a craft from one point to another. *Navigation* usually implies the presence of a human, a navigator, aboard the craft. Compare **guidance**. Also see **celestial navigation, dead reckoning, pilotage, radio navigation,** inertial **navigation.**

navigational planets. The four **planets** commonly used in celestial surface and air navigation: Venus, Mars, Jupiter, and Saturn.

navigational stars. The 57 stars included in the main listing of the Nautical Almanac and Polaris. The navigational stars include almost all the stars with common names.

navigational triangle. The spherical triangle solved in computing **altitude** and **azimuth**. See **celestial triangle**. The celestial triangle is formed on the celestial sphere by the great circles connecting the elevated pole, zenith of the assumed position of the observer, and a celestial body. The terrestrial triangle is formed on the earth by the great circles connecting the pole and two places on the earth. The term *astronomical triangle* applies to either the celestial or terrestrial triangle used for solving celestial observations.

navigation dome. Astrodome.

navigation lights. Lights, specified by regulations, to be shown by all aircraft in flight during the hours of darkness.

navigation, radio. See **radio navigation.**

navigator. See **flight navigator.**

NDB. See **non-directional radio beacon (NDB).**

near miss (aircraft). Any circumstance in flight where the degree of seperation between two aircraft is considered by either pilot to have constituted a hazardous situation involving potential risk of collision.

need to know. A criterion used in security procedures which requires the custodians of classified information to establish, prior to disclosure, that the intended recipient must have access to the information to perform his official duties.

negative. Cancel, or no. Normally used in radio communications.

negative acceleration. Same as deceleration.

negative dihedral. A downward inclination of a wing or other surface.

negative divergence. Same as convergence.

negative feedback. **Feedback** which results in decreasing the amplification.

negative g. In designating the direction of acceleration on a body, the opposite of *positive g,* for example, the effect of flying an outside loop in the upright seated position. See **physiological acceleration.**

nephometer. A general term for instruments designed to measure the amount of cloudiness. An early type consists of a convex hemispherical mirror mapped into six parts. The amount of cloud coverage on the mirror is noted by the observer. Also called *nephelometer*. Compare **nephoscope.**

nephoscope. An instrument for determining the direction of cloud motion. There are two basic designs of nephoscope: the direct-vision nephoscope and the mirror nephoscope. Also called *nepheloscope.*

Neptune. See **planet.**

net (communications). An organization of stations capable of direct communications on a common channel or frequency.

net radiation factor. The fraction of the total **energy** emitted by one surface or volume that is absorbed by another surface or volume directly and indirectly.

net thrust. The gross **thrust** of a jet engine minus the **drag** due to the **momentum** of the incoming air.

network. 1. A combination of **electrical elements. 2.** A group of parts or systems combined to provide a closed information loop, i.e., one that provides for inquiry or command, response, and interpretation of response in relation to inquiry or command.

neutral. Without an electrical charge; neither positive nor negative.

neutral point. 1. In atmospheric optics, one of several points in the sky for which the degree of **polarization** of **diffuse sky radiation** is zero. 2. In aircraft, that location of the center of gravity at which the aircraft would exhibit neutral aerodynamic **stability. 3. Lagrangian point.**

neutron. An uncharged *elementary particle* with a mass slightly greater than that of the *proton,* and found in the *nucleus* of every atom heavier than hydrogen. A free neutron is unstable and decays with a half-life of about 13 minutes into an electron, proton, and neutrino. Neutrons sustain the fission *chain reaction* in a *nuclear reactor.*

neutrosphere. The **atmospheric shell** from the earth's surface upward in which the atmospheric constituents are for the most part not ionized, i.e., it is electrically neutral. The region of transition between 40 and 55 miles depending on latitude and season.

new moon. The moon at **conjunction,** when little or none of it is visible to an observer on the earth because the illuminated side is away from him. Also called *change of the moon.* See **phases of the moon.**

newton. The unit of force in the MKSA system; that force which gives to a mass of 1 kilogram an acceleration of 1 meter per second squared.

Newtonian mechanics. The system of mechanics based upon **Newton laws of motion** in which **mass** and **energy** are considered as separate, conservative, mechanical properties, in constrast to their treatment in relativistic mechanics.

Newtonian telescope. A **reflecting telescope** in which a small plane mirror reflects the convergent beam from the objective to an eyepiece at one side of the telescope. After the sound reflection the rays travel approximately perpendicular to the longitudinal axis of the telescope. See **Cassegrain telescope.**

Newtonian universal constant of gravitation. Gravitional constant.

Newton law of gravitation. Every particle of matter in the universe attracts every other particle with a force, F, acting along the line joining the two particles, proportional to the product of the masses $m_1 m_2$ of the particles and inversely proportional to the square of the distance r between the particles, or $F = Gm_1m_2/r^2$, where $G =$ gravitational constant.

Newton laws of motion. A set of three fundamental postulates forming the basis of the mechanics of rigid bodies, formulated by Newton in 1687. The first law is concerned with the principle of inertia and states that if a body in motion is not acted upon by an external force, its momentum remains constant (law of conservation of momentum). The second law asserts that the rate of change of momentum of a body is proportional to the force acting upon the body and is in the direction of the applied force. A familiar statement of this is the equation $F = ma$ where F is vector sum of the applied forces, m is the mass, and a is the vector acceleration of the body. The third law is the principle of action and reaction, stating that for every force acting upon a body there exists a corresponding force of the same magnitude exerted by the body in the opposite direction.

night. The time between the end of evening civil twilight and the beginning of morning civil twilight, as published in the American Air Almanac, converted to local time.

night cap. Night combat air patrol (written NCAP).

night effect. An effect mainly caused by variations in the state of polarization of reflected waves, which sometimes result in errors in direction finding bearings. The efffect is most frequent at night fall.

nimbostratus. A form of **middle cloud. Also see cloud, nimbus** and **stratiform clouds.**

nimbus. Nimbus means "raincloud". When added to stratus, as in **nimbostratus** or to cumulus, as in **cumulonimbus,** it means precipitation is produced by these clouds. See **cloud.**

nitrogen cycle. The exchange of nitrogen between animals and plants, in which plants convert urea or nitrates to protein, animals digest protein and excrete its nitrogen content as urea, which is taken up again by plants.

nitrogen desaturation. The reduction of the nitrogen content of the tissues of the body by breathing gases not containing nitrogen.

noctilucent clouds. Clouds of unknown composition which occur at great heights, 45 to 60 miles. They resemble thin cirrus, but usually with a bluish or silverish color, although sometimes orange to red, standing out against a dark night sky. Sometimes called *luminous clouds.* These clouds have been seen rarely, and then only during twilight, especially with the sun between 5^0 and 13^0 below the horizon; they have been observed only during summer months in both hemispheres (between latitudes 50^0 to 75^0 N and 40^0 to 60^0S), and only in some parts of these latitude belts.

nocturnal. Occurring during the hours between sunset and sunrise.

nocturnal radiation. Same as **effective terrestrial radiation.**

nodal point. Same as **node.**

node. 1. One of the two points of intersection of the **orbit** of a planet, planetoid, or comet with the **ecliptic,** or of the orbit of a **satellite** with the plane of the orbit of its **primary.** Also called *nodal point.* See **regression of the nodes.** That point at which the body crosses to the north side of the reference plane is called the *ascending node;* the other, the *descending node.* The line connecting the node is called *line of nodes.* **2.** A point, line, or surface in a **standing wave** where some characteristic of the wave field has essentially zero amplitude. The appropriate modifier should be used before the word *node* to signify the type that is intended; e.g., *displacement node, velocity node, pressure node.***3.**A terminal of any branch of a *network* or a terminal common to two or more branches of a network. Also called *junction point, branch point,* or *vertex.*

nodical period. The interval between successive passages of a **satellite** through the **ascending node.**

noise. 1. Any undesired **sound.** By extension, noise is any unwanted disturbance within a useful frequency band, such as undesired electric waves in a transmission channel or device. When caused by natural electrical discharges in the atmosphere, noise may be called *static.* **2.** An erratic, intermittent, or statistically random **oscillation. 3.** In electrical circuit analysis, that portion of the unwanted **signal** which is statistically random, as distinguished from hum, which is an unwanted signal occurring at multiples of the power-supply frequency. If ambiguity exists as to the nature of the noise, a phrase such as *acoustic noise* or *electric noise* should be used. Since the above definitions are not mutually exclusive, it is usually necessary to depend on context for the distinction.

nominal weapon. A nuclear weapon producing a yield of approximately 20 kilotons. See also **kiloton weapon; subkiloton weapon.**

nonair transportable. That which is not transportable by air by virtue of dimension, weight and/or special characteristics or restrictions.

nonalignment. The political attitude of a state which does not associate, or identify itself with the political ideology or objective espoused by other states, groups of states, or international causes, or with the foreign policies stemming therefrom. It does not preclude involvement, but expresses the attitude of no precommitment to a particular state (or bloc) or policy before a situation arises.

noncondensable gas. A gas whose temperature is above its **critical temperature** so that it cannot be liquefied by increase of pressure alone.

nondestructive inspection. A family of methods for investigating the quality, integrity, properties, and dimensions of materials and components without damaging or impairing their serviceability, by the use of optic, penetrant, magnetic, eddy current, ultrasonic, radiographic, infrared, etc., devices.

nondimensional number. A pure number not involving any physical dimensions, e.g., a ratio of two velocities or two lengths. Such numbers are fundamental descriptive quantities of a physical system. Nondimensional numbers involving several variables often are interpreted as estimates of characteristic velocity ratios, force ratios, heat transfer ratios, frequency ratios, etc. Usually several different ratio interpretations are possible and useful for the same number. (See **Mach number, Reynolds number, Prandtle number, Rayleigh number, Nusselt number, Grashof number.**)

nondimensional parameter. Any parameter of a problem which has the dimensions of a pure number, usually rendered so deliberately. See **nondimensional number.**

non-directional radio beacon (NDB). 1. A low or medium-frequency radio beacon transmits nondirectional signals whereby the pilot of an aircraft equipped with a loop antenna can determine his bearing and "home" on the station. These facilities normally operate in the frequency band of 200 to 415 kHz and transmit a continuous carrier with 1,020-cycle modulation keyed to provide identification except during voice transmission. **2.** When a radio beacon is used in conjunction with the **Instrument Landing System** markers, it is called a **Compass Locator.** Also see **beacon.**

noneffective sortie. Any aircraft dispatched which for any reason fails to carry out the purpose of the mission. Abortive sorties are included.

nonimpinging injector. An **injector** used in rocket engines

*Most airplanes weighing less than 12,500 pounds such as the twin engine Piper Aztec are type certificated by the Federal Aviation Administration (FAA) in the **normal category**. Acrobatic maneuvers are not permitted. Some airplanes like the Cessna 150 Aerobat are type certificated for acrobatic maneuvers.*

which employs parallel streams of propellant usually emerging normal to the face of the injector. In this injector, mixing is usually obtained by turbulence and diffusion. The V-2 used a nonimpinging injector.

nonlinear damping. Damping due to a damping **force** that is not proportional to **velocity.** The words *force* and *velocity* should be treated in the generalized sense For example, they can be replaced by *voltage* and *current,* respectively.

nonlinear distortion. Distortion caused by a deviation from a proportional relationship between specified measures of the output and input of a system. The related measures need not be output and input values of the same quantity; e.g., in a linear detector, the desired relation is between the output signal voltage and the input modulation envelope.

nonnuclear electromagnetic radiation. See **nuclear radiation.**

non-operating active aircraft (U.S.Air Force). An allowance of air-frames over and above the authorized unit equipment to permit heavy maintenance, modifications, and inspect and repairs as necessary without reduction of numbers of units available for operations. Ten percent of unit equipment is the normal allowance. No funds, manpower, or flying hours are allocated for these aircraft in the Air Force budget. Included are other aircraft in nonflyable status when they are programmed to eventually be returned to active flying. See also **operating active aircraft.**

non-precision approach procedure. A standard instrument approach procedure in which no electronic glide slope is provided.

nonprogram aircraft (military). All aircraft, other than active and reserve categories, in the total aircraft inventory, including X-models; aircraft for which there is no longer a requirement either in the active or reserve category; and aircraft in the process of being dropped from the total aircraft inventory. See also **aircraft.**

non-rigid airship. An airship in which the internal pressure alone maintains the designed shape of the envelope.

nonscheduled service. Revenue flights that are not operated in regular scheduled service such as charter flights, and all nonrevenue flights incident to such flights.

nonvolatile. Of a **computer** or computer component. The ability to retain information in the absence of power as nonvolatile memory, nonvolatile storage.

nonvolatile storage. Same as permanent memory.

noon. The instant at which a time reference is over the **upper branch** of the **reference meridian.** Noon may be *solar* or *sidereal* as either the sun or vernal equinox is over the upper branch of the reference meridian. Solar noon may be further classiffied as *mean* or *apparent* as the mean or apparent sun is the reference. Noon may also be classiffied according to the reference meridian, either the local or Greenwich meridian or additionally in the case of mean noon, a designated time zone meridian.

normal. 1. Equivalent to usual, regular, rational, or standard conditions. **2.** Perpendicular. A line is normal to another line or a plane when it is perpendicular to it. A line is normal to a curve or curved surface when it is perpendicular to the tangent line or plane at the point of tangency. **3.** Referring to thermal radiation properties, in a direction perpendicular to the surface. **4.** The line **normal,** sense **2,** to a surface.

normal category. As applied to an aircraft **type certificate** issued by the **FAA,** normal **category** is limited to aircraft intended for non-**acrobatic** operation.

normal distribution. The fundamental frequency distribution of statisical analysis.

normal emittance. Emittance in a direction perpendicular to the surface or in a small solid angle whose axial ray is perpendicular to the surface.

normal gravity. See **acceleration of gravity.**

normal mode of vibration. A mode of **free vibration** of an undamped system. In general, any composite motion of a vibrating system can be analyzed into a summation of its normal modes. Also called *natural mode, characteristic mode,* and *eigenmode.*

normal plane. In aerodynamics, a plane at right angles to the **longitudinal axis** of an **aerodynamic vehicle.**

normal shock wave. A **shock wave** perpendicular, or substantially so, to the direction of flow in a **supersonic flow** field. Sometimes shortened to *normal shock.*

north. See **north pole, compass north, magnetic north, true north, magnetic pole.**

northerly turning error. The error caused in a magnetic compass by the vertical component of the earth's magnetic field which is at its maximum when an aircraft on a northerly or southerly heading banks to turn off that heading.

northern lights. Same as **aurora borealis.**

north polar sequence. A list of stars near the north celestial pole arranged in order of photographic **magnitude,** used as reference stars in stellar **photometry.**

north pole. 1. In astronomy, that end of the axis of rotation of a celestial body at which, when viewed from above, the body appears to rotate in a clockwise direction. See **celestial pole, ecliptic pole, geographical pole, geomagnetic pole, magnetic pole. 2.** The north-seeking end of a magnet.

nose cone. The cone-shaped leading end of a **rocket vehicle,** consisting of (**a**) of a chamber or chambers in which a satellite, instruments, animals, plants or auxiliary equipment may be carried, and (**b**) of an outer surface built to withstand high temperatures generated by **aerodynamic heating.** In a satellite vehicle, the nose cone may become the satellite itself after separating from the final stage of the rocket or it may be used to shield the satellite until orbital speed is accomplished, then separating from the satellite.

nose gear. That part of **a landing gear** which is located at the forward end of the vehicle.

nose wheel. A **landing gear** wheel located under the nose of those airplanes which make use of a **tricycle landing gear.**

NOTAM. See **notice to airmen.**

notation. A manner of representing quantities. See **positional notation, binary notation, decimal notation, fixed point, floating point.**

NOT circuit. In computers, a device or **circuit** which inverts the **polarity** of a **pulse.** Also called *inverter.*

notice to airmen. A notice identified either as a **NOTAM** or Airmen Advisory containing information concerning the establishment, condition, or change in any component of, or hazard in, the National Airspace System, the timely knowledge of which is essential to personnel concerned with flight operations. (**1**) *NOTAM.* A Notice to Airmen in message form requiring expeditions and wide dissemination by telecommunications means. (**2**) *Airmen Advisory.* A Notice to

Airmen normally only given local dissemination, during preflight or in-flight briefing or otherwise during contact with pilots.

nova (*plural* **novae**). A star which suddenly becomes many times brighter than previously, and then gradually fades.

nozzle. 1. A duct, tube, pipe, spout, or the like through which a **fluid** is directed and from the open end of which the fluid is discharged, designed to meter the fluid or to produce a desired direction, velocity, or shape of discharge. See **de Laval nozzle, jet nozzle, supersonic nozzle. 2.** Specifically, that part of a **rocket thrust chamber** assembly in which the gases produced in the chamber are accelerated to high velocities.

nozzle blade. Any one of the blades or vanes in a **nozzle diaphragm.** Also called a *nozzle vane.*

nozzle-contraction area ratio. Ratio of the cross-sectional area for gas flow at the nozzle inlet to that at the throat.

nozzle diaphram. A ring of stationary, equally spaced blades or vanes, forming an annulus of nozzles through which fluid is directed onto a turbine wheel. Sometimes called a *nozzle ring.*

nozzle efficiency. The efficiency with which a **nozzle** converts **potential energy** into **kinetie energy,** commonly expressed as the ratio of the actual change in kinetic energy to the ideal change at the given pressure ratio.

nozzle exit area. The cross-sectional area of a rocket **nozzle** available for gas flow measured at the nozzle exit.

nozzle-expansion area ratio. Ratio of the cross-sectional area for gas flow at the exit of a **nozzle** to the cross-sectional area available for gas flow at the throat.

nozzle throat. The portion of a **nozzle** with the smallest cross section.

nozzle throat area. The area of the minimum coss section of a **nozzle.**

nozzle thrust coefficient. A measure of the amplification of **thrust** due to gas expansion in a particular nozzle as compared with the thurst that would be exerted if the **chamber pressure** acted only over the **throat** area. Also called *thrust coefficient.*

NTSB. National Transportation Safety Board.

nuclear battery. A radioisotopic generator.

nuclear cross section. A measure of the probability that the reaction will take place. See **barn.**

nuclear defense. The methods, plans, and procedures involved in establishing and exercising defensive measures against the effects of an attack by nuclear weapons or radiological warfare agents. It encompasses both the training for, and the implementation of, these methods, plans, and procedures. See also **radiological defense.**

nuclear detonation detection and reporting system. A system deployed to provide surveillance coverage of critical friendly target areas and indicate place, height of burst, yield, and ground zero of nuclear detonations. See also **bomb alarm system.**

nuclear device. See **device, nuclear.**

nuclear energy. The energy liberated by a nuclear reaction (fission or fusion) or by radioactive decay. (See **decay, radioactive, fission, fusion, nuclear explosive, nuclear reactor.**)

nuclear explosive. An explosive based on *fission* or *fusion* of atomic *nuclei.* (See **device, nuclear, nuclear weapons.**)

nuclear incident. An unexpected event involving a nuclear weapon, facility, or component resulting in any of the following, but not consituting a nuclear weapon(s) accident: **a.** an increase in the possibility of explosion or radioactive contamination; **b.** errors committed in the assembly, testing, loading, or transportation of equipment, and/or the malfunctioning of equipment and material which could lead to an unintentional operation of all or part of the weapon arming and/or firing sequence, or which could lead to a substantial change in yeild, or increased dud probability; and **c.** any act of God, unfavorable environment or condition resulting in damage to the weapon, facility, or component.

nuclear nations Military nuclear powers and civil nuclear powers.

nuclear parity. A condition at a given point in time when opposing forces possess nuclear offensive and defensive systems approximately equal in overall combat effectiveness.

nuclear power plant. Any device, machine, or assembly that converts nuclear energy into some form of useful power, such as mechanical or electrical power. In a nuclear electric power plant, heat produced by a *reactor* is generally used to make steam to drive a turbine that in turn drives an electric generator.

nuclear radiation. Particulate and electromagnetic radiation emitted from atomic nuclei in various nuclear processes. The important nuclear radiations, from the weapons standpoint, are alpha and beta particles, gamma rays, and neutrons. All nuclear radiations are ionizing radiations, but the reverse is not true; X-rays, for example, are included among ionizing radiations, but they are not nuclear radiations since they do not originate from atomic nuclei.

nuclear reaction. A reaction involving a change in an atomic nucleus, such as *fission, fusion, neutron capture,* or *radioactive decay,* as distinct from a chemical reaction, which is limited to changes in the electron structure surrounding the nucleus. (Compare **thermonuclear reaction.**)

nuclear reactor. A device in which a fission *chain reaction* can be initiated, maintained, and controlled. Its essential component is a *core* with fissionable *fuel.* It usually has a *moderator,* a *reflector, shielding, coolant,* and control mechanisms. Sometimes called an atomic "furnace", it is the basic machine of *nuclear energy.* (See **fission.**)

Nuclear Regulatory Commission. The independent civilian agency of the federal government with statutory responsibility for nuclear energy matters.

nuclear rocket. A rocket powered by an engine that obtains energy for heating a propellant fluid (such as hydrogen) from a nuclear reactor, rather than from chemical combustion.

nuclear stalemate. A concept which postulates a situation wherein the relative strength of opposing nuclear forces results in mutual deterrence against employment of nuclear forces.

nuclear warfare. Warfare involving the employment of nuclear weapons.

nuclear weapons. A collective term for *atomic bombs* and *hydrogen bombs.* Any weapons based on a *nuclear explosive.* (Compare **device, nuclear.**)

nuclear weapon employment time. The time required for delivery of a nuclear weapon after the decision to fire has been made.

nuclear weapon exercise. An operation not directly related to immediate operational readiness. It includes removal of a weapon from its normal storage location, preparing for use, delivery to an employment unit, the movement in a ground training exercise to include loading aboard an aircraft or missile and return to storage. It may include any or all of the operations listed above, but does not include launching or flying operations. Typical exercises include aircraft generation exercises, ground readiness exercises, ground tactical exercises, and various categories of inspections designed to evaluate the capability of the unit to perform its prescribed mission. See also **nuclear weapon maneuver.**

nuclear weapon maneuver. An operation not directly related to immediate operational readiness. It may consist of all those operations listed for a nuclear weapon exercise but is extended to include fly-away in combat aircraft not to include expenditure of the weapon. Typical maneuvers include nuclear operational readiness maneuvers and tactical air operations. See also **nuclear weapon exercise.**

nuclear weapon(s) accident. Any unplanned occurrence involving loss or destruction of, or serious damage to, nuclear weapons or their components which results in an actual or potential hazard to life or property.

nuclear yields. The energy released in the detonation of a nuclear weapon, measured in terms of the kilotons or megatons of trinitrotoluene (TNT) required to produce the same energy release. Yields are categorized as:

very low — less than 1 kiloton.
low — 1 kiloton to 10 kilotons.
medium — over 10 kilotons to 50 kilotons.
high — over 50 kilotons to 500 kilotons.
very high — over 500 kilotons.

See also **nominal weapon; subkiliton weapon.**

nucleonics. The science and technology of nuclear energy and its applications.

nucleus. 1.The positively charged core of an **atom** with which is associated practically the whole mass of the atom but only a minute part of its volume. A nucleus is composed of one or more protons and an approximately equal number of neutrons. The atomic number Z of the element indicates the number of protons in the nucleus. The mass number A of the element is the sum of the protons and neutrons. **2.** In biology, a definitely delineated body within a cell containing the chromsomes.

nuclide. A general term applicable to all atomic forms of the elements. The term is often erroneously used as a synonym for "isotope", which properly has a more limited definition. Whereas isotopes are the various forms of a single element (hence are a family of nuclides) and all have the same *atomic number* and number of protons, nuclides comprise *all* the elements. Nuclides are distinguished by their *atomic number, atomic mass,* and energy state. (Compare **element, isotope.**)

null. In direction-finding systems wherein the output amplitude is a function of the direction of arrival of the signal, the minimum output amplitude (ideally zero). The null is frequently employed as a means of determining bearing. The term *minimum* is often used to indicate an imperfect null.

number. In computer operations, (**a**) amount of units by count, (**b**) a magnitude or quantity represented by group of **digits.** The term *quantity* is preferred to *number* in sense (**b**).

number system. A scheme for representing magnitudes or quantities by a group of **digits.** See **numeric coding, positional notation.**

numeric. Composed wholly or partly of **digits.** See **alphanumeric, number.**

numeric coding. A system of coding in which information is represented by **digits.** See **alphabetic coding, alphanumeric.**

Nusselt number (*symbol* N$_{Nu}$). (After Wilhelm Nusselt, German engineer.) A number expressing the ratio of convective to conductive **heat transfer** between a solid boundary and a moving fluid, defined as hl/k where h is the heat-transfer coefficient, l is the characteristic length, and k is the thermal conductivity of the fluid.

nutating feed. In a **tracking radar** an oscillating antenna **feed** for producing an oscillating deflection of the **beam** in which the plane of **polarization** remains fixed.

nutation. 1. The **oscillation** of the axis of any rotating body, as a gyroscope rotor. **2.** Specifically, in astronomy, irregularities in the precessional motion of the **equinoxes** because of varying positions of the moon and, to lesser extent, of other celestial bodies with respect to the **ecliptic.** Because of nutation, the earth's axis nods like a top, describing a slightly wavy circle about the ecliptic pole. The maximum displacement is about 9.21 seconds (constant of nutation) and the period of a complete cycle is 18.60 tropical years (period of moon's node, nutation period).

nutator. A mechanical device for gyrating the antenna **feed** horn or dipole of a **radar** about the axis of the reflector without changing its **polarization.**

O

O. Military mission designation for **observation** aircraft.

objective. The lens or combination of lenses which receives light rays from an object and refracts them to form an image in the focal plane of the eyepiece of an optical instrument, such as a telescope. Also called *object glass.*

objective. (Military). The physical object of the action taken, e.g., a definite tactical feature, the seizure and/or holding of which is essential to the commander's plan.

oblate spheroid. An **ellipsoid** of **revolution,** the shorter **axis** of which is the axis of revolution. An ellipsoid of revolution, the longer axis of which is the axis of rotation, is called a *prolate spheroid.* The earth is approximately an oblate spheroid.

oblique. Pertaining to, or measured on, an oblique projection, as *oblique equator, oblique pole, oblique latitude.*

oblique air photograph. An air photograph taken with the camera axis directed between the horizontal and vertical planes. Commonly referred to as an "oblique": **a. high oblique:** One in which the apparent horizon appears; and **b. low oblique:** One in which the apparent horizon does not appear.

oblique air photograph strip. Photographic strip composed of oblique air photographs.

oblique coordinates. Magnitudes defining a point relative to two intersecting nonperpendicular lines, called *axes.* See **Cartesian coordinates.** The magnitudes indicate the distance from each axis, measured along a parallel to the other axis. The horizontal distance is called the *abscissa* and the other distance, the *ordinate.*

oblique projection. A map **projection** with an **axis** inclined at an oblique angle to the plane of the **equator.**

oblique shock wave. A **shock wave** inclined at an oblique angle to the direction of flow in a **supersonic flow** field. Sometimes shortened to *oblique shock.* Compare **normal shock.**

obliquity of the ecliptic. The angle between the plane of the **ecliptic** (the plane of the earth's orbit) and the plane of the **celestial equator.** The obliquity of the ecliptic is computed from the following formula: 23 degrees 27 minutes 08.26 seconds—0.4684 (t — 1900) seconds, where t is the year for which the obliquity is desired.

observation (O), (Military designation). Aircraft designed to observe (through visual or other means) and report tactical information concerning composition and disposition of enemy forces, troops and supplies in an active combat area. See **military aircraft types.**

observation post. A position from which military observations are made, or fire directed and adjusted, and which possesses appropriate communications; may be airborne.

observed. In astronomy and navigation, pertaining to a value which has been measured in contrast to one which is computed.

observed altitude. True altitude.

observation. A collection of measured weather information taken at a certain time.

obscuration. In weather observing practice, the designation for sky cover when more than nine-tenths of the sky is hidden by surface based obscuring phenomena and vertical visibility in restricted overhead. Coded X in aviation weather observations. When an obscuration is reported, a ceiling is also reported. Ceiling height ascribed to an obscuration is vertical visibility into the obscuration. Compare **obscuration, partial. 2.** Surface based obscuration phenomena.

obscuration (partial). In surface aviation weather observations, a designation of sky cover when (a) part (0.1 to 0.9) of the sky is completely hidden by surface based phenomena, or (b) ten tenths (to the nearest tenth) of the sky is hidden by surface based phenomena and the vertical visibility is not restricted; compare **obscuration.**

occluded front, (metgy). The **front** that is formed when and where the **cold front** overtakes the **warm front** and it may be either, the warm or cold type depending upon whether the air behind the **cold front** is warmer or colder than that ahead of the **warm front.** This front marks the position of an upper **trough** of warm air, originally from the warm sector, which has been forced aloft by the action of the converging cold and warm fronts. Also see **front, cold front, warm front, stationary front.**

occlusion (metgy). The process whereby a cold front overtakes a warm front or a stationary front and forces warm air aloft. See **occluded front.**

occlusion. Specifically, the trapping of undisolved **gas** in a solid during solidification.

occultation. The disappearance of a body behind another body of larger apparent size. When the moon passes between the observer and a star, the star is said to be *occulted.* The three associated terms, *occultation, eclipse,* and *transit,* are exemplified by the motions of the satellites of Jupiter. An eclipse occurs when a satellite passes into the shadow cast by the planet; an occultation occurs when a satellite passes directly behind the planet, so that it could not be seen even if it were illuminated; and a transit occurs when a satellite passes between the observer and the planet, showing against the disk of the planet.

oceanography. The study of the sea, embracing and integrating all knowledge pertaining to the sea's physical boundaries, the chemistry and physics of sea water, and marine biology.

ocean station ship. A ship assigned to operate within a specified area to provide several services including search and rescue, meteorological information, navigational aid and communications facilities.

o'clock. In clock code section and at range indicated (heading of own aircraft being twelve o'clock).

octant. See **sextant.**

octave. The interval between any two **frequencies** having the ratio of 1:2. The interval in octaves between any two frequencies is the logarithm to the base 2 (or 3.322 times the logarithum to the base 10) of the frequency ratio.

ocular. Pertaining to or in relation with the eye.

oersted. The **centimeter-gram-second** electromagnetic unit of **magnetic intensity.** See **gauss.**

OEW. Operating empty weight.

ogive. A body of revolution formed by rotating a circular arc about an axis that intersects the arc; the shape of this body; also, a nose of a projectile or the like so shaped. Typically, an ogive has the outline of a Gothic arch, although by definition it may be rounded rather than pointed.

ohm. The unit of electrical resistance; the resistance between two points of a conductor when a constant difference of potential of 1 **volt,** applied between these two points, produces in the conductor a current of 1 **ampere** (the conductor not being the source of any electromotive force).

oilcan. Of a sheet-metal skin or of other covering, to snap in and out between rows of rivets or between other places of support in a fashion like that of the bottom of an oilcan.

oleo. A shock-absorbing strut in which the spring action is dampened by oil.

omega/VLF. Basically, Omega/VLF is a long range navigation system utilizing very low frequency signals transmitted between 10 and 24 KHz. There are two independent worldwide networks; the Omega network operated by the U.S. Coast Guard and the VLF Communications network operated by the U.S. Navy. The computerized airborne equipment automatically selects the appropriate station signals and provides navigation information to the aircrew. Omega/VLF provides fixing accuracy within plus or minus 2 nautical miles anywhere on the surface of the earth.

omni. 1. A prefix meaning *all,* as in *omnidirectional.* **2.** Short for *omnirange.*

omnirange. Radio aid to air navigation which creates an infinite number of paths in space throughout 360 degrees azimuth. See **VHF omnidirectional range (VOR), VOR receiver, VORTAC.**

on station. 1. In Air intercept usage, I have reached my assigned station. **2.** In Close Air Support and Air interdiction, Airborne aircraft in position to attack targets, or to perform the mission designated by control agency.

on station time (Military). The time an aircraft can remain on station. May be determined by endurance or orders.

on target. My fire control director(s)/system(s) have acquired the indicated contact and is (are) tracking successfully.

on the deck. At minimum altitude.

opaque plasma. A **plasma** through which an **electromagnetic wave** cannot propagate and is either absorbed or reflected.

open-cycle reactor system (Nucleonics). A reactor system in which the *coolant* passes through the reactor core only once and is then discarded. (Compare **closed-cycle reactor system.**)

open loop. A **system** operating without **feed-back,** or with only partial feedback. See **closed loop.**

open system. A system that provides for the body's metabolism in an aircraft or spacecraft cabin by removal of respiratory products and of waste from the cabin and by use of stored food and oxygen. Compare **closed system.**

operand. In **computer** operations, a word on which an operation is to be performed.

operate. With respect to U.S. Civil aircraft, means use, cause to use or authorize to use aircraft, for the purpose of air navigation including the piloting of aircraft, with or without the right of legal control (as owner, lessee, or otherwise).

operating active aircraft. (Military) The authorized aircraft (unit equipment, training test, test support, or special activity) for which funds, manpower, and flying hours are allocated in the military budget. See also **non-operating active aircraft.**

operating empty weight (OEW). The **manufacturer's empty weight** plus the weight of oil, water, unusable fuel, crew, crew baggage, passenger equipment and emergency equipment.

operating weight. (Military) The term used for transport or tactical aircraft with includes basic weight, crew, oil, crewmen's baggage, stewards' equipment, and emergency and extra equipment that may be required. This does not include the weight of fuel, ammunition, bombs, cargo, antidetonation injection fluid, or external auxiliary fuel tanks if such tanks are to be disposed of in flight.

operation. A military action or the carrying out of a strategic, tactical, service, training, or administrative military mission; the process of carrying on combat, including movement, supply, attack, defense, and maneuvers needed to gain the objectives of any battle or campaign.

operational aircraft damage/loss. The damage or loss of an aircraft from all causes other than combat damage or loss.

operational characteristics. Those characteristics which pertain primarily to the functions to be performed by equipment, e.g., for electronic equipment, operational characteristics include such items as frequency coverage, channeling, type of modulation, and character of emission.

operational command. The authority granted to a commander to assign missions or tasks to subordinate commanders, to deploy units, to reassign forces, and to retain or delegate operational and/or tactical control as may be deemed necessary. It does not of itself include administrative command or logistical responsibility. May also be used to denote the forces assigned to a commander. See also **command.**

operational control. With respect to a flight, means the exercise of authority over initiating, conducting, or terminating a flight.

operational evaluation. (Military) The test and anylsis of a specific end item or system, insofar as practicable under Service operating conditions, in order to determine if quantity production is warranted considering: **a.** the increase in military effectiveness to be gained; and **b.** its effectiveness as compared with currently available items or systems, consideration being given to: (**1**) personnel capabilities to maintain and operate the equipment; (**2**) size, weight, and location considerations; and (**3**) enemy capabilities in the field. See also **technical evaluation.**

operationally ready. 1. as applied to a unit, ship or weapon system. Capable of performing the missions or functions for which organized or designed. Incorporates both equipment readiness and personnel readiness. **2. as applied to equipment.** Available and in condition for serving the functions for which designed. **3. as applied to personnel.** Available and qualified to perform assigned missions or functions.

operational missile. A missile which has been accepted by the using Services for tactical and/or strategic use.

operational phase. (Military). The period from acceptance by the user of the first operating unit until the elimination of the system from the inventory. See also **system life cycle.**

operational status. The attainment by a system, subsystem, or component of the capability to perform its intended mission.

operational test and evaluation. (Military). The term denotes any program or project designed to obtain, verify and provide data for conclusions about the suitability of operational systems, sub-systems, equipment, concepts, tactics, techniques, and procedures. The term *test* denotes the conduct of physical activity in pursuit of prescribed data objectives. The term *evaluation* denotes a review and analysis of quantitative or qualitative data produced by current test projects or programs, by previous testing, or by data provided from other sources; i.e., operational, research and development, supporting activities or from combinations of any of the foregoing.

operative temperature. In the study of human bioclimatology, one of several parameters devised to measure the air's cooling effect upon a human body. It is equal to the temperature at which a specified hypothetical environment would support the same heat loss from an unclothed, reclining human body as the actual environment. In the hypothetical environment, the wall and air temperatures are equal and the air movement is 7.6 centimeters per second.

opposed-cylinder engine. An engine with its cylinders arranged opposite each other in the same plane, their connecting rods working on the same crankshaft.

optical axis. Of an antenna, a line parallel to, but offset from, the electrical axis of an **antenna.** This axis is offset by the distance necessary to have the optical sighting device removed from the electrical center of the antenna.

optical instrumentation. The use of optical systems coupled with television and/or photographic recording devices to record scientific and engineering phenomena for the purposes of technical measurement and evaluation. It may include recording of correlation data to relate images to time, space position or other recorded data.

optical landing system (Military). A shipboard gyrostabilized or shore-based device which indicates to the pilot his displacement from a preselected glide path. See also **ground-controlled approach.**

optical line of sight. The generally curved path of visible light through the atmosphere. Often used erroneously for *geometrical line of sight.*

optical maser. Same as **laser.**

optical path. 1. line of sight. **2.** The path followed by a ray of light through an optical system.

optical pyrometer. A device for measuring the **temperature** of an incandescent radiating body by comparing its **brightness** for a selected wavelength interval within the **visible spectrum** with that of a standard source; a monochromatic radiation **pyrometer.** Temperatures measured by optical pyrometers are known as *brightness temperatures* and except for black bodies are less than the true temperatures.

optical slant range. The horizontal distance in a **homogeneous atmosphere** for which the **attenuation** is the same as that actually encountered along the true oblique path.

optical turbulence. Irregular and fluctuating gradients of optical **refractive index** in the atmosphere. Optical tur-

bulence is caused mainly by mixing of air of different temperatures, and particularly by thermal gradients which are sufficient to reverse the normal decrease in density with altitude, so that convection occurs.

optimal. Pertaining to a **trajectory,** path, or control motion, one that minimizes or maximizes some quantity or combination of quantities such as fuel, time, energy, distance, heat transfer, etc. This optimum condition, or path, is commonly calculated by a type of mathematics known as *calculus of variations.*

oranges (sour) (Military). Weather is unsuitable for aircraft mission.

oranges (sweet) (Military). Weather is suitable for aircraft mission.

orbit. 1. The path of a body or particle under the influence of a **gravitational** or other force. For instance, the orbit of a **celestial body** is its path relative to another body around which it revolves. *Orbit* is commonly used to designate a closed path and *trajectory* to denote a path which is not closed. Thus, the *trajectory of a sounding rocket, the orbit of a satellite.* **2.** To go around the earth or other body in an orbit, sense **1.**

orbital. Taking place in orbit, as *orbital refueling, orbital launch,* or pertaining to an orbit as *orbital plane.*

orbital elements. A set of seven parameters defining the **orbit** of a body attracted by a central, inverse-square force. Several different sets of parameters have been used. For artificial satellites the elements usually given are: longitude of the ascending node, inclination of the orbit plane, argument of perigee, eccentricity, semimajor axis, mean anomaly, and epoch.

orbital glider. See **hypersonic glider.**

orbital motion. Continuous motion in a closed path such as a circle or an ellipse.

orbital period. The interval between successive passages of a **satellite** through the same point in its **orbit.** Often called *period.* See **anomalistic period, nodical period, sidereal period.**

orbital velocity. 1. The average velocity at which an earth **satellite** or other orbiting body travels around its **primary.** Compare **separation velocity. 2.** The velocity of such a body at any given point in its orbit, as in *its orbital velocity at the apogee is less than at the perigee.* **3. circular velocity.**

orbiting. Of a **spacecraft,** in orbit about the earth or other spatial body, as in *an orbiting astronomical laboratory.*

orbiting. Circling, or circle and search.

orbit point. A geographically or electronically defined location over land or water, used in stationing airborne aircraft.

order. (Military) A communication, written, oral, or by signal, which conveys instructions from a superior to a subordinate. In a broad sense, the terms "order" and "command" are synonymous. However, an order implies discretion as to the details of execution whereas a command does not.

order of magnitude. A factor of 10. Compare **octave, magnitude.**

Two quantities of the same kind which differ by less than a factor of 10 are said to be of the same *order of magnitude.*

Order of magnitude is used loosely by many writers to mean a pronounced difference in quantity but the difference may be much less or much more than a factor of 10.

order of reflection. The number of hops, or trips, to the ionosphere and back to earth, that a **radio wave** makes in traveling from one point to another.

ordnance. Explosives, chemicals, pyrotechnic, and similar stores, e.g., bombs, guns and ammunition, flares, smoke, napalm.

organ. A portion or subassembly of a **computer** which constitutes the means of accomplishing some inclusive operation or function (e.g., arithmetic organ).

organic-cooled reactor. (Nucleonics). A reactor that uses organic chemicals, such as mixtures of polyphenyls (diphenyls and terphenyls), as coolant.

organization. A group of individuals with responsibilities, authorities, and relationships defined for the purpose of effective accomplishment of a mission.

orientation. The act of fixing position or attitude by visual or other reference.

orographic. Of, pertaining to, or caused by mountains as in orographic clouds, orographic lifting or orographic precipitation.

orthogonal. Originally, at right angles; later generalized to mean the vanishing of a sum (or integral) of products.

orthogonal antennas. In radar, a pair of transmitting and receiving **antennas**, or a single transmitting-receiving antenna, designed for the detection of a difference in **polarization** between the transmitted energy and the energy returned from the target.

oscillation. 1. Fluctuation or vibration on each side of a mean value or position. **2.** Half an oscillatory cycle, consisting of a fluctuation or vibration in one direction; half a vibration. **3.** The variation, usually with time, of the magnitude of a quantity with respect to a specified reference when the magnitude is alternately greater and smaller than the reference.

oscillator. A nonrotating device for producing alternating current.

oscillatory wave. A **wave** in which only the form advances, the individual particles moving in closed orbits, as ocean waves in deep water.

oscilloscope. 1. An instrument for producing a visual representation of **oscillations** or changes in an electric current. **2.** Specifically, a *cathode-ray oscilloscope*.

The face of the cathode-ray tube used for this representation is called a *scope* or *screen*.

osculating orbit. The **ellipse** that a satellite would follow after a specific time *t* (the **epoch** of osculation) if all forces other than central inverse-square forces ceased to act from time *t* on.

An osculating orbit is tangent to the real, perturbed, orbit and has the same velocity at the point of tangency.

otolith organs. Structures of the inner ear (utricle and saccule) which respond to linear acceleration and tilting.

outbound traffic. (Military) Traffic originating in continental United States destined for overseas or overseas traffic moving in a general direction away from continental United States.

outer atomsphere. Very generally, the atmosphere at a great distance from the earth's surface; an approximate synonym for **exosphere.**

outer fix. A fix in the destination terminal area, other than the approach fix, to which aircraft are normally cleared by an air route traffic control center or a terminal area traffic control facility, and from which aircraft are cleared to the approach fix, or final approach course.

outer marker (OM). An **ILS** marker which is located on a

localizer course line at a recommended distance (normally about 4½ miles) from the approach end of the runway. Also see **beacon, marker beacon, compass locator.**

outer planets. The **planets** with orbits larger than that of Mars: Jupiter, Saturn, Uranus, Neptune, and Pluto.

outgassing. The evolution of gas from a material in a vacuum.

out of phase. The condition of two or more **cyclic** motions which are not at the same part of their cycles at the same instant. Also called *out of step.* Compare **in phase.** Two or more cyclic motions which are at the same part of their cycles at the same instant are said to be *in phase* or *in step.*

outlook. A long range aviation weather forecast stated in general terms.

output. 1. The yield or product of an activity furnished by man, machine, or a system. **2.** Power or energy delivered by an engine, generator, etc. **3.** The electrical **signal** which emanates from a **transducer** and which is a function of the applied stimulus. Compare **input.** The quantity represented by the signal may be given in terms of electrical units, frequency, or time.

output unit. In computer terminology, a unit which delivers information from the computer to an external device or from internal **storage** to external storage.

outside loop. A flight maneuver in which the airplane describes an approximately circular path in the vertical plane and the upper parts of the airplane remaining on the outside of the circle producing negative g effects.

overall heat-transfer coefficient *(symbol U).* The value U, in British thermal units per hour per square foot per °F in the equation $Q = UA (t_1 - t_2)$ where Q is heat flow per unit time; A is area; and t is temperature.

overcast. Sky covered with clouds. See **ceiling.**

overcontrol. To move the aircraft's controls more than is necessary for the desired performance causing the aircraft's attitude to be displaced farther than intended.

overexpanding nozzle. A **nozzle** in which the fluid is expanded to a lower pressure than the external pressure. An overexpanding nozzle has an exit area larger than the optimum.

overhead approach/360 overhead. A series of predetermined maneuvers prescribed for VFR arrival of military aircraft (often in formation) for entry into the VFR traffic pattern and to proceed to a landing.

overpressure; Nucleonics. The transient pressure over and above atmospheric pressure caused by a shock wave from a nuclear explosion.

overseas. (Military) All locations, including Alaska and Hawaii, outside the continental United States.

overseas air commerce. The carriage by aircraft of persons or property for compensation or hire, or the carriage of mail by aircraft, or the operation or navigation of aircraft in the conduct or furtherance of a business or vocation, in commerce between a place in any State of the United States, or the District of Columbia, and any place in a territory or possession of the United States; or between a place in a territory or possession of the United States, and a place in any other territory or possession of the United States.

overseas air transportation. The carriage by aircraft of persons or property as a common carrier for compensation or hire, or the carriage of mail by aircraft, in commerce—

(1) Between a place in a State or the District of

Columbia and a place in a possession of the United States; or

(2) Between a place in a possession of the United States and a place in another possession of the United States;

whether that commerce moves wholly by aircraft or partly by aircraft and partly by other forms of transportation.

overseas (SAR) (search and rescue) region. Overseas unified command areas, including the inland area of Alaska, which are not included within the Inland Region or Maritime Region as defined by the **National SAR Plan.**

overshoot. To fly beyond a designated mark or area, such as a landing field, while attempting to land on the mark or within the area.

over-the-horizon radar. A radar system that makes use of the ionosphere to extend its range of detection beyond line-of-sight. Over-the-horizon radars may be either forward scatter or back-scatter systems.

over-the-shoulder bombing. A special case of loft bombing where the bomb is released past the vertical in order that the bomb may be thrown back to the target. See also **loft bombing; toss bombing.**

over-the-top. Above the layer of clouds or other obscuring phenomena forming the ceiling.

oxidant. Same as **oxidizer.**

oxidizer. Specifically, a substance (not necessarily containing oxygen) that supports the combustion of a **fuel** or **propellant.**

oximeter. An instrument for measuring the oxygen saturation of the blood.

oxygen bottle. A small container for pressurized oxygen used in life-support systems. See **bailout bottle.**

oxygen mask. A covering for the nose and lower face fitted with special attachments for breathing oxygen or a mixture of oxygen and other gases.

The oxygen mask has provision for separating the expired breath from the incoming oxygen.

ozone. A nearly colorless (but faintly blue) gaseous form of oxygen with a characteristic odor like that of weak chlorine. It is found in trace quantities in the atmosphere, primarily above the tropopause.

ozone converter (catalytic). Flight at high altitudes in the higher latitudes is subject to high **ozone** concentrations above Occupational Safety and Health Administration (OSHA) standards. Catalytic converters installed in transport aircraft air-conditioning and pressurization systems, burn-off ozone particles before ozone-laden air enters the crew and passenger compartments.

ozone layer. Same as **ozonosphere.**

ozonosphere. The general stratum of the **upper atmosphere** in which there is an appreciable ozone concentration and in which ozone plays an important part in the radiation balance of the atmosphere. This region lies roughly between 10 and 50 kilometers (6 to 30 miles), with maximum ozone concentration at about 20 to 25 kilometers (12 to 16 miles). Also called *ozone layer.* See **atmospheric shell.**

P

P. Military mission designation for patrol aircraft.

package. Any assembly or apparatus, complete in itself or practically so, identifiable as a unit and readily available for use or installation. See **power package.**

packaged bulk petroleum. Bulk petroleum which because of operational necessity is packaged and supplied (stored, transported, and issued) in 5-gallon cans or 55-gallon drums. See also **petroleum.**

packaged petroleum products. Petroleum products (generally lubricants, oils, greases and specialty items) normally packaged by a manufacturer and procured, stored, transported and issued in containers having a fill capacity of 55 United States gallons (45 Imperial gallons) or less.

package power reactor. A small nuclear power plant designed to be crated in packages small enough to be conveniently transported to remote locations.

pad. See **launch pad.**

pad deluge. Water sprayed upon certain **launch pads** during the launch of a rocket so as to reduce the temperatures of critical parts of the pad or the rocket.

pair production, nucleonics. The transformation of the kinetic energy of a high-energy *photon* or *particle* into mass, producing a *particle* and its *antiparticle,* such as an *electron* and *positron.* See **mass-energy equivalence.**

pallet. A flat base for combining stores or carrying a single item to form a unit load for handling, transportation and storage by materials handling equipment.

PAM (*abbr*). **Pulse amplitude modulation.**

pan (Military). Calling station has a very urgent message to transmit concerning the safety of a ship, aircraft, or other vehicle, or of some person on board or within sight.

pancake (Military). Land, or I wish to land (reason may be specified; e.g., "pancake ammo." "pancake fuel").

panchromatic. See **color sensitive.**

panoramic air camera. An air camera which through a system of moving optics or mirrors, scans a wide area of the terrain, usually from horizon to horizon. The camera may be mounted vertically or obliquely within the aircraft, to scan across or along the line of flight.

pants. A set of teardrop-shaped fairings around the wheels of a fixed landing gear on certain airplanes. Also called spats.

PAR. Precision Approach Radar.

parabola. An open curve all points of which are equidistant from a fixed point, called the *focus,* and a straight line. See **conic section.** The limiting case occurs when the point is on the line, in which case the parabola becomes a straight line.

parabolic. Pertaining to, or shaped like, a **parabola.**

parabolic orbit. An **orbit** shaped like a **parabola;** the orbit representing the least **eccentricity** (that of 1) for **escape** from an attracting body.

parabolic reflector. A reflecting surface having the cross section along the axis in the shape of a **parabola.** See **corner reflector, radar reflector, scanner.** Parallel rays striking the reflector are brought to a focus at a point, or if the source of the rays is placed at the focus, the reflected rays are parallel.

paraboloid. A surface of revolution generated by revolving a section of a **parabola** about its **major axis.**

parabrake. Same as **deceleration parachute.**

parachute. An umbrella-like device used to retard the descent of a falling body by offering resistance to its motion through the air.

parachute rigger. A person who packs and/or maintains a **parachute.** See **parachute rigger certificate, airman certificates.**

parachute rigger certificate. A certificate of competency issued by the FAA to a person meeting the requirements of the applicable **Federal Aviation Regulations.** Also see **parachute rigger, airman certificates.**

paradrop. Delivery by parachute of personnel or cargo from an aircraft in flight. See also **airdrop.**

parallax. The difference in the apparent direction or position of an object when viewed from different points expressed as an angle. For bodies of the solar system, parallax is measured from the surface of the earth and its center and is called *geocentric parallax,* varying with the body's altitude and distance from the earth. The geocentric parallax when a body is in the horizon is called *horizontal parallax* and is the angular semidiameter of the earth as seen from the body. Parallax of the moon is called *lunar parallax.* For stars, parallax is measured from the earth and the sun, and is called *annual, heliocentric,* or *stellar parallax.* Compare **aberration.**

parallax error. 1. The error in measurement between two pairs of **antenna** caused by the fact that the center of the two **baselines** do not coincide. This error is a function of the distance of the target from the baseline, as well as its relative direction. **2.** The error that may exist when a person is reading the indications of a pointer on a dial because of the observer's eyes not being in a plane which includes the graduations on the dial, and the indicator.

parallel. A circle on the surface of the earth, parallel to the plane of the **equator** and connecting all points of equal **latitude,** or a circle parallel to the **primary great circle** of a sphere or spheroid; also a closed curve approximately such a circle. Also called *parallel of latitude, circle of longitude.* See **coordinate.** An astronomical parallel is a line connecting points having the same astronomical latitude. A geodetic parallel is a line connecting points of equal geodetic latitude. Geodetic and sometimes astronomical parallels are also called geographic parallels. Geodetic parallels are shown on charts. A standard parallel is one along which the scale of a chart is as stated. A fictitious, grid, transverse, inverse, or oblique parallel is parallel to a fictitious, grid, transverse, inverse, or oblique equator, respectively. A magnetic parallel is a line connecting points of equal magnetic dip.

parallel of declination. A circle of the **celestial sphere** parallel to the **celestial equator.** Also called *circle of equal declination.* See **diurnal circle.**

parallel of latitude. 1. A circle (or approximation of a circle) on the surface of the earth, parallel to the **equator,** and connecting points of equal **latitude.** Also called *parallel.* **2.** A circle of the **celestial sphere,** parallel to the **ecliptic,** and connecting points of equal **celestial latitude.** Also called *circle of longitude.*

paramagnetic. Having a magnetic **permeability** greater than unity.

parameter. 1. In general, any quantity of a problem that is not an **independent variable.** More specifically, the term is often used to distinguish, from dependent variables, quantities which may be assigned more or less arbitrary values for purposes of the problem at hand. **2.** In statistical terminology, any numerical constant derived from a population or a probability distribution. Specifically, it is an arbitrary constant in the mathematical expression of a probability distribution.

parameterization. The representation, in a mathematical model, or physical effects in terms of admittedly oversimplified **parameters,** rather than realistically requiring such effects to be consequences of the dynamics of the system. Parameterization is often used in systems analysis to determine the effect on the system of changing one parameter while holding other parameters constant.

parametric equations. A set of equations in which the **independent variables** or coordinates are each expressed in terms of a parameter.

parasol monoplane. A monoplane in which the wings are united in a seperate structure above the fuselage.

pararescue team. Specially trained personnel qualified to penetrate to the site of an incident by land or parachute, render medical aid, accomplish survival methods, and rescue survivors.

parasitic element. A radiating element, not coupled directly to the **feed** line of the **antenna,** which materially affects the pattern of the antenna.

parcel. Same as **fluid parcel.**

parent. A radionuclide that upon radioactive decay or disintegration yields a specific nuclide (the daughter), either directly or as a later member of a radioactive series. (See **radioactive series.**)

parking orbit. An **orbit** of a **spacecraft** around a celestial body, used for assembly of components or to wait for conditions favorable for departure from the orbit.

parallel of altitude. A circle of the **celestial sphere** parallel to the **horizon** connecting all points of equal altitude. Also called *altitude circle, almucantar.* See **circle of equal altitude.**

parrot. Identification Friend or Foe transponder equipment.

parsec (abbr **pc**). A unit of length equal to the distance from the sun to a point having a **heliocentric parallax** of 1 second (1″), used as a measure of stellar distance. The name parsec is derived from the words *parallax second.*

1 parsec = pc = 3.084 X 10^{13} kilometers
= 206,265 astronomical units
= 3,262 light years

part. 1. One of the constituents into which a thing may be divided. Applicable to a major **assembly, subassembly,** or the smallest individual piece in a given thing. **2.** Restrictive. The least subdivision of a thing; a piece that functions in interaction with other elements of a thing, but is itself not ordinarily subject to disassembly.

partial-admission turbine. A type of **turbine** in which the working substance is directed only through part of the annular area swept by the rotating turbine blades.

partial pressure. The **pressure** exerted by a designated component or components of a gaseous mixture. This may be separately measured in some cases by suitable selection of gages, traps, or analytical trains. When the percentage composition of the mixture is known, the partial pressure may be calculated from the total

pressure by **Dalton law of partial pressures.**

partial pressure suit. A skintight suit which does not completely enclose the body but which is capable of exerting pressure on the major portion of the body in order to counter-act an increased oxygen pressure in the lungs.

particle. 1. An elementary subatomic particle such as proton, electron, neutron, etc. **2.** A very small piece of matter. **3.** In celestial mechanics, a hypothetical entity which responds to gravitational forces but which exerts no appreciable gravitational force on other bodies, thus simplifying orbital computations.

particle accelerator. (Nucleonics) An **accelerator.**

particle beam. Streams of highly energetic atomic or subatomic size **particles** like **electrons, protons,** hydrogen **atoms** or **ions,** traveling at near the speed of light (186,000 miles per second). If there are enough particles hitting the target rapid transfer of energy to the material of the target cannot be dissipated, thus the beam can burn a hole through the material, melt it or fracture it from the thermal stress caused by this rapid energy deposit. Secondary radiation is also generated that can disable the target. The **accelerators** used for **particle beam** weapons technology are similar to those used in particle physics research except that currents in the beam are much higher.

particle beam weapon. A technology still in the research and development phase requiring a control system to aim the beam at the target and determine that the beam has hit the target. See **particle beam.**

pass. 1. A single circuit of the earth by a **satellite.** Passes start at the time the satellite crosses the equator from the southern hemisphere into the northern hemisphere (the ascending node). See **orbit. 2.** The period of time the satellite is within **telemetry** range of a **data acquisition station.**

pass. (Military) A short tactical run or dive by an aircraft at a target; a single sweep through or within firing range of an enemy air formation.

passenger mile. One passenger transported one mile. For air and ocean transport, use nautical miles; for rail, highway, an inland waterway transport in the continental United States, use statute miles.

passive. Containing no power sources to augment output power, e.g., *passive electrical network, passive reflector* (as in the Echo satellite). Applied to a device that draws all its power from the **input signal.** Compare **active.**

passive air defense. All measures, other than active defense, taken to minimize the effects of hostile air action. These include the use of cover, concealment, camouflage, dispersion, and protective construction. See also **air defense.**

passive communication satellite. A satellite which reflects communications signals between stations. See also **communications satellite.**

passive electronic countermeasures. The search, intercept, direction finding, range estimation, and signal analysis of electromagnetic radiations performed in direct support of operations conducted for other than intelligence purposes.

passive homing. The **homing** of an aircraft or spacecraft wherein the craft directs itself toward the target by means of energy waves transmitted or radiated by the target. See **active homing.**

passive homing guidance. Guidance in which a craft or missile is directed toward a destination by means of natural radiations from the destination.

path. 1. Of a satellite, the projection of the orbital plane on the earth's surface, the locus of the satellite **subpoint.** Since the earth is turning under the satellite, the path of a single orbital pass will not be a closed curve. *Path* and *track* are used interchangeably. On a cylindrical map projection the path is a sine-shaped curve. **2.** Of a meteor, the projection of the **trajectory** on the **celestial sphere,** as seen by the observer. **3. flightpath.**

pathfinder. (Military) An aircraft with a specially trained crew carrying dropping zone/landing zone marking teams, target markers, or navigational aids, which precedes the main force to the dropping zone/landing zone, or target.

patrol. A detachment of ground, sea or air forces sent out by a larger unit for the purpose of gathering information or carrying out a destructive, harassing, mopping-up or security mission See also **combat air patrol.**

patrol (P), (Military designation). Long range, all weather, multi-engine aircraft operating from land and/or water bases, designed for independent accomplishment of the following functions; antisubmarine warfare, maritime reconnaissance, and mining. See **military aircraft types.**

pattern. The configuration or form of a flight path flown by an aircraft, or prescribed to be flown, as in making an approach to a landing.

pattern bombing. The systematic covering of a target area with bombs uniformly distributed according to a plan.

payload. 1. Originally, the revenue producing portion of an aircraft's load, e.g., passengers, cargo, mail, etc. **2.** By extension, that which an aircraft, rocket, or the like carries over and above what is necessary for the operation of the vehicle for its flight.

payload build-up (missile and space). The process by which the scientific instrumentation (sensors, detectors, etc.) and necessary mechanical and electronic sub-assemblies are assembled into a complete operational package capable of achieving the scientific objectives of the mission.

payload integration (mission and space). The compatible installation of a complete payload package into the spacecraft and space vehicle.

P-band. A frequency band used in radar extending approximately from 225 to 390 megacycles per second. See **frequency bands.**

PCM *(abbr). p*ulse *c*ode *m*odulation.

PDM *(abbr). p*ulse *d*uration *m*odulation.

peak overpressure. The maximum value of overpressure at a given location which is generally experienced at the instant the shock (or blast) wave reaches that location. See also **shock wave.**

pebble bed reactor. A reactor in which the fissionable fuel (and sometimes also the moderator) is in the form of packed or randomly placed pellets, which are cooled by gas or liquid.

Peltier effect. The production or absorption of heat at the junction of two metals on the passage of electrical current. Heat generated by current flowing in one direction will be absorbed if the current is reversed. This effect is presently being extensively studied as a possible energy conversion method for space vehicles.

pencil beam. Emission, from an **antenna,** having the form of a narrow **conical beam.**

pencil-beam antenna. A unidirectional antenna, so designed that cross sections of the **major lobe** by planes perpendicular to the direction of maximum radiation are approximately circular, and having a very small angular cross section.

penetration. A form of offensive maneuver which seeks to break through the enemy's defensive position, widen the gap created, and destroy the continuity of his positions.

penetration aids. Techniques and/or devices employed by aerospace systems to increase the probability of weapon system penetration of an enemy defense. Examples are: low altitude flight profiles, trajectory adjustments, reduced radar cross-sections of attack vehicles, improved vehicle hardness to effects of defense engagements, terrain avoidance radar, bomber defense missiles, decoys, chaff, electronic countermeasures, etc. Penetration aids are used by an offensive system to penetrate more effectively enemy defenses.

penetration (air traffic control) (Military) That portion of a published high altitude instrument approach procedure which prescribes a descent path from the fix on which the procedure is based to a fix or altitude from which an approach to the airport is made.

penetration area. A general area within which appropriate enemy defenses are scheduled to be neutralized to a degree which will assist succeeding aircraft to proceed to their assigned targets.

penetration probability. The probability that a penetrating delivery vehicle will survive the enemy defense through the point of weapon or warhead delivery.

penetrometer. A simple device for measuring the penetrating power of a beam of X rays or other penetrating radiation by comparing transmission through various absorbers. (See **absorber.**)

perfect gas. A gas which has the following characteristics: (**a**) it obeys the **Boyle-Mariotte law** and the **Charles-Gay-Lussac law,** thus satisfying the equation of state for perfect gases; (**b**) it has internal energy as a function of temperature alone; and (**c**) it has specific heats with values independent of temperature. Also called *ideal gas.* Compare **perfect fluid.** The normal volume of a perfect gas is 2.24136 X 10^4 centimeters cubed per mole.

perfect fluid. In simplifying assumptions, a **fluid** chiefly characterized by lack of viscosity and, usually by incompressibility. Also called an *ideal fluid, inviscid fluid.* See **perfect gas.** A perfect fluid is sometimes further characterized as homogeneous and continuous.

perfect-gas laws. Same as **gas laws.**

perfect radiator. A black body.

performance. The flying properties of an aircraft which can be expressed quantitatively (e.g., maximum speed, rate of climb, ceiling, range, loads and runway requirements.

periapsis. The orbital point nearest the center of attraction. See **orbit.**

pericynthian. That point in the **trajectory** of a vehicle which is closest to the moon.

perifocus. The point on an **orbit** nearest the dynamical center **focus).** The pericenter is at one end of the **major axis** of the orbital **ellipse.**

perigee. That orbital point nearest the earth when the earth is the center of attraction. See **orbit.** That orbital point farthest from the earth is called *apogee.* Perigee and apogee are used by some writers in referring to orbits of satellites, especially artificial satellites, around any planet or satellite, thus avoiding coinage of new terms for each planet and moon.

perigee propulsion. A programed-thrust technique for **escape** from a planet, which uses intermittent applications of thrust at perigee (when vehicle velocity is high) and coasting periods.

perigee speed. The speed of an orbiting body when at **perigee.**

perihelion. That point in a solar **orbit** which is nearest the sun. That orbital point farthest from the sun is called *aphelion.* The term *perihelion* should not be confused with *parhelion,* a form of halo.

period. 1. The interval needed to complete a **cycle. 2.** = **orbital period. 3.** Specifically, the interval between passages at a fixed point of a given **phase** of a simple **harmonic** wave; the reciprocal of **frequency. 4.** The time interval during which the power level (flux) of a **reactor** changes by a factor of *e* (2.718, the base of natural logarithms).

periodic inspections. Servicings/inspections repeated at regular intervals of calendar time or hours of operation.

periodic table (periodic chart). A table or chart listing all the *elements,* arranged in order of increasing *atomic numbers* and grouped by similar physical and chemical characteristics into "periods". The table is based on the chemical law that the physical or chemical properties of the elements are periodic (regularly repeated) functions of their *atomic weights,* first proposed by the Russian chemist, Dmitri I. Mendeleev, in 1869.

periodic quantity. In mathematics, an oscillating quantity whose values recur for certain increments of the **independent variable.**

periscope. An optical instrument which displaces the **line of sight** parallel to itself to permit a view which may otherwise be obstructed.

permanent magnetism. Magnetism which is retained for long periods without appreciable reduction, unless the magnet is subjected to demagnetizing force. See **induced magnetism.** Because of the slow dissipation of such magnetism, it is sometimes called *subpermanent magnetism,* but the expression *permanent magnetism* is considered preferable.

permanent memory. In computer terminology, **storage** of information which remains intact when the power is turned off. Also called *nonvolatile storage.*

permeability. 1. Of a magnetic material, the ratio of the **magnetic induction** to the **magnetic-field intensity** in the same region. **2.** The ability to permit penetration or passage. In this sense the term is applied particularly to substances which permit penetration or passage of **fluids. 3. permeability coefficient.**

permeability coefficient. The steady-state rate of flow of **gas** through unit area and thickness of a solid barrier per unit pressure differential at a given temperature. Also called *permeability.* Usually expressed in volume or mass per unit time, per unit area of cross section, per unit thickness, per unit pressure differential across the barrier.

permeation. As applied to gas flow through solids, the passage of **gas** into, through, and out of a solid barrier having no holes large enough to permit more than a small fraction of the gas to pass through any one hole. The process always involves **diffusion** through the solid and may involve various surface phenomena, such as **sorption, dissociation, migration,** and **desorption** of the gas molecules.

permissive action link. A device included in or attached to a nuclear weapon system to preclude arming and/or launching until the insertion of a prescribed discrete code or combination. It may include equipment and cabling external to the weapon or weapon system to activate components within the weapon or weapon system.

permissible dose. (See **maximum permissible dose, radiation protection guide.**)

personnel locator beacon. A portable, lightweight beacon, manually operated, which is designed to be carried on the person, in the cockpit of an aircraft, or attached to a parachute, which operates from its own power source on 121.5 MHz and/or 243 MHz, preferably on both emergency frequencies, transmitting a distinctive downward swept audio tone for homing purposes, which may or may not have voice capability, and which is capable of operation by unskilled persons.

perturbation. 1. Any departure introduced into an assumed **steady state** of a system, or a small departure from a nominal path such as a desired trajectory. Usually used as equivalent to **small perturbation. 2.** Specifically, a disturbance in the regular motion of a **celestial body,** the result of a force additional to that which causes the regular motion, specifically, a gravitational force.

perturbation quantity. Any parameter of a system, e.g., velocity components or temperature, which may or may not have been assumed to be small perturbations from a mean or **steady-state** value.

petroleum. An oil, liquid solution of hydro-carbons which, when fractionally distilled, yields paraffin, kerosene fuel oil, gasoline, etc.

PFM *(abbr). *pulse *frequency *modulation.

phase. 1. Of a **periodic quantity,** for a particular value of the independent variable, the fractional part of a **period** through which the independent variable has advanced, measured from an arbitrary reference. The arbitrary reference is generally so chosen that the fraction is less than unity. In case of a **simple harmonic quantity,** the reference is often taken as the last previous passage through zero from the negative to positive direction. Thus, if two waves crest one-fourth cycle apart, they are said to be *90° apart in phase,* or *90° out of phase.* The moon is said to be at *first quarter* when it has completed one-fourth of its cycle from new moon. **2.** The state of aggregation of a substance, for example solid, liquid, or gas. **3.** The extent to which the disk of the moon or a planet, as seen from the earth, is illuminated or not illuminated by the sun. **4.** In astronomy, **configuration.**

phase angle. 1. The **phase** difference of two periodically recurring phenomena of the same **frequency,** expressed in angular measure. **2.** The angle at a **celestial body** between the sun and earth.

phase detector. A device that continuously compares the **phase** of two signals and provides an output proportional to their difference in phase.

phase deviation. The peak difference between the instantaneous **phase** of the **modulated** wave and the **carrier** frequency. The extent of deviation is proportional to the amplitude of the modulating signal.

phase front. A surface of constant **phase** (or phase angle) of a propagating **wave** disturbance. Also called *wave front.* Generally, phase fronts spread out spherically from their source but in cases where energy is assumed to travel in parallel rays (as in many radiation problems), phase fronts may be approximated as plane surfaces oriented perpendicularly to the rays.

phase lock. The technique of making the **phase** of an **oscillator** signal follow exactly the phase of a reference signal by comparing the phases between the two signals and using the resultant difference signal to adjust the frequency of the reference oscillator. See **correlation detection.**

phase-lock loop. An electronic **servo** system incorporating **phase lock** and used either as a tracking filter or as a frequency discriminator.

phase modulation *(abbr* **PM).** **Angle modulation** in which the angle of a sine-wave carrier is caused to depart from the **carrier** angle by an amount proportional to the instantaneous value of the **modulating wave.** Combinations of phase and frequency modulation are commonly referred to as *frequency modulation.*

phases of military government. 1. assult— That period which commences with first contact with civilians ashore and extends to the establishment of military government control ashore by the landing force. **2. consolidation—** That period which commences with the establishment of military government control ashore by the landing force and extends to the establishment of control by occupation forces. **3. occupation—** That period which commences when an area has been occupied in fact, and the military commander within that area is in a position to enforce public safety and order.

phases of the moon. The various appearances of the moon during different parts of the **synodical month.** The cycle begins with new moon or change of the moon at **conjunction.** The visible part of the waxing moon increases in size during the first half of the cycle until full moon appears at **opposition,** after which the visible part of the waning moon decreases for the remainder of the cycle. First quarter occurs when the waxing moon is at east **quadrature**; last quarter when the waning moon is at west quadrature. From last quarter to new and from new to first quarter, the moon is crescent; from first quarter to full and from full to last quarter, it is **gibbous.** The elapsed time, usually expressed in days since the last new moon, is called *age of the moon.*

Philips gage. A cold-cathode type of **vacuum gage** wherein an electrical discharge is maintained in the presence of a superposed magnetic field in order to increase the ionization current. See **cold-cathode ionization gage.**

phonetic alphabet. A list of standard words used to identify letters in a message transmitted by radio or telephone.

phosphor. A luminescent substance; a material capable of emitting light when stimulated by radiation. (See **scintillation.**)

phosphor. A phosphorescent substance, such as zinc sulfide, which emits light when excited by radiation, as on the scope of a cathode-ray tube. See **phosphorescence.**

phosphorescence. Emission of light which continues after the exciting mechanism has ceased. See **luminescence.** Compare **fluorescence.** An example of phosphorescence is the glowing of an oscilloscope screen after the exciting beam of electrons has moved to another part of the screen.

photochemical reaction. A chemical reaction which involves either the **absorption** or **emission** of **radiation.**

photoelectric. 1. Pertaining to the **photoelectric effect. 2.** Using a **photoelectric cell,** as a *photoelectric photometer.*

photoelectric cell. A **transducer** which converts **electromagnetic radiation** in the infrared, visible, and ultraviolet regions into electrical quantities such as voltage, current, or resistance. Also called *photocell.*

See **photoelectric effect.**

photoelectric effect. The **emission** of an **electron** from a surface as the surface absorbs a **photon** of electromagnetic radiation. Electrons so emitted are termed *photoelectrons.* The effectiveness of the process depends upon the surface metal concerned and the wavelength of the radiant energy to which it is exposed. Cesium, for example, will emit electrons when exposed to visible radiation. The energy of the electron produced is equal to the energy of the incident photon minus the amount of work needed to raise the electron to a sufficient energy level to free it from the surface. The resultant energy of the electron, therefore, is proportional to the frequency (i.e., inversely proportional to the wavelength) of the incident radiation.

photoelectric photometry. Photometry in which a **photoelectric cell** is used as the sensing element.

photoelectric transducer. A **transducer** which converts changes in light energy to changes in electrical energy.

photoelectron. An **electron** which has been ejected from its parent atom by interaction between that atom and a high-energy **photon.** Photoelectrons are produced when electromagnetic radiation of sufficiently short wavelength is incident upon metallic or other solid surfaces (photoelectric effect) or when radiation passes through a gas.

photoflash cartridge. A pyrotechnic cartridge designed to produce a brief and intense illumination for low altitude night photography.

photogrammetry. The science or art of obtaining reliable measurements from photographic images.

photographic magnitude. Stellar **magnitudes** measured from a photographic plate exposed without filters.

photographic scale. The ratio of a distance measured on a photograph or mosaic to the corresponding distance on the ground, classified as follows: **very large scale** — 1:6,000 and larger   **large scale** —1:6,000 to 1:12,000   **medium scale** —1:12,000 to 1:30,000   **small scale** —1:30,000 to 1:70,000 **very small scale** — 1:70,000 and smaller

photographic sortie. (**1**) One flight by one aircraft for the purpose of doing air photography. (**2**) All the photographs obtained by one air vehicle on one photographic sortie.

photographic strip. A series of successive overlapping photographs made from an aircraft flying a seclected course or direction.

photology. The study of **light.**

photometer. An instrument for measuring the **intensity** of light or the relative intensity of a pair of lights. Also called *illuminometer.* If the instrument is designed to measure the intensity of light as a function of wavelength, it is called a *spectrophotometer.* Photometers may be divided into two classes: photoelectric photometers in which a photoelectric cell is used to compare electrically the intensity of an unknown light with that of a standard light, and visual photometers in which the human eye is in sensor.

photometry. The study of the measurement of the **intensity** of light. At one time *photometry* referred only to the measurement of **luminous intensity,** intensity of light in the wavelength to which the eye is sensitve. This restriction has proved difficult to maintain in practice.

photon. The carrier of a quantum of electromagnetic *energy.* Photons have an effective mementum but no mass or electrical charge. (See **radiation, quantum.**)

photon engine. A projected type of **reaction engine** in which thrust would be obtained from a stream of **electromagnetic radiation.** Compare **ion engine.** Although the thrust of this engine would be minute, it may be possible to apply it for extended periods of time. Theoretically, in space, where no resistance is offered by air particles, very high speeds may be built up.

photosynthesis. A process operating in green plants in which carbohydrates are formed under the influence of light with chlorophyl serving as a catalyst. See **closed ecological system.**

phototheodolite. An instrument or device incorporating one or more cameras for taking and recording angular measurements. The phototheodolite, sometimes in conjunction with radar equipment, is used to track rockets and to measure and record attitude, altitude, azimuth and elevation angles, etc.

photovoltaic cell. A **transducer** which converts **electromagnetic radiation** into electric current. The solar cells used on satellites and space probes are photovoltaic cells employing a semiconductor such as silicon which releases electrons when bombarded by photons from solar radiation.

phugoid oscillation. In a flightpath, a long-period longitudinal **oscillation** consisting of shallow climbing and diving motions about a median **flightpath** and involving little or no change in angle of attack.

physical constant. An abstract number or physically dimensional quantity having a fixed or approximately fixed value; a universal and permanent value, as the constant of gravitation; a characteristic of a substance, as the refractive index of a liquid.

physical meteorology. That branch of meteorology which deals with optical, electrical, acoustical, and thermodynamic phenomena of atmospheres, their chemical composition, the laws of radiation, and the explanation of clouds and precipitation. As generally accepted, it does not include mathematical theory of the motions of the atmosphere and the forces responsible therefore (which matters fall in the field of **dynamic meteorology**). Also called *atmospheric physics.* Subdivisions of physical meteorology include **atmospheric electricity, cloud physics, precipitation physics, atmospheric acoustics,** and **atmospheric optics.**

physical system. Same as **cgs system.**

physiological acceleration. The **acceleration** experienced by a human or an animal test subject in an accelerating vehicle. Several different terminologies have been used to describe physiological acceleration. Since the terminology may be based either on the action of the accelerating vehicle or the reaction of the passenger, the terms used are often confusing to a reader without prior knowledge of the system of terminology used. Probably the most easily understood system is the *eyeballs in, eyeballs out, eyeballs down, eyeballs up,* etc., terminology used by test pilots, which refers to the sensations experienced by the person being accelerated. Thus, the acceleration experienced in an aircraft pullout or inside loop is *eyeballs down.* Some physiological-acceleration terminologies designate accelerations in terms of the equivalent displacement acceleration of the subject as if he were starting from rest. In such terminologies a man standing up or sitting down on the surface of the earth is experienceing 1 g of *headward* acceleration because of gravity. Other descriptive terms used in this way are *footward, forward* (the acceleration

experienced by a man pressed into the seat back by an accelerating vehicle), *rearward, leftward, rightward, spineward, sternumward,* and *tailward.* One terminology based on reaction uses the terms *head-to-foot* (the acceleration generated by a pullout in an aircraft), *chest-to-back. foot-to-head,* and *back-to-chest.*

picket ship. One of the ocean-going ships used on a missile range to provide added instrumentation for tracking or recovering the missiles. The picket ship may be used to extend the length of the range.

pickoff. A sensing device that responds to angular movement to create a signal or to effect some type of control, as a *pickoff on a gyro in an automatic pilot.* A pickoff may be a potentiometer, a photoelectric device, a kind of valve controlling the fluid flows and pressures in a system, or one of various other devices.

pictorial symbolization. A method of representing prominent map details by means of stylized symbols.

pickup. 1. A device that converts a sound, scene, or other form of intelligence into corresponding electric **signals** (e.g., a microphone, a television camera, or a phonograph pickup). **2.** The minimum current, voltage, power, or other value at which a **relay** will complete its intended function. **3. Interference** from a nearby circuit or electrical system.

piezoelectricity. The property exhibited by some asymmetrical crystalline materials which when subjected to strain in suitable directions develop electric **polarization** proportional to the strain.

pile. Old term for *nuclear reactor.* This name was used because the first reactor was built by piling up graphite blocks and natural uranium.

pilot. 1. A person who handles the **controls** of an aircraft or spacecraft from within the craft, and in so doing, guides or controls it in three-dimensional flight. **2.** A mechanical system designed to exercise **control** functions in an aircraft or spacecraft. **3.** To operate, control, or guide an aircraft or spacecraft from within the vehicle so as to move in three-dimensional flight through the air or space. See **airman certificates.**

pilotage. A form of **navigation,** whereby an aircraft is directed with respect to visible landmarks.

pilot certificates. See **student pilot certificate, private pilot certificate, commercial pilot certificate, airline transport pilot certificate, airman certificates.**

piloted. Of an aircraft or spacecraft, under, or subject to, continuous **control** by a person inside the vehicle. This term is more specific than the term *manned.*

pilot in command. The pilot responsible for the operation and safety of an aircraft during flight time.

pilot report (PIREP). A communication received from a pilot concerning details of his flight (meteorological, navigation) or status of his aircraft.

pilot's trace. (Military) A rough overlay to a map made by the pilot of a photographic reconnaissance aircraft during or immediately after a sortie. It shows the locations, direction, number, and order of photographic runs made, together with the camera(s) used on each run.

pinpoint. 1. A precisely identified point, especially on the ground, that locates a very small target, a reference point for rendezvous or for other purposes; the coordinates that define this point. **2.** The ground position of aircraft determined by direct observation of the ground.

pip. Signal indication on the **oscilloscope** screen of an electronic instrument, produced by a short, sharply peaked pulse of voltage. Also called **blip.**

pipeline. In logistics, the channel of support or a specific portion thereof by means of which materiel or personnel flow from sources of procurement to their point of use.

piston engine. An engine in shich the working fluid is expanded in a cylinder against a reciprocating piston.

pitch. 1. Of a vehicle, an angular displacement about an axis parallel to the **lateral axis** of the vehicle. **2.** In **acoustics,** that attribute of auditory sensation in terms of which sounds may be ordered on a scale extending from low to high. Pitch depends primarily upon the frequency of the sound stimulus, but it also depends upon the sound pressure and waveform of the stimulus. The pitch of a sound may be described by the frequency or frequency level of that simple tone having a specified sound pressure level which is judged by listeners to produce the same pitch.

pitch attitude. The **attitude** of an aircraft, rocket, etc., referred to the relationship between the longitudinal **body axis** and a chosen reference line or plane as seen from the side.

pitch axis. A lateral axis through an aircraft, missile, or similar body, about which the body pitches. It may be a *body, wind,* or *stability* axis. Also called a *pitching axis.* See **axis,** sense **2** and note.

pitch, collective. See **collective pitch control.**

pitch, cyclic. See **cyclic pitch control.**

pitching moment. A **moment** about a **lateral axis** of an aircraft, rocket, airfoil, etc. This moment is positive when it tends to increase the angle of attack or to nose the body upward.

pitchover. 1. The programmed turn from the **vertical** that a **rocket** takes as it describes an arc and points in a direction other than vertical. **2.** The point-in-space of this action.

pitch setting. The propeller blade setting as determined by the blade angle measured in a manner, and at a radius, specified by the instruction manual for the propeller.

pitot-static tube. A device consisting essentially of a unit combination of a **pitot tube** and a **static tube** arranged coaxially or otherwise parallel to one another, used principally in measuring **impact** and **static pressures;** also called **pitot-static head.** The difference between impact and static pressures is used to measure the velocity of flow past the tube by means of a differential-pressure gage. The static pressure from a pitot-static tube may in addition, be used in the operation of an altimeter and similar instruments.

pitot tube. (Pronounced pee-toe. After Henri Pitot, 1695—1771, French scientist.) An open-ended tube or tube arrangement which, when immersed in a moving **fluid** with its mouth pointed upstream, may be used to measure the **stagnation pressure** of the fluid for subsonic flow; or the stagnation pressure behind the tube's **nor-mal shock** wave for supersonic flow.

plain flap. A flap forming the rear portion of the airfoil and moving as a whole.

Planck law. An expression for the variation of monochromatic radiant flux per unit area of source as a function of wavelength of black-body radiation at a given temperature; it is the most fundamental of the **radiation laws.**

plane. (aerodynamics). An airfoil section for deflection of air.

plane. (geometrical). A surface, infinite in scope (width and length), and having no thickness. A plane is determined by a point and a line, three points, or two intersecting parallel lines.

planet. A **celestial body** of the **solar system,** revolving around the sun in a nearly circular orbit, or a similar body revolving around a star. See table II. See also **astronomical constant.** The larger of such bodies are sometimes called *principal planets* to distinguish them from asteroids, planetoids, or minor planets which are comparatively very small. The larger planets are accompanied by satellites, such as the moon. An inferior planet has an orbit smaller than that of the earth; a superior planet has an orbit larger than that of the earth. The four planets nearest the sun are called *inner planets;* the others, *outer planets.* The four largest planets are called *major planets.* The four planets commonly used for celestial observations are called *navigational planets.* The word *planet* is of Greek origin, meaning, literally, *wanderer,* applied because the planets appear to move relative to the stars.

planetary aberration. A displacement in the apparent position of a **planet** in the **celestial sphere** due to the relative movement of the observer and the planet. See **aberration.**

planetary boundary layer. That layer of the atmosphere from a planet's surface to the **geostrophic wind level** including, therefore, the **surface boundary layer** and the **Ekman layer.** Above this layer lies the **free atmosphere.** Also called *friction layer, atmospheric boundary layer.*

planetary circulation. 1. The system of large-scale disturbances in a planet's **troposphere** when viewed on a hemispheric or world-wide scale. **2.** The mean or time-averaged hemispheric circulation of a planetary **atmosphere;** also called *general circulation.*

planetary configurations. Apparent position of the **planets** relative to each other and to other bodies of the **solar system,** as seen from the earth.

planetographic. Referring to positions on a **planet** measured in **latitude** from the planet's **equator** and in **longitude** from a reference meridian.

planetoid. Same as **asteroid.** See **planet.**

planform. The shape or form of an object, such as an *airfoil,* as seen from above, as in a *plan view.*

plan position indicator *(abbr PPI).* **1.** A **cathode-ray indicator** in which a **signal** appears on a radial line. Distance is indicated radially and bearing as an angle. **2.** In radar technique, a cathode-ray indicator on which **blips** produced by signals from reflecting objects and **transponders** are shown in plan position, thus forming a maplike display.

plan range. In air photographic reconnaissance, the horizontal distance from the point below the aircraft to an object on the ground.

plasma. An electrically conductive gas comprised of neutral particles, **ionized** particles, and **free electrons** but which, taken as a whole, is electrically neutral. A plasma is further characterized by relatively large intermolecular distances, large amounts of energy stored in the internal energy levels of the particles, and the presence of a **plasma sheath** at all boundaries of the plasma. Plasmas are sometimes referred to as a fourth state of matter.

plasma cloud. Specifically, a mass of **ionized** gas flowing out of the sun.

plasma engine. A **reaction engine** using magnetically

TABLE II. —PLANETS[a]
Mean elements of planetary orbits
(for epoch 1960 January 1.5 E.T.)

	Inclination i	Mean longitude of node Ω	Mean longitude of perihelion w	Mean longitude at epoch L	Eccentricity e
	°	°	°	°	
Mercury	7.00399	47.85714	76.83309	222.62165	0.205627
Venus	3.39423	76.31972	131.00831	174.29431	0.006793
Earth	0.0	0.0	102.25253	100.15815	0.016726
Mars	1.84991	49.24903	335.32269	258.76729	0.093368
Jupiter	1.30536	100.04444	13.67823	259.83112	0.048435
Saturn	2.48991	113.30747	92.26447	280.67135	0.055682
Uranus	0.77306	73.79630	170.01083	141.30496	0.047209
Neptune	1.77375	131.33980	44.27395	216.94090	0.008575
Pluto[b]	17.1699	109.88562	224.16024	181.64632	0.250236

	Mean distance from Sun, AU	10^6 km	Sidereal period (tropical years)	Synodic period	Mean daily motion n	Orbital velocity (km/sec)
				d	°	
Mercury	0.387099	57.9	0.24085	115.88	4.092339	47 8
Venus	0.723332	108.1	0.61521	583.92	1.602131	35.0
Earth	1.000000	149.5	1.00004		0.985609	29.8
Mars	1.523691	227.8	1.88089	779.94	0.524033	24.2
Jupiter	5.202803	778	11.86223	398.88	0.083091	13.1
Saturn	9.538843	1426	29.45772	378.09	0.033460	9.7
Uranus	19.181951	2868	84.01331	369.66	0.011732	6.8
Neptune	30.057779	4494	164.79345	367.48	0.005981	5.4
Pluto[b]	39.43871	5896	247.686	366.72	0.003979	4.7

Dimensions and rotations of the planets

	Semi-diameter at unit distance	Radius[e] on scale Earth = 1	Reciprocal of flattening	Mass on scale Earth = 1	Density, g/cm³	Surface gravity Earth = 1	Rotation period	Inclination of Equator to orbit
	″							
Mercury	3.34	0.39	∞	0.056	5.13	0.36	88ᵈ	?
Venus	8.41	0.97	∞	0.817	4.97	0.87	?	32°
							h m s	° ′
Earth	8.80	1.00	2.9825ᵈ	1.000	5.52	1.00	23 56 04	23 27
Mars	4.68	0.53	192	0.108	3.94	0.38	24 37 23	23 59
Jupiter	98.47	11.19	16.1	318.0	1.33	2.64	9 50 30	3 04
Saturn	83.33	9.47	10.4	95.2	0.69	1.13	10 14	26 44
Uranus	34.28	3.69	16	14.6	1.56	1.07	10 49	97 53
Neptune	36.56	3.50	50	17.3	2.27	1.41	14 ?	28 48
Pluto	10 ?	1.1 ?	?	0.9 ?	4 ?	?	6ᵈ.39 ?	?

[a] From the *Explanatory Supplement to the Astronomical Ephemeris and the American Ephemeris and Nautical Almanac, 1961.*
[b] The elements for Pluto are osculating values for epoch 1960 September 23.0 E.T. = J.D. 2,437,200.5.
[e] The radii of the planets are based on recent values for the angular semi-diameters; the equatorial radius of the earth is 6,378 km = 3,963 miles.
[d] Adopted by IAU, 1963.

accelerated **plasma** as **propellant.** A plasma engine is a type of electrical engine.

plasma generator. 1. A machine, such as an electric-arc chamber, that will generate very high heat fluxes to convert neutral gases into **plasma. 2.** A device which uses the interaction of a plasma and electrical field to generate a current.

plasma physics. The study of the properties of **plasmas.**

plasma rocket. A rocket using a **plasma engine.** Also called *electromagnetic rocket.*

plasma sheath. 1. The boundary layer of charged particles between a **plasma** and its surrounding walls, electrodes, or other plasmas. The sheath is generated by the interaction of the plasma with the boundary material. Current flow may be in only one direction across the sheath (single sheath), in both directions across the sheath (double sheath), or when the plasma is immersed in a magnetic field, it may flow along the sheath surface at right angles to the magnetic field (magnetic current sheath). **2.** An envelope of **ionized** gas that surrounds a body moving through an atmosphere at **hypersonic** velocities. The plasma sheath affects transmission, reception, and diffraction of radio waves; thus it is important in operational problems of spacecraft.

plasticity. The tendency of a loaded body to assume a deformed state other than its original state when the load is removed.

plastic range. The stress range in which a material will not fail when subjected to the action of a force, but will not recover completely, so that a permanent deformation results when the force is removed.

plate. 1. A planar body whose thickness is small compared with its other dimensions. **2.** A common name for the principal **anode** in an **electron tube.**

plot. 1. Map, chart, or graph representing data of any sort. **2.** Represent on a diagram or chart the position or course of a target in terms of angles and distances from known positions; locate a position on a map or chart. **3.**

The visual display of a single geographical location of an airborne object at a particular instant of time. **4.** A portion of a map or overlay on which are drawn the outlines of the areas covered by one or more photographs.

plowshare. The Atomic Energy Commission program of research and development on peaceful uses of *nuclear explosives*. The possible uses include large-scale excavation, such as for canals and harbors, crushing ore bodies, and producing heavy transuranic isotopes. The term is based on a Biblical reference: *Isaiah 2:4.*

plus count. In the **launch** of a **rocket,** a count in seconds (plus 1, plus 2, etc.) that immediately follows **T-time,** used to check on the sequence of events after the action of the **countdown** has ended.

Pluto. See **planet.**

plutonium. A heavy, radioactive, man-made, metallic element with atomic number 94. Its most important isotope is fissionable plutonium—239, produced by neutron irradiation of uranium—238. It is used for reactor fuel and in weapons.

PM *(abbr). phase modulation.*

PMR *(abbr). Pacific Missle Range.*

pneumatic system. 1. A particular method of transmitting power by means of air under pressure. **2.** Of an aircraft: The complete pneumatic installation.

pod. An enclosure, housing, or detachable container of some kind, as an *engine pod.*

pogo. (Military) Switch to communications channel number preceding "pogo". If unable to establish communications, switch to channel number following "pogo".

point defense. Point defense has as its purpose the defense of specified geographical areas, cities, and vital installations. One distinguishing feature of point defense missiles is that their guidance information is received from radars located near the launching sites. See also **area defense.**

point designation grid. A system of lines, having no relation to the actual scale or orientation, drawn on a map, chart, or air photograph, dividing it into squares so that points can be more readily located.

point of no return. A point along an aircraft track beyond which its endurance will not permit return to its own or some other associated base on its own fuel supply.

point target 1. A target which requires the accurate placement of bombs or fire. **2.** (nuclear) A target in which the ratio of radius of damage to target is equal to or greater than 5.

poise. The unit of viscosity in the cgs system equal to 1 dyne second per square centimeter.

polar coordinates. 1. In a plane, a system of **curvilinear coordinates** in which a point is located by its distance from the origin (or pole) and by the angle which a line (radius vector) joining the given point and the origin makes with a fixed reference line, called the polar axis. **2.** In three dimensions, short for **space polar coordinates.**

polar distance. Angular distance from a **celestial pole;** the arc of an **hour circle** between a celestial pole, usually the elevated pole, and a point on the **celestial sphere,** measured from the celestial pole through 180°.

polarimeter. An instrument for determining the degree of **polarization** of **electromagnetic radiation,** specifically the polarization of light.

polariscope. An instrument for detecting **polarized** radia-

tion and investigating its properties.

polarity. The sign of the electric discharge associated with a given object, as an **electrode** or an **ion.**

polarization. 1. The state of **electromagnetic radiation** when transverse vibrations take place in some regular manner, e.g., all in one plane, in a circle, in an ellipse, or in some other definite curve. Radiation may become polarized because of the nature of its emitting source, as is the case with many types of radar antennas, or because of some processes to which it is subjected after leaving its source as that which results from the scattering of solar radiation as it passes through the earth's atmosphere. **2.** With respect to particles in an **electric field,** the displacement of the charge centers within a particle in response to the electric force acting thereon. **3.** The response of the molecules of a **paramagnetic** medium (such as iron) when subjected to a **magnetic field.**

polar orbit. The **orbit** of an earth **satellite** that passes over or near the earth's poles.

pole 1. The origin of a system of **polar coordinates. 2.** For any circle on the surface of a sphere, the point of intersection of the surface of the sphere and the normal line through the center of the circle. See **geographical pole, celestial pole, ecliptic pole. 3.** A point of concentration of electric charge. See **dipole. 4.** A point of concentration of magnetic force. See **magnetic pole.**

pole of the Milky Way. The **pole** in the **galactic system of coordinates.**

polytropic atmosphere. A **model atmosphere** in **hydrostatic equilibrium** with a constant nonzero lapse rate.

pontoon. See **float.**

population. In statistical usage, any definite class of individuals or objects. Also called *universe.* Compare **sample.**

pop-up maneuver. A maneuver utilized by tactical aircraft when transitioning from the low level approach phase of an attack mission to an altitude and point from which the target can be identified and attacked.

porpoising. In seaplanes, dynamic pitching while planing.

port. 1. A place of access to a **system** where energy may be supplied or withdrawn or where system variables may be observed or measured. In any particular case, the ports are determined by the way in which the system is used, and not by the structure alone. A designated pair of terminals is an example of a port. **2.** An opening, as the *port* in a solid rocket.

posigrade rocket. An auxiliary **rocket** which fires in the direction in which the **vehicle** is pointed, used, for example, in separating two stages of a vehicle.

position. 1. A point in space. **2.** A point defined by stated or implied **coordinates,** particularly one on the surface of the earth. **3. attitude. 4.** A crew member's station aboard an aircraft or spacecraft.

positional notation. Any scheme for representing quantities characterized by the arrangement of **digits** in sequence with the understanding that successive digits are to be interpreted as coefficients of successive **powers** of an integer called the **base** or radix. The base determines the name of the notation, as, *binary* (base 2), *decimal* (base 10), or *duodecimal* (base 12).

position lights. See **navigation lights.**

positive acceleration. 1. Acceleration such that speed increases. **2.** Accelerating force in an upward sense or direction, e.g., from bottom to top, seat to head, etc.;

acceleration in the direction that this force is applied. See **physiological acceleration.**

positive control. Control of all air traffic, within designated airspace, by air traffic control.

positive feedback. Feedback which results in increasing the amplification.

positive G. In a gravitational field or during an acceleration, when the human body is so positioned that the force of inertia acts on it in a head-to-foot direction, i.e., the footward inertial force produced by a headward acceleration. See **physiological acceleration.**

positron. An *elementary particle* with the mass of an electron but charged positively. It is the "antielectron". It is emitted in some radioactive disintegrations and is formed in pair production by the interaction of high-energy gamma rays with matter. (See **antimatter, electron, pair production.**)

post-strike reconnaissance. Missions undertaken for the purpose of gathering information used to measure results of a strike.

potential. 1. A function of space, the **gradient** of which is equal to a **force. 2.** Applied to the value that an atmospheric thermodynamic variable would attain if processed adiabatically from its initial pressure to the standard pressure of 1000 millibars. **3.** Short for **electric potential.**

potential energy. Energy possessed by a body by virtue of its position in a **gravity** field in contrast with **kinetic energy,** that possessed by virtue of its motion.

potentiometer. 1. An instrument for measuring differences in **electric potential** by balancing the unknown voltage against a variable known voltage. If the balancing is accomplished automatically, the instrument is called a *self-balancing potentiometer.* **2.** A variable electric resistor.

pound *(abbr* **lb). 1.** A unit of mass equal in the United States to 0.45359237 kilogram, exactly. **2.** Specifically, a unit of measurement of the thrust or force of a reaction engine representing the weight the engine can move, as *an engine with 100,000 pounds of thrust.* See **poundal, pound mass. 3.** The force exerted on 1 **pound mass** by the standard **acceleration of gravity.** See **gravity,** sense **2.**

poundal. A unit of force; that unbalanced force which, acting on a body of 1 pound mass, produces an **acceleration** of 1 foot per second squared. See **pound, pound mass.**

pound force. Pound, sense **3.**

pound mass. 1. A **mass** equal to 0.45359237 kilogram. **2.** A unit of measure of the inertial property equal to the mass of a body weighing 1 pound at the standard acceleration of **gravity** (980.665 centimeters per second squared).

pound weight. A **force** equal to the earth's attraction for a mass of 1 pound. This force, acting on a 1-**pound mass,** will produce an acceleration of 32.1747 feet per second squared.

powder metallurgy. The science or art involving the production of powdered metals or metallic objects by compressing a powdered metal or alloy with or without other materials, and heating without thoroughly melting to solidify and strengthen.

power. 1. *(symbol* **P).** Rate of doing **work. 2.** Luminous intensity. **3.** The number of times an object is magnified by an optical system, such as a telescope. Usually called *magnifying power.* **4.** The result of multiplying a number by itself a given number of times, as *the third power of a number is its cube;* the superscript which indictates this process as in $2^3 = 2 \times 2 \times 2$.

power gain. 1. The ratio of the **power** that a **transducer** delivers to a specified **load,** under specified operating conditions, to the power absorbed by its input circuit. If the input and/or output power consist of more than one component, such as multifrequency signal or noise, then the particular components used and their weighting must be specified. This gain is usually expressed in decibels. **2.** Of an antenna, in a given direction, 4π times the ratio of the radiation intensity in that direction to the total power delivered to the **antenna.**

power loading. The ratio of the gross weight of a propeller-driven **aircraft** to its power, usually expressed as the gross weight of the aircraft divided by the rated horsepower of the power plant corrected for air of standard density. With turboprop engines, the equivalent shaft horsepower is used. Compare **thrust loading.**

power package. An engine, especially a **reciprocating engine** together with its accessories, lines, cowling, etc., ready for quick installation on an aircraft.

powered control system. A flight control system in which a power amplifier is placed between the flying control and the control surface.

powered lift flight regime. That flight regime of any aircraft in which controlled, level flight is possible below the power off stall speed and in which part or all of the lift and/or control moments are derived directly from power plant(s).

power rating. The power permitted by the relevant regulations for a certain specified use; e.g. Maximum continuous rating.

power plant. 1. The complete assemblage or installation of **engine** or engines with accessories (induction system, cooling system, ignition system, etc.) that generates the motive power for a self-propelled vehicle or vessel such as an aircraft, rocket, etc. **2.** An engine or engine installation regarded as a source of **power.**

powerplant mechanic. A person who repairs and/or maintains an aircraft powerplant. When applied to U. S. Civil aviation, powerplant mechanic may be a **rating** to an **aviation mechanic certificate.**

powerplant unit. A complete aircraft engine package including accessories and cowling, and which is designed as a single unit for quick installation or removal.

PPI *(abbr). p*lan *p*osition *i*ndicator.

Prandtl number. (After Ludwig Prandtl, 1875—1953, German scientist.) A dimensionless number representing the ratio of **momentum transport** to **heat** transport in a flow.

preamplifier. 1. An **amplifier,** the primary function of which is to raise the output of a low-level **source** to an intermediate level so that the **signal** may be further processed without appreciable degradation in the signal-to-noise ratio of the system. A preamplifier may include provision for equalizing and/or mixing. **2.** In radar an amplifier separated from the remainder of the receiver and located so as to provide the shortest possible input circuit path from the antenna so as to avoid deterioration of the signal-to-noise ratio.

precession. Change in the direction of the axis of rotation of a spinning body, as a gyro, when acted upon by a **torque.** See **apparent wander, precession of the equinoxes.** The direction of motion of the axis is such that it causes the direction of spin of the gyro to tend to coincide with that of the impressed torque. The horizontal component of precession is called *drift,* and the vertical component is called *topple.*

precession of the equinoxes. The conical motion of the earth's **axis** about the normal to the plane of the **ecliptic,** caused by the attractive force of the sun, moon, and other planets on the equatorial protuberance of the earth.

precipitation. A general term for the forms in which water may fall from the atmosphere.

precipitation attenuation. The loss of radio **energy** due to the passage through a volume of the atmosphere containing precipitation. Part of the energy is lost by **scattering** and part by **absorption.** See **cloud attenuation, range attenuation.** Radars operating at wavelengths of 10 centimeters and higher are generally unaffected, whereas even the smallest precipitation rates will seriously attenuate radar energy of wavelengths less than 1 centimeter. For rain and snow diameter-to-wavelength ratios less than 0.07, the loss is due primarily to absorption. Scattering becomes important for ratios near 0.1 and greater. Attenuation by dry snow is small for most radar wave-lengths.

precision. The quality of being exactly or sharply defined or stated. A measure of the precision of a representation is the number of distinguishable alternatives from which it was selected, which is sometimes indicated by the number of **significant digits** it contains. Compare **accuracy.**

precision approach radar. (PAR). Radar displaying range, azimuth, and elevation (in relation to a glide slope) normally encompassing an area from 10 miles on final approach to a position on the runway intercepted by the glide slope.

precision approach procedure. Means a standard instrument approach procedure in which an electronic glide slope is provided, such as ILS or PAR.

precision bombing. Bombing directed at a specific target.

precombustion chamber. In a **rocket,** a chamber in which the **propellants** are ignited and from which the burning mixture expands torchlike to ignite the mixture in the main **combustion chamber.**

preemptive attack. An attack initiated on the basis of incontrovertible evidence that an enemy attack is imminent.

preferred routes. Airways designated between major terminals on which IFR flight in one direction only is preferred for air traffic control reasons. These routes have been developed to increase the efficiency of the air traffic control system.

presentation. In electronics, the act or process of displaying **radar** echoes on a cathode-ray screen; the **echo** or images displayed on a **cathode-ray screen.**

preset guidance. A type of **guidance** in which devices in the aircraft or spacecraft, adjusted before launching, **control** the path of the missile.

pressure *(symbol* **p).** **1.** In a gas, the net rate of transfer of momentum in the direction of the positive normal to an imaginary plane surface of specified area located in a specified position in the gas by molecules crossing the surface in both directions, momentum transmitted in the opposite direction being counted as negative, divided by the area of the surface. In general, it is assumed that the area of the imaginary plane surface is small enough so that the pressure with respect to any part of the surface is equal (within narrow limits) to the pressure based on the whole surface. Different kinds of pressure (static, dynamic, partial, total, vapor, etc.) are distinguished by the orientation of the surface with respect to mass-flow velocity vectors or by the restriction to a specified set of molecular species crossing the imaginary surface. **2.** On a boundary surface, the force applied per unit area and equal to the pressure in the gas as determined by molecules crossing an imaginary surface located at a fixed distance of molecular magnitude in front of the real surface, the positive normal being drawn from the imaginary surface toward the real surface. The term *pressure* when used alone can be assumed to refer to the total pressure in a gas at rest or else to refer to the static pressure in a gas flowing under steady-state condition. **3. atmospheric pressure. 4.** As measured in a vacuum system, the quantity measured at a specified time by a so-called **vacuum gage,** whose sensing element is located in a cavity (gage tube) with an opening oriented in a specified direction at a specified point within the system, assuming a specified calibration factor. The sensitivity of the sensing element is, in general, not the same for all molecular species, but the gage reading is frequently reported using the calibration factor for air regardless of the composition of the gas. The opening to the gage tube is often carelessly oriented with respect to mass-flow vectors in the gas (which is seldom at rest), and errors due to variations in wall temperatures of tube and system are frequently neglected. The actual total pressure in a high-vacuum system cannot usually be measured by a single gage, but in vacuum technology the term *total pressure* is sometimes used to refer to the reading of a single untrapped gage which responds to condensable vapors as well as permanent gases.

pressure altimeter. An **altimeter** that utilizes the change of **atmospheric pressure** with height to measure **altitude.** It is commonly an aneroid altimeter. Also called *barometric altimeter.* See **aneroid,** sense **1.**

pressure altitude. 1. Altitude in the earth's atmosphere above the standard datum plane, standard sea level pressure, measured by a pressure altimeter. **2.** The altitude in a standard atmosphere corresponding to atmospheric pressure encountered in a real atmosphere. **3.** The simulated altitude created in an altitude chamber.

pressure breathing. The breathing of oxygen or of a suitable mixture of gases at a **pressure** higher than the surrounding pressure. See **continuous pressure breathing, intermittent pressure breathing.**

pressure-breathing system. An oxygen system in which oxygen is injected inside the respiratory ducts through a pressure higher than the surrounding **pressure.**

pressure-demand oxygen system. A **demand oxygen system** that furnishes oxygen at a **pressure** higher than atmospheric pressure above a certain altitude.

pressure pattern navigation. A navigation technique which makes use of the characteristics of the atmospheric pressure difference to obtain lines of position, drift or minimal flight path.

pressure stabilized. Referring to membrane-type structures that require internal pressure for maintenance of a stable structure; for example, the Atlas missile structure.

pressure suit. A garment designed to provide pressure upon the body so that respiratory and circulatory functions may continue normally, or nearly so, under low-pressure conditions, such as occur at high altitudes or in space without benefit of a **pressurized cabin.** A pressure suit is distinguished from a pressurized suit, which inflates, although it may be fitted with inflating parts that tighten the garment as ambient pressure decreases. Compare **g-suit.**

pressure thrust. In rocketry, the product of the cross-section area of the exhaust **jet** leaving the **nozzle exit** and the difference between the **exhaust pressure** and the **ambient pressure.**

pressure transducer. A **transducer** which produces an output related to imparted pressure.

pressure wave. In meteorology, a short-period **oscillation** of **pressure** such as that associated with the propagation of **sound** through the atmosphere.

pressurization. The process of producing **pressures** higher than ambient, as in a **pressurized cabin.**

pressurized. Containing air, or other gas, at a **pressure** higher than **ambient.**

pressurized cabin. The occupied space of an aircraft in which the air pressure has been increased above that of the ambient atmosphere by compression of the ambient atmosphere into the space.

pressurized suit. A suit designed to be inflated so as to provide pressure directly upon the body, not to air surrounding the body. Compare **pressure suit.**

pressurizing gas. Specifically, a gas used to expel propellant from a fuel tank.

prestage. 1. A step in the action of igniting a large liquid rocket taken prior to the ignition of the full flow, and consisting of igniting a partial flow of **propellants** into the **thrust chamber. 2.** The partial flow thus ignited. Also called *preliminary stage.*

prestrike reconnaissance. Missions undertaken for the purpose of obtaining complete information about known targets for use by the strike force.

prevailing visibility. The horizontal distance at which targets of known distance are visible over at least half of the horizon. It is determined by an observer viewing selected dark objects against the horizon sky during the day and moderate intensity unfocused lights at night. It is reported in statute miles and fractions thereof and does not necessarily represent the visibility along the runway.

prevailing wind. The wind direction most frequently observed during a given period.

preventive maintenance. The care and servicing by personnel for the purpose of maintaining equipment and facilities in satisfactory operating condition by providing for systematic inspection, detection, and correction of incipient failures either before they occur or before they develop into major defects.

preventive war. A war initiated in the belief that military conflict, while not imminent, is inevitable, and that to delay would involve greater risk.

primary alerting system. A leased-circuitry voice communications system employed by HQ SAC as the alert and execution medium for the SAC forces.

primary body. The **celestial body** or **central force field** about which a **satellite** or other body orbits, or from which it is escaping, or towards which it is falling. The primary body of the moon is the earth; the primary body of the earth is the sun.

primary circulation. In meteorology, the prevailing fundamental atmospheric circulation on a planetary scale which must exist in response to radiation differences with latitude, to the rotation of the planet, and to the particular distribution of land and oceans; and which is required from the viewpoint of **conservation of energy.**

primary configuration. The configuration in which a weapon system is delivered or in which its primary mission capability is contained.

primary cosmic rays. High-energy **particles** originating outside the earth's **atmosphere.** Primary cosmic rays appear to come from all directions in space. Their energy appears to range from 10^9 to more than 10^{17} electron volts.

primary great circle. A **great circle** used as the origin of measurement of a **coordinate;** particularly, such a circle 90° from the poles of a system of **spherical coordinates,** as the equator. Also called *primary circle, fundamental circle.*

primary radar. **Radar** using reflection only, in contrast with secondary radar which uses automatic retransmission on the same or a different radio frequence.

primary standard. A unit directly defined and established by some authority, against which all secondary standards are calibrated, as the prototype kilogram.

prime airlift. (Military) The number of aircraft of a force that can be continuously maintained in a flow from home base to onload base to off load base, thence to the recycle base. Spare and self-support aircraft are not included.

prime meridian. 1. The **meridian** of longitude 0°, used as the origin for measurement of **longitude.** The meridian of Greenwich, England, is almost universally used for this purpose. **2.** Any meridian in any **coordinate system** used as an origin for measurement of longitude.

primer (Eng.) A device for spraying fuel into the induction system or the combustion chamber of an engine to facilitate starting.

prime vertical. The **vertical circle** through the east and west points of the **horizon.** It may be *true, magnetic, compass,* or *grid* depending upon which east or west points are involved. Also called *prime vertical circle.*

primitive atmosphere. The **atmosphere** of a **celestial body** as it existed in the early stages of its formation; specifically, the earth's atmosphere of 3 billion or more years ago, thought to consist of water vapor, carbon dioxide, methane, and ammonia gas.

principal planets. The larger bodies revolving about the Sun in nearly circular orbits. See **planet.** The known principal planets, in order of their distance from the Sun are; Mercury, Venus, Earth, Mars, Jupiter, Saturn, Uranus, Neptune, and Pluto.

principal stresses. The normal stresses on three mutually perpendicular planes on which there are no shear stresses.

prior permission (air). Permission granted by the appropriate national authority prior to the commencement of a flight or a series of flights landing in or flying over the territory of the nation concerned.

private pilot. A person authorized by the **FAA** to **pilot** an aircraft without payment for his services and cannot carry passengers for hire. See **private pilot certificate, airmen certificates.**

private pilot certificate. A certificate of competency issued by the **FAA** to a person meeting the requirements of the applicable **Federal Aviation Regulations.** See **private pilot, instrument rating.**

process lapse rate. The rate of decrease of the **temperature** T of an air **parcel** as it is lifted. The concept

190

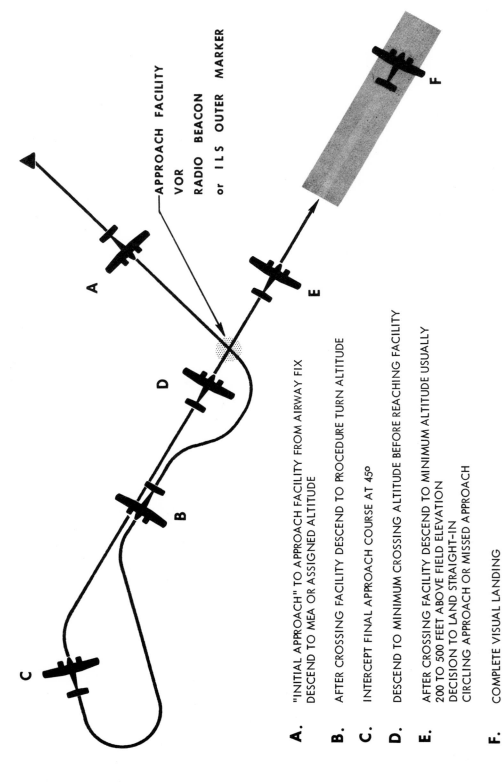

APPROACH FACILITY

VOR

RADIO BEACON

or ILS OUTER MARKER

A. "INITIAL APPROACH" TO APPROACH FACILITY FROM AIRWAY FIX
DESCEND TO MEA OR ASSIGNED ALTITUDE

B. AFTER CROSSING FACILITY DESCEND TO PROCEDURE TURN ALTITUDE

C. INTERCEPT FINAL APPROACH COURSE AT 45°

D. DESCEND TO MINIMUM CROSSING ALTITUDE BEFORE REACHING FACILITY

E. AFTER CROSSING FACILITY DESCEND TO MINIMUM ALTITUDE USUALLY
200 TO 500 FEET ABOVE FIELD ELEVATION
DECISION TO LAND STRAIGHT-IN
CIRCLING APPROACH OR MISSED APPROACH

F. COMPLETE VISUAL LANDING

*A **procedure turn** is an integral part of an instrument approach procedure accomplished without radar vectors.*

may be applied to other atmospheric variables, e.g., the process lapse rate of density. The process lapse rate is determined by the character of the fluid processes and should be carefully distinguished from the environmental lapse rate, which is determined by the distribution of temperature in space. In the atmosphere the process lapse rate is usually assumed to be either the dry-adiabatic lapse rate or the saturation-adiabatic lapse rate.

probability. The chance that a prescribed event will occur, represented as a pure number P in the range $0 \leq P \leq 1$. The probability of an impossible event is zero and that of an inevitable event is unity.

Probability is estimated empirically by relative frequency, that is, the number of times the particular event occurs divided by the total count of all events in the class considered.

probably destroyed (aircraft). A damage assessment on an enemy aircraft seen to break off combat in circumstances which lead to the conclusion that it must be a loss although it is not actually seen to crash.

probe. 1. Any device inserted in an **environment** for the purpose of obtaining information about the environment. 2. In geophysics, a device used to make a **sounding.** 3. Specifically, an instrumented vehicle moving through the **upper atmosphere** or **space** or landing upon another celestial body in order to obtain information about the specific environment. In sense 3, almost any instrumented spacecraft can be considered a probe. However, earth satellites are not usually referred to as *probes.* also , almost any instumented rocket can be considered a probe. In practice, rockets which attain an altitude of less than 1 earth radius (4000 miles) are called *sounding rockets,* those which attain an altitude of *more* than 1 earth radius are called *probes* or *space probes.* Spacecraft which enter into orbit around the sun are called *deep-space probes.* Spacecraft designed to pass near or land on another celestial body are often designated *lunar probe, Martian probe, Venus probe, etc.* . Specifically, a slender device or apparatus projected into a moving **fluid,** as for measurement pusposes; a **pitot tube.** 5. Specifically, a slender projecting pipe on an aircraft which is thrust into a **drogue** to receive fuel in in-flight refueling.

procedure turn. A maneuver, normally part of an instrument approach procedure, in which a turn is made away from a designated track followed by a turn in the opposite direction, both turns being executed so as to permit the aircraft to intercept and proceed along the reciprocal of the designated track. **Note 1.** Procedure turns are designated left or right according to the direction of the initial turn as follows: a) Procedure turn left. (standard). a procedure turn in which the initial turn is to the left; b) Procedure turn right (non-standard). a procedure turn in which the initial turn is to the right. **Note 2.** Procedure turns may be designated as being made either in level flight or while descending, according to the circumstances of each individual instrument approach procedure.

procurement. The process of obtaining personnel, services, supplies, and equipment.

procurement lead time. The interval in months between the initiation of procurement action and receipt into the supply system of the production model (excludes prototypes) purchased as the result of such actions, and is composed of two elements, production lead time and

administrative lead time.

production. The conversion of raw materials into products and/or components thereof, through a series of manufacturing processes. It includes functions of production engineering, controlling, quality assurance, and the determination of resources requirements.

production lead time. The time interval between the placement of a contract and receipt into the supply system of material purchases. Two entries are provided; **a.(initial)** — The time interval if the item is not under production as of the date of contract placement. **b.(reorder)** — The time interval if the item is under production as of the date of contract placement. See also **procurement lead time.**

proficiency training aircraft. (Military) Aircraft required to maintain the proficiency of pilots and other aircrew members who are assigned to nonflying duties.

profile. 1. Of a **variable,** a curve representing corresponding values or two or more variables which may occur. A profile accounts for the correlation from point to point on the curve and has some possibility, not necessarily specified, of actual occurrence. 2. The contour or form of a body, expecially in a cross section; specifically, an airfoil profile. 3. Something likened to a profile (sense 1), such as a line on a graph, as *a flight profile.*

profile drag. For shock-free flow, the sum of the surface-friction drag and the form drag.

program. 1. In **computer** operations, a plan for the solution of a problem. 2. To create a plan for the solution of a problem. A complete program includes plans for the transcription of data, coding for the computer, and plans for the absorption of the results into the system. The list of coded instructions, called a *routine,* plans a computation or process from the asking of a question to the delivery of the result, including the integration of the operation into an exisiting system. Thus, programming consists of planning and coding, including numerical analysis, systems analysis, specification of printing formats, and any other functions necessary to the integration of a computer in a system.

programmer. A person who prepares problem-solving procedures and writes computer routines.

progress payment. Payment made as work progresses under a contract, upon the basis of costs incurred, of percentage of completion accomplished, or of a particular stage of completion. The term does not include payments for partial deliveries accepted by the Government under a contract, or partial payments on contract termination claims.

prohibited area. Airspace of defined dimensions identified by an area on the surface of the earth within which flight is prohibited.

project. A planned undertaking of something to be accomplished, produced, or constructed, having a finite beginning and a finite ending.

Project Blue Book. The Air Force project for the investigation of unidentified flying objects.

projection; map or chart. The system of reference lines representing the earth's meridians and parallels on a chart. The projection is usually designed to retain some special property such as true directions, true distances, true shape, or true area of certain portions of the sphere.

project officer. An individual, military or civilian, who is responsible for planned undertaking or assignment to accomplish something specific. The project assigned is usually of limited life and not normally a responsibility already established within organizational and supervisory channels.

projectile. 1. Any object, especially a **missile,** fired, thrown, launched, or otherwise projected in any manner, such as a bullet, a guided rocket missile, a sounding rocket, a pilotless airplane, etc. 2.Originally, an object, such as a bullet or artillery shell, projected by an applied external force.

prolate spheroid. An **ellipsoid of revolution,** the longer **axis** of which is the axis of revolution. An ellipsoid of revolution, the shorter axis of which is the axis of revolution, is called *oblate spheroid.*

prominence. A filamentlike protuberance from the **chromosphere** of the sun. See **flocculi** Compare **flare.** Prominences can be observed visually (optically) whenever the sun's disk is masked, as during an eclipse or by using a **coronagraph;** and can be observed instrumentally by filtering in certain wavelengths, as with a **spectroheliograph.** A typical prominence is 6,000 to 12,000 kilometers thick, 60,000 kilometers high, and 200,000 kilometers long.

propaganda. Any form of communication in support of national objectives designed to influence the opinions, emotions, attitudes, or behavior of any group in order to benefit the sponsor either directly or indirectly.

propagation. The spreading abroad or sending forward, as of **radiant energy.**

propellant. Any agent used for consumption or combustion in a **rocket** and from which the rocket derives its thrust, such as a **fuel, oxidizer,** additive, catalyst, or any compound or mixture of these; specifically, a fuel, oxidant, or a combination or mixture of fuel and oxidant used in propelling a rocket. See **fuel.** Propellants are commonly in either liquid or solid form.

propeller. A device for propelling an aircraft that has blades on an engine-driven shaft and that, when rotated, produces by its action on the air, a thrust approximately perpendicular to its plane of rotation. It includes control components normally supplied by its manufacturer, but does not include main and auxiliary rotors or rotating airfoils of engines.

propfan. A new development, highly loaded, multiblade turboprop using advanced blade structure and aerodynamics technology for efficient, high-speed operation. Analysis and tests indicate the propulsive efficiency of the propfan is about 20 percent better at Mach 0.8 than a high **bypass ratio turbofan.**

proportional control. Control of an aircraft, rocket, etc. in which control-surface deflection is proportional to the movement of the remote **controls.** Compare **flicker control.**

provisioned spares. (Military) Those spare parts which are procured under certain special Government procurement procedures at a certain point in the weapon system acquisition cycle.

protective action guide (PAG). (nucleonics) The *absorbed dose* of *ionizing radiation* to individuals in the general population which would warrant protective action following a contaminating event, such as a nuclear explosion. (See **radiation protection guide.**)

proton. An *elementary particle* with a single positive electrical charge and a mass approximately 1837 times that of the *electron. The nucleus of an ordinary or light hydrogen atom.* Protons are constituents of all nuclei. The *atom number (Z) of an atom is equal to the number

of protons in its nucleus.

prototype. 1. Of any mechanical device, a production model suitable for complete evaluation of mechanical and electrical form, design, and performance. 2. The first of a series of similar devices. 3. A physical standard to which replicas are compared, as *the prototype kilogram.*

protractor. A device for measuring angles. Usually used in navigation to determine compass courses on a chart.

proximity fuze. A fuze designed to detonate a projectile, bomb, mine, or charge when activated by an external influence in the close vicinity of a target. The variable time fuze is one type of a proximity fuze. See also **fuze.**

pseudoadiabatic expansion. A saturation-adiabatic process in which the condensed water substance is removed from the system, and therefore best treated by the thermodynamics of open systems. See **adiabatic process.** Meterologically, this process corresponds to rising air from which the moisture is precipitating. Descent of air so lifted becomes by definition a **dry-adiabatic process.**

psychological warfare. The planned use of propaganda and other psychological actions having the primary purpose of influencing the opinions, emotions, attitudes, and behavior of hostile foreign groups in such a way as to support the achievement of national objectives.

psychology. The science which studies the functions of the mind. such as sensation, perception, memory, thought, and, more broadly, the behavior of an organism in relation to its environment.

psychomotor ability. Of or pertaining to muscular action ensuing directly from a mental process, as in the coordinated manipulation of aircraft or spacecraft controls.

psychophysical quantity. A physical measurement, as a **threshold,** dependent on human attributes or perception.

psychrometer. An instrument for measuring atmospheric **humidity,** consisting of a dry-bulb thermometer and wet-bulb thermometer (covered with a muslin wick); used in the calculation of **dew point** and relative humidity.

public affairs. Those public information and community relations activities directed toward the general public by the various elements of the Department of Defense.

public aircraft. Aircraft used only in the service of a government, or a political subdivision.

public information. Information of a military nature, the dissemination of which through public news media is not inconsistent with security, and the release of which is considered desirable by or nonobjectionable to the responsible releasing agency.

pull-up point (close air support). The point at which a pilot must start to climb from a low level approach in order to gain sufficient height from which to execute the attack.

pulmonary. Pertaining to, or affecting, the lungs or any component of the lungs.

pulse. 1. A variation of a quantity whose value is normally constant; this variation is characterized by a rise and a decay, and has a finite duration. The word *pulse* normally refers to a variation in time; when the variation is in some other dimension, it should be so specified, such as *space pulse.* This definition is so broad that it covers almost any transient phenomenon. The only features common to all pulses are rise, finite duration, and decay. It is necessary that the rise, duration, and decay be of a quantity that is constant (not necessarily zero) for some time before the pulse and has the same constant value for some time afterwards. The quantity has a normally constant value and is perturbed during the pulse. No relative time scale can be assigned. **2. Radar,** sense **2. 3.** The intermittent change in the shape of an artery due to an increase in the tension of its walls following the contraction of the heart. The pulse is usually counted at the wrist (radial pulse), but may be taken over any artery that can be felt.

pulse amplitude. A general term indicating the **magnitude** of a **pulse.** For specific designation, adjectives such as *average, instantaneous, peak, root-mean-square* (effective), etc., should be used to indicate the particular meaning intended. Pulse amplitude is measured with respect to the normally constant value unless otherwise stated.

pulse amplitude modulation (*abbr* **PAM**). See pulse **modulation.**

pulse code. 1. A sequence of **pulses** so modulated as to represnet information. **2.** Loosely, a **code** consisting of pulses, such as Morse code, **binary code.**

pulse code modulation *(abbr* **PCM**). Any modulation which involves a **pulse code.** This is a generic term and additional specification is required for a specific purpose.

pulsed Doppler system. A **pulse radar** system which utilizes the **Doppler effect** for obtaining information about the **target** (not including simple resolution from fixed targets).

pulse duration. In radar, mesurement of pulse transmission time microseconds, that is, the time the radar's transmitter is energized during each cycle. Also called *pulse length* and *pulse width.*

pulsejet engine. A type of compressorless **jet engine** in which conbustion takes place intermittently, producing thrust by a series of explosions, commonly occurring at the approximate **resonance frequency** of the engine. Often called a *pulsejet.*

pulsejet. A jet-propulsion engine, containing neither compressor nor turbine. Equipped with valves in the front which open and shut, it takes in air to create thrust in rapid periodic bursts rather than continuously.

pulse modulation. 1. Modulation of a **carrier** by a **pulse train.** Compare **frequency modulation.** In this sense, the term is used to describe the process of generating carrier frequency pulses. **2.** Modulation of one or more characteristics of a **pulse carrier.** In this sense, the term is used to describe methods of transmitting information on a pulse carrier.

pulse radar. A type of **radar,** designed to facilitate range measurement, in which the transmitted energy is emitted in periodic short **pulses.** Also called *pulsed radar.* Compare **continuous-wave radar.** The distance to any target causing a detectable echo can be determined by measuring one-half the time interval between transmitted pulse and received echo and multiplying this number by the speed of light. This is by far the most common type of radar.

pulse repeater. In a **transponder,** a device used for receiving **pulses** from one circuit and transmitting corresponding pulses into another circuit. It may also change the frequency and wave forms of the pulses and perform other functions.

pulse repetition frequency. In radar, the number of pulses that occur each second. Not to be confused with transmission frequency which is determined by the rate at which cycles are repeated within the transmitted pulse.

pulse train. In radio, a sequence of **pulses.**

pulse width. The time interval during which a **pulse** exceeds a reference level. For measuring pulse width, the reference level is generally taken at the **half power points.**

punch (Military). You should very soon be obtaining a contact on the aircraft that is being intercepted. (only use with "air intercept" interceptions.)

purge. To rid a line or tank of residual fluid, especially of fuel or oxygen in the tanks or lines of a **rocket** after a test firing or simulated test firing.

pusher. An airplane in which the propeller is mounted aft of the engine, and pushes the air away from it.

pylon. 1. An object on the surface used as a reference for pilots when performing certain aerial maneuvers. **2.** A prominent landmark outside of which pilots must fly during certain racing events. **3.** An attachment to an aircraft that facilitates carriage of ordnance, fuel tanks, engines, other accessories or integral aircraft systems.

pylon eight. An airplane flight training maneuver, the flight path having the shape of a figure 8 and the turns using a prominant landmark (pylon) as a reference.

pyrheliometer. An actionometer which measures the intensity of **direct solar radiation,** consisting of a radiation sensing element enclosed in a casing which is closed except for a small aperture, through which the direct solar rays enter, and a recorder unit.

pyrolysis. Chemical decomposition by the action of heat.

pyrometer. An instrument for the measurement of **temperatures;** generally applied to instruments measuring temperatures above 600° C.

pyrometry. High-temperature **thermometry,** the technique of measurement of temperatures, generally above 600° C, at a distance.

pyrophoric fuel. A **fuel** that ignites spontaneously in air. Compare **hypergolic propellants.**

Q

Q. Military mission designation for **drone** aircraft.

q. **dynamic pressure,** as *the vehicle encountered maximum q 40 seconds after lift-off.*

Q-band. A **frequency band** used in **radar** extending approximately from 36 to 46 kilomegacycles. See **frequency band.**

quadrant. A quarter part of a circle, centered on a NAVAID, oriented clockwise from magnetic north as follows: NE quadrant 000-089, SE quadrant 090-179, SW quadrant 180-269, NW quadrant 270-359.

quality control. That management function by which conformance to established standards is assured, performance is measured, and in the event of defects, corrective action is initiated.

quanta. See **quantum theory.**

quantum. Unit quantity of energy according to the *quantum theory.* The *photon* is a quantum of electromagnetic energy. The term is applied in energy transfers which are not continuous in nature, that is waves. (See **electromagnetic radiation, radiation.**)

quantum theory. The statement, originated by Max Planck, German physicist, that the energy of *radiation* emitted or absorbed is directly proportional to its frequency, and is concentrated in units, or quanta, each having 6.656×10^{-27} erg.

quarter-chord point. The point on the chord of an airfoil section at one quarter of the chord length behind the leading edge.

quick engine change unit. See **power plant unit.**

quiet sun. The sun when it is free from unusual radio wave or thermal **radiation** such as that associated with **sun spots.**

R

R. Military mission designation for **reconnaissance** aircraft.

rabbit. Video **display** of a beacon's replies to **interrogations** from two or more nonsynchronized **radars.**

racon. (From *radar beacon.*) A **transponder** for **interrogation** by a **primary radar.**

rad. (Acronym for radiation absorbed dose). The basic unit of absorbed dose of *ionizing radiation.* A dose of one rad means the absorption of 100 ergs of radiation energy per gram of absorbing material. (Compare **rem, roentgen;** see **absorbed dose.**)

radar. (From *radio detection and ranging.*) **1.** A method, system, or technique of using beamed, reflected, and timed **radio waves** for detecting, locating, or tracking objects (such as rockets), for measuring altitude, etc., in any of various activities, such as air traffic control or guidance. **2.** The electronic equipment or apparatus used to generate, transmit, receive, and, usually, to display radio scanning or locating **waves** ; a radar set. The terms *primary radar* and *secondary radar* may be used when the return signals are, respectively, by reflection and by the transmission of a second signal as a result of triggering responder **beacon** by the incident signal.

radar advisory. The term used to indicate that the provision of advice and information is based on radar observation. An air traffic control term.

radar altimeter. An altimeter operating on the radar principle to determine the altitude of the air or space vehicle above the terrain directly beneath it. For manned aircraft, altitude information (feet) is usually displayed to the pilot on the instrument face in digital readout form. Output signals can be provided to input to automatic flight control equipment. Also see **absolute altimeter, ground avoidance radar, radar, radio altimeter.**

radar altitude. The altitude of an aircraft or spacecraft as determined by a **radio altimeter;** thus, the actual distance from the nearest terrain feature.

radar approach control. A facility providing radar approach control service by use of airport surveillance radar and precision approach radar equipment.

radar astronomy. The study of celestial bodies within the solar system by means of **radiation** originating on earth but reflected from the body under observation. See **radio astronomy.**

radar band. See **frequency band.**

radar beacon. A **beacon** transmitting a characteristic signal on **radar** frequency, permitting a craft to determine the bearing and sometimes the range of the beacon. A *racon* returns a coded signal when triggered by the proper type of radar pulse; a *ramark* continuously transmits a signal which appears as a radial line on the plan position indicator.

radar beam. See **beam.**

radar clutter. Unwanted signals, echoes, or images on the face of the display tube which interfere with observation of desired signals.

radar contact. The term air traffic controllers use to indicate that an aircraft is identified on the radar display and that radar service can be provided until radar identification is lost or radar service is terminated, and that when the aircraft is informed of "radar contact" it automatically discontinues reporting over compulsory reporting points.

radar control. Term used to indicate that radar-derived information is employed directly in the provision of air traffic control service.

radar controller. A person capable of performing the functions of surveillance controller, traffic director or precision controller.

radar cross section. The ratio of **power** returned in a radar **echo** to power received by the **target** reflecting the signal.

radar echo. See **echo.**

radar flight following. The general observation of the progress of identified aircraft targets sufficiently to retain their identity or the observation of the movement of specific radar targets.

radar frequency. See **frequency band.**

radar horizon. The **angle of elevation** at which the **beam** from a **radar** antenna is intercepted by the earth's horizon. Compare **radio horizon.**

radar indentification. The process of ascertaining that a radar target is the radar return from a particular aircraft.

radar imagery. Imagery produced by recording radar waves reflected from a given target surface.

radar indicator. See **radarscope.**

radar mile. A time unit or 10.75 microseconds duration; the time it takes for the **signal** emitted by a **radar** to travel from the radar to a **target** one mile distant and return to the radar.

radar monitoring. See **radar service.**

radar netting. The linking of several radars to a single center to provide integrated target information.

radar picket. (Military) Any ship, aircraft, or vehicle, stationed at a distance from the force protected, for the purpose of increasing the radar detection range.

radar picket cap. Radar picket combat air patrol.

radar range. 1. The distance from a **radar** to a target as measured by the radar. **2.** The maximum distance at which a radar set is effective in detecting targets. Radar range depends upon variables such as weather conditions, type of target, etc. Radar range, sense **2,** is sometimes given a specific definition, e.g., the range at which the set is effective one-half of the time.

radar ranging. The use of radar transmissions to determine range to the target.

radar reflectivity. In general, the measure of the efficiency of a **radar target** in intercepting and returning a radar **signal.** It depends upon the size, shape, aspect, and the dielectric properties at the surface of the target. It includes the effects of not only reflection (see **reflectivity**) but also **scattering** and **diffraction.**

radar reflector. A device capable of or intended for reflecting **radar** signals. See **parabolic reflector.**

radar return/echo. The signal indication of an object which has reflected energy transmitted by a primary radar.

radar scan. 1. The searching motion of a radar **beam** in any of various path configurations; the pattern of the

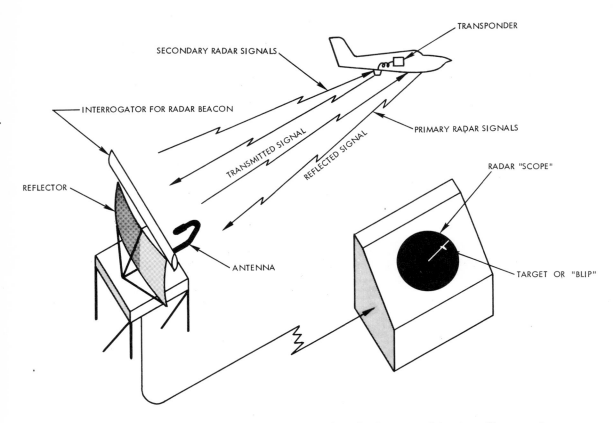

SECONDARY RADAR SIGNALS

TRANSPONDER

INTERROGATOR FOR RADAR BEACON

PRIMARY RADAR SIGNALS

TRANSMITTED SIGNAL

REFLECTED SIGNAL

REFLECTOR

ANTENNA

RADAR "SCOPE"

TARGET OR "BLIP"

A simplified schematic showing the basic principles of **radar** *as used in air traffic control.*

Giant 85 foot **radar** *locates and tracks missiles at the Pacific Missile Range.*

motion of a radar beam. **2. Radar scanning.**

radar scanning. The action or process of moving or directing a searching radar **beam.**

radarscope. The **cathode-ray tube** or oscilloscope in a **radar set,** which displays the received **signal** in such a manner as to indicate range, bearing, etc. Sometimes called a *radar indicator.*

radarscope overlays. Transparent overlays for placing on the radarscope for comparison and identification of radar returns.

radar screen. 1. The face of a **cathode-ray oscilloscope** used in a **radar set. 2.** A network of **radar** installations, or their emanations, serving, e.g., to detect strange aircraft.

radar separation. See **radar service.**

radar service. An air traffic control term which encompasses one or more of the following services based on the use of radar which can be provided by a controller to a pilot of a radar-identified aircraft.   **Radar Separation.** Radar spacing of aircraft in accordance with established minima. **Radar Navigation Guidance** —Vectoring aircraft to provide course guidance. **Radar Monitoring** — The radar flight following of aircraft, whose primary navigation is being performed by the pilot, to observe and note deviations from its authorized flight path airway, or route. As applied to the monitoring of instrument approaches from the final approach fix to the runway, it also includes the provision of advice on position relative to approach fixes and whenever the aircraft proceeds outside the prescribed safety zones.

radar set. An electronic apparatus consisting principally of a transmitter, antenna, receiver, and indicator for sending out scanning beams and receiving and displaying the reflected waves or the waves emitted by a radar beacon. See **radar.**

radar shadow. A condition in which **radar frequency** signals do not reach a region because of an intervening obstruction.

radar silence. An imposed discipline prohibiting the transmission by radar of electromagnetic signals on some or all frequencies.

radar surveillance. The radar observation of a given geographical area for the purpose of performing some radar function.

radar target. An object which reflects a sufficient amount of a **radar** signal to produce an **echo** signal on the radar screen.

radar tracking station. A radar facility which has the capability of tracking moving targets.

radar vector. A heading issued to an aircraft to provide navigational guidance by radar.

radar volume. The volume in space that is irradiated by a given **radar.** For a **continuous-wave radar** it is equivalent to the antenna radiation pattern. For a **pulse radar** it is a function of the cross-section area of the beam of the antenna and the pulse length of the transmitted pulse.

radar wave. A transmitted or reflected **radio wave** used in **radar;** a radio wave in one of the **frequency bands** used for radar.

radial. 1. Motion along a radius. **2.** A magnetic bearing extending from a **very high frequency omnirange/tactical air navigation** station.

radial engine. An engine with a row, or rows, of cylinders spaced radially round a common crankshaft, the cylinders being stationary and the crankshaft revolving.

radial motion. Motion along a radius, or a component in

such a direction, particularly that component of space motion of a **celestial body** in the direction of the line of sight.

radial velocity. In **radar,** that vector component of the **velocity** of a moving **target** that is directed away from or toward the ground station.

radian. The angle subtended at the center of a circle by an arc equal in length to a radius of the circle. It is equal to $360°/2\pi$ or approximately 57 degrees 17 minutes 44.8 seconds.

radiant. 1. Pertaining to the **emission** or the measurement of **electromagnetic radiation.** Compare **luminous. 2.** In astronomy, the apparent location on the **celestial sphere** of the origin of the luminous **trajectories** of **meteors** seen during a **meteor shower.** For convenience, the common meteor showers are named for the constellations of stars in which their radiants appear. **3.** In describing auroras, a projected point of intersection of lines drawn coincident with auroral streamers; that is, the point from which the **aurora** seems to originate.

radiant energy. 1. The **energy** of any type of **electromagnetic radiation,** such as light, infrared, radio and radar. Also called **radiation. 2.** Infrequently, any energy that may be radiated, as, for example, acoustic energy.

radiant energy thermometer. An instrument which determines the **black-body temperature** of a substance by measuring its thermal radiation.

radiant heat. Infrared radiation. This term, still used in certain engineering fields, is to be avoided since it confuses the distinct physical concepts of radiation and heat.

radiant temperature. The **temperature** obtained by use of a total radiation **pyrometer** when sighted upon a non-black body. This is always less than the true temperature.

radiating element. A basic subdivision of **an antenna** which in itself is capable of radiating or receiving **radiofrequency** energy.

radiation. 1. The process by which electromagnetic energy is propagated through free space by virtue of joint undulatory variations in the electric and magnetic fields in space. This concept is to be distinguished from conduction and convection. A group of physical principles known as the **radiation laws** comprise, to a large extent, the current state of practical knowledge of the complex radiative processes. **2. The process by which energy is propagated through any medium by virtue of the wave motion** of that medium, as in the propagation of sound waves through the atmosphere, or ocean waves along the water surface. **3. radiant energy. 4. electromagnetic radiation,** specifically, high-energy radiation such as **gamma rays** and **X-rays. 5.** Corpuscular emissions, such as α or β-radiation. **6. nuclear radiation. 7. radioactivity.**

radiation area. Any accessible area in which the level of radiation is such that a major portion of an individual's body could receive in any one hour a dose in excess of 5 milirem, or in any 5 consecutive days a dose in excess of 150 milirem. (See **absorbed dose, rem.)**

radiation belt. An envelope of **charged particles** trapped in the **magnetic field** of a spatial body. See **Van Allen belt.**

radiation burn. Radiation damage to the skin. Beta burns result from skin contact with or exposure to emitters of beta particles. Flash burns result from sudden thermal radiation. (See **beta particles, flash burn, ionizing radia-**

tion, thermal burn.)

radiation cooled. Of a structure, pertaining to the use of materials able to radiate heat at a rate such that the rate of increase of the temperature of the material is low.

radiation detection instruments. Devices that detect and record the characteristics of *ionizing radiation.* (See **counter, dosimeter, monitor.**)

radiation dosimetry. The measurement of the amount of radiation delivered to a specific place or the amount of radiation that was absorbed there. (See **dosimeter.**)

radiation fogs. Fogs characteristically resulting from the radiational cooling of air near the surface of the ground on relatively calm clear nights. Such fogs which usually occur on calm cloudless nights and are brought about by the cooling of water vapor created by evaporation during the heat of the day. These fogs usually disappear with the coming of the morning sun; their depth is usually less than 300 feet.

radiation illness. An acute organic disorder that follows exposure to relatively sever doses of ionizing radiation. It is characterized by nausea, vomiting, diarrhea, blood cell changes, and in later stages by hemorrhage and loss of hair. (See **ionizing radiation.**)

radiation laws. 1. The four physical laws which, together, fundamentally describe the behavior of black-body radiation: (**a**) the **Kirchoff law** is essentially a thermodynamic relationship between emission and absorption of any given wavelength at a given temperature; (**b**) the **Planck law** describes the variation of intensity of black-body radiation at a given temperature, as a function of wavelength; (**c**) the **Stefan-Boltzmann law** relates the time rate of radiant energy emission from a black body to its absolute temperature; (**d**) the **Wien law** relates the wavelength of maximum intensity emitted by a black body to its absolute temperature. **2.** All the more inclusive assemblage of empirical and theoretical laws describing all manifestations of radiative phenomena; e.g. **lambert law.**

radiation lobe. A portion of the **radiation pattern** bounded by one or two cones of **nulls.**

radiation medicine. That branch of medicine dealing with the effect of **radiation,** specifically high-energy radiation such as X-rays, gamma rays, and energetic particles on the body and with the prevention or cure of physiological injuries resulting from such radiation.

radiation monitoring. Nucleonics. Continuous or periodic determination of the amount of radiation present in a given area. (See **monitor.**)

radiation pattern. A graphical representation of the **radiation** of an **antenna** as a function of direction. Cross sections in which radiation patterns are frequently given are vertical planes and the horizontal plane, or the principal electric and magnetic polarization planes. Also called *antenna pattern, lobe pattern, coverage diagram.* Two types of radiation patterns should be distinguished. They are : (**a**) the free-space radiation pattern which is the complete lobe pattern of the antenna and is a function of the wavelength, feed system, and reflector characteristics, and (**b**) the field radiation pattern which differs primarily from the free-space pattern by the formation of interference lobes whenever direct and reflected wave trains interfere with each other as is found in most surface-based radars. The envelope of these interference lobes has the same shape, but for a perfectly reflecting surface, it has up to twice the amplitude of the free-space radiation pattern.

radiation protection. Nucleonics. Legislation and regulations to protect the public and laboratory or industrial workers against radiation. Also measures to reduce exposure to radiation. (Compare *protection.*)

radiation protection guide. The officially determined radiation doses which should not be exceeded without careful consideration of the reasons for doing so. These standards, established by the Federal Radiation Council, are equivalent to what was formerly called the *maximum permissible dose* or *maximum permissible exposure.*

radiation shield. 1. A device used on certain types of instruments to prevent unwanted **radiation** from biasing the measurement of a quantity. **2.** A device used to protect human beings from the harmful effects of **nuclear radiation,** cosmic radiation, or the like. **3. heat shield.**

radiation shielding. Reduction of radiation by interposing a shield of absorbing material between any radioactive source and a person, laboratory area, or radiation-sensitive device. (See **absorber, shield.**)

radiation sickness. See **radiation illness.**

radiation source. Nucleonics. Usually a man-made, sealed source of *radioactivity* used in teletherapy, radiography, as a power source for batteries, or in various types of industrial gauges. Machines such as accelerators, and radioisotopic generators and natural radionuclides may also be considered as sources.

radiation sterilization. Use of radiation to cause a plant or animal to become sterile, that is, incapable of reproduction. Also the use of radiation to kill all forms of life (especially bacteria) in food, surgical sutures, etc. (Compare **radiation illness, radiomutation.**)

radiation therapy. Treatment of disease with any type of radiation. Often called radiotherapy.

radiator. 1. Any source of radiant energy, especially **electromagnetic radiation. 2.** A device that dissipates the heat from something, as from water or oil, not necessarily by radiation only. Generally, the application of the terms *radiator* (in sense **2**) or *heat exchanger* to a particular apparatus depends upon the point of view: If the emphasis is upon merely getting rid of heat, *radiator* is most often used, or sometimes *cooler;* if the emphasis is upon transferring heat, *heat exchanger* is used — but these distinctions do not always hold true.

radio. 1. Communication by **electromagnetic waves,** without a connecting wire. **2.** Pertaining to **radiofrequency,** as in *radio wave.*

radioactive. Exhibiting radioactivity or pertaining to radioactivity.

radioactive cloud. A mass of air and vapor in the atmosphere carrying radioactive debris from a nuclear explosion. (See **atomic cloud.**)

radioactive contamination. Deposition of radioactive material in any place where it may harm persons, spoil experiments, or make products or equipment unsuitable or unsafe for some specific use. The presence of unwanted radioactive matter. Also radioactive material found on the walls of vessels in used-fuel processing plants, or radioactive material that has leaked into a reactor coolant. Often referred to only as *contamination.* (Compare **background radiation.** See **decontamination.**)

radioactive dating. Nucleonics. A technique for measuring the age of an object or sample of material by determining the ratios of various *radioisotopes* or products of radioactive decay it contains. For example, the ratio of carbon-14 to carbon-12 reveals the approximate age of bones, pieces of wood, or other archeological specimens that contain carbon extracted from the air at the time of their origin. (Compare **atomic clock**; see **decay, radioactive.**)

radioactive decay. nucleonics. (See decay, radioactive.) (disintegration).

radioactive fallout. (See fallout).

radioactive gas. 1. In atmospheric electricity, any one of the three radioactive inert gases, radon, thoron, and actinon, which contribute to atmospheric **ionization** by virtue of the ionizing effect of the alpha particles which each emits on disintegration. These three gases are isotopic to each other, all having atomic number 86. **2.** Any gaseous material containing **radioactive** atoms.

radioactive series. A succession of nuclides, each of which transforms by radioactive disintegration into the next until a stable nuclide results. The first member is called the *parent*, the intermediate members are called *daughters*, and the final stable member is called the *end product*. (See **decay, radioactive.**)

radioactivity. 1. Spontaneous disintegration of atomic nuclei with emission of **corpuscular** or **electromagnetic radiations.**

The principal types of radioactivity are alpha decay, beta decay, and isomeric transition.

To be considered as radioactive a process must have a measureable lifetime between approximately 10^{-10} second and approximately 10^{17} years. Radiations emitted within a time too short for measurement are called *prompt*.

Prompt radiations such as gamma rays and X-rays are often associated with radioactive disintegrations. **2.** The number os spontaneous disintegrations per unit mass and per unit time of a given unstable (radioactive) element, usually measured in curies.

radio altimeter. A device that measures the **altitude** of a craft above the terrain by measuring the elapsed time between transmission of **radio waves** from the craft and the reception of the same waves reflected from the terrain. Also called *radar altimeter*.

radio approach aids. Equipment making use of radio to determine the position of an aircraft with considerable accuracy from the time it is in the vicinity of an airfield or carrier until it reaches a position from which landing can be carried out.

radio astronomy. 1. The study of **celestial** objects through observation of **radiofrequency** waves emitted or reflected by these objects.

In this sense *radio astronomy* includes both the use of radiation emitted by the celestial bodies and of radiation originating on earth and reflected by celestial bodies (radar astronomy).

2. Specifically, the study of celestial objects by measurement of the radiation emitted by them in the radiofrequency range of the electromagnetic spectrum.

Radio astronomy measurements are usually of the **intensity** of the received signal but often include **polarization** of the signal and angular size of the source.

radio beacon. A radio transmitter which emits a distinctive or characteristic signal used for the determination of bearings, courses, or location. See also **beacon, nondirectional radio beacon, homing beacon, marker beacon.**

radio beam. See **beam.**

radiobiology. The study of the effects produced on living organisms by **radiation.**

radio blackout. Same as **blackout,** sense **1.**

radio channel. A **frequency band** comprised of the emission **bandwidth,** the **interference guard bands,** and the **frequency tolerance.**

radiochemistry. The body of knowledge and the study of the chemical properties and reactions of radioactive materials.

radio command. A radio **signal** to which a rocket, satellite, or the like responds.

radio compass. A device employing a fixed-loop antenna and visual indicator, used chiefly for "homing" flight (flight directly toward or away from a radio station). This equipment has been largely superceded by the **automatic direction finder (ADF).**

radio control. 1. Remote **control** of a pilotless airplane, a rocket, etc., by means of radio **signals** that activate controlling devices. **2.** Any radio apparatus used for this kind of control.

radio deception. The employment of radio to deceive the enemy. Radio deception includes sending false dispatches, using deceptive headings, employing enemy call signs, etc. See also **electronic deception.**

radio detection. The detection of the presence of an object by radio location without precise determination of its position.

radio direction finder. A device similar to the radio compass, but employing a rotatable loop antenna; signals are often received both aurally and visually. This equipment has been largely superceded by the **automatic direction finder (ADF).**

radio direction finder, automatic. See **automatic direction finder (ADF).**

radio duct. A rather shallow, almost horizontal layer in the **atmosphere** through which vertical temperature and moisture gradients are such as to produce an **index of refraction** lapse rate of greater than—48 N-units per 1000 feet. Strong temperature, or moisture inversions, or both are necessary for the formation of radio ducts. The resulting **superstandard propagation** is such as to cause the curvature of rays traveling through it to be greater than that of the earth. Radio energy which originates within the duct and leaves the antenna at angles near the horizontal may thus be *trapped* within the layer. See **anomalous propagation, skip effect.**

The effect is similar to that of a mirage (it is sometimes called *radio mirage*), and radar targets may be detected at phenomenally long ranges if both target and radar are in the duct. The greater the elevation angle between radar and target, the less the possibility of serious distortion due to transmission through ducts. Ducts may be surface based or elevated, with thickness ranging from a few tens of feet up to a maximum of 1000 feet. Elevated ducts are generally associated with subsidence or frontal inversions and are seldom found above 15,000 to 20,000 feet.

radioecology. The body of knowledge and the study of the effects of radiation on species of plants and animals in natural communities.

radio energy. Electromagnetic **radiation** of greater wavelength (lower frequency) than **infrared** radiation, that is, of **wavelength** greater than about 1000 microns (0.01 centimeter). The high-frequency end of the radioenergy spectrum is known as *microwave radiation*. See **frequency bands.**

radio fadeout. fadeout.

radio fix. 1. The location of a friendly or enemy radio transmitter, determined by finding the direction of the radio transmitter from two or more listening stations. **2.** The location of a ship or aircraft by determining the direction of radio signals coming to the ship or aircraft from two or more sending stations, the locations of which are known. Also see **fix.**

radiofrequency (*abbr* **RF**). **1.** A **frequency** at which coherent electromagnetic radiation of energy is useful for communication purposes.

Roughly, the radiofrequency of the electromagnetic spectrum lies between 10^4 and 10^{12} cycles per second. See **frequency bands.**

2. Specifically, the frequency of a given radio **carrier wave.**

radiofrequency band. See **frequency band.**

radio goniometer. radio direction finder.

radiography. The use of ionizing radiation for the production of shadow images on a photographic emulsion. Some of the rays (*gamma rays* or *X rays*) pass through the subject, while others are partially or completely absorbed by the more opaque parts of the subject and thus cast a shadow on the photographic film.

radio guard. A ship, aircraft, or radio station designated to listen for and record transmissions, and to handle traffic on a designated frequency for a certain unit or units.

radio guidance system. A **guidance** system that uses radio **signals** to guide an aircraft or spacecraft in flight; the **system** includes both the flight-borne equipment and the guidance station equipment on the ground.

radio interferometer. An **interferometer** operating at radiofrequencies. *Radio interferometers* are used in radio astronomy and in satellite tracking.

radioisotope. A radioactive isotope. An unstable isotope of an element that decays or disintegrates spontaneously, emitting radiation. More than 1300 natural and artificial radioisotopes have been identified. (See **decay, radioactive; isotope.**)

radioisotopic generator. A small power generator that converts the heat released during radioactive decay directly into electricity. These generators generally produce only a few watts of electricity and use thermoelectric or thermionic converters. Some also function as electrostatic converters to produce a small voltage. Sometimes called an "atomic battery". (See **decay, radioactive; SNAP**)

radio horizon. The locus of points at which direct rays from a radio **transmitter** become tangential to the earth's surface. Assuming a smooth surface, the distance of the horizon is given approximately by the equation $r = \sqrt{2h}$ where r is the distance, statute miles, and h is the height, feet, of the **antenna** above the surface. Compare **radar horizon.** The horizon extends beyond (below) the geometrical and visible horizons as the result of normal atmospheric refraction. Beyond the radio horizon, surface targets cannot be detected under normal atmospheric conditions although significant amounts of radio power have been detected in the diffraction zone below the horizon. It is now felt that this represents power scattered by turbulence-produced atmospheric inhomogeneities.

radiological defense. Defensive measures taken against the radiation hazard resulting from the employment of nuclear and radiological weapons.

radiology. The science which deals with the use of all forms of ionizing radiation in the diagnosis and the treatment of disease.

radioluminescence. Visible light caused by radiations from radioactive substances; an example is the glow from luminous paint containing radium and crystals of zinc sulfide, which give off light when struck by alpha particles from the radium. (See **luminescence.**)

radio magnetic indicator (RMI). An instrument installed in an aircraft instrument panel which displays heading information and magnetic bearing to one or two radio range/beacon stations. All inputs to the RMI are from remotely installed **compass, gyro** and **radio** systems.

radio marker beacon. See **marker beacon.**

radiometer. An instrument for detecting and, usually, measuring **radiant energy.** Compare **bolometer.** See **actinometer, photometer.**

radiometric magnitude. The **magnitude** of a **celestial body** measured with reference to the total radiation observable through the atmosphere.

radiometry. The science of measurement of radiant energy. In practice, there is no clear distinction between radiometry and **photometry** although *photometry* usually refers to measurements in the visible and near-visible range.

radiomutation. A permanent, transmissible change in form, quality, or other characteristic of a cell or offspring from the characteristics of its parent, due to radiation exposure. (See **genetic effects of radiation, mutation.**)

radio navigation. Radio location intended for the determination of position or direction or for obstruction warning in navigation. Also see **navigation.**

radio range. A transmitting facility which radiates signals which have a unique characteristic in specific bearings from the transmitting site. The facilities are normally operated on a continuous basis and, in general, are monitored but do not require operating personnel. See **VHF omnidirectional range (VOR), Tacan, LF/MF four course radio range.**

radio range finding. Radio location in which the distance of an object is determined by means of its radio emissions, whether independent, reflected or retransmitted on the same or other wave length.

radio range station. Same as **radio range.**

radioresistance. Nucleonics. A relative resistance of cells, tissues, organs, or organisms to the injurious action of radiation. (Compare **radiosensitivity.**)

radiosensitivity. Nucleonics. A relative susceptibility of cells, tissues, organs or organisms to the injurious action of radiation. (Compare **radioresistance.**)

radio silence. A period during which all or certain radio equipment capable of radiation is kept inoperative. (In combined or United States joint or intra-Service communications the frequency bands and/or types of equipment affected will be specified.)

radiosonde. An instrument, usually balloon-borne, for the simultaneous measurement and transmission of meteorological data while moving vertically through the atmosphere. See **dropsonde.** The instrument consists of transducers for the measurement of pressure, temperature, and humidity; a modulator for the conversion of the output of the transducers to a quantity which controls a property of the radiofrequency signal; a selector switch which determines the sequence in which the parameters are to be transmitted; and a transmitter which generates the radiofrequency carrier.

radiospectrum. The range of frequencies of **electromagnetic radiation** usable for radio communication. The radiospectrum ranges from about 10 kilocycles per second to over 300,000 megacycles per second. Corresponding wavelengths are 30 kilometers to 1 millimeter. See **frequency bands.**

radio telegraphy. The transmission of telegraphic codes by means of radio.

radio telephony. The transmission of speech by means of modulated radio waves.

radio telescope. A device for receiving, amplifying, and measuring the intensity of **radio waves** originating outside the earth's atmosphere or reflected from a body outside the atmosphere. A radio telescope usually includes a source of radiation of known power for calibration of the received signal. A term *radio telescope* is not restricted to devices incorporating a paraboloidal *dish* antenna. A radio telescope can use any antenna or combination of antennas which will accept the radiation being studied.

radio waves. Waves produced by **oscillation** of an electric charge at a **frequency** useful for radiocommunication. Formerly called *Hertizian waves.* See **frequency bands, electromagnetic radiation.**

radium. [Symbol Ra] A radioactive metallic element with atomic number 88. As found in nature, the most common isotope has an atomic weight of 226. It occurs in minute quantities associated with uranium in pitchblende, carnotite and other minerals; the uranium decays to radium in a series of alpha and beta emissions. By virtue of being an alpha- and gamma-emitter, radium is used as a source of *luminescence* and as a *radiation source* in medicine and radiography.

radius of action. (Military) The maximum distance a ship, aircraft, or vehicle can travel away from its base along a given course with normal combat load and return without refueling, allowing for all safety and operating factors.

radius vector. A straight line connecting a fixed reference point or center with a second point, which may be moving; specifically, in astronomy, the straight line connecting the center of a **celestial body** with the center of a body which revolves around it, as *the radius vector of the moon.* See **polar coordinates, spherical coordinates.**

radix. base Same as **base** (of a number system).

radome. (From *radar dome.* Pronounced *raydome.*) A **dielectric** housing for an **antenna.**

raid. An operation, usually small scale, involving a swift penetration of hostile territory to secure information, confuse the enemy, or to destroy his installations. It ends with a planned withdrawal upon completion of the assigned mission.

rain. Liquid **precipitation** in the form of drops of appreciable size such that their individual impact on water surfaces is perceptible.

rainout. Radioactive material in the atmosphere brought down by precipitation.

rain gage. An instrument for measuring rainfall.

ram air. Air entering an **airscoop** or **air inlet** as a result of the high-speed forward movement of a vehicle.

ram drag. The **drag** produced by the **momentum** of air entering an **airscoop** or an **air inlet** of an aeronautical vehicle in flight.

ramjet engine. A type of **jet engine** with no mechanical compressor consisting of a specially shaped tube or duct open at both ends, the air necessary for combustion being shoved into the duct and compessed by the forward motion of the engine, where the air passes through a **diffuser** and is mixed with **fuel** and burned, the exhaust gases issuing in a **jet** from the rear opening. The ramjet engine cannot operate under static conditions. Often called a *ramjet.* Also called *Lorin tube.*

ramp. A defined area, on a land airport, intended to accommodate aircraft for purposes of loading or unloading passengers or cargo, refueling, parking, or maintenance.

ramp weight. The static weight of a mission aircraft determined by adding operating weight, payload, flight plan fuel load, and fuel required for ground turbine power unit, taxi, runup and takeoff.

random. Eluding pricise prediction, completely irregular. Compare **stochastic.** In connection with probability and statistics, the term *random* implies collective or long-run regularity; thus, a long record of the behavior of a random phenomenon presumably gives a fair indication of its general behavior in another long record, although the individual observations have no discernible system of progression.

random error. Errors that are not systematic, are not erratic, and are not **mistakes.** Such random errors are caused by disturbed elements in the measuring instrument and usually are of an approximately normal or **Gaussian distribution.** Such random errors are sometimes called *short-period errors.*

random sample. A **sample** taken at random from a population.

range. 1. The difference between the maximum and minimum of a given set of numbers; in a **periodic** process it is twice the amplitude, i.e., the wave height. **2.** The distance between two objects, usually an observation point and an object under observation. See **slant range. 3.** A maximum distance attributable to some process, as in *visual range* or *the range of a rocket.* **4.** An area in and over which rockets are fired for testing, as *Atlantic Missile Range.* **5. radar range.**

range attenuation. In **radar** terminology, the decrease in **power density** (flux density) caused by the divergence of the flux lines with distance, this decrease being in accordance with the inverse-square law.

range error. The error in **radar range** measurement due to the **propagation** of radio energy through a nonhomogeneous atmosphere. This error is due to the fact that the volocity of radio-wave propagation varies with the **index of refraction** and that ray travel is not in straight lines through actual atmospheres. The resulting range error is generally insignificant. Compare **azimuth error.**

range gating. The use of circuits in **radar** to suppress **signals** from all targets falling outside selected **range** limits.

range-height-indicator scope. A type of **radar** indicator (radar-scope); an intensity-modulated **indicator** on which echoes are displayed in coordinates of slant range and elevation angle, simulating, thereby, a vertical cross section of the atmosphere along some azimuth from the radar. The power of the signal returned from the target is used to modulate the intensity of the electron beam.

range marker. The index marks displayed on **radar** indicators to establish the scale or facilitate determination of the distance of a target from the radar. On the **plan-position-indicator** scope, for example, range markers take the form of concentric circles with the position of the radar at the center. See **azimuth marker.** Also called *distance marker.*

range only measurement of trajectory and recording. A nonambiguous spherical and elliptical, long-baseline, range-only **trajectory measuring system** utilizing phase comparison techniques with range-modulation frequencies. The system consists of three or more receivers which track a **transponder** interrogated by a single transmitter. The reference signal from the ground transmitter is also received by the ground receivers. Simultaneous range measurements are made by the ground receivers, which are correlated with base timing from which space position can be computed by triangulation. The system operates on 387 and 417 megacycles.

range, radio. See **radio range.**

range rate. The rate at which the distance from the measuring equipment to the target or signal source being tracked is changing with respect to time. See **radial velocity.**

range resolution. The ability of the radar equipment to separate two reflecting objects on a similar bearing, but at different ranges from the antenna. The ability is determined primarily by the pulse length in use.

range ring. A circle on a **plan position indicator,** particularly one with an adjustable diameter, to indicate distance from the antenna. See **distance marker.**

range safety officer. An official on a **rocket test range** whose responsibility is to supervise the planning and execution of each test to insure the maximum safety of all personnel and property within the range boundaries.

range strobe. An index mark which may be displayed on some types of radar **indicators** to assist in the determination of the exact **range** of a **target.**

range sweep. See **sweep,** note.

range wind. The component of a **ballistic** wind which is parallel to the longitudinal axis of the range.

ranging. The process of establishing target distance. Types of ranging include echo, intermittent, manual, navigational, explosive echo, optical, radar, etc.

ranging pulse. In a radar system the **pulse** used to measure the **range** of the object being tracked.

ranging system. A **radar** system which measures range (distance).

Rankine cycle. An idealized thermodynamic cycle consisting of two constant-pressure processes and two isentropic processes.

Rankine temperature scale. *(abbr* °**R).** A temperature scale with the degree-interval of the **Fahrenheit temperature scale** and the zero point at **absolute zero.** The **ice point** is thus 491.69 degrees Rankine and the **boiling point** of water is 671.69 degrees Rankine.

raob. (From *radiosonde ob*servation.) An observation of the vertical distribution of temperature, pressure and relative humidity, obtained by means of a **radiosonde.**

rare earths. A group of 15 chemically similar metallic elements, including Elements 57 through 71 on the *Periodic Table* of the Elements, also known as the Lanthanide Series.

rarefied gas dynamics. The study of the phenomena related to the molecular or noncontinuum nature of gas **flow** at densities where $\lambda/l >|0.01$ when λ is molecular mean free path and l is a characteristic dimension of the flow field.

Flow with $\lambda/l > 0.01$ is called *molecular flow.*

Flow with $\lambda/l < 0.01$ is called *continuum flow.*

Flow with $\lambda/l \approx 0.01$ to 0.1 is called *slip flow.*

Flow with $\lambda/l \approx 0.1$ to 10 is called *transition flow.*

Flow with $\lambda/l > 10$ is called *free molecule flow.*

Slip flow and *transition flow* are not always distinguished from each other.

rare gas. inert gas.

raster. The pattern followed by the **electronbeam** exploring element scanning the screen of a television transmitter or receiver.

raster line. One line of a raster, or scanning pattern.

rated maximum continuous power. With respect to reciprocating, turbopropeller, and turboshaft engines, means the approved brake horsepower that is developed statically or in flight, in standard atmosphere at a specified altitude, within the engine operating limitations for unrestricted periods of use.

rated maximum continuous thrust. With respect to turbojet engines, means the approved jet thrust that is developed statically or in flight, in standard atmosphere at a specified altitude, within the engine operating limitations for unrestricted periods of use.

rated takeoff power. With respect to reciprocating, turbopropeller, and turboshaft engines, means the approved brake horsepower that is developed statically under standard sea level conditions, within the engine operating limitations and limited in use to periods of not over 5 minutes for takeoff operation.

rated takeoff thrust. With respect to turbojet engine types, the approved jet thrust that is developed statically under standard sea level conditions, within the engine operating limiations and limited in use to periods of not over 5 minutes for takeoff operation.

rated 30-minute power. With respect to helicopter turbine engines, the maximum brake horsepower, developed under static conditions at specified altitudes and atmospheric temperatures, under the maximum conditions of rotor shaft rotational speed and gas temperature, and limited in use to periods of not over 30 minutes as shown on the engine data sheet.

rated 2½-minute power. With respect to helicopter turbine engines, the brake horsepower, developed statically in standard atmosphere at sea level, or at a specified altitude, for one-engine-out operation of multiengine helicopters for 2½ minutes at rotor shaft rotation speed and gas temperature established for this rating.

rate gyro. A **single-degree-of-freedom** gyro having primarily elastic restraint of its spin axis about the output axis. In this gyro an output signal is produced by gimbal angular displacement, relative to the base, which is proportional to the angular rate of the base about the input axis.

rate integrating gyro. A **single-degree-of-freedom gyro** having primarily viscous restraint of its spin axis about the output axis. In this gyro an output signal is produced by gimbal angular displacement, relative to the base, which is proportional to the integral of the angular rate of the base about the input axis.

rate of decay. 1. Of a sound, the time rate at which the **sound pressure level** (or other stated characteristic) decreases at a given point and at a given time. A commonly used unit is the decibel per second. **2.** Of a **radioactive nuclide,** the number of nuclei of that nuclide changing (or disintegrating) per unit time.

rate-of-climb indicator. See **vertical speed indicator.**

rate of fire. The number of rounds fired per weapon per minute.

rating. A statement that, as a part of a certificate, sets forth special conditions, privileges, or limitations. See **airman certificates.**

RATO, Rato, or **rato.** (From *rocket-assisted take-off*.) **1.** A **take-off** in which a rocket or rockets, commonly of the solid-fuel type, are used to provide additional thrust. Hence, *RATO bottle, Rato bottle, rato unit,* etc., a **rocket** so used. **2.** A RATO bottle or unit; the complete apparatus on an aircraft, comprising rockets, ignition system, etc., for assisted take-off. See **JATO.**

raw data. Data that is in a form ready for processing. Different groups regard data in various forms as *raw,* dependent on their function. A photographic processing group may regard the latent image as raw data, a reading group may regard the photographic image as raw data, a computing group may regard certain digital data as raw data, and so on.

rawin. A measurement of wind direction and speed at altitude by radar tracking of a balloon-borne target.

rawinsonde. A combination **raob** and **rawin;** an observation of temperature, pressure, relative humidity, and winds-aloft by means of **radiosonde** and radio direction finding equipment or **radar** tracking.

ray. 1. An elemental **path** of radiated **energy;** or the energy following this path. It is perpendicular to the **phase fronts** of the **radiation.** See **incident ray, reflected ray, refracted ray. 2.** One of a series of lines diverging from a common point, as radii from the center of a circle. **3.** A long, narrow, light colored streak on the lunar surface originating from a **crater.** Rays range in length to over 150 kilometers and usually several radiate from the same crater, like spokes of a wheel.

Rayleigh atmosphere. An idealized **atmosphere** consisting of only those particles, such as molecules, that are smaller than about one-tenth the **wavelength** of all radiation incident upon that atmosphere. In such an atmosphere, simple **Rayleigh scattering** would prevail. This model atmosphere is amenable to reasonably complete theoretical treatment, and hence has often served as a useful starting point in descriptions of the optical properties of actual atmospheres. The polarization of skylight, for example, exhibits almost none of the complexities found in the real atmosphere.

Rayleigh scattering. Any **scattering** process produced by spherical particles whose radii are smaller than about one-tenth the **wavelength** of the scattered **radiation.** In Rayleigh scattering, the scattering coefficient varies inversely with the fourth power of the wavelength, a relation known as the *Rayleigh law.* The angular intensity polarization relationships for Rayleigh scattering are conveniently simple. For particles not larger than the Rayleigh limit, there is complete symmetry of scattering about a plane normal to the direction of the incident radiation, so that the forward scatter equals the backward scatter.

RBN. See **radio beacon (RBN).**

reaction balance. A type of thrust meter using a balance to measure the static **thrust** of a **rocket** or jet engine.

reaction engine. An engine that develops **thrust** by its reaction to a substance ejected from it; specifically, such an engine that ejects a **jet** or stream of gases created by the burning of fuel within the engine. Also called **reaction motor.** A reaction engine operates in accordance with Newton third law of motion, i.e., to every action (force) there is an equal and opposite reaction. Both rocket engines and jet engines are reaction engines.

reaction propulsion. Propulsion by reaction to a **jet** or jets ejected from one or more **reaction engines.**

reaction time. In human engineering, the interval between an input signal (physiological) or a stimulus (psychophysiological) and the response elicited by the signal.

reaction time. (Military). **1.** The elapsed time between the initiation of an action and the required response. **2.** The time required between the receipt of an order directing an operation and the arrival of the initial element of the force concerned in the designated area. **3.** The time between stimulus and response, for example: (a) Elapsed time to dispatch a weapon system after receipt of executive order. (b) The interval of time between a command to launch a missile and the actual launch. (c) The total time between first warning and dispatch of a weapon system. (d) The time between a pilot's first seeing another aircraft and initiation of maneuvers to avoid collision.

reaction turbine. A type of **turbine** having rotor blades shaped so that they form a ring of **nozzles,** the turbine being rotated by the reaction of the **fluid** ejected from between the blades. Compare **impulse turbine.**

reactor. (See **nuclear reactor.**)

read. In **computer** operations, to acquire information, usually from some form of **storage.** See **write.**

readback. A procedure whereby the receiving station repeats a received message or an appropriate part thereof back to the transmitting station so as to obtain confirmation of correct reception.

read in. In **computer** operations, to introduce information into **storage.**

readout. 1. The action of a radio **transmitter** transmitting data either instantaneously with the acquisition of the data or by playing of a magnetic tape upon which the data have been recorded. See **instantaneous readout. 2.** The data transmitted by the action described in sense **1. 3.** In **computer** operations, to extract information from **storage.**

readout indicators. Any type of indicating instrument from which meaningful information and data can be directly obtained and used.

readout station. A recording or receiving radio station at which data are received from a **transmitter** in a probe, satellite, or other **spacecraft.**

ready cap. Fighter aircraft in condition of "standby."

Ready Reserve. That part of the Reserve components of the Air Force that may be immediately called to active duty in an emergency declared by the Congress, proclaimed by the President, or otherwise authorized by law. The Ready Reserve consists of all members and units of the Air National Guard of the United States and the United States Air Force Reserve not assigned to either the Standby or Retired Reserve. See also **Standby Reserve.**

real time. Time in which reporting on events or recording of events is simultaneous with the events. For example, the real time of a satellite is that time in which it simultaneously reports its environment as it encounters it; the real time of a computer is that time during which it is accepting data.

real-time data. Data presented in usable form at essentially the same time the event occurs. The delay in presenting the data must be small enough to allow a corrective action to be taken if required.

rear echelon (air transport). Elements of a force which are not required in the objective area.

rearward acceleration. See **physiological acceleration**.

rebreather. An **oxygen system** with a circuit closed to the atmosphere, to which oxygen is added to meet the user's needs; carbon dioxide and water vapor are removed from the expired gas.

received power. In **radar**, the power of a **target signal** received at the antenna. This power is normally of the order of microwatts as compared to the megawatts of transmitted power. Also called *scattered power*.

receiver. 1. The initial component or sensing element of a measuring system. For example, the receiver of a thermoelectric thermometer is the measuring thermocouple. 2. An instrument used to detect the presence of and to determine the information carried by **electromagnetic radiation**. A receiver includes circuits designed to detect, amplify, rectify, and shape the incoming **radiofrequency** signals received at the **antenna** in such a manner that the information-containing component of this received energy can be delivered to the desired indicating or recording equipment.

reciprocal. 1. A direction 180° from a given direction 2. The quotient of 1 divided by a given number.

reciprocating engine. An **engine**, especially an internal-combustion engine, in which a piston or pistons moving back and forth work upon a crankshaft or other device to create rotational movement.

recognition. The psychological process in which an observer so interprets the visual stimuli he receives from a distant object that he forms a correct conclusion as to the exact nature of that object. Recognition is a more subtle phenomenon than the antecedent step of detection, for the latter involves only the simpler process of interpreting visual stimuli to the extent of concluding that an object is present at some distance from the observer.

reconnaissance. A mission undertaken to obtain, by visual observation or other detection methods, information about the activities and resources of an enemy or potential enemy; or to secure data concerning the meteorological, hydrographic, or geographic characteristics of a particular area.

reconnaissance (R), (Military designation). Aircraft having equipment permanently installed for photographic and/or electronic reconnaissance missions. See **military aircraft types.**

reconnaissance photography. Photography taken primarily for purposes other than making maps, charts, or mosaics. It is used to obtain information on the results of bombing, or on enemy movements, concentrations, activities, and forces.

recoverable. Of a **rocket** vehicle or one of its parts, so designed or equipped as to be located after flight and recovered with or without damage.

recovery. 1. The procedure or action that obtains when the whole of a **satellite,** or a section, instrumentation package, or other part of a rocket vehicle is retrieved after a launch, as in *recovery was counted upon to give added data.* 2. The conversion of **kinetic energy** to **potential energy** such as in the deceleration of air in the

duct of a **ramjet engine.** Also called *ram recovery.* 3. In flying, the action of a lifting vehicle returning to an equilibrium attitude after a nonequilibrium maneuver.

recovery airfield. (Military) Any airfield, military or civil, at which aircraft might land post H-hour. It is not expected that combat missions would be conducted from a recovery airfield. See also **airfield.**

recovery base. The rear area base used for maintenance and reservicing of aircraft to eliminate the need for those services at airfields in the combat zone.

recovery capsule. A **capsule** designed to be recovered after reentry. See **reentry vehicle.**

recovery gear. The devices and equipment used to mark and locate a nose cone or other part of a **rocket vehicle** after impact.

recovery package. A package attached to a **reentry** or other body designed for recovery, containing devices intended to locate the body after **impact.** This package may, for example, release a balloon that will buoy up a reentry body (if it impacts in water) and serve as a radio beacon or light.

recovery temperature. Short for *adiabatic recovery temperature.*

recrystallization. 1. In metals, the change from one crystal structure to another, as occurs on heating or cooling through a critical temperature. 2. The formation of a new strain-free grain structure from that existing in cold-worked metal, usually accomplished by heating.

rectifier. A static device having an asymmetrical conduction characteristic which is used to convert attending current into direct current. A rotating device for this purpose is called a *converter.* Compare **inverter.**

recycle. 1. In a **countdown** to stop the count and to return to an earlier point in the countdown, as in *we have recycled, now at T minus 80 and counting.* Compare **hold.** 2. To give a completely new **checkout** to a rocket or other object.

recycle base. A base used by returning mission aircraft for servicing and maintenance before next mission launch.

recycling. Nucleonics. The reuse of *fissionable material,* after it has been recovered by chemical processing from spent or depleted reactor fuel, reenriched, and then refabricated into net fuel elements. (See **fuel cycle, spent fuel.**)

redeployment airfield. An airfield not occupied in its entirety in peacetime, but available immediately upon outbreak of war for use and occupation by units redeployed from their peacetime locations. It must have substantially the same standard of operational facilities as a main airfield. See also **airfield; alternative airfield; departure airfield; main airfield.**

redout. The condition occurring under **negative g** in which objects appear to have a red coloration due to uncertain causes possibly venous congestion of engorged eyelids. Compare **blackout,** sense **3.**

redundancy. 1. In **information theory;** of a source, the amount by which the logarithm of the number of symbols available at the source exceeds the average information content per symbol of the source. The term *redundancy* has been used loosely in other senses. For example, a source whose output is normally transmitted over a given channel has been called *redundant,* if the channel utilization index is less than unity. 2. The existence of more than one means for accomplishing a given task, where all means must fail before there is an overall failure to the system. *Parallel redundancy* applies to systems where both means are working at the same time to accomplish the task, and either of the systems is capable of handling the job itself in case of failure of the other system. *Standby redundancy* applies to a system where there is an alternative means of accomplishing the task that is switched in by a malfunction sensing device when the primary system fails.

reentry. The event occurring when a **spacecraft** or other object comes back into the **sensible atmosphere** after being rocketed to higher altitudes; the action involved in this event.

reentry body. That part of a **space vehicle** that reenters the atmosphere after flight above the **sensible atmosphere.**

reentry nose cone. A **nose cone** designed especially for **reentry,** consisting of one or more chambers protected by an outer shield. See **heat sink.**

re-entry system. That portion of a ballistic missile designed to place one or more re-entry vehicles on terminal trajectories so as to arrive at selected targets. Penetration aids, spacers, deployment modules, and associated programming, control and sensing devices are included in the re-entry system.

reentry trajectory. That part of a **rocket's** trajectory that begins at reentry and ends at target or at the surface. If the rocket is unguided at reentry, its reentry trajectory is ballistic in character.

reentry vehicle. Any payload carrying **vehicle** designed to leave the **sensible atmosphere** and then return through it to earth. This term applies both to return vehicles from orbital or space payloads and to boostglide vehicles.

reference frame. Same as **coordinate system.**

reference line. Same as **datum line.**

reference meridian. See **meridian.**

reference plane. Same as **datum plane.**

reference signal. In **telemetry,** the **signal** against which data-carrying signals are compared to measure differences in time, phase, frequency, etc.

refire time (ballistic missile). Time required, after initial firing, to launch a second missile from the same pad or launcher.

reflectance (*symbol* ρ). The ratio of the radiant **flux** reflected by a body to that incident upon it. Also called *reflection factor.* For an opaque body, the sum of the reflectance and the absorptance for the incident radiation is unity $\rho + \alpha = 1$.

reflected ray. A **ray** extending outward from a point of reflection.

reflected wave. 1. A **shock wave, expansion wave,** or **compression wave** reflected by another wave incident upon a wall or other boundary. **2.** In electronics, a radio wave reflected from a surface or object.

reflecting telescope. A telescope which collects light by means of a concave mirror.

reflection. The process whereby a surface of discontinuity turns back a portion of the incident radiation into the medium through which the radiation approached. See **albedo, reflectivity, radar reflectivity.**

reflectivity. 1. A measure of the fraction of **radiation** reflected by a given surface; defined as the ratio of the **radiant energy** reflected to the total that is incident upon that surface. See **radar reflectivity.** The reflectivity of any given substance is, in general, a variable strongly dependent upon the wavelength of the radiation in ques-

tion. The reflectivity of a given surface for a specified broad spectral range, such as the visible spectrum or the solar spectrum, is referred to as the *albedo.* **2.** In thermal radiation, a property of a material, measured as the **reflectance** of a specimen of the material that is thick enough to be completely opaque and has an optically smooth surface.

reflector. 1. In general, any object that reflects incident **energy;** usually it is a device designed for specific reflection characteristics. See **corner reflector, parabolic reflector, radar reflector. 2.** In an antenna, a **parasitic element** located in a direction other than the general direction of the major lobe of radiation. **3.** A material of high **scattering** cross section that surrounds a reactor core to reduce the escape of neutrons, many of which are reflected back into the core.

refracted ray. A **ray** extending onward from the point of **refraction.**

refracted wave. A **wave** that has had its direction of motion changed by **refraction.**

refracting telescope. A telescope which collects light by means of a lens or system of lenses. Also called *refractor.*

refraction. The process in which the direction of energy **propagation** is changed as the result of a change in density within the propagating medium, or as the energy passes through the interface representing a density discontinuity between two media. In the first instance the rays undergo a smooth bending over a finite distance. In the second case the **index of refraction** changes through an interfacial layer that is thin compared to the wavelength of the radiation; thus, the refraction is abrupt, essentially discontinuous. Compare **reflection, diffraction, scattering.**

refraction index. Same as **index of refraction.**

refractive index. Same as **index of refraction.**

refractivity. The algebraic difference between an **index of refraction** and unity.

refractometer. An instrument for measuring the **index of refraction** of a liquid, gas or solid.

refractory. A material, usually **ceramic,** that resists the action of heat, does not fuse at high temperatures, and is very difficult to break down.

refractory metal. A metal with melting point above 4000°F. Usually refers to columbium, molybdenum, tantalum, or tungsten.

regeneration. 1. positive feedback. 2. In **computer** operations, the process of restoring a **storage** device, whose information storing state may deteriorate, to its latest underteriorated state. See **rewrite.**

regenerative cooling. The cooling of a part of an engine by the fuel or propellant being delivered to the combustion chamber; specifically, the cooling of a rocket-engine **combustion chamber** or **nozzle** by circulating the **fuel** or **oxidizer,** or both, around the part to be cooled.

regenerative engine. A **liquid rocket engine** cooled by **regenerative cooling.**

regenerator. A device used in a **thermodynamic** process for capturing and returning to the process **heat** that would otherwise be lost. Also called a *heat exchanger* (which see).

region. A portion of the **ionosphere** usually characterized by a particular altitude or range of altitudes, in which concentrations of **free electrons** tend to form.

register. A device capable of retaining information, often that contained in a small subset (e.g., one word) of the aggregate information in a digital **computer**. See **storage**.

registration certificate. A document which must be displayed in a U.S. **civil aircraft** showing the owner as registered with the **FAA**. An aircraft is eligible for registration only if it is owned by a citizen of the United States, and not registered under the laws of a foreign country.

regression of the nodes. Precessional motion of a set of **nodes**. See **precession**. The expression is used principally with respect to the moon, the nodes of which make a complete westerly revolution in approximately 18.6 years.

regulating rod. Nucleonics. A reactor *control rod* used for making frequent fine adjustments in reactivity.

reheat. Same as **afterburning**.

reheating. 1. The addition of heat to a **working fluid** in an engine after a partial expansion. **2.** The retention of heat in a fluid, as after passing through a turbine stage, owing to the inefficiency of the stage.

relative. Of angle measurements in navigation, measured from the **heading** of a craft, as *relative bearing*.

relative bearing. The horizontal angle at the aircraft measured clockwise from the true heading (or longitudinal axis) of the aircraft to the great circle containing the aircraft and object. The term is used in conjunction with navigation by means of a radio compass or **Automatic Direction Finding (ADF)** equipment. Relative bearing is simply the direction of a straight line between the aircraft longitudinal axis and the radio station or vice versa.

relative biological effectiveness (RBE). Nucleonics. A factor used to compare the biological effectiveness of different types of *ionizing* radiation, required to produce a given effect, to a standard (or reference) radiation required to produce the same effect. (See **absorbed dose, rad, rem.**)

relative coordinate system. Any **coordinate system** which is moving with respect to an **inertial coordinate system**. Referred to a relative system, various apparent forces arise in Newton laws owing to motion of the system. See, e.g., **centrifugal force, coriolis force**.

relative humidity. The ratio of the amount of moisture in the air to the amount which the air could hold at the same temperature if it were saturated; usually expressed in percent. See **humidity**.

relative motion. relative movement. See **motion**.

relative movement. Motion of one object or body measured relative to another. Usually called *apparent motion* when applied to the change of position of a celestial body as observed from the earth. Also called *relative motion*. The expression is usually used in connection with problems involving motion of one craft or vehicle relative to another, the direction of such motion being called *direction of relative movement* and the speed of such motion being called *speed of relative movement* or *relative speed*. Distance relative to a specified reference point, usually one in motion, is called *relative distance*.

relative position. A point defined with reference to another **position**, either fixed or moving. The coordinates of such a point are usually **bearing**, true or relative, and distance from an identified reference point.

relative sunspot number. A measure of **sunspot** activity.

relative wind. The direction of an airflow with respect to an airfoil.

relativistic. In general, pertaining to material, as a particle, moving at speeds which are an appreciable fraction of the **speed of light** thus increasing the mass.

relativistic velocity. A velocity sufficiently high that some properties of a particle of this velocity have values significantly different from those obtaining when the particle is at rest. See **rest mass**. The property of most interest is the mass. For many purposes, the velocity is relativistic when it exceeds about one-tenth the velocity of light.

relativity. A principle that postulates the equivalence of the description of the universe, in terms of physical laws, by various observers, or for various frames of reference.

relaxation time. 1. In general, the time required for a system, object, or fluid to recover to a specified condition or value after disturbance. **2.** Specifically, the time taken by an exponentially decaying quantity to decrease in amplitude by a factor of $1/e = 0.3679$.

release altitude. Altitude of an aircraft above the ground at the time of release of bombs, rockets, missiles, tow targets, etc.

reliability. The probability that a system, subsystem, or equipment will perform a required function under specified conditions, without failure, for a specified period of time.

remaining body. That part of a **rocket** or **vehicle** that remains after the separation of a **fallaway section** or **companion body**. In a multistage rocket, the remaining body diminishes in size successively as each section or part is cast away and successively becomes a different body.

remark. A fixed **radar frequency** facility which continuously emits a signal so that a bearing indication appears on a radar **display**. See **radar beacon**.

remote control. Control of an operation from a distance, especially by means of electricity or electronics; a controlling switch, lever, or other device used in this kind of control; as in *remote-control armament, remote-control switch*, etc.

remote indicating. Of an instrument, displaying indications at a point remote from its sensing element, often by electrical or electronic means.

remote indicating compass. A magnetic compass, the magnetic detecting element of which is installed in an aircraft in a position as free as possible from causes of deviation. A transmitter system is included to enable compass indications to be read on a number of repeater dials suitably positioned in the aircraft.

remotely piloted vehicle (RPV). An unmanned aircraft used for military missions such as reconnaissance, target or surveillance purposes and remotely controlled from a ground base or from a *director* aircraft.

rendezvous. 1. The event of two or more objects meeting with zero **relative** velocity at a preconceived time and place. **2.** The point in **space** at which such an event takes place, or is to take place. A rendezvous would be involved, for example, in servicing or resupplying a space station.

repair. Repair of materiel encompasses the inspection, servicing, overhaul, test, calibration, and preservation applied to restoring the item to servicable condition and extending its servicable life.

repairman certificate. An **airman certificate** issued by the **FAA** to a person authorized to perform or supervise the maintenance of aircraft or components appropriate to the job for which he is employed and certificated, in connection with duties for the repair station, commercial operator or air carrier by whom he is employed. See **aviation mechanic.**

report. Any transmission or presentation of data or information, whether in tabular, graphic, narrative, questionnaire, punch-card or other form regardless of the method of transmission.

reporting point. A specified geographical location in relation to which the position of an aircraft can be reported.

report, weather. See **weather report.**

rescue coordination center. A primary search and rescue (SAR) facility suitably staffed by supervisory personnel and equipped for coordinating and controlling SAR operations in a region, sub-region, or sector as defined by the National SAR Plan.

research (X), (Military designation). Aircraft designed for testing configurations of radical nature. These aircraft are not normally intended for use as tactical aircraft. See **military aircraft types.**

research and development categories. Major categories are: (1) Research and development: (a) *research*. Includes all efforts directed toward increased knowledge of natural phenomena and environment and efforts directed toward the solution of problems in physical, behavioral, biological and social sciences that have no clear, direct military application. (b) *exploratory development*. includes all efforts directed toward the solution of specific military problems, short of major development projects. (c) *advanced development*. Includes all projects which have moved into the development of hardware for experimental or operational test. It is characterized by line item projects, and program control is exercised on a project basis. (d) *engineering development*. Includes those development programs being engineered for service use but which have not yet been approved for procurement or operation. (e) *management and support*. Includes research and development effort directed toward support of installations or operations required for general research and development use. Included are test ranges, military construction not included elsewhere, and operation and maintenance of test aircraft. (2) Operational systems development. Includes a development effort directed toward development, engineering and test of systems, vehicles, weapons, and support programs that have been approved for production and service employment.

research and development test. An experiment or operation designed to measure, verify, assess, and provide data for evaluation of: research investigations or experiments carried on beyond the laboratory bench; progress in attainment of accomplishment of development objectives; and performance capability and/or operational suitability of systems, subsystems, components, and equipment items.

research rocket. A rocket propelled vehicle used to collect scientific data as a part of a research effort.

research testing. Operations performed as a part of research experiments and investigations that are carried on beyond the laboratory bench to measure, verify, or assess phenomena, hypotheses, results of experimentation, and to gain new knowledge.

reserve aircraft. Those aircraft which have been accumulated in excess of immediate needs for active aircraft and are retained in the inventory against possible future needs. See also **aircraft.**

Reserve components. Reserve components of the Armed Forces of the United States are: **a.** the Army National Guard of the United States; **b.** the Army Reserve; **c.** the Naval Reserve; **d.** The Marine Corps Reserve; **e.** the Air National Guard of the United States; **f.** the Air Force Reserve; and **g.** the Coast Guard Reserve. In each reserve component there are three reserve categories, namely: a Ready Reserve, a Standby Reserve, and a Retired Reserve. Each reservist shall be placed in one of these categories.

reset. 1. To restore a **storage** device to a prescribed state. **2.** To place a **binary cell** in the initial or *zero* state. See **clear.**

residence time. Nucleonics. The time during which radioactive material remains in the atmosphere following the detonation of a nuclear explosive. It is usually expressed as a *halftime* since the time for all material to leave the atmosphere is not well known. (Compare **half-life; see fallout.**)

residual load. Of a **vehicle,** the sum of the payload, all items directly associated with the payload, and other relatively fixed weights of the overall vehicle; calculated as the difference between gross weight and the sum of propellant, tank, structure, and power-plant weights.

residual nuclear radiation. Lingering radiation, or radiation emitted by radioactive material remaining after a nuclear explosion. Residual radiation is arbitrarily designated as that emitted more than one minute after the explosion. (Compare **fallout, initial nuclear radiation.**)

residual stress. In structures, any **stress** in an unloaded body. These stresses arise from local yielding of the material due to machining, welding, quenching, cold work, etc.

resistance *(symbol* **R).** **1.** In electricity, the factor by which the square of the instantaneous conduction current must be multiplied to obtain the power lost by heat dissipation or other permanent radiation of energy away from the electrical current. **2.** In mechanics, the opposition by frictional effects to forces tending to produce motion.

resolution 1. The ability of a film, a lens, a combination of both, or a **vidicon** system to render barely distinguishable a standard pattern of black and white lines. When the resolution is said to be 10 lines per millimeter, it means that the pattern whose line plus space width is 0.1 millimeter is barely resolved, the finer patterns are not resolved, and the coarser patterns are more clearly resolved. In satellite television systems the limiting element is the television scanning pattern. **2.** In **radar,** the minimum angular separation at the antenna at which two targets can be distinguished (a function of **beamwidth**); or the minimum range at which two targets at the same azimuth can be separated (equal to one-half the **pulse** length). **3.** Of a **gyro,** a measure of response to small changes in input; the maximum value of the minimum input change that will casue a detectable change in the output for inputs greater than the threshold, expressed as a percent of one half the input range.

resolving power. 1. resolution, senses **1** and **2.** In a unidirectional **antenna,** the reciprocal of its beam width measured in degrees. The **resolution** of a directional radio system can be different from the resolving power of its antenna, since the resolution is affected by other factors.

resonance. **1.** The phenomenon of amplification of a **free wave** or **oscillation** of a system by a **forced wave** or oscillation of exactly equal **period.** The forced wave may arise from an impressed force upon the system or from a boundary condition. The growth of the resonant amplitude is characteristically linear in time. **2.** Of a system in **forced oscillation,** the condition which exists when any change, however small, in the **frequency** of excitation causes a decrease in the response of the system.

resonance frequency. A **frequency** at which **resonance** exists. Also called *resonant frequency.* In case of possible confusion, the type of resonance must be indicated, as *velocity resonance frequency.*

resonator. In radio and radar applications, a **circuit** which will **resonate** at a given **frequency,** or over a range of frequencies, when properly excited. A very important type of resonator is the cavity resonator, a closed hollow volume having conducting walls. The frequency at which these cavities will resonate is a function of their volume and shape; thus, they are used for making accurate frequency comparisons and for generating radio frequencies, usually in the microwave region.

respiration. The interchange of gases of living organisms and the gases of the medium in which they live. Respiration applies to the interchange by any channel as *pulmonary respiration, cutaneous respiration,* etc.

responder. **1.** In general an instrument that indicates reception of an electric or electromagnetic **signal. 2.** A **transponder.**

response. Of a device or **system,** the motion (or other output)resulting from an **excitation** under specified conditions. Modifying phrases must be prefixed to the term *response* to indicate what kinds of input and output are being utilized. The response characteristic, often presented graphically, gives the response as a function of some independent variable such as frequency or direction. For such purposes it is customary to assume that other characteristics of the input (for example, voltage) are held constant.

responsor. A radio **receiver** which receives the reply from a **transponder** and produces an output suitable for feeding to a **display** system. A responsor is usually combined in a single unit with an interrogator, which sends out the pulse that triggers a transponder, the combined unit being called an *interrogator-responsor.*

restart. Specifically, the act of firing a **stage** of a **rocket** after a previous powered flight and a coast phase in a **parking orbit.**

rest mass. According to **relativistic** theory, the mass which a body has when it is at absolute rest. Mass increases when the body is in motion according to

$$m = m_0/\sqrt{1 - (v^2/c^2)}$$

where m is its mass in motion; m_0 is its rest mass; v is the body's speed of motion; and c is the speed of light. Newtonian physics, in contrast with relativistic physics, makes no distinction between rest mass and mass in general.

restricted area. (Military) **1** An area (land, sea, or air) in which there are special restrictive measures employed to prevent or minimize interference between friendly forces. **2.** An area under military jurisdiction in which special security measures are employed to prevent unauthorized entry. See also **restricted areas (air).**

restricted areas (air). Designated areas established by appropriate authority over which flight of aircraft is restricted. They are shown on aeronautical charts and published in notices to airmen, and publications of aids to air navigation. See also **restricted area.**

restricted propellant. A **solid propellant** having only a portion of its surface exposed for burning, the other surfaces being covered by an **inhibitor.**

restrictor. In **solid-propellant rockets,** a layer of fuel containing no **oxidizer,** or of noncombustible material, adhered to the surface of the propellant so as to prevent burning in that region.

resultant. The sum of two or more **vectors.**

reticle. A system of lines, wires, etc., placed in the focal plane of an optical instrument to serve as a reference. Also called *reticule.* A crosshair is a hair, thread, or wire constituting part of a reticle.

retractable landing gear. A type of **landing gear** which may be withdrawn into the body or wings of an airplane while it is in flight, in order to reduce the parasite drag.

retrofire. To ignite a **retrorocket.**

retrofit action. Action taken to modify inservice equipment.

retrograde motion. **1.** Motion in an **orbit** opposite to the usual orbital direction of **celestial bodies** within a given system. Specifically, of a **satellite,** motion in a direction opposite to the direction of rotation of the **primary. 2.** The apparent motion of a planet westward among the stars. Also called *retrogression.*

retropack. A rocket unit built into or strapped to a **spacecraft** that provides **retrothrust.**

retrorocket. (From *retro*acting.) A **rocket** fitted on or in a spacecraft, satellite, or the like to produce **thrust** opposed to forward motion.

retrothrust. **Thrust** used for a braking maneuver; reverse thrust.

retrosequence. The sequence of events preparatory to, and programmed to follow, the **retrofiring** for **spacecraft** reentry.

return to base. (Military). Proceed to the point indicated by the displayed information. This point is being used to return the aircraft to a place at which the aircraft can land. Command heading, speed and altitude may be used, if desired.

reverberation. **1.** The persistence of **sound** in an enclosed space, as a result of multiple reflections after the sound source has stopped. **2.** The sound that persists in an enclosed space, as a result of repeated reflection or scattering, after the source of the sound has stopped.

reverse pitch. See **reversible propeller.**

reverse thrust. Thrust applied to a moving object in a direction opposite to the direction of the object's motion. Reverse thrust is sometimes used to reduce the landing roll of turbo-jet aircraft. Deflector vanes are incorporated in the jet engine exhaust nozzle to provide reverse thrust.

reversible propeller. A propeller whose blades may be entirely reversed while the airplane is in flight. Reversible propellers are normally used to reduce the landing roll of propeller driven aircraft.

revetment. A wall of concrete, earth, sandbags, or the like installed for protection, as against the blast of ex-

*The powerful F-1 **rocket engine** undergoes final inspection prior to delivery from the plant of the Rocketdyne division of Rockwell International. The F-1 is used in the Saturn V first stage (S-1C), in a cluster of five developing 7.5 million pounds of lift-off thrust.*

ploding fuel during a rocket **abort.**

revolution. 1. Motion of a **celestial body** in its orbit; circular motion about an **axis** usually external to the body. In some contexts, the terms *revolution* and *rotation* are used interchangeably but, with reference to the motions of a celestial body, *revolution* refers to motion in an orbit or about an axis external to the body, whereas *rotation* refers to motion about an axis within the body. Thus, the earth revolves about the sun annually and rotates about its axis daily. **2.** One complete cycle of the movement of a **celestial body** in its orbit, or of a body about an external axis, as *a revolution of the earth about the sun.*

revolve. To move in a path about an axis, usually external to the body accomplishing the motion, as in *the planets revolve about the sun.* Hence **revolution.** See **rotate.**

rewrite. In a **storage** device whose information storing state may be destroyed by **reading,** the process of restoring the device to its state prior to reading.

Reynolds number. The product of a typical length and the fluid speed divided by the kinematic viscosity of the fluid. It expresses the ratio of the inertial forces to the viscous forces.

RF *(abbr).* **radio***frequency.*

rho-theta system. 1. Any electronic navigation system in which position is defined in terms of distance, or radius ρ and **bearing** θ with respect to a transmitting station. Also called an *R-theta system.* **2.** Specifically, a **polar-coordinate** navigation system providing data with sufficient accuracy to permit the use of a computer which will provide arbitrary course lines anywhere within the coverage area of the system.

rhumb line (navigation). A line on a chart which intersects all meridians at the same angle.

rib. A fore-and-aft member which maintains the required contour of the covering material of wings or control surfaces, and which may also act as a structural member.

ribbon parachute. A type of parachute having a canopy consisting of an arrangement of closely spaced tapes. This parachute has high porosity with attendant stability and slight opening shock.

rich. Of a combustible mixture; having a relatively high proportion of **fuel** to **oxidizer;** more precisely, having a value greater than **stoichiometric.**

ridge. (metgy) An elongated area of relatively high pressure extending from the center of a high pressure region.

rig. Adjustment of the airfoils of an airplane to produce desired flight characteristics.

rigging. The relative adjustment or alignment of the different components of an aircraft.

right ascension. Angular distance east of the **vernal equinox;** the arc of the **celestial equator,** or the angle at the celestial pole, between the **hour circle** of the vernal equinox and the hour circle of a point on the **celestial sphere,** measured eastward from the hour circle of the vernal equinox through 24 hours. Angular distance west of the vernal equinox, through 360°, is sidereal hour angle.

rill. (from German *rille* meaning *groove*). A deep, narrow, depression on the lunar surface which cuts across all other types of lunar topographic features.

rime ice. Rime ice is a milky, opaque, and granular deposit with a rough surface. Rime ice is formed by the instantaneous freezing of small supercooled water droplets upon contact with exposed aircraft surfaces. This instant freezing traps a large amount of air, giving the ice its opaqueness and making it very brittle. Rime ice usually forms on leading edges and protrudes forward into the airstream as a sharp nose. It has little tendency to spread over and take the shape of the airfoil.

rip cord. A cord or flexible cable on a parachute which, when pulled, opens the pack and allows the parachute to deploy.

RMI. Radio Magnetic Indicator.

Rnav. Area navigation.

rockair. A high-altitude sounding system consisting of a small **solid-propellant** research **rocket** carried aloft by an aircraft. The rocket is fired while the aircraft is in vertical ascent.

rocket. 1. A projectile, pyrotechnic device, or flying **vehicle** propelled by a **rocket engine. 2.** A **rocket engine;** any one of the *combustion chambers* or tubes of a multichambered **rocket engine.**

rocket airplane. 1. Any airplane using rocket propulsion for its main or only propulsive power. **2.** An airplane fitted out to carry and fire rocket ammunition.

rocket ammunition. Rocket-powered projectiles of relatively small size fired from an aircraft or other platform, mobile or stationary.

rocket-assisted take-off. The full term for **RATO.**

rocket booster. A booster, senses **2** and **3.**

rocket engine. A **reaction engine** that contains within itself, or carries along with itself, all the substances necessary for its operation or for the consumption or combustion of its **fuel,** not requiring intake of any outside substance and hence capable of operation in outer space. Also called *rocket motor.* Chemical rocket engines contain or carry along their own fuel and oxidizer, usually in either liquid or solid form, and range from simple motors consisting only of a combustion chamber and exhaust nozzle to engines of some complexity incorporating, in addition, fuel and oxygen lines, pumps, cooling systems, etc., and sometimes having two or more combustion chambers. Experimental rocket motors have used neutral gas, ionized gas, and plasmas as propellants. See **liquid-propellant rocket engine, solid-propellant rocket engine, ion rocket, plasma rocket.**

rocket fuel. A **fuel,** either liquid or solid, developed for, or used by a **rocket.**

rocket launcher. A device for launching a **rocket.** See **launcher.** Rocket launchers are wheel mounted, motorized, or fixed for use on the ground; or they are mounted on aircraft, as under the wings; or they are installed below or on the decks of ships.

rocket nozzle. The exhaust **nozzle** of a **rocket.**

rocket propellant. 1. Any agent used for consumption or combustion in a rocket and from which the rocket derives its thrust, such as a **fuel,** oxidizer, additive, catalyst, or any compound or mixture of these. **2.** The ejected **fluid** in a nuclear rocket.

rocket propulsion. Reaction propulsion by a **rocket engine.**

rocket ramjet. A **ramjet engine** having a **rocket** mounted within the ramjet duct, the rocket being used to bring the ramjet up to the necessary operating speed. Sometimes called a *ducted rocket.*

rocketry. The science or study of **rockets,** including theory, research, development, experimentation, and

Bell Helicopter Textron's Model 222 twin turbine helicopter is a **rotorcraft.**

application; the art or science of using rockets.

rocket ship. An aircraft, space-air vehicle, or spacecraft using **rocket propulsion.**

rocket sled. A sled that runs on a rail or rails and is accelerated to high velocities by a **rocket engine.** This sled is used in determining g-tolerances and for developing crash survival techniques. Rocket sleds are at Edwards Air Force Base, Holloman Air Force Base, and the Naval Ordnance Test Station.

rocketsonde. Same as **meteorological rocket.**

rocket thrust. The **thrust** of a **rocket engine** usually expressed in pounds. On a test stand, rocket thrust may be measured by use of strain gages, thrust-balancing pistons, dynamometers, or spring scales, each calibrated in pounds to represent the static weight moved by the engine.

rocket vehicle. A **vehicle** propelled by a **rocket engine,** used to place a satellite in orbit, place a missile upon target, carry a passenger over a rail as on a rocket sled, etc.

rockoon. A high-altitude **sounding** system consisting of a small **solidpropellant** research **rocket** carried aloft by a large plastic balloon.

The rocket is fired near the maximum altitude of the balloon flight. It is a relatively mobile sounding system and has been used extensively on shipboard.

rod. A type of photoreceptive cell in the retina of the mammalian eye. Rods are involved in detection of movement and scotopic vision (night vision).

roentgen. A unit of exposure to *ionizing radiation.* It is that amount of gamma or X rays required to produce ions carrying 1 electrostatic unit of electrical charge (either positive or negative) in 1 cubic centimeter of dry air under standard conditions. Named after Wilhelm Roentgen, German scientist who discovered X rays in 1895. (Compare **curie, rad, rem.**)

roentgen equivalent man (rem). The unit of dose of any ionizing radiation which produces the same biological effect as a unit of *absorbed dose* of ordinary X rays. The RBE dose (in *rems) = RBE x* absorbed dose (in *rads).* (Compare **curie, roentgen.**)

roentgen rays, X rays.

roger. I have received all of your last transmission. It should not be used to answer a question requiring a yes or no answer. (See **affirmative, negative**)

roll. 1. The act of rolling; rotational or oscillatory movement of an aircraft or similar body about a **longitudinal axis** through the body — called *roll* for any degree of such rotation. **2.** The amount of this movement, i.e., the angle of roll.

roll. An airplane acrobatic maneuver consisting of displacement around the longitudinal axis. See **aileron roll, snap roll.**

roll axis. A **longitudinal axis** through an aircraft, rocket, or similar body, about which the body rolls. A roll axis may be a body, wind, or stability axis, or any other lengthwise axis.

roll cloud. Sometimes called Rotor Cloud.) A turbulent altocumulus-type cloud formation found in the lee of some large mountain barriers, particularly in the Sierra Nevadas near Bishop, Calif. The air in the cloud rotates around an axis parallel to the range. Also sometimes refers to part of the cloud base along the leading edge of a cumulonimbus cloud; it is formed by rolling action in the wind shear region between cool downdrafts within the cloud and warm updrafts outside the cloud. Also see **cloud.**

rolling moment. A **moment** that tends to rotate an aircraft, a rocket, etc., about a **longitudinal axis.** This moment is considered positive when it tends to depress the starboard side of the body.

rolling vertical takeoff. The ability of an aircraft to take off in the vertical mode after a ground roll of 50 feet or less. This capability represents a compromise between true vertical takeoff and landing and short takeoff and landing.

rollout RVR. The RVR readout values obtained from RVR equipment located nearest the runway end (applicable to Category II ILS runways only).

root chord. In aerodynamics, the **chord** of a lifting surface at the intersection of that surface with its supporting body, e.g., *wing root chord.*

root-mean-square error. In statistics, the square root of the **arithmetic mean** of the squares of the deviations of the various items from the arithmetic mean of the whole. Also termed *standard deviation.*

rope. An element of chaff consisting of a long roll of metallic foil or wire which is designed for broad, low-frequency response. See also **chaff.**

rotate. To turn about an internal axis. Said especially of **celestial bodies.** Hence **rotation.** Compare **revolve.**

rotation. 1. Turning of a body about an axis within the body, as *the daily rotation of the earth.* See **revolution. 2.** One turn of a body about an internal axis, as *a rotation of the earth.*

rotation. (flight operations). During the takeoff run of a fixed wing airplane, when the airspeed reaches a predetermined value, the pilot increases the angle-of-attack in order to become airborne, by a definite and positive rearward motion of the pilot control (control column).

rotational speed. Revolutions per unit time.

rotor. See **gyro.**

rotor. A system of rotating aerofoils.

rotor angular momentum. Of a **gyro,** the product of spin **angular velocity** and **rotor moment of inertia,** usually expressed in gram centimeters squared per second. It is a measure of the ability of a gyrorotor to maintain its **spin axis** fixed in space.

rotorcraft. A heavier-than-air aircraft that depends principally for its support in flight on the lift generated by one or more rotors.

round. A single munition, missile or device to be loaded on or in a delivery platform, vehicle, or device for purposes of expenditure. The configuration of the round may vary.

round off. To delete less significant **digits** from a number and possibly apply some rule or correction to the part retained.

route. A defined path, consisting of one or more courses, which an aircraft traverses in a horizontal plane over the surface of the earth.

routine. A set of **instructions** arranged in proper sequence to cause a **computer** to perform a desired operation, such as the solution of a mathematical problem.

RP *(abbr). rocket propellant.* Used with a number in designations of different propellants, as in *RP-1.*

RP-1. A rocket **fuel** consisting essentially of kerosene.

RPV. remotely piloted vehicle

R-T unit. The receiver-transmitter portion of a **radar beacon** system.

rubber-base propellant. A **solid propellant** mixture in which the oxygen supply is obtained from a perchlorate and the fuel is provided by a synthetic rubber latex.

rudder. A hinged vertical control of an aerodynamic vehicle surface used to induce or overcome yawing moments about the vertical axis.

rudder pedals. Pedals by which the rudder is operated.

rumble. A form of **combustion instability,** especially in a **liquid-propellant rocket engine,** characterized by a low-pitched, low-frequency rumbling noise; the noise made in this kind of combustion.

run. That part of flight of one photographic reconnaissance aircraft during which photographs are taken.

running fix. The intersection of two or more position lines, not obtained simultaneously, adjusted to a common time.

runway. A strip, either paved or improved, on which takeoffs and landings are effected.

runway direction number. A whole number to the nearest one tenth of the magnetic bearing of the runway and measured in degrees clockwise from magnetic north.

runway lights. Lights defining a runway to indicate the area of taking-off and landing.

runway visual range. (RVR). See **visibility**

rupture disk. Same as **burst disk.**

RVR. runway visual range.

S

S. Military mission designation for **antisubmarine** aircraft.

SAC. Strategic Air Command.

safety belt. See **lap belt.**

safety pilot. A pilot who accompanies another pilot (e.g., a student pilot or a pilot practicing instrument flying) to warn of other traffic or to take over the controls if need be.

safing. As applied to weapons and ammunition the changing from a state of readiness for initiation to a safe condition.

sailing. In seaplanes, the use of wind and current conditions to produce the desired track while taxiing on the water.

sailplane. A glider designed for sustained flight utilizing atmospheric currents.

St. Elmo's fire. A luminous brush discharge of electricity from elevated objects, such as the masts and yardarms of ships, lightning rods, steeples, etc., occurring in stormy weather. Also called corposant. See **corona discharge.**

salvo. 1. In naval gunfire support, a method of fire in which a number of weapons are fired at the same target simultaneously. **2.** In close air support/air interdiction operations, a method of delivery in which the release mechanisms are operated to release or fire all ordnance of a specific type simultaneously.

salvo launch. Act of launching two or more **rockets** simultaneously.

SAM. Surface-to-Air Missile.

sample. In statistics, a group of observations selected from a statistical **population** by a set procedure. See **random sample.** Samples may be taken at random or systematically. The sample is taken in an attempt to estimate the population.

sandwich construction. A type of construction in which two sheets, sides, or plates are separated by a core of stiffening material, generally lightweight. See **honeycomb core.**

SAR. Search and Rescue.

Sarah. (From *search and rescue and homing.*) A radio **homing** device originally designed for personnel rescue and now used in spacecraft **recovery** operations at sea.

satellite. 1. An attendant body that revolves about another body, the **primary;** especially in the solar system, a secondary body, or moon, that revolves about a planet. **2.** A manmade object that revolves about a spatial body, such as Explorer I orbiting about the earth. **3.** Such a body intended and designed for orbiting, as distinguished from a companion body that may incidentally also orbit, as in *the observer actually saw the orbiting rocket rather than the satellite.* **4.** An objet not yet placed in orbit, but designed or expected to be launched into an orbit.

satelloid. A **vehicle** that revolves about the earth or other body, but at such altitudes as to require sustaining **thrust** to balance **drag.**

saturated air. Air that contains the maximum amount of water vapor it can hold at a given pressure and temperature (relative humidity of 100 percent).

saturation-adiabatic lapse rate. A special case of **process lapse rate,** defined as the rate of decrease of temperature with height of an air **parcel** lifted in a saturation-adiabatic process through an atmosphere in **hydrostatic equilibrium.** Also called *moist-adiabatic lapse rate.* Owing to the release of latent heat, this lapse rate is less than the **dry-adiabatic lapse rate,** and the differential equation representing the process must be intergrated numerically. Wet-bulb potential temperature is constant with height in an atmosphere with this lapse rate.

saturation vapor pressure. 1. The **vapor pressure** of a system, at a given temperature, wherein the vapor of a substance is in equilibrium with a plane surface of the pure liquid or solid phase of that substance; that is, the vapor pressure of a system that has attained saturation but not supersaturation. Compare **equilibrium vapor pressure, vapor tension.** The saturation vapor pressure of any pure substance, with respect to a specified parent phase, is an intrinsic property of that substance and is a function of temperature alone. **2. equilibrium vapor pressure.**

Saturn. See **planet.**

saturnographic. Referring to positions on Saturn measured in **latitude** from Saturn's **equator** and in **longitude** from a reference **meridian.**

saunter. (Military) Fly at best endurance.

S-band. A **frequency band** used in **radar** extending approximately from 1.55 to 5.2 kilomegacycles per second.

scalar. Any physical quantity whose field can be described by a single numerical value at each point in space. A scalar quantity is distinguished from a **vector** quantity by the fact that a scalar quantity possesses only **magnitude,** whereas a vector quantity possesses both magnitude and direction.

scale effect. Any variation in the nature of the **flow** and in the force coefficients associated with a change in value of the **Reynolds number,** i.e., caused by change in size without change in shape.

scale model. A model of a different size from its prototype and having dimensions in some constant ratio to the dimensions of the prototype, especially such a model of smaller size than its prototype.

scaler. Nucleonics. An electronic instrument for rapid counting of radiation-induced *pulses* from Geiger counters or other radiation detectors. It permits rapid counting by reducing (by a definite scaling factor) the number of pulses entering the counter. (See **counter, Geiger-Muller counter.**)

scan. The motion of an electronic beam through space searching for a target. Scanning is produced by the motion of the antenna or by lobe switching. See **scanning.**

scanner. A radar mechanism incorporating a rotatable antenna, or radiator, motor drives, mounting, etc., for directing a searching **radar beam** through space and imparting target information to an indicator. See **parabolic reflector.**

scanning. In **radar,** the motion of the radar **antenna** assembly when searching for **targets.** Scanning usually follows a systematic pattern involving one or more of the following: (**a**) In horizontal scanning (or searchlighting), the antenna is continuously rotated in azimuth around the horizon or in a sector (sector scanning); used to generate plan-position-indicator-scope displays. (**b**) Vertical scanning is accomplished by holding the azimuth constant but varying the elevation angle of the antenna; used in height-finding radars to generate the relative-height-indicatorscope display. (**c**) For conical scanning, a somewhat offcenter radiating element is rotated while its parabolic reflector is fixed in position so that the radiated beam generates a conically shaped volume with the antenna at the apex; used to determine accurate bearing and elevation angle of targets and employed in automatic tracking radars. (**d**) In helical scanning (or spiral scanning) the azimuth and elevation angle of the antenna are constantly varied so that at a given distance from the radar the radiated beam generates the surface of a hemisphere; used for radio direction finding, in certain types of search radars, and in tracking radars to search areas for targets.

scanning radioisotope. A method of determining the location and amount of radioactive isotopes within the body by measurements taken with instruments outside the body; usually the instrument, called a scanner, moves in a regular pattern over the area to be studied, or over the whole body, and makes a visual record. (Compare **whole-body counter**).

scan type. The path made in space by a point on the radar beam; for example: circular, helical, conical, spiral, or sector.

scatter. 1. scattering. 2. The relative dispersion of points on a graph, especially with respect to a mean value, or any curve used to represent the points. See dispersion. **3.** To accomplish **scattering.**

scattering. The process by which small particles suspended in a medium of a different **index of refraction** diffuse a portion of the incident **radiation** in all directions. In scattering, no energy transformation results, only a change in the spatial distribution of the radiation. Also called *scatter.*

scattering. Nucleonics. A process that changes a particle's trajectory. Scattering is caused by *particle* collisions with atoms, nuclei, and other particles or by interactions with fields of magnetic force. If the scattered particle's internal energy (as contrasted with its kinetic energy) is unchanged by the collision, elastic scattering prevails; if there is a change in the internal energy, the process is called inelastic scattering. (See **collision,.**)

scattering loss. That part of the **transmission loss** which is due to **scattering** within the medium or due to roughness of the reflecting surface.

scattering power. In radar terminology, the ratio of the total power scattered by a **target** to the power in the incident wave, independent of the direction of **scattering.** The scattering power measures the loss of energy by **absorption** in the scatterers. Also called *total scattering cross section.* Compare **radar reflectivity.**

scatter propagation. Specifically, the longrange **propagation** of **radio** signals by **scattering** due to **index of refraction** inhomogeneities in the lower atmosphere. Also called *tropospheric propagation.* Recognition of this process and the development of specialized equipment (basically, more powerful transmitters and sensitive

receivers) has greatly increased the range of VHF and UHF communications. The over-all technique is known as *scatter communication.*

scavenging. In chemistry, the use of a nonspecific precipitate to remove one or more undesirable radionuclides from solution by absorption or coprecipitation. In atmospheric physics, the removal of radionuclides from the atmosphere by the action of rain, snow or dew. (See **fallout.**)

scheduled maintenance. Periodic prescribed inspection and/or servicing of equipment accomplished on a calendar, or hours of operation basis.

scheduled service (air transport). A routine air transport service operated in accordance with a timetable.

schedule target (nuclear). A planned target on which a nuclear weapon is to be delivered at a specific time during the operation of the supported force. The time is specified in terms of minutes before or after a designated time or in terms of the accomplishment of a predetermined movement or task. Coordination and warning of friendly troops and aircraft are mandatory.

schlieren. (German, *steaks, striae*). **1.** Regions of different density in a **fluid,** especially as shown by special apparatus. **2.** Pertaining to a method or apparatus for visualizing or photographing regions of varying density in a field of **flow.** See **schlieren photography, scintillation.** Used in compounds, such as *schlieren lens, schlieren method, schlieren photograph,* etc.

schlieren photography. A method of photography for flow patterns that takes advantage of the fact that light passing through a density gradient in a *gas* is **refracted** as though it were passing through a prism. Compare **shadow-graph.**

scientific and technical intelligence. The product resulting from the collection, evaluation, analysis, and interpretation of foreign scientific and tehnical information which covers: **a.** foreign developments in basic and applied research and in applied engineering techniques; and **b.** scientific and technical characteristics, capabilities, and limitations of all foreign military systems, weapons, weapon systems, and material, the research and development related thereto, and the production methods employed for their manufacture.

scintillation. 1. Generic term for rapid variations in apparent position, brightness, or color of a distant luminous object viewed through the **atmosphere.** If the object lies outside the earth's atmosphere, as in the case of stars and planets, the phenomenon is termed *astronomical scintillation;* if the luminous source lies within the atmosphere, the phenomenon is termed *terrestrial scintillation.* As one of the three principal factors governing astronomical seeing, scintillation is defined as variations in **luminance** only. Parcels of the order of only centimeters to decimeters are believed to produce most of the scintillatory irregularities in the atmosphere. **2.** A flash of light produced in a **phosphor** by an **ionizing** event. See **scintillation counter. 3.** On a radar display, a rapid apparent displacement of the **target** from its mean positon. Also called *target glint* or *wander.* This includes but is not limited to shift of effective reflection point on the target.

scintillation counter. An instrument that detects and measures ionizing radiation by counting the light flashes (scintillations) caused by radiation impinging on certain materials *(phosphors).*

scintillometer. A type of photoelectric **photometer** used in a method of determinging high-altitude winds on the assumption that stellar **scintillation** is caused by atmospheric inhomogeneities (**schlieren**) being carried along by the wind near **tropopause** level. Also called *scintillation meter.*

scope. The general abbreviation for an instrument of viewing, such as telescope, microscope, and oscilloscope. In radar installations, the **cathode-ray oscilloscope** indicators are commonly referred to as *scopes* or *radarscopes.* Because of possible ambiguity this term should be avoided in formal reports.

scram. The sudden shutdown of a nuclear reactor, usually by rapid insertion of the *safety rods.* Emergencies or deviations from normal reactor operation cause the reactor operator or automatic control equipment to scram the reactor.

scramble. (Military) Take off as quickly as possible (usually followed by course and altitude instructions).

screaming. A form of **combustion instability,** especially in a **liquid-propellant rocket engine,** of relatively high frequency and characterized by a high-pitched noise.

screeching. A form of **combustion instability,** especially in an **afterburner,** of relatively high frequency and characterized by a harsh, shrill noise.

screen. 1. A device to shield or separate one part of an apparatus from other parts, or to separate the effects of one part on others. **2.** A surface on which images are displayed, as the face of a **cathode-ray tube.**

scrub. To cancel a scheduled firing, either before or during **countdown.**

sea breeze. An on-shore wind during the day, caused by the more rapid heating of the air over land than over water. See **land and sea breezes.**

sea-launched ballistic missile. A missile launched from a submarine or surface ship.

sealed cabin. The occupied space of an aircraft or **spacecraft** characterized by walls which do not allow any gaseous exchange between the inner atmosphere and its surrounding atmosphere and containing its own mechanisms for maintenance of the inside atmosphere.

sea level. Same as **mean sea level.**

sea-level pressure. The **atmospheric pressure** at **mean sea level,** either directly measured or, most commonly, empirically determined from the observed station pressure.

seaplane. Any airplane designed to rise from and alight on the water. This general term applies to both boat and float types, although the boat type is usually designated as a **flying boat.**

search and rescue. The use of aircraft, surface craft, submarines, specialized rescue teams and equipment to search for and rescue personnel in distress on land or at sea.

search/rescue (H), (Military designation). Aircraft having special equipment for performance of search and rescue missions. See **military aircraft types.**

search and rescue facility. A facility responsible for maintaining and operating a search and rescue service to render aid to persons and property in distress.

search and rescue region. See **inland search and rescue region; maritime search and rescue region; overseas search and rescue region.**

searchlighting. Horizontal **scanning,** in which the antenna beam is continuously rotated in **azimuth.**

search mission (air). An air reconnaissance by one or more aircraft dispatched to locate an object or objects known or suspected to be in a specific area.

search radar. A **radar** designed for the approximate location of (usually airborne) objects. Search radar beams are usually wide, wider in the vertical than in the horizontal, making it possible to scan large volumes of space quickly. Compare **tracking radar.**

sea return. See **ground return.**

seat belt. Same as **lap belt.**

seat-to-head acceleration. See **physiological acceleration.**

second *(abbr s).* See **ephemeris second.**

second in command. A pilot who is designated to be second in command of an aircraft during flight time.

second law of thermodynamics. An inequality asserting that it is impossible to transfer **heat** from a colder to a warmer system without the occurrence of other simultaneous changes in the two **systems** or in the **environment.** It folows from this law that during an adiabatic process, entropy cannot decrease. For reversible adiabatic processes entropy remains constant, and for irreversible adiabatic processes it increases. Another equivalent formulation of the law is that it is impossible to convert the heat of a system into work without the occurrence of other simultaneous chagnes in the system or its environment. This version, which requires an engine to have a cold source as well as a heat source, is particularly useful in engineering applications. See **first law of thermodynamics.**

second officer. In airline service the second officer is third in command of an aircraft after the **captain** and **first officer.** The second officer normally performs the duties of **flight engineer.**

second pilot. A pilot, not necessarily qualified on type, who is responsible for assisting the first pilot to fly the aircraft and is authorized as second pilot.

secondary instrument. An intrument whose calibration is determined by comparison with an **absolute instrument.**

secondary radar. See **radar, air traffic control radar beacon system (ATCRBS).**

Secor/DME. *(Se*quential *c*ollation *of r*ange/*di*stance *m*easuring equipment). A distance-measuring system used in **rocket** tracking.

secret. See **defense classification.**

section. One of the cross-section parts that a **rocket vehicle** is divided into, each adjoining another at one or both of its ends. Usually described by a designating word, as in *nose section, aft section, center section, tail section, thrust section, tank section, etc..*

section. As applied to ships or naval aircraft, a tactical subdivision of a division. It is normally one-half of a division in the case of ships, and two aircraft in the case of aircraft.

sectionalized vertical antenna. A vertical **antenna** which is insulated at one or more points along its length. The insertion of suitable reactances or applications of a driving voltage across the insulated points results in a modified current distribution giving a more desired **radiation pattern** in the vertical plane.

secular. Pertaining to long periods of time on the order of a century, as *secular perturbations, secular terms.*

secular terms. In the mathematical expression of an orbit, terms for very long period **perturbations,** in contrast to *periodic terms,* terms of short period.

security clearance. An administrative determination by competent authority that an individual is eligible, from a security standpoint, for access to classified information.

Seebeck effect. The establishment of an electric potential difference tending to produce a flow of current in a circuit of two dissimilar metals the junctions of which are at different temperatures.

seeing. A blanket term long used by astronomers for the disturbing effects produced by the **atmosphere** upon the image quality of an observed celestial body. Also called *astronomical seeing.* Recent studies show that *seeing* is a combination of three principal and distinct effects that the human eye is not capable of distinguishing: (**a**) scintillation, i.e., fluctuations in brightness; (**b**) transverse displacements of the image; and (**c**) **variations of the radius of curvature of the wavefront rendering the image in and out of focus.**

seeker. A guidance system which homes on energy emanating or reflected from a target or station.

segmented circle. A basic marking device used to aid pilots in locating airports, and which provides a central location for such indicators and signal devices as may be required.

select code. (Air traffic control). That code displayed when the ground interrogator and the airborne transponder are operating on the same mode and code simultaneously. See **air traffic control radar beacon system (ATCRBS).**

selective identification feature. A capability which, when added to the basic **Identification Friend or Foe** system, provides the means to transmit, receive, and display selected coded replies.

selectivity. The quality of a radio receiver which permits it to receive only signals of one frequency, or signals in a band of frequencies, and simultaneously reject signals of undesired frequencies.

selenocentric. Relating to the center of the moon; referring to the moon as a center.

selenographic. 1. Of or pertaining to the physical geography of the moon. **2.** Specifically, referring to positions on the moon measured in **latitude** from the moon's **equator** and in **longitude** from a **reference meridian.**

selenology. That branch of astronomy that treats of the moon, its magnitude, motion, consitution, and the like. *Selene* is Greek for *moon.*

self-adaptive control system. A particular type of stability augmentation system which changes the response of a given **control** input by constantly sampling response and adjusting its **gain,** rather than having a fixed or selective gain system.

self-induced vibration. Vibration of a mechanical **system** resulting from conversion, within the system, of-nonoscillatory excitation to oscillatory excitation. Also called *self-excited vibration.*

selsyn. (A trade name, from *self-syn*chronous; often capitalized.) An electrical remote-indicating instrument operating on direct current, in which the angular position of the transmitter shaft, carrying a contact arm moving on a resistance strip, controls the pointer on the **indicator** dial.

semiactive homing guidance. Guidance in which a craft or vehicle is directed toward a destination by means of information received from the destination in response to transmissions from a source other than the craft. In active homing guidance the information received is in response to transmissions from the craft. In passive homing guidance natural radiations from the destina-tion are utilized.

semiactive tracking system. A trajectory measuring system which tracks a signal source normally aboard the target for other purposes, or a system that illuminates the target by use of a ground transmitter but requires no special electronics on board the missile.

semiautomatic ground environment. Air defense system in which air surveillance data are processed for transmission to computers at direction centers.

semicircular canals. Structures of the **inner ear,** the primary function of which is to register movement of the body in space. They respond to change in the rate of movement.

semiconductor. An electronic **conductor,** with **resistivity** in the range between metals and insulators, in which the electrical charge carrier concentration increases with increasing temperature over some temperature range. Certain semiconductors possess two types of **carriers,** namely, negative **electrons** and positive **holes.**

semiconductor device. An **electron device** in which the characteristic distinguishing electronic conduction takes place within a **simiconductor.**

semidiameter. 1. The radius of a closed figure. **2.** Half the angle at the observer subtended by the visible disk of a **celestial body.**

semidiameter correction. A correction due to **semidiameter,** particularly that sextant altitude correction resulting from observation of the upper or lower **limb** of a celestial body, rather than the center of that body.

semimonocoque. A structural concept in which longitudinal members as well as **formers** reinforce the skin and help carry the stresses. Compare with **monocoque.**

sensation level. The level of psychophysiologic stimulation above the **threshold.**

sense antenna An antenna used to resolve a 180° **ambiguity** in a **directional antenna.**

sensible atmosphere. That part of the **atmosphere** that offers resistance to a body passing through it.

sensible temperature. The temperature at which average indoor air of moderate **humidity** would induce, in a lightly clothed person, the same sensation of comfort as that induced by the actual environment. Compare **effective temperature.** Sensible temperature depends on the air temperature; radiation from the sun, sky, and surrounding objects; relative humidity and air motion. The wet-bulb temperature is often taken as an approximate measure.

sensing element. Same as **sensor.**

sensitivity. 1. The ability of electronic equipment to amplify a **signal,** measured by the minimum strength of signal **input** capable of causing a desired value of **output.** The lower the input signal for a given output, the higher the sensitivity. **2.** In measurements, the derivative representing the change in instrument indication produced by a change in the variable being measured.

sensor. 1. The component of an instrument that converts an **input** signal into a quantity which is measured by another part of the instrument. Also called *sensing element.* **2.** The nerve endings or sense organs which receive information from the environment, from the organism, or from both.

sensor. (Military) A technical means to extend man's natural senses; an equipment which detects and indicates terrain configuration, the presence of military targets, and other natural and man-made objects and activities by means of energy emitted or reflected by such targets or objects. The energy may be nuclear, electromagnetic, including the visible and invisible portions of the spectrum, chemical, biological, thermal, or mechanical, including sound, blast, and earth vibration.

separation. (Air traffic control) Spacing of aircraft to achieve their safe and orderly movement in flight and while landing and taking off.

separation. 1. The action of a **fallaway section** or **companion body** as it casts off from the remaining body of a **vehicle,** or the action of the remaining body as it leaves a fallaway section behind it. **2.** The moment of this action.

separation minima. (Air traffic control) The minimum longitudinal, lateral, or vertical distances by which aircraft are spaced through the application of air traffic control procedures.

separation velocity. The velocity at which a **space vehicle** is moving when some part or section is separated from it; specifically, the velocity of a space probe or satellite at the time of separation from the **launch vehicle.**

sequencer. A mechanical or electronic device that may be set to initiate a series of events and to make the events follow in a given sequence. See **program.**

sequence report. The weather report transmitted hourly to all teletype stations, and available at all Flight Service Stations and **National Weather Service** airport stations. Sometimes called *hourly report.*

sequential collation of range*(abbr* **Secor).** A spherical, long-baseline, phase-comparison **trajectory-measuring system** utilizing three or more ground stations, time sharing a single **transponder,** to provide nonambiguous range measurements to determine the instantaneous position of a vehicle in flight.

sequential control. Control by completion of a series of one or more events.

service ceiling. The altitude at which the maximum rate of climb has a defined value approximating to the lowest practicable for a service operation.

service engineering. The function of determining the integrity of materiel and services in order to measure and maintain operational reliability, approve design changes, and assure their conformance with established specifications and standards.

service test. A test of an item, system of materiel, or technique conducted under simulated or actual operational conditions to determine whether the specified requirements or characteristics are satisfied. See also **test.**

servo. 1. servomechanism. 2. Pertaining to or incorporating a servomechanism.

servomechanism. A **control** system incorporating **feedback** in which one or more of the system signals represent mechanical motion. It should be noted that *servomechanism* and *regulator* are not mutually exclusive terms; their application to a particular system will depend on the method of operation of that system.

servotab. A **balance tab** directly operated by the pilot to produce forces which in turn move the main surface.

set. 1. To place a **storage** device in a prescribed state. **2.** To place a **binary cell** in the *one* state.

sextant. A double-reflecting instrument for measuring angles, primarily **altitudes** of **celestial bodies.** As originally used, the term applied only to instruments

having an arc of 60°, a sixth of a circle, from which the instrument derived its name. Such an instrument had a range of 120°. In modern practice the term applies to a similar instrument, regardless of its range, very few modern instruments being sextants in the original sense.

sextant altitude. The **altitude** of a **celestial body** as actually measured by a **sextant.** See **altitude difference.**

sferics. 1. (Also called *spherics).* The study of **atmospherics,** especially from a meteorological point of view. This involves techniques of locating and tracking atmospherics sources and evaluating received signals (waveform, frequency, etc.) in terms of source. **2. atmospherics.**

sferics fix. The estimated location of a source of **atmospherics,** presumably a lightning discharge.

sferics observation. An evaluation, from one or more **sferics receivers,** of the location of weather conditions with which lightning is associated. Such observations are more commonly obtained from networks of two or three widely spaced stations. Simultaneous observations of the azimuth of the discharge are made at all stations and the location of the storm is determined by triangulation.

sferics receiver. An instrument which measures, electronically, the direction of arrival, intensity, and rate of occurrence of **atmospherics.** In its simplest form the instrument consists of two orthogonally crossed antennas. Their output signals are connected to an oscillograph so that one loop measures the north-south component whereas the other measures the east-west component. These are combined vertically to give the azimuth. Also called **lightning recorder.**

shadow. Darkness in a region, caused by an obstruction between the source of light and the region. By extension, the term is applied to a similar condition when any form of radiant energy is cut off by an obstruction as a *radar shadow.* The darkest part of a shadow in which light is completely cut off is called the *umbra;* a lighter part surrounding the umbra, in which the light is only partly cut off, is called the *penumbra.*

shadowgraph. 1. A picture or image in which steep density **gradients** in the flow about a body are made visible, the body itself being presented in silhouette. **2.** The optical method or technique by which this is done. A shadowgraph differs from a **schlieren photograph** in that the schlieren method depends on the first derivative of the refractive index while the shadow method depends on the second derivitive. Interference measurements give the refractive index directly.

shadow shield. A **shield** that is interposed between a **radiation** source and a specific area to be protected. Useful in space, a shadow shield is less effective in the earth's atmosphere because air scattering deflects radiation around it.

shaker. An electromagnetic device capable of imparting known **vibratory acceleration** to a given object.

shake-table test. A laboratory test for **vibration** tolerance, in which the device to be tested is placed in a vibrator.

shaped-beam antenna. A **unidirectional antenna** whose major **lobe** differs materially from that obtainable from an aperture of uniform phase. Also called *phase-shaped antenna.*

shear strength. In materials, the **stress** required to produce fracture in the plane of cross section, the conditions of loading being such that the directions of force and of resistance are parallel and opposite although their paths are offset a specified minimum amount.

shell. A body one of whose dimensions is small compared with the others.

shield (shielding); nucleonics. A body of material used to reduce the passage of radiation. (See **logical shield, radiation shielding.**)

shimmy. An oscillation of a castoring wheel such as an aircraft nosewheel about the castor axis excited when traveling on a surface the coefficient of friction of which exceeds a critical value.

shimmy damper. A damper designed for suppressing shimmy.

shock. 1. Same as **shock wave. 2.** A blow, impact, collision, or violent jar. **3.** A sudden agitation of the mental or emotional state or an event causing it. **4.** The sudden stimulation caused by an electrical discharge on the animal or human organism (e.g., electric shock).

shock absorber. A device for the dissipation of energy used to modify the response of a mechanical system to applied **shock.** Also see **oleo.**

shock front. 1. A **shock wave** regarded as the forward surface of a **fluid** region having characteristics different from those of the region ahead of the wave. **2.** The front side of a shock wave.

shock isolator. A resilient support that tends to isolate a system from applied **shock.** Also called *shock mount.*

shock mount. Same as **shock isolator.**

shock spectrum. A plot of the maximum **acceleration** expreienced by a **single-degree-of-freedom** system as a function of its own **natural frequency** in response to an applied **shock.**

shock tube. A relatively long tube or pipe in which very brief high-speed gas flows are produced by the sudden release of gas at very high pressure into a low-pressure portion of the tube; the high-speed flow moves into the region of low pressure behind a shock wave.

shock tunnel. A **shock tube** used as a **wind tunnel.**

shock wave. A surface or sheet of discontinuity (i.e., of abrupt changes in conditions) set up in a **supersonic** field of **flow,** through which the **fluid** undergoes a finite decrease in velocity accompanied by a marked increase in pressure, density, temperature, and entropy, as occurs, e.g., in a supersonic flow about a body. Sometimes called a *shock.* See **attached shock wave, bow wave, condensation shock wave, detached shock wave, Mach wave, normal shock wave, oblique shock wave.**

shooting star. Same as **meteor.**

shoran. A precise short-range electronic navigation system which uses the time of travel of pulse-type transmission from two or more fixed stations to measure slant-range distance from the stations. Also, in conjunction with suitable computer, used in precision bombing. (This term is derived from the words *"short-range navigation."*)

short range attack missile. An air-to-surface missile designed to be launched at ranges sufficient to attack targets effectively without exposing the launch aircraft to terminal defenses.

short-range ballistic missile. A ballistic missile with a range capability up to about 600 nautical miles.

short-range navigation. See **shoran.**

short takeoff and landing (STOL). The ability of an aircraft to clear a 50-foot obstacle within 1500 feet of commencing takeoff, or in landing, to stop within 1500 feet after passing over a 50-foot obstacle.

short-wave radiation. In meteorology, a term used loosely to distinguish radiation in the visible and near-visible portions of the **electromagnetic spectrum** (roughly 0.4 to 1.0 micron in wavelength) from long-wave radiation (**infrared radiation**).

shot. 1. An act or instance of firing a **rocket,** especially from the earth's surface, as, *the shot carried the rocket 200 miles.* **2.** The flight of a rocket, as, *the rocket made a 200-mile shot.*

shoulder harness. A harness that fastens over a person's shoulders to prevent his being thrown forward in his seat. See **lap belt.**

shower. Precipitation from a convective cloud. Characterized by the suddenness with which it starts and stops, by the rapid changes of intensity, and usually by rapid changes in the appearance of the sky.

shutdown. The process of decreasing engine **thrust** to zero.

shuttle bombing. Bombing of objectives, utilizing two bases. By this method, a bomber formation bombs its target, flies on to its second base, reloads, and returns to its home base, again bombing a target if required.

SID. Standard Instrument Departure.

sideband. 1. Either of the two **frequency bands** on both sides of the **carrier** frequency within which fall the frequencies of the wave produced by the process of **modulation. 2.** The wave components lying within such a band.

side lobe. See **lobe.**

side-looking airborne radar. An airborne radar, viewing at right angles to the axis of the vehicle, which produces a presentation of terrain or moving targets. Commonly referred to as SLAR.

side oblique air photograph. An oblique photograph taken with the camera axis at right angles to the longitudinal axis of the aircraft.

sidereal. Of or pertaining to the stars. Although *sidereal* generally refers to the stars and *tropical* to the vernal equinox, sidereal time and the sidereal day are based upon the position of the vernal equinox relative to the meridian. The sidereal year is based upon the stars.

sidereal day. The duration of one **rotation** of the earth on its axis, with respect to the **vernal equinox.** It is measured by successive transits of the vernal equinox over the **upper branch** of a **meridian.** Because of the precession of the equinoxes, the sidereal day thus defined is slightly less than the period of rotation with respect to the stars, but the difference is less than 0.01 second. The length of the mean sidereal day is 24 hours of sidereal time or 23 hours 56 minutes 4,09054 seconds of mean solar time.

sidereal hour angle *(abbr* **SHA).** Angular distance west of the **vernal equinox;** the arc of the **celestial equator,** or the angle at the celestial pole, between the **hour circle** of the vernal equinox and the hour circle of a point on the **celestial sphere,** measured westward from the hour circle of the vernal equinox through 360°. Angular distance east of the vernal equinox, through 24 hours, is *right ascension.*

sidereal month. Tha average **period** of **revolution** of the moon with respect to the stars, a period of 27 days 7 hours 43 minutes 11.5 seconds, or approximately 27 1/3 days.

sidereal period. 1. The time taken by a **planet** or **satellite** to complete one **revolution** about its **primary** as seen from the primary and as referred to a fixed star. **2.** Specifically, the interval between two successive returns of an earth satellite in orbit to the same geocentric **right ascension.**

sidereal time. Time based upon the **rotation** of the earth relative to the **vernal equinox.** Sidereal time may be designated as *local* or *Greenwich* as the local or Greenwich meridian is used as the reference. When adjusted for nutation, to eliminate slight irregularities in the rate, it is called *mean sidereal time.*

sidereal year. The period of one apparent **revolution** of the earth around the sun, with respect to the stars, averaging 365 days 6 hours 9 minutes 9.55 seconds in 1955, and increasing at the rate of 0.000095 second annually. Because of the precession of the equinoxes this is about 20 minutes longer than a tropical year.

sideslip. Similar to a **forward slip** except the flight path of the airplane over the ground (track) is not a straight line. Also see **slip.**

sight. Same as **celestial observation.**

sighting. Actual visual contact. Does not include other contacts, which must be reported by type, e.g., radar and sonar contacts.

sigma. See **standard deviation.**

SIGMET and AIRMET. A weather advisory service to warn airmen of potentially hazardous weather such as squall lines, thunderstorms, fog, icing, and turbulence. SIGMET's concern severe and extreme conditions of importance to all aircraft. AIRMET's concern less severe conditions which may be hazardous to some aircraft or to relatively inexperienced pilots.

signal. 1. A visible, audible, or other, indication used to convey **information. 2.** The information to be conveyed over a communication system. **3.** Any carrier of information; opposed to **noise.**

signal strength. In radio, a measure of the received radiofrequency **power,** generally expressed in decibels relative to some standard value, normally either 1 milliwatt or that power which would have resulted at the same distance under free-space transmission. Also called *field strength.*

signal-to-noise ratio *(abbr* **S/N).** A ratio which measures the comprehensibility of a data source or transmission link, usually expressed as the root-mean-square **signal** amplitude divided by the root-mean-square **noise** amplitude. The higher the S/N ratio, the less the interference with reception.

signal transmission level. In a transmission system, the signal **level,** of a kind to be specified, at a designated position in the system. The signal level at some specified position near the source may be taken as the zero reference level. In an acoustic system the signal level is often in the form of a sound pressure level; either the reference sound pressure or the reference sound pressure level must be specified.

silo. A missile shelter that consists of a hardened vertical hole in the ground with facilities either for lifting the missile to a launch position, or for direct launch from the shelter.

silver-cell battery. A type of short-duration, high-power-density **battery** of light weight used for single-time, high-power applications in vehicles where weight is critical.

silver-disk pyrheliometer. An instrument used for the measurement of **direct solar radiation.** See **pyrheliometer.** It is constructed in the following manner. A silver disk located at the lower end of a diaphragmed tube serves as the radiation receiver for a calorimeter. Radiation falling on the silver disk is periodically intercepted by means of a shutter located in the tube, causing temperature fluctuations of the calorimeter which are proportional to the intensity of the radiation. The instrument is normally used as a secondary instrument and is calibrated against the water-flow pyrheliometer. It is used by the National Weather Service as a standard instrument.

simple harmonic motion. A motion such that the **displacement** is a **sinusoidal** function of time.

simplex. A method in which telecommunication between two stations takes place in one direction at a time. **Note.** In application to Aeronautical Communications this method may be subdivided as follows: (a) **single channel simplex;** (b) **double channel simplex.**

simplified directional facility/SDF. A NAVAID used for nonprecision instrument approaches. The final approach course is similar to that of an ILS localizer except that the SDF course may be offset from the runway, generally not more than 3 degrees, and the course may be wider than the localizer, resulting in a lower degree of accuracy.

simulator. See **flight simulator.**

sine wave. A **wave** which can be expressed as the sine of a linear function of time, or space, or both.

single channel simplex. Simplex using the same frequency channel in each direction.

single-degree-of-freedom system. A mechanical **system** for which only one **coordinate is** required to define completely the configuration of the system at any instant. See **degree of freedom.**

single-entry compressor. A **centrifugal compressor** that takes in air or fluid on only one side of the **impeller,** the impeller being faced with vanes only on that side.

single-sideband modulation. Modulation whereby the spectrum of the modulating wave is translated in **frequency** by a specified amount either with or without inversion.

single-sideband transmission. That method of operation in which one **sideband** is transmitted and the other sideband is suppressed. The **carrier wave** may be either transmitted or suppressed.

single-stage compressor. A **centrifugal compressor** having a single **impeller** wheel with vanes either on one or on both sides of the wheel; also, an **axial-flow compressor** with one row of **rotor** blades and one row of **stator** blades. Axial-flow compressors are normally multistage.

single-stage rocket. A **rocket vehicle** provided with a single rocket propulsion system. See **stage.**

single-stage turbine. A turbine having one set of **stator** blades followed by a set of **rotor** blades.

sink. 1. In the mathematical representation of fluid **flow,** a hypothetical point or place at which the fluid is absorbed. **2.** A **heat sink.** See **source.**

sintered ceramic. A **ceramic** body or coating prepared by heating a ceramic powder below its melting point but at a sufficiently high temperature to cause interdiffusuion of ions between contacting particles and subsequent adherence at the points of contact.

sintering. The **bonding** of adjacent surfaces of particles in a mass of powders, usually metal, by heating.

sinusoidal. Having the form of a sine wave.

skiatron. 1. A dark trace oscilloscope tube. See **dark trace tube. 2.** A **display** employing an optical system with a dark trace tube.

skid. Sideward motion of an airplane in flight produced by centrifugal force.

ski landing gear. A landing gear incorporating skis.

skin. The covering of a body, of whatever material, such as the covering of a fuselage, of a wing, of a hull, of an entire aircraft, etc.; a body shell, as of a rocket; the surface of a body.

skin paint. A radar indication caused by the reflected radar signal from an object.

skin temperature. The outer surface **temperature** of a body.

skin tracking. The **tracking** of an object by means of **radar** without using a **beacon** or other signal device on board the object being tracked.

skip bombing. A method of aerial bombing in which the bomb is released from such a low altitude that it slides or glances along the surface of the water or ground and strikes the target at or above water level or ground level. See also **minimum-altitude bombing.**

skip effect. A phenomenon in which **sound** or **radio energy** may be detected only at various distance intervals from the energy source as the result of the presence of an energy reflecting or refracting layer in the atmosphere. See **radio duct.** For long radio waves, the ionosphere acts as the reflecting layer. For shorter wavelengths, the effects may be produced by strong superstandard propagation in elevated layers of the troposphere. Skip effects make it possible on occasion to detect targets at distances far greater than the normal radio horizon, while closer targets remain undetected.

skirt. The lower outer part of a **rocket vehicle;** specifically, the half-stage of an Atlas.

skirt fog. The cloud of steam and water that surrounds the engines of a **rocket** being launched from a **wet emplacement.**

sky screen. An optical device used to detect the departure of a **rocket** from its intended **trajectory.**

sky wave. In radio, **radio energy** that is received after having been reflected by the **ionosphere.** Compare **wave.**

slant range. The **line-of-sight** range of a radar or radio. See **range.**

skyhook balloon. (Originally a code name for a U.S. Navy project.) A large free balloon having a plastic envelope, used especially for constant-level meteorological observations at very high altitudes.

slash. In radar terminology, a radar beacon reply displayed as an elongated target.

slat. The forward portion of a slotted aerofoil with forward located slot.

slave. 1. slave station. 2. Device that follows an order given by a **master** through remote control.

slave antenna. A **directional antenna** that is positioned in azimuth and elevation by a **servoystem.** The information controlling the servosystem is supplied by a **tracking** or positioning system.

slave station. In a hyperbolic navigation system, a station whose transmissions are controlled by a **master station.** Often shortened to *slave.* See **hyperbolic navigation.**

slaving. Of a **gyro,** the use of a **torquer** to maintain the orientation of the **spin axis** relative to an external reference such as a pendulum or magnetic compass.

slenderness ratio. A dimensionless number expressing the ratio of a **rocket vehicle** length to its diameter.

sleet. Generally transparent, globular, solid grains of ice which have formed from the freezing of raindrops, or the refreezing of largely melted snowflakes when falling through a below-freezing layer of air near the Earth's surface.

slew. To change the position of an **antenna** or range gear assembly by injecting a synthetic error signal into the positioning **servo-amplifier.**

slewing. 1. Of a **gyro,** the rotation of the **spin axis** caused by applying **torque** about the axis of rotation. **2.** In **radar,** changing the scale on the **display.**

slip. The controlled flight of an airplane in a direction not in line with its longitudinal axis. Also see **forward slip, sideslip and skid.**

slipstream. The stream of air discharged aft by a rotating propeller.

slope angle. The angle in the vertical plane between the **flightpath** and the **horizontal.**

slope of the lift curve. The slope of the curve of the wing lift coefficient against the angle of attack.

sloshing. The back-and-forth movement of a **liquid** fuel in its tank, creating problems of stability and control in the **vehicle.**

slotted airfoil. An airfoil having one or more air passages (or slots) connecting its upper and lower surfaces, to modify the normal force.

slotted aileron. An aileron whose leading edge is so shaped that the slot between it and the wing improves the flow over its upper surface when the aileron is deflected downwards.

slotted flap. A flap whose leading edge is so shaped that the slot, or slots, between it and the wing improves the flow over its upper surface when the flap is deflected downwards.

slow roll. See **aileron roll.**

slug. A unit of **mass;** the mass of a free body which if acted upon by a force of 1 pound would experience an acceleration of 1 foot per square second; thus approximately 32.17 pounds.

slurry. A **suspension** of fine solid particles in a liquid.

slurry fuel. A **fuel** consisting of a **suspension** of fine solid particles in a liquid.

small aircraft. Aircraft of 12,500 pounds or less, maximum certificated takeoff weight. An FAA definition.

small arms. All arms, including automatic weapons, up to and including .60 caliber and shotguns.

small circle. The intersection of a sphere and a plane which does not pass through the center of the sphere, as a **parallel** of latitude.

small ion. An atmospheric **ion,** apparently a singly charged atmospheric molecule (or, rarely, an atom) about which a few other neutral molecules are held by the electrical attraction of the central ionized molecule. Estimates of the number of satellite molecules range as high as 12. Also called *light ion, fast ion.* Small ions may disappear either by direct recombination with oppositely charged small ions or by combination with neutral **Aitken nuclei** to form new large ions, or by combination with large ions of opposite sign. The small ion, collectively, is the principal agent of atmospheric conduction.

small perturbation. A disturbance imposed on a system in steady state, with amplitude assumed small, i.e., the square of the amplitude is negligible in comparison with the amplitude, and the derivatives of the perturbation are assumed to be of the same order of magnitude as the perturbation. See **perturbaton.**

smog. A natural fog contaminated by industrial pollutants; a mixture of smoke and fog.

snaking. An uncontrolled oscillation in yaw, of an aircraft, the amplitude of which remains approximately constant.

SNAP. (Acronym for *S*ystems for *N*uclear *A*uxiliary *P*ower.) An Atomic Energy Commission program to develop small auxiliary nuclear power sources for specialized space, land, and sea uses. Two approaches are employed: the first uses heat from radioisotope decay to produce electricity directly by thermoelectric or thermionic methods; the second uses heat from small reactors to produce electricity by thermoelectric or thermionic methods or by turning a small turbine and electric generator. (See **radioisotopic generator, thermionic conversion, thermoelectric conversion.**)

snap roll. An acrobatic maneuver in which the airplane is made to effect a quick, complete roll about its longitudinal axis. It is in effect, a horizontal **spin.** Also see **roll.**

snow. Precipitation in the form of white or translucent ice crystals, chiefly in complex branched hexagonal form and often clustered into snowflakes.

snow. See **grass.**

snubber. A device used to increase the stiffness of an elastic system, usually by a large factor, whenever the **displacement** becomes larger than a specified amount.

soaring. A sailplane's use of rising air columns (thermals) and/or large scale upgliding motions of air for motive power.

sodium-graphite reactor. Nucleonics. A reactor that uses liquid sodium as coolant and graphite as moderator.

sofar. (From *s*ound *f*ixing *a*nd *r*anging.) A system of navigation providing hyperbolic **lines of position** determined by shore listening stations which receive sound signals produced by depth charges dropped at sea and exploding in a sound channel which is at a considerable depth in most areas. This system was used in Project Mercury for locating spacecraft down at sea.

softening range. An arbitrarily defined temperature range below the crystal melting point where a **ceramic** becomes soft and noticeably viscous; a softening range rather than a sharp melting point occurs in ceramics containing a glass base.

soft hail. White, opaque, round pellets of snow.

soft landing. The act of landing on the surface of a **planet** without damage to any portion of the vehicle or payload except possibly the landing gear.

soft missile base. A launching base that is not protected against a nuclear explosion.

soft radiation. Radiation absorbable by an absorber equivalent to 10 centimeters of lead or less. Radiation which can penetrate more than 10 centimeters of lead is termed *hard radiation.*

software. The programs and routines used to extend the capability of automatic data processing equipment. The types of software are as follows: **a. Basic Software** comprises those routines and programs designed to extend or facilitate the use of particular automatic data processing equipment, the requirement for which takes into account the design characteristics of such equipment. This software is usually provided by the original equipment manufacturer and is normally essential to and a part of the system configuration furnished by him. Examples of basic software are executive and operating programs; diagnostic programs; compilers;

assemblers, utility routines, such as soft-merge and input/output conversion routines; file management programs, and data management programs. Data management programs are commonly linked to, and/or under the control of, the executive or operating programs. **b. Application Software** consists of those routines and programs designed by or for automatic data processing equipment users to accomplish specific, mission-oriented tasks, jobs or functions using the automatic data processing equipment and basic software available. Applications software may be either general purpose packages, such as demand-deposit accounting, payroll, machine tool control, etc.; or specific application programs tailored to accomplish a single or limited number of users' functions, such as base level personnel, depot maintenance, missile or satellite tracking, etc. Except for general purpose packages which are acquired directly from software vendors or from the original equipment manufacturers, this type of software is normally developed by the user, either with in-house resources or through contract services.

solar. **1.** Of or pertaining to the sun or caused by the sun, as *solar radiation, solar atmospheric tide.* **2.** Relative to the sun as a *datum* or *reference,* as *solar time.*

solar activity. Any type of variation in the appearance or energy output of the sun. See **faculae, flare, flocculi, prominence, spicules, sunspot.**

solar apex. The point on the **celestial sphere** toward which the sun is traveling. Also called *apex of the sun's way.* The solar apex is at approximately right ascension 270° declination 34° N. The point diametrically opposite the solar apex on the celestial sphere is the solar antapex, right ascension 90° declination 34° S.

solar cell. A **photovoltaic cell** that converts sunlight into electrical energy.

solar constant. The rate at which **solar radiation** is received outside the earth's atmosphere on a surface normal to the incident radiation and at the earth's mean distance from the sun. Measurements of solar radiation at the earth's surface by the Smithsonian Institution for several decades give a best value for the solar constant of 1.934 calories per square centimeter per minute. Measurements from rockets of the intensity of the ultraviolet end of the spectrum have corrected this value to 2.00 calories per square centimeter per minute with a probable error of + 2 percent.

solar cycle. The periodic increase and decrease in the number of **sunspots.** The cycle has a period of about 11 years.

solar day. **1.** The duration of one **rotation** of the earth on its axis, with respect to the sun. This may be either a *mean solar day,* or an *apparent solar day,* as the reference is the mean or apparent sun, respectively. **2.** The duration of one **rotation** of the sun on its axis.

solar eclipse. The obscuration of the light of the sun by the moon. A solar eclipse is *partial* if the sun is partly obscured, *total* if the entire surface is obscured, or *annular* if a thin ring of the sun's surface appears around the obscuring body.

solar flare. See **flare.**

solar protons. Protons emitted by the sun, especially during solar **flares.**

solar radiation. The total **electromagnetic radiation** emitted by the sun. See **insolation, direct solar radiation, diffuse sky radiation, global radiation, extraterrestrial radiation, solar constant.** To a first ap-

223

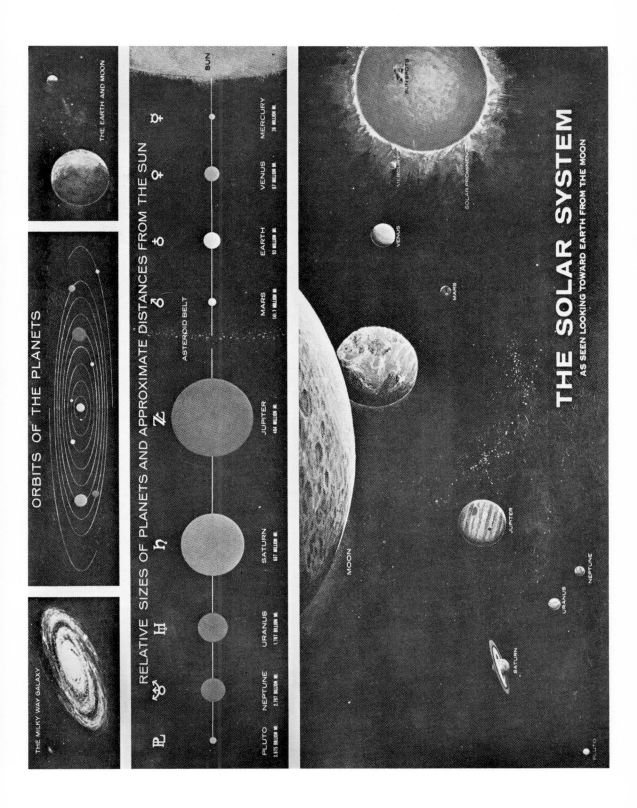

proximation, the sun radiates as a black body at a temperature of about 5700° K; hence about 99.9 percent of its energy output falls within the wavelength interval from 0.15 micron to 4.0 microns, with peak intensity near 0.47 micron. About one-half of the total energy in the solar beam is contained within the visible spectrum from 0.4 to 0.7 micron, and most of the other half lies in the near infrared, a small additional portion lying in the ultraviolet.

solar-radiation observation. An evaluation of the radiation from the sun that reaches the observation point. The observing instrument is usually a **pyrheliometer** or **pyranometer.** Two types of such observations are taken. The more common consists of measurements of the radiation reaching a horizontal surface, consisting of both radiation from the sun (direct solar radiation and that reaching the instrument indirectly by scattering in the atmosphere (diffuse sky radiation). The other type of observation involves the use of an equatorial mount that keeps the instrument pointed directly at the sun at all times. The sensitive surface of the instrument is normal to the path of the radiation and is shielded from indirect radiation from the sky.

solar radio burst. A sudden increase in the **flux** from the sun at **radio frequencies.**

solar radio waves. Radiation at **radio frequencies** originating in the sun or its **corona.**

solar simulator. A device which produces thermal energy, equivalent in intensity and spectral distribution to that from the sun, used in testing materials and space vehicles.

solar system. The sun and other celestial bodies within its gravitational influence, including **planets, asteroids, satellites, comets,** and **meteors.**

solar time. Time based upon the **rotation** of the earth relative to the sun. Solar time may be designated as *mean* or *astronomical* if the mean sun is the reference, or *apparent* if the apparent sun is the reference. The difference between mean and apparent time is called *equation of time.* Solar time may be further designated according to the reference meridian, either the *local* or *Greenwich* meridian or additionally in the case of mean time, a designated zone meridian. Standard or daylight-saving are variations of zone time. Time may also be designated according to the timepiece, as *chronometer time* or *watch time,* the time indicated by these instruments.

solar wind. Streams of **plasma** flowing approximately radially outward from the sun.

solar year. Same as **tropical year.**

solenoid. 1. A tube formed in space by the intersection of unit-interval **isotimic** surfaces of two scalar quantities. Solenoids formed by the intersection of surfaces of equal pressure and density are frequently referred to in meteorology. A barotropic atmosphere implies the absence of solenoids of this type, since surfaces of equal pressure and density coincide. **2.** A hollow or tubular shaped electric coil, and made up of many turns of fine insulated wire and possessing the same properties as an electromagnet. The hollow core will impart lineal motion to an iron core placed within the hollow core of the solenoid. Solenoids are often used to operate remote-controlled switches or mechanisms.

solidity (rotor). The ratio of the total blade area of a rotor to the disk area.

solid propellant. Specifically, a **rocket propellant** in solid form, usually containing both fuel and oxidizer combined or mixed, and formed into a monolithic (not powdered or granulated) **grain.**

solid-propellant rocket engine. A **rocket engine** fueled with a **solid propellant.** Such motors consist essentially of a **combustion chamber** containing the propellant, and a **nozzle** for the exhaust jet, although they often contain other components, as grids, liners, etc.

solid rocket. A rocket that uses a **solid propellant.**

solid rocket fuel. A **solid propellant.**

solid-state devices. Devices which utilize the electric, magnetic, and photic properties of solid materials, e.g., binary magnetic cores, transistors, etc.

solo. A flight during which a pilot is the only occupant of the airplane.

solstice. 1. One of the two points of the **ecliptic** farthest from the **celestial equator;** one of the two points on the **celestial sphere** occupied by the sun at maximum **declination.** That in the northern hemisphere is called the *summer solstice* and that in the southern hemisphere the *winter solstice.* Also called *solstitial point.* **2.** That instant at which the sun reaches one of the solstices, about June 21 (summer solstice) or December 22 (winter solstice).

somatic effects of radiation. Effects of radiation limited to the exposed individual, as distinguished from genetic effects (which also affect subsequent, unexposed generations). Large radiation doses can be fatal. Smaller doses may make the individual noticeably ill, may merely produce temporary changes in blood-cell levels detectable only in the laboratory, or may produce no detectable effects whatever. Also called physiological effects of radiation. (Compare **genetic effects of radiation;** see **radiation illness.**)

sonar. (From sound, *na*vigation, and *r*anging.) A method or system, analogous to **radar** used under water, in which high-frequency **sound waves** are emitted so as to be reflected back from objects, and used to detect the objects of interest. Called *asdic* by the British.

sonar capsule. A device designed to reflect high-frequency **sound waves.** See **sonar.** The sonar capsule, if attached to a reentry body, may be used to locate the reentry body in case of a water landing.

sone. A unit of **loudness.** A simple tone of frequency 1000 cycles per second, 40 decibels above a listener's **threshold,** produces a loudness of 1 sone. The loudness of any sound that is judged by the listener to be *n* times that of the 1-sone tone is *n* sones. A millisone is equal to 0.001 sone. The loudness scale is a relation between loudness and level above threshold for a particular listener. In presenting data relating loudness in sones to sound pressure level, or in averaging the loudness scales of several listeners, the thresholds (measured or assumed) should be specified.

sonic. 1. In aerodynamics, of or pertaining to the speed of sound; that which moves at **acoustic velocity** as in *sonic flow;* designed to operate or perform at the speed of sound as in *sonic leading edge.* **2.** Of or pertaining to sound, as in *sonic amplifier.* In sense **2,** *acoustic* is preferred to *sonic.*

sonic barrier. A popular term for the large increase in **drag** that acts upon an aircraft approaching **acoustic velocity;** the point at which the speed of sound is attained and existing subsonic and supersonic flow theories are rather indefinite. Also called *sound barrier.*

sonic boom. A noise caused by a **shock wave** that

emanates from an aircraft or other object traveling at or above **sonic velocity.** A shock wave is a pressure disturbance and is received by the ear as a noise or clap.

sonic drilling. The process of cutting or shaping materials with an abrasive **slurry** driven by a reciprocating tool attached to an **electromechanical transducer** operating at **ultrasonic** frequencies.

sonics. The technology of **sound** in processing and analysis. *Sonics* includes the use of sound in any non-communication process.

sonic speed. Acoustic velocity; by extension, the speed of a body traveling at a **Mach number** of 1.

sonobuoy. A buoy equipped for detecting underwater sounds and transmitting them by radio. It can be dropped from an **antisubmarine warfare** aircraft.

sophisticated. Complex and intricate; making use of advanced art; requiring special skills to operate.

sorb. To take up gas by **sorption.**

sorbent. The material which takes up gas by **sorption.**

sorption. The taking up of gas by **absorption, adsorption, chemisorption,** or any combination of these processes. See **absorption.**

sortie (air). An operational flight by one aircraft.

sound. 1. An **oscillation** in pressure, stress, particle displacement, particle velocity, etc., in a medium with internal forces (e.g., elastic, viscous), or the superposition of such propagated oscillations. **2.** A sensation evoked by the oscillation described above in the human ear. In case of possible confusion, the term *sound wave* or *elastic wave* may be used for concept **1** and the term can evoke an auditory sensation, e.g., ultrasound. The medium in which the sound exists is often indicated by an appropriate adjective; e.g., *airborne, water borne, structure borne.*

sound absorption. Sound **absorption** is the change of **sound energy** into some other form, usually heat, in passing through a medium or on striking a surface.

sound barrier. Same as **sonic barrier.**

sound energy. The energy which **sound waves** contribute to a particular medium.

sound level. Specifically, a weighted **sound pressure level,** obtained by the use of metering characteristics and the weightings A, B, or C specified in American Standard Publication Z24.3-1944: *Sound Level Meters for Measurement of Noise and Other Sounds.* The weighting employed must always be stated. The reference pressure is 0.0002 microbar. A suitable method of stating the weighting is, for example, *The A-sound level was 43 decibels.*

sound pressure. At a point, the total instantaneous pressure at that point in the presence of a **sound wave** minus the **static pressure** at that point.

sound pressure level. In decibels, 20 times the logarithm to the base 10 of the ratio of the **sound pressure** to the reference pressure. The reference pressure must be explicitly stated. The following reference pressures are in common use: (a) 2×10^{-4} microbar, (b) 1 microbar. Reference pressure (a) is in general use for measurements concerned with hearing and with sound in air and liquids, whereas (b) has gained widespread acceptance for calibration of transducers and various kinds of sound measurements in liquids. Unless otherwise explicitly stated, it is to be understood that the sound pressure is the effective (root-mean-square) sound pressure. It is to be noted that in many sound

fields the sound pressure ratios are not the square roots of the corresponding power ratios.

sound probe. A device that responds to some characteristic of an **acoustic wave** (e.g., sound pressure, particle velocity) and that can be used to explore and determine this characteristic in a **sound field** without appreciably altering that field.

sound suppressor. A device for reducing to within levels tolerable to those within hearing, noises produced by operating equipment, such as jet type aircraft. The device may consist of an arrangement for absorbing a portion of the noise or an arrangement to deflect the noise in a direction where it is not expected to impinge on sensitive receptors, including ears and mechanical or electronic equipment which could be adversely affected.

sound wave. A mechanical disturbance advancing with infinite velocity through an elastic medium and consisting of longitudinal displacements of the medium, i.e., consisting of compressional and rarefactional displacements parallel to the direction of advance of the disturbance; a longitudinal wave. Sound waves are small-amplitude **adiabatic** oscillations. As defined, this includes waves outside the frequency limits of human hearing, which limits customarily define sound. Also called *acoustic wave, sonic wave.* See **ultrasonic, infrasonic, pressure wave.** Gases, liquids, and solids transmit sound waves, and the propagation velocity is characteristic of the nature and physical state of each of these media. In those cases where a steadily vibrating sound generator acts as a source of waves, one may speak of a uniform wave train; but in other cases (explosions, lightning discharges) a violent initial disturbance sends out a principal wave, followed by waves of more or less rapidly diminishing amplitude.

sounding. 1. In **geophysics,** any penetration of the natural environment for scientific observation. **2.** In meteorology, same as **upper air observation.** However, a common connotation is that of a single complete **radiosonde** observation. **3. air sounding.**

sounding balloon. A free, unmanned balloon carrying a set of self-registering meteorological instruments.

sounding rocket. A **rocket** that carries aloft equipment for making observations of or from the **upper atmosphere.** See **air sounding.** Compare **probe,** sense **3.** Usually a sounding rocket has a near vertical trajectory.

source. 1. The location or device from which energy emanates as a *sound source, heat source,* etc. **2** Specifically, in the mathematical representation of fluid, a hypothetical point or place from which fluid emanates. **3.** Specifically, the device which supplies **signal power** to a transducer.

source material. In atomic energy law any material, except *special nuclear material,* which contains 0.05% or more of uranium, thorium, or any combination of the two. (See **licensed material, special nuclear material.**)

South Tropical Disturbance. An elongated dark band in the cloud surface of Jupiter at about the latitude of the **Great Red Spot.** It was first seen in 1901 as a dark spot which then spread rapidly. It has at times exceeded 180° of longitude in length and, like the Red Spot, appears and disappears intermittently.

space. 1. Specifically, the part of the universe lying outside the limits of the earth's atmosphere. **2.** More

The space shuttle *on the launch pad at the Kennedy Space Center*

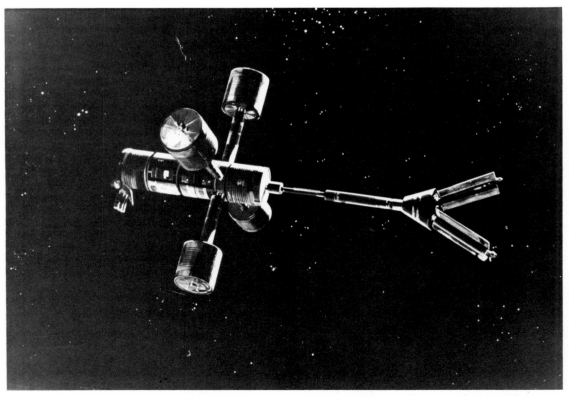

An artist's concept depicting a possible manned, modularized **space station** *in earth orbit. The modules could be carried to earth orbit by the* **space shuttle.**

generally, the volume in which all celestial bodies, including the earth, move.

space-air vehicle. A vehicle operable either within or above the **sensible atmosphere.** Also called *aerospace vehicle.*

space biology. Same as **bioastronautics.**

space capsule. A container used for carrying out an experiment or operation in space. A capsule is usually assumed to carry an organism or equipment.

space charge. 1. The electric charge carried by a cloud or stream of **electrons** or **ions** in a vacuum or a region of low gas pressure when the charge is sufficient to produce local changes in the potential distribution. **2.** The net electric charge within a given volume.

space coordinates. A three-dimensional system of **Cartesian coordinates** by which a point is located by three **magnitudes** indicating distance from three planes which intersect at a point.

spacecraft. Devices, manned and unmanned, which are designed to be placed into an **orbit** about the earth or into a **trajectory** to another **celestial body.**

space base. Same as a **space station** except considerably larger in size and capability.

space defense. All measures designed to reduce or nullify the effectiveness of hostile acts by vehicles (including missiles) while in space.

space equivalent. A condition within the earth's atmosphere that is virtually identical, in terms of a particular function, with a condition in outer space. For example, at 50,000 feet, the drop in air pressure and the scarcity of oxygen creates a condition, so far as respiration is concerned, that is equivalent to a condition in outer space where no appreciable oxygen is present; thus, a physiological space equivalent is present in the atmosphere.

space medicine. A branch of **aerospace medicine** concerned specifically with the health of persons who make, or expect to make, flights into space beyond the **sensible atmosphere.**

space modulation. The combining of signals outside of an electronic device or conductor to form a signal of desired characteristics. See **modulation.**

space motion. Motion of a **celestial body** through **space.** That component perpendicular to the line of sight is termed *proper motion* and that component in the direction of the line of sight, *radial motion.*

space polar coordinates. A system of **coordinates** by which a point on the surface of a sphere is located in three dimensions by (**a**) its distance from a fixed point at the center, called the pole; (**b**) the colatitude or angle between the polar axis (a reference line through the pole) and the radius vector (a straight line connecting the pole and the point); and (**c**) the longitude or angle between a reference plane through the polar axis and a plane through the radius vector and polar axis. See **polar coordinates, spherical coordinates.**

space probe. See **probe,** note, and **spacecraft,** note.

space reddening. The observed reddening, or absorption of shorter **wavelengths,** of the light from distant **celestial bodies** due to **scattering** by small particles in interstellar **space.** Compare **red shift.**

space shuttle. A space transportaion system designed to carry passengers, supplies and equipment to and from orbit on a routine, aircraft-like basis.

space simulator. 1. Any device used to simulate one or more parameters of the space **environment** used for testing space systems or components. **2.** Specifically, a closed chamber capable of approximately the vacuum and normal environments of space.

space station. A permanent structure, generally operating continuously in a planetary orbit for purposes of supporting manned space activities. Periodic crew replacement and resupply are accomplished by a **space shuttle** system.

space suit. A **pressure suit** for wear in space or at very low ambient pressures within the atmosphere, designed to permit the wearer to leave the protection of a **pressurized** cabin.

Spacetrack. A global system of radar, optical and radiometric sensors linked to a computation and analysis center in the North American Air Defense Command combat operations center complex. The Spacetrack mission is detection, tracking, and cataloging of all man-made objects in orbit of the earth. It is the Air Force portion of the North American Air Defense Command Space Detection and Tracking System. See also **Spadats; Spasur.**

Spadats. A *space d*etection *t*racking *s*ystem capable of detecting and tracking space vehicles from the earth, and reporting the orbital characteristics of these vehicles to a central control facility. See also **Spacetrack; Spasur.**

span. 1. The dimension of a craft measured between lateral extremities; the measure of this dimension. **2.** Specifically, the dimension of an **airfoil** from tip to tip measured in a straight line. *Span is not usually applied to vertical airfoils.*

spar. A principal spanwise structural member of an airfoil or control surface.

spark discharge. That type of gaseous **electrical discharge** in which the charge transfer occurs intermittently along a relatively constricted path of high ion density, resulting in high **luminosity.** It is of short duration and to be contrasted with the nonluminous **point discharge,** with the diffuse **corona discharge,** and also with the continuous **arc discharge.** The exact meaning to be attached to the term *spark discharge* varies somewhat in the literature. It is frequently applied to just the transient phase of the establishment of any arc discharge. A lightning discharge is a large-scale spark discharge, though its very length introduces certain details not found in laboratory short-spark processes.

Spasur. An opertional space surveillance system with the mission to detect and determine the orbital elements of all man-made objects in orbit of the earth. The mission is accomplished by means of a continuous fan of continuous wave energy beamed vertically across the continental United States and an associated computational facility. It is the Navy portion of the North American Air Defense Command/Continental Air Defense Command Space Detection and Tracking System. See also **Spacetrack; Spadats.**

spatial. Pertaining to **space.**

special electronic installation (E), (Military designation). Aircraft equipped with electronic devices for employment in the following roles: (**a**) Electronic countermeasures. (**b**) Airborne early warning radar. (**c**) Airborne command and control including communications relay aircraft. (**d**) Tactical data communications link for all nonautonomous modes of flight. See **military aircraft types.**

special nuclear material. In atomic energy law, this term refers to plutonium-239, uranium-233, uranium containing more than the natural abundance of uranium-235, or any material artifically enriched in any of these substances. (Compare **source material; enriched material, licensed material.**)

special (or restricted) theory of relativity. A theory developed by Albert Einstein in 1905 that is of great importance in atomic and nuclear physics. It is especially useful in studies of objects moving with speeds approaching the speed of light. Two of the results of the theory with specific application in nuclear physics are statements (a) that the mass of an object increases with its velocity and (b) that mass and energy are equivalent. (See **mass-energy equation.**)

special perturbations. A method of **orbit** determination by numerical integration which takes into account the perturbing forces which are causing the orbit to depart from the orbit as calculated by **Kepler laws.**

special VFR conditons (special VFR minimum weather conditions). Weather conditions which are less than basic VFR weather conditions and which permit flight under Visual Flight Rules.

special VFR operations. Aircraft operating in accordance with clearances within control zones in weather conditions less than the basic VFR weather minima.

specific. A modifier generally implying *per unit mass.*

Specific fuel consumption. 1. Of engines driving propellers; the quantity of fuel consumed per horsepower per hour. **2.** Of jet reaction engines: the weight of the propellant or fuel consumed per pound of thrust per hour.

specific heat. The ratio of the **heat** absorbed (or released) by unit mass of a system to the corresponding **temperature** rise (or fall).

specific humidity. In a system of moist air, the (dimensionless) ratio of the mass of **water vapor** to the total mass of the system. See **absolute humidity, relative humidity, dew point, humidity.**

specific impulse *(symbol Isp).* A performance parameter of a **rocket propellant,** expressed in seconds, equal to the **thrust** F in pounds divided by the weight flow rate w in pounds per second: $Isp = F/w.$ Specific impulse is equivalent to the effective exhaust velocity divided by the gravitational acceleration.

specific propellant consumption. The reciprocal of the **specific impulse,** i.e., the required **propellant** flow to produce one pound of thrust in an equivalent rocket.

specific speed. Of a pump, a parameter used to predict pump performance.

specific thrust. Same as **specific impulse.**

specific volume *(symbol v).* Volume per unit mass of a substance. The reciprocal of density.

spectra. Plural of *spectrum.*

spectral. 1. Of or pertaining to a spectrum. **2.** Referring to thermal radiation properties, for ratios such as **emittance, reflectance,** and **transmittance,** at a specified wavelength; for powers, such as **emissive power,** within a narrow wavelength band centered on a specified wavelength.

spectral line. A bright, or dark, line found in the **spectrum** of some radiant source. See **absorption line, emission line.** Bright lines indicate emission, dark lines indicate absorption.

spectroheliograph. An instrument for taking photographs (spectroheliograms) of the image of the sun in **monochromatic** light. The wavelength of light chosen for this purpose corresponds to one of the **Fraunhofer**

lines, usually the light of hydrogen or ionized calcium. A similar instrument used for visual, instead of photographic, observations is a spectrohelioscope.

spectroscope. An apparatus to effect **dispersion** of radiation and visual display of the **spectrum** obtained. A spectroscope with a photographic recording device is called a *spectrograph.*

spectrum. 1. In physics, any series of energies arranged according to **wavelength** (or **frequency**). **2.** The series of images produced when a **beam** of radiant energy is subject to **dispersion. 3.** Short for **electromagnetic spectrum** or for any part of it used for a specific purpose as the *radio spectrum* (10 kilocycles to 300,000 megacycles). **4.** In mathematics, = **function. 5.** In acoustics, the distribution of effective **sound pressures** or intensities measured as a function of frequency in specified frequency bands.

speed. Rate of motion. Rate of motion in a straight line is called *linear speed,* whereas change of direction per unit time is called *angular speed. Speed* and *velocity* are often used interchangeably although some authorities maintain that *velocity* should be used only for the *vector* quantity.

speed brakes/dive brakes. Moveable aerodynamic devices on aircraft that reduce airspeed during descent and landing.

speed of light. The speed of propagation of **electromagnetic radiation** through a perfect vacuum; a universal dimensional constant equal to $299,792.5 \pm 0.4$ kilometers (approximately 186,000 miles) per second. Also called velocity of light.

speed of sound *(symbol c_s).* The speed of propagation of sound waves. At sea level in the standard atmosphere, the speed of sound is 340.294 meters per second (1116.45 feet per second). The concept of the speed of sound in the atmosphere loses its applicability at about 90 kilometers where the mean free path of air molecules approaches the wavelengths of sound waves.

spent (depleted) fuel. Nucleonics. Nuclear reactor fuel that has been irradiated (used) to the extent that it can no longer effectively sustain a chain reaction. (Compare **depleted uranium.**)

sphere of influence. The surface in space about a **planet** where the ratio of the force with which the sun perturbs the motion of a particle about the planet, to the force of attraction of the planet equals the ratio of the force with which the planet perturbs the motion of a particle about the sun, to the force of attraction of the sun on the particle. The volume inside this surface defines the region where the attracting body exerts the primary influence on a particle.

spherical angle. The angle between two intersecting **great circles.**

spherical coordinates. 1. A system of **coordinates** defining a point on a sphere or spheroid by its angular distances from a **primary great circle** and from a reference secondary great circle, as latitude and longitude. See **celestial coordinates. 2. space polar coordinates.**

spherical system. A **trajectory measuring system,** whose locus of the measured ranges is a sphere with the ground equipment at the center. A unique point in space is determined by the intersection of three or more spheres. The term *spherical system* has been applied to systems using three or more **slant ranges** to determine space position.

spherical triangle. A closed figure having arcs of three **great circles** as sides.

spheroid. An **ellipsoid;** a figure resembling a sphere. Also called *ellipsoid* or *ellipsoid of revolution* from the fact that it can be formed by revolving an ellipse about one of its axes. If the shorter axis is used as the axis of revolution, an oblate spheroid results, and if the longer axis is used, a prolate spheroid results. The earth is approximately an oblate spheroid.

spicules. Bright spikes extending into the **chromosphere** of the sun from below. They are several hundred miles in diameter and extend outward 5000 to 10,000 miles. Spicules have a lifetime of several minutes and may be related to **granules.**

spin. A prolonged stall in which an airplane rotates about its center of gravity while it descends, usually with its nose well down. Also see **autorotation.**

spin axis. The axis of rotation of the **rotor** of a **gyro.**

spineward acceleration. See **physiological acceleration.**

spinner. A streamlined fairing, fitted coaxially and rotating with the propeller, enclosing the hub or boss.

spin rocket. A small **rocket** that imparts spin to a larger **rocket vehicle** or **spacecraft.**

spin stabilization. Directional stability of a spacecraft obtained by the action of gyroscopic forces which result from spinning the body about its axis of symmetry.

spin table. A flat round platform on which human and animal subjects can be placed in various positions and rapidly rotated, much as on a phonograph record, in order to simulate and study the effects of prolonged tumbling at high rates. Complex types of tumbling can be simulated by mounting the spin table on the arm of a centrifuge.

spiral. (flight maneuver). A prolonged gliding or climbing turn during which at least 360° change of direction is effected.

splashed. Enemy aircraft shot down (followed by number and type).

split cameras. An assembly of two cameras disposed at a fixed overlapping angle relative to each other.

split flap. A flap inset into the lower surface of the wing.

split vertical photography. Photographs taken simultaneously by two cameras mounted at an angle from the vertical, one tilted to the left and one to the right, to obtain a small sidelap.

spoiler. A plate, series of plates, comb, tube, bar, or other device that projects into the **air-stream** about a body to break up or spoil the smoothness of the **flow**, especially such a device that projects from the upper surface of an airfoil, giving an increased **drag** and a decreased **lift**. Compare **deflector**, sense (a). Spoilers are normally movable and consist of two basic types: the flap spoiler, which is hinged along one edge and lies flush with the airfoil or body when not in use, and the retractable spoiler, which retracts edgewise into the body.

sponson. A projection from the hull of a flying boat to give lateral stability on the water.

spontaneous fission. Fission that occurs without an external stimulus. Several heavy isotopes decay mainly in this manner; examples: californium-252 and californium-254. The process occurs occasionally in all fissionable materials, including uranium-235.

spontaneous-ignition temperature. In testing fuels, the lowest temperature of a plate or other solid surface adequate to cause ignition in air of a fuel upon the surface.

spot jamming. The jamming of a specific channel or frequency. See also **barrage jamming; electronic jamming.**

spring feel system. An artificial feel system in which the load required to move a flying control in the absence of air forces is dependent on the displacement from the trimmed condition.

spring tab. A balance tab the angular movement of which is geared to the compression or extension of a spring embedded in the main control circuit. The primary purpose is to reduce the pilot's effort at high airspeeds. The spring may be preloaded so that the tab does not move until that effort exceeds the preset value.

spurious radiation 1. Any undesired **emission** from a radio **transmitter. 2.** Any **electronmagnetic radiation** from a radio **receiver.** Also called **spurious emission.**

spurious response. Output from a **receiver** due to a signal or signals having frequencies other than that to which the receiver is tuned.

spurious transmitter output. Any part of the **radio frequency** output of a transmitter which is not a component of the theoretical **output** as determined by the type of modulation and specified bandwidth limitations.

squadron. 1. An organization consisting of two or more divisions of ships, or two or more divisions (Navy) or flights of aircraft. It is normally, but not necessarily, composed of ships or aircraft of the same type. **2.** The basic administrative aviation unit of the Army, Navy, Marine Corps, and Air Force.

squall. A strong wind which rises and dies away rapidly lasting only for a few minutes, frequently associated with a temporary change in wind direction.

squall line. Any nonfrontal line or narrow band of active thunderstorms (with or without squalls); a mature instability line.

square wave. 1. An **oscillation,** the amplitude of which shows periodic discontinuities between two values, remaining constant between jumps. **2.** Specifically, in radar a **pulse** initiated by a rapid rise to peak power, maintained at a constant peak power over the finite pulse length, and terminated by rapid decrease from peak power.

squawk. A code meaning switch civil transponder or military **Identification Friend or Foe** master control to "normal" (Mode and Code as directed) position.

squawk flash. A code meaning actuate **Identification Friend or Foe** I/P switch.

squawking. Showing civil **transponder** or military **Identification Friend or Foe** in Mode (and Code) indicated.

sqawk may day. A code meaning switch civil transponder or military **Identification Friend or Foe** master control to "emergency" position.

sqawk standby. A code meaning switch civil **transponder** or military **Identification Friend or Foe** master control to "standby" position.

squib. 1. Any of various small explosive devices. **2.** An explosive device used in the **ignition** of a rocket. Usually called an *igniter.*

SSB *(abbr).* **single sideband.**

SS loran. Sky-wave synchronized **loran,** or loran in which the **sky wave** rather than the ground wave from the **master** controls the **slave.** SS loran is used with unusually long baselines.

SSM. Surface-to-Surface Missile.

stabilator. A movable horizontal surface of an airplane used to increase longitudinal stability as well as provide a pitching moment. A stabilator combines the actions of a **stabilizer** and **elevator.**

stability. 1. The property of a body, as an aircraft or rocket, to maintain its attitude or to resist displacement, and, if displaced, to develop forces and moments tending to restore the original condition. **2.** Of a fuel, the capability of a **fuel** to retain its characteristics in an adverse **environment,**e.g. extreme temperature.

Regarding stability of aircraft, static stability is defined as the initial tendency to return to equilibrium conditions following some disturbance. The term "static" is applied to this form of stability since resulting motion is not considered. Dynamic stability is defined by the resulting motion with time. If an aircraft is displaced from equilibrium, the time history of the resulting motion indicates the dynamic stability of its aircraft.

stability, of air; metgy. The degree of stability of an air layer, especially as regards to its internal vertical air currents. It is a state in which the vertical distribution of temperature is such that an air particle will resist displacement from its level. In the case of **unsaturated air,** the **lapse rate** for stability will be less than the **dry adiabatic lapse rate;** in that of **saturated air,** less than the **saturated adiabatic lapse rate.**

stability augmentation system. An auxiliary system to the basic manual **vehicle control system** whereby response of the contol surfaces to inputs by the pilot can be adjusted to give a preselected vehicle response by selection of certain fixed gains in a standard **feedback** loop on control-surface output.

stabilizer. The fixed **airfoil** of an airplane used to increase stability; usually, the aft fixed horizontal surface to which the elevators are hinged (horizontal stabilizer) and the fixed vertical surface to which the rudder is hinged (vertical stabilizer). Also see **stabilator, elevator, rudder, tail plane, tail unit, empennage.**

stable. nucleonics. Incapable of spontaneous change. Not radioactive.

stable isotope. An isotope that does not undergo radioactive decay. (Compare **radioisotope.**)

stable platforms. A gyroscopic device so designed as to maintain a plane of reference in space regardless of the movement of the vehicle carrying the stable platform.

staff (V), (military designation). Aircraft having accommodations such as chairs, tables, lounge, berths, etc., for the transportation of staff personnel. See **military aircraft types.**

stage. 1. A self-propelled separable element of a **rocket vehicle.** See **multistage rocket. 2.** A step or process through which a **fluid** passes, especially in compression or expansion. **3.** A set of **stator** blades and a set of **rotor** blades in an axial-flow compressor or in a turbine; an **inpeller wheel** in a radial-flow compressor. See **multistage compressor, single-stage compressor, single-stage turbine.**

stage-and-a-half. A liquid-propellant **rocket** of which only part of the propulsion unit falls away from the **rocket vehicle** during flight, as in the case of **booster rockets** falling away to leave the **sustainer engine** to consume remaining fuel.

staging. The process or operation during the flight of a **rocket vehicle** whereby a full **stage** or **half stage** is disengaged from the remaining body and made free to decelerate or be propelled along its own flightpath. See **separation.**

stagnation point. A point in a field of **flow** about a body where the **fluid** particles have zero velocity with respect to the body.

stagnation pressure. 1. The pressure at a **stagnation point.**

2. In compressible flow, the pressure exhibited by a moving gas or liquid brought to zero velocity by an isentropic process. **3. total pressure. 4. impact pressure.** Because of the lack of a standard meaning, *stagnation pressure* should be defined when it is used.

stagnation region. Specifically, the region at the front of a body moving through a fluid where the fluid has negligible relative velocity.

stall. The abrupt loss of lift when the angle of attack increases to the point at which the flow of air tends to tear away from a wing or other airfoil.

stalling speed. The equivalent air speed corresponding to the maximum lift-coefficient of an aircraft.

standard. 1. An exact value, or a concept, that has been established by authority or agreement, to serve as a model or rule in the measurement of a quantity or in the establishment of a practice or procedure. **2.** A document that establishes engineering and technical limitations and applications for items, materials, processes, methods, design, or engineering practices.

standard atmosphere. 1. A hypothetical vertical distribution of atmospheric temperature, pressure, and density which, by international agreement, is taken to be representative of the **atmosphere** for purposes of pressure altimeter calibrations, aircraft performance calculations, aircraft and rocket design, ballistic tables, etc. The air is assumed to be devoid of dust, moisture, and water vapor and to obey the perfect gas law and the hydrostatic equation (the air is static with respect to the earth). Standard atmospheres, sense **1,** which have been used are: (**a**) The NACA standard atmosphere, also called *U.S. standard atmosphere,* prepared in 1925, which was supplanted by (**b**) The ICAO standard atmosphere, adopted in 1952, which was extended to greater altitudes by (**c**) The ARDC model atmosphere 1956, and (**d**) The U.S. extension to the ICAO standard atmosphere, adopted in 1956, which has been revised by (**e**) The ARDC model atmosphere, 1959, which incorporated some satellite data which has been supplanted by (**f**) The U.S. Standard Atmosphere — 1962. **2.** *(abbr* **atm).** A standard unit of **atmospheric pressure,** defined as that pressure exerted by a 760-millimeter column of mercury at standard gravity (980.665 centimeters per second per second) at temperature 0° C. 1 standard atmosphere = 760 millimeters of mercury
= 29.9213 inches of mercury
= 1013.250 millibars

standard conditions. standard temperature and pressure.

standard deviation *(symbol* σ*).* A measure of the **dispersion** of data points around their mean value.

standard instrument departure (SID). A preplanned coded air traffic control IFR departure routing, preprinted for pilot use in graphic and textual or textual form only.

standardization. 1. The act or process of reducing something to, or comparing it with, a **standard. 2.** A measure of uniformity. **3.** A special case of **calibration** whereby a known input is applied to a device or system for the purpose of verifying the output or adjusting the output to a desired level or scale factor. Applied to **transducers, standardization** indicates adjustment of the output to a standard value within specified limits of error.

standard pressure. 1. In meteorology, usually a pressure of 1000 millibars, but other pressures may be used as standard for specific purposes. **2.** In physics, a pressure of 1 **standard atmosphere.**

standard propagation. The **propagation** of **radio** energy over a smooth spherical earth of uniform dielectric constant and conductivity under conditons of standard **refraction** in the atmosphere, i.e., an atmosphere in which the index of refraction decreases uniformly with height at a rate of 12 N-units per 1000 feet. See **superstandard propagation, substandard propagation, standard atmosphere.** Standard propagation results in a ray curvature due to refraction which has a value approximately one-fourth that of the earth's curvature, giving a radio horizon which is about 15 percent greater than the distance to the geometrical horizon. This is equivalent to straight-line propagation over a fictitious earth whose radius is four-thirds the radius of the actual earth.

standard rate turn. A turn conducted at a constant rate of 3 degrees per second for low/medium speed airplanes and 1½ degrees per second for high speed airplanes. Standard rate turns are normally conducted during **instrument flight.** A 3 degree per second turn is also known as a **two minute turn** since a 360 degree turn requires 2 minutes. A 1½ degree per second turn is also called a **four minute turn.**

standard refraction. The **refraction** which would occur in an idealized atmosphere in which the **index of refraction** decreases uniformly with height at the rate of 39×10^{-6} per kilometer. See **standard propagation.** Standard refraction may be included in ground-wave calculations by use of an effective earth radius of 8.5×10^{-6} meters, or four-thirds the geometrical radius of the earth.

standard temperature. 1. A **temperature** that depends upon some characteristic of some substance, such as the melting, boiling, or freezing point, that is used as a reference standard of temperature. **2.** In physics, usually the ice point (0° C); less frequently, the temperature of maximum water density (4° C). **3.** In meteorology, this has no generally accepted meaning, except that it may refer to the temperature at zero altitude in the standard atmosphere (15° C).

standard temperature and pressure. Usually a temperature of 0° C but also used to designate a temperature of 15° C and 1 **standard atmosphere.** (See **standard pressure**).

standard terminal arrival route (STAR). A preplanned coded air traffic control IFR arrival routing, preprinted for pilot use graphic and textual or textual form only.

standard time. See **time.**

standby reserve. That part of the United States Air Force Reserve that is normally called to active duty only in time of war, of national emergency declared by the Congress, or when otherwise authorized by law. See also **ready reserve.**

standing operating procedure. A set of instructions covering those features of operations which lend themselves to a definite or standardized procedure without loss of effectiveness. The procedure is applicable unless prescribed otherwise in a particular case. Thus, the flexibility necessary in special situations is retained.

standing wave. A periodic **wave** having a fixed distribution in space which is the result of **interference** of progressive waves of the same frequency and kind. Such waves are characterized by the existence of **nodes** or partial nodes and **antinodes** that are fixed in space.

standoff weapon. A weapon which may be launched at a distance sufficient to allow attacking personnel to evade defensive fire from the target area.

stand talker. A person on a static **test stand** responsible for coordinating and timing the preparations for a **static test.**

star. 1. A self-luminous celestial body exclusive of nebulas, comets, and meteors; any one of the suns seen in the heavens. Distinguished from **planets** or planet **satellites** that shine by reflected light. See **navigational stars. 2.** Any luminous body seen in the heavens. The star (sense **1**) of our solar system is the sun. In sense **2.** *star* sometimes excludes the sun, the moon, and man-made satellites from the category.

STAR. Standard Terminal Arrival Route.

star catalogue. A listing of stars giving positions for a specified mean equinox and equator. Stars are often identified by catalogue numbers.

star classification. Stars are classified by their spectra, designated by letters, sometimes with numerical subdivisions, as *the sun is a G1-type star.* The seven main types with their principal spectral characteristics are, in order of decreasing temperature:

O He *II* absorption
B He *I* absorption
A H absorption
F Ca *II* absorption
G strong metallic lines
K bands developing
M very red

Also, the letters, P, W, Q, R, N, and S are used to designate comparatively rare types of stars which do not fall into the main series.

star cluster. A group of **stars** physically close together in space.

star perforated grain. A hollow rocket-propellant **grain** with the cross section of the hole having a multipointed shape.

starting pressure. In rocketry, the minimum **chamber pressure** required to establish shock-free flow in the exit plane of a supersonic **nozzle.**

star tracker. A telescopic instrument on a **rocket** or other flight borne vehicle that locks onto a **celestial body** and gives **guidance** reference to the vehicle during flight. See **astro-tracker, celestial guidance, sun tracker.**

state and regional defense airlift. The program for use during an emergency of civil aircraft other than air carrier aircraft.

state chicken. (Military) I am at a fuel state requiring recovery, tanker service, or diversion to an airfield.

state lamb. (Military) I do not have enough fuel for an intercept plus reserve required for carrier recovery.

state of the art. The level to which technology and science have at any designated cutoff time been developed in a given industry or goup of industries.

state tiger. (Military) I have sufficient fuel to complete my mission as assigned.

static. 1. Involving no variation with time. **2.** Involving no movement, as in *static test.* **3.** Any **radio interference** detectable as **noise** in the audio stage of a receiver.

static balance. The condition of a control surface in which the mass-balance is such that the center of mass lies on the hinge axis.

static conversion. Energy conversion in which no moving parts or equipment are utilized.

static firing. The firing of a **rocket engine** in a hold-down position to measure **thrust** and accomplish other tests.

static jet thrust. In jet propulsion engines; the net thrust in pounds with no translational motion at specified ambient conditions.

static line (air transport). A line attached to a parachute pack and to a strop or anchor cable in an aircraft so that when the load is dropped the parachute is deployed automatically.

static pressure *(symbol p).* **1.** The **pressure** with respect to a stationary surface tangent to the **mass-flow velocity vector. 2.** The pressure with respect to a surface at rest in relation to the surrounding fluid.

static stability. See **stability.**

static testing. The testing of a **rocket** or other device in a stationary or hold-down position, either to verify structural design criteria, structural integrity, and the effects of limit loads or to measure the **thrust** of a rocket engine.

static vent. A small aperture in a plate fixed to form part of the fuselage and located appropriately for measuring the ambient static air pressure.

station. A location where measurements are made, e.g., along an **airfoil** in a **wind tunnel** test.

station. (Military) **1.** A general term meaning any military or naval activity at a fixed land location. **2.** A particular kind of activity to which other activities or individuals may come for a specific service, often of a technical nature, e.g., aid station. **3.** An assigned or prescribed position in a naval formation or cruising disposition; or an assigned area in an approach, contact, or battle disposition. **4.** Any place of duty or post or position in the field to which an individual, or group of individuals, or a unit may be assigned. **5.** One or more transmitters or receivers or a combination of transmitters and receivers, including the accessory equipment necessary at one location, for carrying on radio communication service.

stationary front (meteorology). A **front** where neither air mass is replacing the other. See **warm front, cold front, occluded front.**

stationary orbit. An orbit in which the **satellite** revolves about the **primary** at the angular rate at which the primary rotates on its axis. From the primary, the satellite thus appears to be stationary over a point on the primary. A stationary orbit with respect to the earth is commonly called a *24-hour orbit.*

station keeping. The sequence of maneuvers that maintains a **vehicle** in a predetermined **orbit.**

station pressure. The **atmospheric pressure** computed for the level of the station elevation. This may or may not be the same as either the climatological station pressure or the actual pressure, the difference being attributable to the difference in reference elevations. Station pressure usually is the base value from which sea-level pressure and altimeter setting are determined.

stator. In machinery, a part or assembly that remains stationary with respect to a rotating or moving part or assembly such as the field frame of an electric motor or generator, or the stationary casing and blades surrounding an axial-flow-compressor rotor or turbine wheel; a stator blade.

statute mile. 5280 feet = 1.6093 **kilometers** = 0.869 **nautical mile.** Also called *land mile.*

steady. (Military) Am on prescribed heading, or straighten out immediately on present heading or heading indicated.

steady flow. A **flow** whose velocity vector components at any point in the **fluid** do not vary with time. See **streamline flow.**

steady state. 1. The condition of a substance or system whose local physical and chemical properties do not vary with time. **2.** Specifically, the stable operating condition of a **reactor** in which the neutron inventory remains constant.

steerable antenna. A **directional antenna** whose major **lobe** can be readily shifted in direction.

steering function. An empirical relation based on the **relative** distance and velocity of the target, used in **guidance** of rockets and spacecraft.

Stefan-Boltzmann law. One of the **radiation laws** which states that the amount of energy radiated per unit time from a unit surface area of an ideal **black body** is proportional to the fourth power of the **absolute temperature** of the black body. The law is written: $E = \sigma T^4$ where E is the emittance of the black body; σ is the Stefan-Boltzmann constant; and T is the absolute temperature of the black body. Also called *Stefan law.* This law was established experimentally by Stefan and was given theoretical support by thermodynamic reasoning due to Boltzmann. This law may be deduced by integrating **Planck law** over the entire frequency spectrum.

stellar. Of or pertaining to **stars.**

stellar guidance. Same as **celestial guidance.**

stellar inertial guidance. The guidance of a flight-borne vehicle by a combination of **celestial** and **inertial guidance;** the equipment which accomplishes the guidance.

stellar map matching. A process during the flight of a **vehicle** by which a chart of the stars set into the **guidance** system is automatically matched with the position of the stars observed through telescopes so as to give guidance to the vehicle. See **mapmatching guidance.**

St. Elmo's fire. Same as **corona discharge.** Also see **Saint Elmo's fire.**

step. A "break" in the bottom of a float of a seaplane's hull to improve planing characteristics on the water.

step-climb profile. The aircraft climbs a specified number of feet whenever its weight reaches a predetermined amount, thus stepping to an optimum altitude as gross weight decreases.

stereochemistry. Chemistry dealing with the arrangement of atoms and molecules in three dimensions.

stern attack. In air interception, an attack by an interceptor aircraft which terminates with a heading crossing angle of 45 degrees or less. See also **heading crossing angle.**

sternward acceleration. See **physiological acceleration.**

steward/stewardess. A member of the cabin crew responsible for the feeding, general comfort and safety of the passengers.

stick. 1. A succession of missiles fired or released separately at predetermined intervals from a single aircraft. **2.** A number of parachutists who jump from one aperture or door of an aircraft during one run over a dropping zone.

stick commander (air transport). A designated individual who controls parachutists from the time they enter the aircraft until their exit.

stiffener. A member attached to a sheet to restrain its movement normal to the surface.

stiffness. The ratio of change of force (or torque) to the corresponding change in translational (or rotational) displacement of an elastic element.

stimulus. 1. Same as **excitation. 2.** Same as **measurand.**

Stirling cycle. A theoretical **heat engine** cycle in which heat is added at constant volume, followed by **isothermal** expansion with heat addition. The heat is then rejected at constant volume, followed by isothermal compression with heat rejection. If a regenerator is used so that heat rejected during the constant-volume process is recovered during heat addition at constant volume, the **thermal efficiency** of the Stirling cycle is the same as for the **Carnot cycle** with less compressive work needed.

stochastic. Conjectural; in statistical analysis, = **random.**

stoichiometric. Of a mixture of chemicals, having the exact proportions required for complete chemical combination, applied especially to combustible mixtures used as **propellants.**

STOL. See **Short Take-Off and Landing.**

stop squawk. Turn **Identification Friend or Foe** master control to "off".

stopway. An area beyond the take-off runway, no less wide than the runway and centered upon the extended centerline of the runway, able to support the airplane during an aborted takeoff, without causing structural damage to the airplane, and designated by the airport authorities for use in decelerating the airplane during an aborted takeoff.

storable. Of a liquid; subject to being placed and kept in a tank without benefit or special measures for temperature or pressure control, in *storable propellant.*

storage. 1. The act of storing **information.** See **store. 2.** Any device in which information can be stored. Also called a *memory device.* **3.** In a **computer,** a section used primarily for storing information. Such a section is sometimes called a *memory* or a *store.* The physical means of storing information may be electrostatic, ferroelectric, mangetic, acoustic, optical, chemical, electronic, electrical, mechanical, etc., in nature.

storage capacity. The amount of **information,** usually expressed in **bits** (i.e., the $\log_2$ of the number of distinguishable states in which the storage can exist), that can be retained in **storage.** Also called *memory capacity.*

store. 1. To retain information in a device from which it can later be withdrawn. **2.** To introduce information into such a device. **3.** A container, rocket, bomb, or vehicle carried externally in a craft.

strafing. The delivery of automatic weapons fire by aircraft on ground targets.

straight-in approach—IFR. An instrument approach wherein final approach is begun without first having executed procedure turn.

straight-in approach—VFR. Entry of the traffic pattern by interception of the extended runway centerline without executing any other portion of the traffic pattern.

strain. The deformation produced by a **stress** divided by the original dimension.

strain gage. An instrument used to measure the **strain** or distortion in a member or test specimen (such as a structural part) subjected to a **force.**

stranger (bearing, distance, altitude) (Military) An unidentified aircraft, bearing, distance, and altitude as indicated relative to you.

strangle. (Military) A code meaning switch off equipment indicated.

strangle parrot. (Military) A code meaning switch off Identification Friend or Foe equipment.

Strategic Air Command (SAC). See **major command.**

strategic airlift. The continuous, sustained air movement of units, personnel, and material in support of all Department of Defense agencies between area commands; the Contintental United States and overseas areas; and within an area command when directed. Strategic airlift resources possess a capability to airland, or airdrop troops, supplies and equipment for augmentation of tactical forces when required

strategic air transport. The movement of personnel and materiel by air in accordance with a strategic plan.

strategic air warfare. Air combat and supporting operations designed to effect, through the systematic application of force to a selected series of vital targets, the progressive destruction and disintegration of the enemy's war-making capacity to a point where he no longer retains the ability or the will to wage war. Vital targets may include key manufacturing systems, sources of raw material, critical material, stockpiles, power systems, transportation systems, communication facilities, concentrations of uncommitted elements of enemy armed forces, key agricultural areas, and other such target systems.

strategic attack. An attack by means of aerospace forces directed at selected vital targets of an enemy nation so as to destroy its war-making capacity or will to fight.

strategic material (critical). A material required for essential uses in a war emergency, the procurement of which is adequate quantity, quality, or time, is sufficiently uncertain, for any reason, to require prior provision of the supply thereof.

strategic mission. A mission directed against one or more of a selected series of enemy targets with the purpose of progressive destruction and disintegration of the enemy's war-making capacity and his will to make war. Targets include key manufacturing systems, sources of raw material, critical material, stockpiles, power systems, transportation systems, communication facilities, and other such target systems. As opposed to tactical operations, strategic operations are designed to have a long-range rather than immediate, effect on the enemy and his military forces.

strategic nuclear weapon. A nuclear weapon which is programmed primarily for use against strategic targets in strategic nuclear war.

strategic transport aircraft. Aircraft designed primarily for the carriage of personnel and/or cargo over long distances.

strategic warning. A notification that enemy-initiated hostilities may be imminent. This notification may be received from minutes to hours, to days, or longer, prior to the initiation of hostilities.

strategy. The art and science of developing and using political, economic, psychological, and military forces as necessary during peace and war, to afford the maximum support to policies, in order to increase the probabilities and favorable consequences of victory and to lessen the chances of defeat.

stratiform clouds. Stratiform clouds are formed by cooling air in stable layers such as advection cooling whereby moist air moves over a colder surface. However, stratiform clouds are sometimes formed during warm front conditions whereby the air is lifted so slowly that clouds form in stable, stratiform layers. Stratiform clouds occur in sheets or layers and often cover the entire sky. Although these cloud layers have some thickness, they do not have the vertical development of cumuliform clouds. Stratiform clouds are indicative of stable air not associated with the extreme vertical lifting of unstable air forming cumuliform clouds. Precipitation from stratiform clouds is generally in the form of light continuous rain, drizzle, or snow. While flying conditions are usually smooth, the extensive precipitation areas can produce low visibility and possibly IFR conditions even beneath the cloud bases. Also see **cloud.**

stratocumulus. A form of **low cloud.** See also **cloud, cumuliform clouds, stratiform clouds.**

stratosphere. The layer of the atmosphere above the troposphere in which the change of temperature with height is relatively small. See also **atmospheric shell.**

stratus. A form of **low cloud.** Also see **cloud, stratiform clouds.**

streamline. A line whose tangent at any point in a **fluid** is parallel to the instantaneous velocity vector of the fluid at that point. In steady-state flow the streamlines coincide with the trajectories of the fluid particles; otherwise, the streamline pattern changes with time. See **free streamline.** Compare **trajectory.**

streamline flow. Same as **laminar flow.**

stream takeoff. Aircraft taking off in trail/column formation.

stress. 1. The **force** per unit area of a body that tends to produce a deformation. **2.** The effect of a physiological, psychological, or mental load on a biological organism which causes fatigue and tends to degrade proficiency.

stress concentration. In structures, a localized area of high **stress.** See **stress raiser.**

stress corrosion. Chemical corrosion that is accelerated by stress concentrations, either built into or resulting from a load.

stressed-skin structure. A structure covered with sheet which contributes substantially to its strength and stiffness.

stress raisers. Changes in contour or discontinuities in a structure that cause local increases in **stress.**

stress ratio. The ratio of the minumum **stress** to the maximum stress occurring in one **stress cycle.**

stretchout. An action whereby the time for completing an action, especially a contract, is extended beyond the time originally programed or contracted for.

strike. An attack which is intended to inflict damage on, seize or destroy an objective.

strike control and reconnaissance. A United States Air Force mission flown for the primary purpose of acquiring and reporting air interdiction targets and controlling air strikes against such targets.

strike force. A force composed of appropriate units necessary to conduct strikes, attack or assult operations. See also **task force.**

strike photography. Air photographs taken during an air attack.

stringer. A slender, lightweight, lengthwise fill-in structural member in a rocket body, or the like, serving to reinforce and give shape to the skin.

structural weight. Same as **construction weight.**

strut. A compression or tension member in a truss structure. In airplanes, usually applied to an external major structural member.

student pilot. A student pilot is a person authorized by the **FAA** to pilot an aircraft under the supervision of an appropriately rated **flight instructor.** A student pilot can-

not carry passengers and cannot accept payment for his services. See **student pilot certificate.**

student pilot certificate. A certificate of competency issued by the **FAA** to a person meeting the requirements of the applicable **Federal Aviation Regulations.** See **student pilot, airmen certificates.**

subassembly. An **assembly** that is a component part of a larger assembly.

subaudio frequency. A **frequency** below the **audiofrequency range,** below about 15 cycles per second.

subcarrier. A **carrier** which is applied as a **modulating wave** to modulate another carrier or an intermediate subcarrier.

subcarrier oscillator. In a **telemetry** system, the **oscillator** which is directly modulated by the **measurand** or by the equivalent of the measurand in terms of changes in the transfer elements of a **transducer.**

subchannel. In a **telemetry** system, the route required to convey the **magnitude** of a single subcommutated **measurand.**

subcommutation. In telemetry, **commutation** of additional **channels** with output applied to individual channels of the primary **commutator.** Subcommutation is called *synchronous* if its rate is a submultiple of that of the primary commutator. Unique identification must be provided for the subcommutation frame pulse.

subcritical mass. Nucleonics. An amount of fissionable material insufficient in quantity or of improper geometry to sustain a fission chain reaction. (See **critical mass, criticality.**)

subgravity. A condition in which the resultant ambient acceleration is between 0 and one G.

subkiloton weapon. A nuclear weapon producing a yield below one kiloton. See also **kiloton weapon; nominal weapon.**

subframe. In telemetry, a complete sequence of **frames** during which all subchannels of a specific **channel** are sampled once.

sublimation. The transition of a substance directly from the solid state to the vapor state, or vice versa, without passing through the intermediate liquid state. See **condensation, evaporation.**

submarine launched missile. A sea-launched missile launched from a submarine or surface ship.

subroutine. A set of instructions necessary to direct a computer to carry out a well-defined mathematical or logical operation; a subunit of a **routine,** usually coded in such a manner that it can be treated as a **black box** by the routine using it.

subsidence. (Metgy) An extensive sinking motion of air, most frequently occurring in polar highs. The subsiding air is warmed by compression and becomes more stable.

subsolar point. The **georgraphical position** of the sun; that point on the earth at which the sun is in the **zenith** at a specified time.

subsonic. In aerodynamics, of or pertaining to, or dealing with speeds less than **acoustic velocity** as in *subsonic aerodynamics.*

subsonic flow. Flow of a **fluid,** as air over an airfoil, at speeds less than **acoustic velocity.** Aerodynamic problems of subsonic flow are treated with the assumption that air acts as an incompressible fluid.

substandard propagation. The **propagation** of radio energy under conditions of substandard refraction in the atmosphere, that is, refraction by an atmosphere or

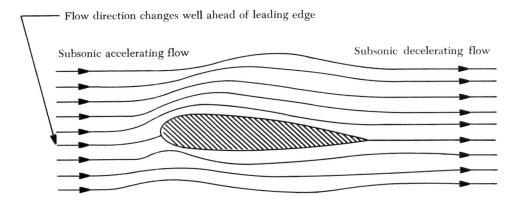

Flow direction changes well ahead of leading edge

Subsonic accelerating flow

Subsonic decelerating flow

Typical subsonic flow pattern, subsonic wing.

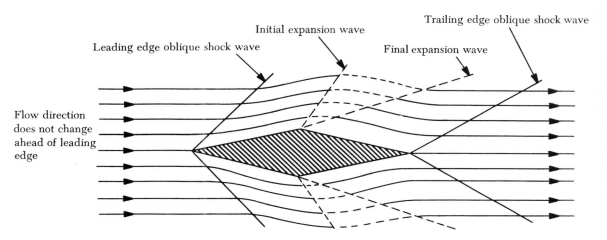

Trailing edge oblique shock wave

Leading edge oblique shock wave

Initial expansion wave

Final expansion wave

Flow direction does not change ahead of leading edge

Typical supersonic flow pattern, supersonic wing.

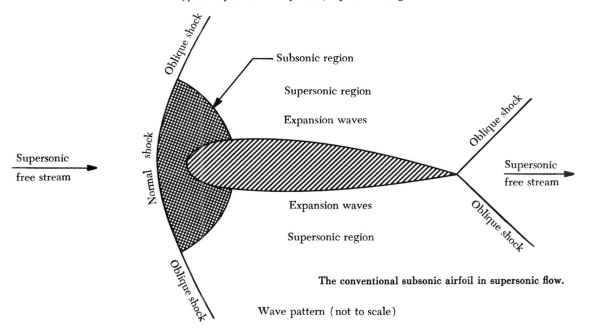

Oblique shock

Subsonic region

Supersonic region

Expansion waves

Normal shock

Supersonic free stream

Oblique shock

Oblique shock

Supersonic free stream

Expansion waves

Supersonic region

The conventional subsonic airfoil in supersonic flow.

Oblique shock

Oblique shock

Wave pattern (not to scale)

Supersonic flow *patterns relative to airfoils.*

section of the atmosphere in which the **index of refraction** decreases with height at a rate of less than 12 N-units per 1000 feet. See **standard propagation, superstandard propagation.** Substandard propagation produces a less-than-normal downward bending or even an upward bending of radio waves as they travel through the atmosphere, giving closer radio horizons and decreased radar and radio coverage. It results primarily when propagation takes place through a layer in which moisture remains constant or increases with height.

substrate (of a microcircuit or integrated circuit). The supporting material upon or within which the elements of a microcircuit or integrated circuit are fabricated or attached.

suction flap. A flap whose effectiveness is increased by boundary-layer suction.

sudden ionospheric disturbance. A complex combination of sudden changes in the condition of the **ionosphere** and the effects of these changes. A sudden ionospheric disturbance usually occurs a few minutes after a solar flare and is noted only on the sunlit side of the earth. The return of the ionosphere to its *normal* condition following a pronounced sudden ionospheric disturbance usually takes from half an hour to an hour, sometimes longer.

SUM. Surface-to-Underwater Missile.

summer solstice. 1. That point on the **ecliptic** occupied by the sun at maximum northerly declination. Sometimes called *June solstice, first point of Cancer.* **2.** That instant at which the sun reaches the point of maximum northerly **declination**, about June 21.

sun. The star at the center of the **solar system,** around which the planets, planetoids, and comets revolve. It is a G-type star. The sun visible in the sky is called *apparent* or *true* sun. A fictitious sun conceived to move eastward along the celestial equator at a rate that provides a uniform measure of time equal to the average apparent time is called *mean sun;* a fictitious sun conceived to move eastward along the ecliptic at the average rate of the apparent sun is called *dynamical mean sun.*

sunrise. The crossing of the visible **horizon** by the **upper limb** of the ascending sun.

sunset. The crossing of the visible **horizon** by the **upper limb** of the descending sun.

sunspot. A relatively dark area on the surface of the sun consisting of a dark central umbra surrounded by a penumbra which is intermediate in brightness between the umbra and the surrounding **photosphere.** Sunspots usually occur in pairs with opposite magnetic polarities. They have a lifetime ranging from a few days to several months. Their occurrence exhibits approximately an 11-year period (the **sunspot cycle**).

sunspot cycle. A cycle with an average length of 11.1 years but varying between about 7 and 17 years in the number and area of **sunspots,** as given by the **relative sunspot number.** This number rises from a minimum of 0 to 10 to a maximum of 50 to 140 about 4 years later, and then declines more slowly. An approximate 11-year cycle has been found or suggested in geomagnetism, frequency of aurora, and other ionospheric characteristics. The u-index of geomagnetic intensity variation shows one of the strongest known correlations to solar activity. Eleven-year cycles have been suggested for various tropospheric phenomena, but none of these has been substantiated.

sun tracker. A species of **star tracker** designed to lock onto the sun to afford **guidance** to a rocket or other flight-borne object. See **star tracker.**

super high frequency *(abbr* **SHF).** See **frequency bands.**

superadiabatic lapse rate. An environmental **lapse rate** greater than the **dry-adiabatic lapse rate,** such that potential temperature decreases with height.

supercharged engine. An engine in which the charge pressure in the induction system may be increased by mechanical means above that produced by normal aspiration.

supercharger. A compressor used to increase the density of the air or mixture supplied to an engine. Normally driven either by the engine or by an exhaust turbine.

supercritical mass. Nucleonics. A mass of fuel whose effective multiplication factor is greater than one. (See **critical mass.)**

supercritical reactor. Nucleonics. A reactor in which the effective multiplication factor is greater than one; consequently a reactor that is increasing its power level. If uncontrolled, a supercritical reactor would undergo an excursion. (See **criticality, excursion.)**

supercritical wing. The supercritical wing employs a supercritical airfoil section with a relatively flat upper surface which develops a weak shock wave at transonic speeds in contrast to the very strong upper surface shock associated with the more conventional airfoil. A substantial increase in drag rise mach number results from use of a supercritical wing.

superior planets. The **planets** with **orbits** larger than that of the earth; Mars, Jupiter, Saturn, Uranus, Neptune, and Pluto.

supersonic. Of or pertaining to, or dealing with, speeds greater than the **acoustic velocity.** Compare with **ultrasonic.**

supersonic compressor. A **compressor** in which a **supersonic** velocity is imparted to the fluid relative to the **rotor** blades, the **stator** blades, or to both the rotor and stator blades, producing oblique **shock waves** over the blades to obtain a high pressure rise.

supersonic diffuser. A **diffuser** designed to reduce the velocity and increase the pressure of **fluid** moving at **supersonic** velocities.

supersonic flow. In aerodynamics, flow of a **fluid** over a body at speeds greater than the **acoustic velocity** and in which the **shock waves** start at the surface of the body. Compare **hypersonic flow.**

supersonic nozzle. A converging-diverging **nozzle** designed to accelerate a **fluid** to **supersonic** speed.

supersonics. Specifically, the study of **aerodynamics** of **supersonic** speeds. See **hypersonics.**

superstandard propagation. The **propagation** of radio waves under conditions of super-standard refraction (superrefraction) in the atmosphere, that is, **refraction** by an atmosphere or section of the atmosphere in which the **index of refraction** decreases with height at a rate of greater than 12 N-units per 1000 feet. See **standard propagation, substandard propagation.** Superstandard propagation produces a greater-than-normal downward bending of radio waves as they travel through the atmosphere, giving extended radio horizons and increased radar coverage. It results primarily from propagation through layers near the earth's surface in which the moisture lapse rate is greater than normal, or the temperature lapse rate less than normal, or both. A condition in which warm dry air moves out over a cool water surface is an example of superrefraction. A layer in which the downward bending is greater than the curvature of the earth is called a *radio duct*. Frequently, the general term, *anomalous propagation*, is used for superstandard propagation.

supplementary angles. Two angles whose sum is 180°.

support. (Military) **1.** The action of a force which aids, protects, complements, or sustains another force in accordance with a directive requiring such action. **2.** A unit which helps another unit in battle. Aviation, artillery, or naval gunfire may be used as a support for infantry. **3.** A part of any unit held back at the beginning of an attack as a reserve. **4.** An element of a command that assists, protects, or supplies other forces in combat.

support equipment. See **ground support equipment.**

supporting aircraft. (Military) All active aircraft other than unit aircraft. See also **aircraft.**

supporting arms. Air, sea, and land weapons of all types employed to support ground units.

supporting surfaces. Surfaces the primary function of which is to provide lift for an aircraft.

surface. 1. A two-dimensional extent; the outside or superficies of any body; especially, the surface of the earth, either land or water, used in combinations as surface-to-air, etc. **2.** A wing, rudder, propeller blade, vane, hydrofoil, or the like—applied in this sense to the entire structure or body.

surface boundary layer. That thin layer of air adjacent to the earth's surface, extending up to the so-called anemometer level (the base of the **Ekman layer**). Within this layer the wind distribution is determined largely by the vertical temperature gradient and the nature and contours of the underlying surface; shearing stresses are approximately constant. Also called *surface layer, friction layer, atmospheric boundary layer, ground layer*. See **planetary boundary layer, free atmosphere.**

surface-to-air. (SAM). A surface-launched missile designed to operate against a target above the surface.

surface-to-air missile installation. A surface-to-air missile site with the surface-to-air missile system hardware installed.

surface-to-surface missile. (SSM) A surface-launched missile designed to operate against a target on the surface.

surge. A transient rise in power, pressure, etc., such as a brief rise in the discharge pressure of a rotary compressor.

surveillance. The systematic observation of aerospace, surface, or subsurface areas, places, persons, or things by visual, aural, electronic, photographic, or other means. See also **air surveillance.**

surveillance approach. An instrument approach conducted in accordance with directions issued by a controller referring to the surveillance radar display.

survey. The process of determining accurately the position, extent, contour, etc., of an area, usually for the purpose of preparing a chart.

survey meter. Nucleonics. Any portable *radiation detection instrument* especially adapted for surveying or inspecting an area to establish the existence and amount of radioactive material present. (Compare **counter,** **monitor.**)

survivability. The capability of a system to withstand a man-made hostile environment without suffering an abortive impairment of its ability to accomplish its designated mission.

survival radio equipment. A self-buoyant, water resistant, portable emergency radio signaling device which operates from its own power source on 121.5 and/or 243 MHz, perferably on both emergency frequencies, transmitting a distinctive downward swept audio tone for homing purposes, which may or may not have voice capability, and which is capable of operation by unskilled persons. This type equipment is agreed upon internationally for extended overwater operations and is presently required for all carriers engaged in extended overwater operations.

suspension. In physical chemistry, a **system** composed of one substance (suspended phase, suspensoid) dispersed throughout another substance (suspending phase) in a moderately finely divided state, but not so finely divided as to acquire the stability of a **colloidal system.** Given sufficient time, a suspension will, by definition, separate itself by gravitational action into two visibly distinct portions, whereas a colloidal system, by definition, is stable. Dust in the atmosphere is an example of a suspension of a solid in a gas.

suspensoid. See **suspension.**

sustainer engine. A **rocket engine** that maintains the velocity of a rocket vehicle once it has achieved its programed velocity by use of **booster** or other engine. This term is applied, for example, to the remaining engine of the Atlas after the two booster engines have been jettisioned. The term is also applied to a rocket engine used on an orbital glider to provide the small amount of thrust now and then required to compensate for the drag imparted by air particles in the upper atmosphere.

sweat cooling. Same as **transpiration cooling.**

sweep. The motion of the visible dot across the face of a **cathode-ray tube,** as a result of deflections of the **electron beam.**

sweepback. A wing design in which the leading edge (and sometimes the trailing edge) slope in plan form, is such that the wing tips are further aft than the wing root.

swing-around trajectory. A planetary round trip **trajectory** which requires no propulsion at the destination planet, but uses the planet's **gravitational** field to effect the necessary orbit change to return to earth.

swinging the compass. Checking the indications of an installed compass by comparing them with an accurate compass rose laid out on the ground. Also see **compass compensation.**

swing-wing aircraft. See **variable geometry aircraft.**

synchronous. Coincident in time, phase, rate, etc.

synchronous orbit. 1. An orbit in which a satellite makes a limited number of equatorial crossing points which are then repeated in synchronism with some defined reference (usually earth or sun). **2.** This term in common usage, is applied to the equatorial, circular, 24 hour case in which the satellite appears to hover over a specific point of the earth.

synchronous satellite. An equatorial west-to-east **satellite** orbiting the earth at an altitude of approximately 35,-900 kilometers (22,300 miles) at which altitude it makes one **revolution** in 24 hours, synchronous with the earth's **rotation.**

synergic curve. A curve plotted for the ascent of a **rocket vehicle** calculated to give the vehicle an optimum economy in fuel with an optimum velocity. This curve, plotted to minimize air resistance, starts off vertically, but bends towards the horizontal between 20 and 60 miles altitude to minimize the thrust required for vertical ascent.

synodical month. The average period of **revolution** of the moon about the earth with respect to the sun, a period of 29 days 12 hours 44 minutes 2.8 seconds. This is sometimes called the *month of the phases,* since it extends from new moon to the next new moon. Also called *lunation.*

synoptic. Pertaining to or affording an overall view. In meteorology, this term refers to meteorological data obtained simultaneously over a wide area for the purpose of presenting a comprehensive and nearly instantaneous picture of the state of the atmosphere. Thus, to a meteorologist, *synoptic* takes on the additional connotation of simultaneity.

synoptic meteorology. The study and analysis of weather information gathered at the same time. See **synoptic.**

synoptic weather chart. A chart showing the weather conditions prevailing at a given time over a wide area.

system. **1.** Any organized arrangement in which each component part acts, reacts, or interacts in accordance with an overall design inherent in the arrangement. **2.** Specifically, a major component of a given **vehicle** such as a *propulsion system* or a *guidance system.* Usually called a *major system* to distinguish it from the systems subordinate or auxiliary to it. The system of sense **1** may become organized by a process of evolution, as in the solar system, or by deliberate action imposed by the designer, as in a missile system or an electrical system. In sense **2,** the system embraces all its own subsystems including checkout equipment, servicing equipment, and associated technicians and attendants. When the term is proceded by such designating nouns as *propulsion* or *guidance,* it clearly refers to a major component of the missile. Without the designating noun, the term may become ambiguous. When modified by the word *major,* however, it loses its ambiguity and refers to a major component of the missile.

system management. A concept for the technical and business management of a particular system program based on the use of a designated centralized management authority who is responsible for planning, directing, and controlling the definition, development, production and deployment of a system, and for assuring that planning is accomplished by the support, personnel training, operational testing, activiation, or deployment. The centralized management authority is supported by functional organizations, which are responsible to the centralized management authority for the execution of specifically assigned system tasks.

system/project. Equipment and/or skills together with any related facilities, services, information, and techniques, that form a complex or an entity capable of performing specific operational tasks in support of an identifiable objective.

system/project management. A concept for the technical and business management of particular systems/projects based on the use of a designated, centralized management authority who is responsible for planning, directing, and controlling the definition, development, and production of a system/project: and for assuring that planning is accomplished by the organizations responsible for the complementary functions of logistic and maintenance support, personnel training, operational testing, activation, or deployment. The centralized management authority is supported by functional organizations, which are responsible to the centralized management authority for the execution of specifically assigned system/project tasks.

systems engineering. The process of applying science and technology to the study and planning of an overall aerospace vehicle system, whereby the relationships of various parts of the system and the utilization of various subsystems are fully planned and comprehended prior to the time that hardware designs are committed.

T

T. Military mission designation for **trainer** aircraft.

tab. A small auxiliary airfoil, usually attached to a movable control surface to aid in its movement, or to effect a slight displacement of it for the purpose of **trimming** the airplane for varying conditions of power, load, or airspeed.

TAC. Tactical Air Command.

Tacan *(abbr). tactical air navigation.*

tachometer. An instrument that measures and indicates the revolutions per unit of time of a rotating mechanism.

Tactical Air Command. 1. An Air Force organization designed to conduct offensive and defensive air operations in conjunction with land or sea forces. **2.** A designation of one of the subordinate commands of the Air Force. See **major command.**

tactical air commander. See **tactical air force.**

tactical air control center. The principal air operations installation (land or ship-based) from which all aircraft and air warning functions of tactical air operations are controlled.

tactical air control group. 1. (land-based)— A flexible administrative and tactical component of a tactical air organization which provides aircraft control and warning functions ashore for offensive and defensive missions within the tactical air zone of responsibility. **2.** (ship-based)— An administrative and tactical component of an amphibious force which provides aircraft control and warning facilities afloat for offensive and defensive missions within the tactical air command area of responsibility.

tactical air control squadron. 1. (land-based) — A flexible administrative component of a tactical air control group, known as tacron, which provides the control mechanism for a land-based tactical air control center, a tactical air direction center, or tactical air control parties. **2.** (ship-based) — An administrative and tactical air control group, known as tacron, which provides the control mechanism for the shipbased tactical air direction center or the ship-based tactical air control center.

tactical air coordinator (airborne). An officer who coordinates, from an aircraft, the action of combat aircraft engaged in close support of ground or sea forces.

tactical air direction center. An air operations installation under the overall control of the tactical air control center (afloat)/tactical air command center, from which aircraft and air warning service functions of tactical air operations in an area of responsibility are directed. See also **tactical air director.**

tactical air director. The officer in charge of all operations of the tactical air direction center. He is responsible to the tactical air controller for the direction of all aircraft and air warning facilities assigned to his area of responsibility. When operating independently of a tactical air control center (afloat), the tactical air director assumes the functions of the tactical air controller. See also **fighter director; tactical air direction center.**

tactical air force. An air force charged with carrying out tactical air operations in coordination with ground or naval forces.

tactical air navigation (Tacan). A military navigation aid which provides distance and direction information to appropriately equipped aircraft. Tacan ground equipment consists of either a fixed or mobile transmitting unit. The airborne unit in conjunction with the ground unit reduces the transmitted signal to a visual presentation of both azimuth and distance information. Tacan is a pulse system and operates in the UHF band of frequencies. Its use requires Tacan airborne equipment and does not operate through conventional VOR equipment.

tactical air observer. An officer trained as an air observer whose function is to observe from airborne aircraft and report on movement and disposition of friendly and enemy forces, on terrain, weather, and hydrography and to execute other missions as directed.

tactical air operations. An air operation involving the employment of air power in coordination with ground or naval forces to: **a.** gain and maintain air superiority; **b.** prevent movement of enemy forces into and within the objective area and to seek out and destroy these forces and their supporting installations; and **c.** join with ground or naval forces in operations within the objective area, in order to assist directly in attainment of their immediate objective.

tactical air reconnaissance. The use of air vehicles to obtain information concerning terrain, weather, and the disposition, composition, movement, installations, lines of communications, electronic and communication emissions of enemy forces. Also included are artillery and naval gunfire adjustment, and systematic and random observation of ground battle area, targets, and/or sector of airspace.

tactical air support. Air operations carried out in coordination with surface forces which directly assist the land or naval battle. See also **air support.**

tactical air transport. The use of air transport in direct support of: **a.** airborne assults; **b.** carriage of air-transported forces; **c.** tactical air supply; **d.** evacuation of casualties from forward airdromes; and **e.** clandestine operations.

tactical aircraft shelters. A shelter to house fighter-type aircraft and to provide protection to the aircraft from attack by conventional weapons, or damage from high winds or other elemental hazards. Levels of protection may be extended to include protection against nuclear blast effects, radiation penetration, chemical, or biological warfare.

tactical airlift. That airlift which provides the immediate and responsive air movement and delivery of combat troops and supplies directly into objective areas through air landing, extraction, airdrop, or other delivery techniques; and the air logistic support of all theater forces, including those engaged in combat operations, to meet specific theater objectives and requirements.

tactical electronic warfare. That application of electronic warfare to tactical air operations. Tactical electronic warfare encompasses the three major subdivisions of electronic warfare: electronic warfare support measures, electronic countermeasures, and electronic counter-countermeasures.

tactical nuclear weapon employment. The use of nuclear weapons by land, sea, or air forces against opposing forces, supporting installations or facilities, in support of operations which contribute to the accomplishment of a military mission of limited scope, or in support of the military commander's scheme of maneuver, usually limited to the area of military operations.

tactical transport aircraft. Aircraft designed primarily for the carriage of personnel and/or cargo over short or medium distances.

tactics. 1. The employment of units in combat. **2.** The ordered arrangement and maneuver of units in relation to each other and/or to the enemy in order to utilize their full potentialities.

tail. 1. The rear part of a body, as of an aircraft, a rocket, etc. **2.** The tail surfaces of an aircraft or rocket.

tail booms. A cantilever carrying the tail unit of an aircraft in which the fuselage does not perform this function.

tail fin. A **fin** at the rear of a **rocket** or other body.

tailheavy. A condition of trim in an airplane in which the tail tends to sink when the elevator control is released.

tail plane. An airfoil fixed, movable or adjustable in flight, located aft of the main plane, contributing to longitudinal control and/or stability. See **stabilizer, empennage.**

tail unit. The combination of stabilizing and controlling surfaces situated at the rear of an aircraft. See **stabilizer, empennage.**

tailless aircraft. An aircraft with its longitudinal control and stabilizing surfaces incorporated in the main plane.

tailspin. Same as **spin.**

tailward acceleration. See **physiological acceleration.**

tailwheel. A wheel located at the aft end of the fuselage and aft of the main landing gear. Airplanes with such wheels are referred to as tail wheel-type airplanes. Also see **tricycle landing gear.**

takeoff. 1. The action of a **rocket vehicle** departing from its **launch pad.** See **liftoff. 2.** The action of an aircraft as it becomes airborne. **3.** To perform the action of a takeoff Said of a rocket vehicle or aircraft.

takeoff power. 1. With respect to reciprocating engines, means the brake horsepower that is developed under standard sea level conditions, and under the maximum conditions of crankshaft rotational speed and engine manifold pressure approved for the normal takeoff, and limited in continuous use to the period of time shown in the approved engine specification; and **2.** With respect to turbine engines, means the brake horsepower that is developed under static conditions at a specified altitude and atmospheric temperature, and under the maximum conditions of rotorshaft rotational speed and gas temperature approved for the normal takeoff, and limited in continuous use to the period of time shown in the approved engine specification.

takeoff power rating (Eng.). The brake horsepower developed under standard sea level conditions, under the maximum conditions of crankshaft rotational speed and engine manifold pressure approved for use in normal takeoff.

takeoff thrust. With respect to turbine engines, means the jet thrust that is developed under static conditions at a specific altitude and atmospheric temperature under the maximum conditions of rotorshaft rotational speed and gas temperature approved for the normal takeoff, and limited in continuous use to the period of time shown in the approved engine specification.

takeoff weight. The weight of an aircraft or rocket vehicle ready for takeoff. (This weight includes the weight of the vehicle, the fuel, and the payload).

tally ho. (Military) Target visually sighted (presumably the target I have been ordered to intercept.) This should be followed by initial contact report as soon as possible. The sighting should be amplified if possible (e.g., "tally-ho pounce," or "tally-ho heads up").

tangential acceleration. The **acceleration** acting at the periphery of a system rotating about an axis.

tangential wave path. For a direct **radio wave,** that path of **propagation** tangential to the surface of the earth. This path is curved slightly by **atmosphere refraction.**

tank. 1. A container incorporated into the structure of a **liquid propellant rocket** from which a liquid **propellant** or propellants are fed into the firing chamber or chambers. **2.** A container for storage of liquid oxygen, liquid fuel, or other liquid propellant until transferred to the rocket's tanks or some other receptacle. **3.** In **computers,** a container of mercury, or other liquid, and associated components used as **delay-line** storage.

tankage. Of a **liquid propellant rocket,** the aggregate of the **tanks** carried by the rocket.

tanker (K), (Military designation). Aircraft having special equipment to provide in-flight refueling of other aircraft. See **military aircraft types.**

tapered wing. A wing in which there is progressive decrease in the chord length from root to tip.

target. 1. Any object, point, etc., toward which something is directed. **2.** An object which reflects a sufficient amount of a radiated signal to produce an **echo** signal on detection equipment. See **radar target.**

target. (Military) 1. A geographical area, complex, or installation planned for capture or destruction by military forces. **2.** In intelligence usage — A country, area, installation, agency, or person against which intelligence operations are directed. **3.** An area designated and numbered for future firing. **4.** In gunfire support usage — An impact burst which hits the target.

target. Nucleonics. Material subjected to particle bombardment (as in an *accelerator*) or irradiation (as in a research reactor) in order to induce a *nuclear reaction;* also a nuclide that has been bombarded or irradiated.

target acquisition. The process of optically, manually mechanically, or electronically orienting **tracking** system in direction and range to **lock on** a target.

target combat air patrol. A patrol of fighters maintained over an enemy target area to destroy enemy aircraft and to cover friendly shipping in the vicinity of the target area in amphibious operations. See also **combat air patrol.**

target discrimination. The ability of a surveillance or guidance system to identify or engage any one target when multiple targets are present.

target of opportunity. 1. A target visible to a surface or air sensor or observer, which is within range of available weapons and against which fire has not been scheduled or requested. **2.** (nuclear) A nuclear target observed or detected after an operation begins that has not been previously considered, analyzed or planned for a nuclear strike. Generally fleeting in nature, it should be attacked as soon as possible within the time limitations imposed for coordination and warning of friendly troops and aircraft.

target pattern. The flight path of aircraft during the attack phase. Also called **attack pattern.**

target signal. The radar **energy** returned to a **radar** by a **target.** Also called *echo signal, video signal.* The amount of this energy is termed *received power.*

TAS. True Air Speed.

task force. 1. A temporary grouping of units under one commander, formed for the purpose of carrying out a specific operation or mission. **2.** A semipermanent organization of units under one commander for the purpose of carrying out a continuing specific task. **3.** A component of a fleet organized by the commander of a task fleet or higher authority for the accomplishment of a specific task or tasks. See also **force(s).**

taxi. To operate an airplane under its own power on the ground, except that movement incident to actual takeoff and landing.

taxiway. A specially prepared or designated path on a land airfield for the use of taxing aircraft.

taxiway lights. Aeronautical ground lights arranged along a taxiway to indicate the route to be followed by taxiing aircraft.

TCA. Terminal control area.

teardrop balloon. A **sounding** balloon which, when operationally inflated, resembles an inverted teardrop. This shape was determined primarily by aerodynamic considerations of the problem obtaining maximum stable rates of balloon ascension.

technical characteristics. Those characteristics of equipment which pertain primarily to the engineering principles involved in producing equipment, e.g., for electronic equipment, technical characteristics include such items as circuitry, and types and arrangement of components.

technical evaluation. The study and investigations by a developing agency to determine the technical suitability of material, equipment, or a system.

technical intelligence. See **scientific and technical intelligence.**

technical specifications. A detailed description of technical requirements stated in terms suitable to form the basis for the actual design development and production processes of an item having the qualities specified in the operational characteristics.

tektite. Small glassy bodies containing no crystals, composed of at least 65 percent silicon dioxide, bearing no relation to the geological formations in which they occur, and believed to be of extraterrestrial origin. Tektites are found in certain large areas called *strewn fields.* They are named, as are minerals, with the suffix *ite,* as *australite,* found in Australia, *billitonite, indochinite,* and *rizalite,* found in Southeast Asia, *bediasite* from Texas, and *moldavite* from Bohemia and Moravia.

telecommunications. Any transmission, emission, or reception of signs, signals, writing, images, and sounds or information of any nature by wire, radio, visual, or other electromagnetic systems.

telemeter. 1. To measure at a distance. See **telemetering, telemetry. 2.** The electronic unit which transmits the **signal** in a telemetering system.

telemetering. 1. A measurement accomplished with the aid of intermediate means which allows perception, recording, or interpretation of data at a distance from a primary **sensor.** The most widely employed interpretation of telemetering restricts its significance to data transmitted by means of electromagnetic propagation. **2.** Automatic radio **communication** intended to indicate or record a measurable variable quantity at a distance.

telemetry. The science of measuring a quantity or quantities, transmitting the results to a distant station, and there interpreting, indicating, and/or recording the quantities measured.

teletherapy. Radiation treatment administered by using a *source* that is at a distance from the body, usually employing gamma-ray beams from radioisotope sources. (See **radiation therapy.**)

television imagery. Imagery acquired by a television camera and recorded or transmitted electronically.

temperature. 1. In general, the **intensity** of **heat** as measured on some definite temperature scale by means of any of various types of thermometers. **2.** In statistical mechanics, a measure of translational molecular **kinetic energy** (with three degrees of freedom). **3.** In thermodynamics, the integrating factor of the differential equation referred to as the **first law of thermodynamics.**

temperature inversion. See **inversion.**

temperature stress. A stress induced when a structure embodying materials with different coefficients of linear expansion is exposed to a temperature other than that prevailing at the time of assembly.

teracycle. 1000 gigacycles.

terminal. 1. A point at which any **element** in a **circuit** may be directly connected to one or more other elements. **2.** Pertaining to a final condition or the last division of something, as *terminal ballistics.*

terminal area. A general term used to describe airspace in which approach control service or airport traffic control service is provided.

terminal ballistics. That branch of **ballistics** dealing with the motion and behavior of **projectiles** at the termination of their flight, or in striking and penetrating a target.

terminal control area (TCA). (See controlled airspace)

terminal forecasts. Weather forecasts available each six hours at all FAA **Flight Service Stations** and **National Weather Service** airport stations covering the airways weather at major terminal airports. See area **forecast, weather forecast.**

terminal guidance. Guidance from an arbitrary point, at which **midcourse guidance** ends, to the destination.

terminal phase. The period of flight of a missile between the end of midcourse guidance and impact. See also **terminal guidance.**

terminal velocity. The maximum velocity attainable, especially by a freely falling body, under the given conditions.

terminal VOR (TVOR). Very high frequency terminal omnirange station (located on or near an airport and used as an approach aid).

terminator. The line separating illuminated and dark portions of a celestial body, as the moon, which is not self luminous.

terrain following. The flight of a military aircraft maintaining a constant AGL altitude above the terrain or the highest obstruction. The altitude of the aircraft will constantly change with the varying terrain and/or obstruction.

terrestrial. Of or pertaining to the earth.

terrestrial coordinates. Same as **geographical coordinates.**

terrestial environment. The earth, including its interface with the atmosphere and the oceans, including the ocean floor seaward to the five fathom line, and the influence of those man-made and natural features which affect the solid earth.

terrestrial equator. Same as **astronomical equator.**

terrestrial latitude. Latitude on the earth; angular distance from the **equator.** See **coordinate.** *Terrestrial latitude* is named for the datum used to measure angular distance from the equator. *Astronomical latitude* is the angular distance between the direction of gravity and the plane of the equator. *Geodetic* or *topographical latitude* is the angular distance between the plane of the equator and a normal to the spheroid. *Geodetic* and sometimes *astronomical* latitude are also called *geographic latitude.* *Geocentric latitude* is the angle between a line to the center of the earth and the plane of the equator. *Geodetic latitude* is used for charts.

terrestrial longitude. **Longitude** on the earth; the arc of a **parallel,** or the angle at the pole, between the **prime meridian** and the meridian of a point on the earth. See **coordinate.** *Terrestrial longitude* is the angle between the plane of the reference meridian and the plane of the celestial meridian. *Geodetic longitude* is the angle between the plane of the reference meridian and the plane through the polar axis and the normal to the spheroid. *Geodetic* and sometimes *astronomical longitude* are also called *geographic longitude.* *Geodetic longitude* is used for charts.

terrestrial magnetism. The magnetism of the earth. Also called *geomagnetism.*

terrestrial meridian. Same as **astronomical meridian.**

terrestrial pole. One of the poles of the earth. See **geographical pole, geomagnetic pole, magnetic pole.**

terrestrial radiation. 1. The total **infrared** radiation emitted from the earth's surface; to be carefully distinguished from **effective terrestrial radiation, atmospheric radiation** (which is sometimes erroneously used as a synonym for terrestrial radiation), and **insolation.** Also called *earth radiation, eradiation.*

terrestrial-reference guidance. See **guidance,** note.

test. 1. A procedure or action taken to determine under real or simulated conditions the capabilities, limitations, characteristics, effectiveness, reliability, or suitability of a material, device, system, or method. 2. A similar procedure or action taken to determine the reactions, limitations, abilities, or skills of a person, other animal or organism.

test bed. 1. A base, mount, or frame within or upon which a piece of equipment, especially an **engine,** is secured for testing. 2. A flying test bed.

test chamber. A place, section, or room having special characteristics where a person or object is subjected to experiment, as *an altitude chamber;* specifically, the **test section** of a **wind tunnel.**

test firing. The **firing** of a **rocket engine,** either live or static, with the purpose of making controlled observations of the engine or of an engine component.

test flight. A **flight** to make controlled observations of the operation or performance of an aircraft or rocket, of an aircraft or rocket component, of a system, etc.

test point. A convenient safe access to a circuit or system so that a significant quantity can be measured or introduced to facilitate maintenance, repair, calibration, alignment, and checkout (test).

test section. The section of a **wind tunnel** where objects are tested to determine their aerodynamic characteristics. Also called a *test chamber.*

test stand. A stationary platform or table, together with any testing apparatus attached thereto, for testing or proving engines, instruments, etc. Compare **launch stand.**

tetrahedron. A device with four triangular sides which indicates wind direction and which may be used as a landing direction indicator.

T-hangar. An aircraft hangar in which aircraft are parked alternately tail to tail, each in the T-shaped space left by the other row of aircraft or aircraft compartments.

theodolite. An optical instrument which consists of a sighting telescope, mounted so that it is free to rotate around horizontal and vertical axes, and graduated scales so that the angle of **rotation** may be measured. The telescope is usually fitted with a right-angle prism so that the observer continues to look horizontally into the eyepiece, whatever the variation of the **elevation** angle.

thermal. 1. Of or pertaining to **heat** or **temperature.** 2. A vertical air current caused by differential heating of the terrain.

thermal barrier. A popular term for speed limitations within an atmosphere imposed by **aerodynamic heating.** Also called the *heat barrier.*

thermal burn. Nucleonics. A burn of the skin or other organic material due to radiant heat, such as that produced by the detonation of a nuclear explosive. (See **flash burn, radiation burn, radiation illness.**)

thermal breeder reactor. Nucleonics. A breeder reactor in which the fission chain reaction is sustained by **thermal neutrons.**

thermal conductivity. An intrinsic physical property of a substance, describing its ability to conduct heat as a consequence of molecular motion. Also called *heat conductivity, coefficient of thermal conduction, coefficient of heat conduction.*

thermal efficiency. 1. The **efficiency** with which a **heat engine** transforms the potential heat of its fuel into work or output, expressed as the ratio of the useful work done by the engine in a given time interval to the total heat energy contained in the fuel burned during the same time interval, both work and heat being expressed in the same units. 2. thermodynamic efficiency.

thermal emission. The process by which a body emits **electromagnetic radiation** as a consequence of its temperature only.

thermal fatigue. In metals, fracture resulting from the presence of temperature gradients which vary with time in such a manner as to produce cyclic **stresses** in a structure.

thermal imagery (infrared). Imagery produced by measuring and recording electronically the thermal radiation of objects.

thermal instability. The conditions of temperature **gradient, thermal conductivity,** and **viscosity** which lead to the onset of **convection** in a fluid. Such gross phenomena as atmospheric winds are an example of this type of instability. In general, if the fluid is conducting, as a plasma, the application of a magnetic field tends to reduce these thermal instabilities.

thermal jet engine. A **jet engine** that utilizes heat to expand gases for rearward ejection. This is the usual form of aircraft jet engine.

thermal (slow) neutron. A neutron in thermal equilibrium with its surrounding medium. Thermal neutrons are those that have been slowed down by a *moderator* to an average speed of about 2200 meters per second (at room temperature) from the much higher intial speeds they had when expelled by fission. This velocity is similar to that of gas molecules at ordinary temperatures.

thermal radiation. The **electromagnetic radiation** emitted by any substance as the result of the **thermal** excitation of its molecules. Thermal radiation ranges in wavelength from the longest infrared radiation to the shortest ultraviolet radiation.

thermal reactor. Nucleonics. A reactor in which the fission chain reaction is sustained primarily by thermal neutrons. Most reactors are thermal reactors.

thermal shock. The development of a steep temperature **gradient** and accompanying high **stresses** within a structure.

thermal stresses. **Stresses** in metal, resulting from non-uniform temperature distribution.

thermionic. Of or pertaining to the emission of **electrons** by **heat.**

thermionic conversion. The process whereby **electrons** released by **thermionic emission** are collected and utilized as electric current. The simplest example of this is provided by a vacuum tube, in which the electrons released from a heated anode are collected at the cathode or plate. Used as a method of producing electrical power for spacecraft.

thermionic emission. Direct ejection of **electrons** as the result of heating the material, which raises electron energy beyond the **binding energy** that holds the electron in the material.

thermionic tube. An **electron tube** in which one or more of the **electrodes** is heated to cause electron or ion **emission.**

thermistor. An **electron device** employing the temperature-dependent change of resistivity of a **semiconductor.**

thermochemical. Pertaining to a chemical change induced by **heat.**

thermochemistry. A branch of chemistry that treats of the relations of **heat** and chemical changes.

thermocouple. A device which converts **thermal** energy directly into **electrical** energy. In its basic form it consists of two dissimiliar metallic electrical conductors connected in a closed loop. Each junction forms a thermocouple. See **thermopile.** If the junctions are at different temperatures, an electrical potential proportional to the temperature difference will exist in the circuit; the value of the potential generated is different for various combinations of materials. For meteorological purposes couples of copper and constantan are frequently used; these generate approximately 40 microvolts per °C of couple temperature difference.

thermodynamic. Pertaining to the flow of **heat** or to **thermodynamics.**

thermodynamic efficiency. In thermodynamics, the ratio of the work done by a **heat engine** to the total **heat** supplied by the heat **source.** Also called *thermal efficiency, Carnot efficiency.*

thermodynamic energy equation. The mathematical statement of the concept of **conservation of energy** embodied in the **first law of thermodynamics.** See **energy equation.**

thermodynamics. The study of the **flow of heat.**

thermoelectric conversion. The conversion of heat into electricity by the use of thermocouples. (Compare **thermionic conversion,** see **thermocouple.**)

thermograph. A self-registering thermometer.

thermometer. A device for measuring **temperature.**

thermonuclear bomb (device). A **hydrogen bomb** (device).

thermonuclear reaction. A reaction in which very high temperatures bring about the fusion of two light nuclei to form the nucleus of a heavier atom, releasing a large amount of energy. In a *hydrogen bomb,* the high temperature to initiate the thermonuclear reaction is produced by a preliminary *fission* reaction. (See **fusion.**)

thermopile. 1. A **transducer** for converting **thermal** energy directly into **electrical** energy, composed of pairs of **thermocouples** which are connected either in series or in parallel. **2.** A battery of thermocouples connected in series to form a single compact unit.

thermosphere. See **atmospheric shell.**

thermoswitch. A temperature-activated switch.

thickness/chord ratio. The ratio of the maximum thickness of an airfoil section measured perpendicular to the chord, to the chord length.

third law of thermodynamics. The statement that every substance has a finite positive **entropy,** and that the entropy of a crystalline substance is zero at the temperature of absolute zero.

three-body problem. That problem in classical **celestial mechanics** which treats the motion of a small body, usually of negligible mass, relative to and under the gravitational influence of two other finite point masses.

threshold. Generally, the minimum value of a **signal** that can be detected by the **system** or **sensor** under consideration.

threshold. The beginning of that portion of the runway usable for landing.

threshold contrast. The smallest contrast of **luminance** (or brightness) that is perceptible to the human eye under specified conditions of **adaptation luminance** and target visual angle. Also called *contrast threshold, liminal contrast.* Compare **threshold illuminance.** Psychophysically, the existence of a threshold contrast is merely a special case of the general rule that for every sensory process there is a corresponding lowest detectable intensity of stimulus, i.e., a **limen.**

threshold dose. Nucleonics. The minimum dose of radiation that will produce a detectable biological effect. (See **absorbed dose, biological dose.**)

threshold illuminance. The lowest value of illuminance which the eye is capable of detecting under specified conditions of **background luminance** and degree of **dark adaptation** of the eye. Also called *flux-density threshold.* Compare **threshold contrast.**

This threshold, which controls the visibility of point light sources, especially at night, cannot be assigned any universal value, but nonflashing lights can generally be seen by a fully dark-adapted eye when the lights yield an illuminance of the order of 10^{-1} lumen per square kilometer at the eye.

threshold lights. Lights indicating the longitudinal limits of that portion of the runway or strip which is usable for landing.

threshold of audibility. For a specified **signal**, the minimum effective **sound pressure level** of the signal that is capable of evoking an auditory sensation in a specified fraction of the trials. The characteristics of the signal, the manner in which it is presented to the listener, and the point at which the sound pressure level is measured must be specified. Also called *threshold of detectability.* Unless otherwise indicated, the ambient noise reaching the ears is assumed to be negligible. The threshold is usually given as a sound pressure level in decibels, relative to 0.0002 microbar.

threshold sensitivity. Of a **transducer**, the lowest level of the **input** signal which produces desired response at the **output**. The term applies equally to psychophysics.

throat. The narrowest portion of a constricted duct, as in a diffuser, a venturi tube, etc., specifically, a **nozzle throat.**

throttling. The varying of the **thrust** of a **rocket engine** during powered flight by some technique. Tightening of fuel lines, changing of thrust chamber pressure, pulsed thrust, and variation of nozzle expansion are methods to achieve throttling.

thrust. 1. The pushing or pulling force developed by an aircraft engine or a **rocket engine. 2.** The force exerted in any direction by a fluid jet or by a powered screw, as, *the thrust of an antitorque rotor.* **3.** *(symbol F).* Specifically, in rocketry, $F = mv$ where m is propellant mass flow and v is exhaust velocity relative to the vehicle. Also called *momentum thrust.*

thrust augmentation. The increasing of the **thrust** of an engine or power plant, especially of a **jet engine** and usually for a short period of time, over the thrust normally developed. The principal methods of thrust augmentation are the introduction of additional air into the induction system, liquid injection, and afterburning. With a piston engine, *thrust augmentation* usually refers to the direction of exhaust gases so as to give additional thrust.

thrust augmenter. Any device used to increase the thrust of a piston, jet, or rocket engine, such as an **afterburner.** See **augmenter tube.**

thrust axis. A line or **axis** through an aircraft, rocket, etc., along which the **thrust** acts; an axis through the longitudinal center of a jet or rocket engine along which the thrust of the engine acts; a center of thrust. Also called *axis of thrust.*

thrust chamber. Same as **firing chamber.**

thrust horsepower. 1. The force-velocity equivalent of the **thrust** developed by a jet or rocket engine. **2.** The **thrust** of an engine-propeller combination expressed in horsepower. It differs from the shaft horsepower of the engine by the amount the propeller efficiency varies from 100 percent.

thrust loading. The weight-thrust ratio of a jet or rocket-propelled aircraft or other vehicle expressed as gross weight in pounds divided by **thrust** in pounds. See **power loading.**

thrust meter. An instrument for measuring static **thrust**, especially of a jet engine or rocket.

thrust power. The power usefully expended on **thrust**, equal to the thrust (or **net thrust**) times airspeed.

thrust reverser. A device or apparatus for reversing **thrust**, especially of a **jet engine.** See **reverse thrust.**

thrust section. 1. A section in a **rocket vehicle** that houses or incorporates the **combustion chamber** or chambers and **nozzles. 2.** In loose usage, a propulsion system.

thrust terminator. A device for ending the **thrust** in a **rocket engine,** either through propellant **cutoff** (in the case of a liquid) or through diverting the flow of gases from the **nozzle.**

thrust-weight ratio. A quantity used to evaluate engine performance, obtained by dividing the **thrust** output by the engine weight less fuel. If the pound is used as the unit of measure for thrust and weight, the result is pounds of thrust per pound of engine.

thunder. (Meteorology) The sound emitted by rapidly expanding gases along the channel of a lightning discharge.

thunderstorm. A storm, invariably produced by a **cumulonimbus cloud,** and always accompanied by lightning and thunder; usually attended by strong wind gusts, heavy rain, and sometimes hail. It is usually of short duration, seldom over 2 hours for any one storm. Also see **anvil cloud.**

TIAS. True Indicated Air Speed.

tide. The periodic rising and falling of the earth's oceans and **atmosphere.** It results from the **gravitational** forces of the moon and sun acting upon the rotating earth. The disturbance actually propagates as a wave through the atmosphere and along the surface of the waters of the earth. Atmospheric tides are always so designated, whereas the term *tide* alone commonly implies the oceanic variety.

tied on. (Military) The aircraft indicated is in formation with me.

tiedown. See **aircraft tiedown.**

tie-down point. 1. An attachment point provided on or within a vehicle for securing cargo. **2.** A ring or hook attached to the aircraft structure used for securing aircraft when parked in the open. Usually a minimum of three tiedown points are installed. See **aircraft tiedown.**

tilt table. A device used to calibrate linear **accelerometers** with rated ranges of, or below, $\pm 1.0 \ g$. It allows the accelerometer to be positioned at different angles in reference to a surface perpendicular to the direction of the earth's gravity, so that the applied values of acceleration are equal to the cosine of the angle between the reference surface and the direction of the earth's gravity.

time. *(symbol t).* The hour of the day reckoned by the position of a celestial reference point relative to the reference **celestial meridian.** Time may be designated *solar, lunar,* or *sidereal* as the reference is the sun, moon, or vernal equinox, respectively. Solar time may be further classified as *mean* or *astronomical* if the mean sun is the reference, or as *apparent* if the apparent sun is the reference. Time may also be designated according to the reference meridian, either the *local* or *Greenwich* meridian or, additionally, in the case of mean solar time, a designated zone meridian. Standard and daylight-saving time are variations of zone time. Time may also be designated according to the timepiece, as *chronometer time* or *watch time*, the time indicated by these instruments.

time change items. Accessories and components of weapon systems which have been identified as having some fixed service life expectancy and which must be replaced with a new or overhauled item after accrual of a specified number of hours or cycles of operation or at the expiration of a given calender time period.

time division multiplex. A system for the transmission of information about two or more quantities (**measurands**) over a common **channel** by dividing available time intervals among the measurands to form a composite **pulse train.** Information may be transmitted by variation of pulse duration, pulse amplitude, pulse position, or by a pulse code. (Abbreviations used are PDM, PAM, PPM, and PCM, respectively.)

time in service. With respect to maintenance time records the time from the moment an aircraft leaves the surface of the earth until it touches it at the next point of landing.

time lag. The total time between the application of a **signal** to a measuring instrument and the full indication of that signal within the uncertainty of the instrument.

time-of-flight spectrometer. Nucleonics. A device for separating and sorting neutrons (or other particles) into categories of similar energy, measured by the time it takes the particles to travel a known distance. (Compare **mass spectrometer.**)

time over target. The time at which an aircraft or formation of aircraft arrives over a designated point for the purpose of conducting an air mission on a target.

time signal. 1. An accurate **signal** marking a specified time or time interval. It is used primarily for determining errors of timepieces. Such signals are usually sent from an observatory by radio or telegraph. **2.** In photography, a time indication registered on the film to serve as a time reference for interpretation of the data recorded on the film.

time tic. Markings on **telemetry** records to indicate time intervals.

time zone. See **zone time.**

timing pulse. In telemetry, a **pulse** used as a time reference.

tip speed. The mean angular velocity of the rotor multiplied by the rotor radius.

tip tanks. External fuel tanks. See **external load.**

titanium. A silver-gray, metallic element found combined with ilmenite and rutile. Alloys of titanium are used extensively in aircraft. It is light, strong and resistant to stress-corrosion cracking. Titanium is approximately 60 percent heavier than aluminum and about 50 percent lighter than stainless steel.

TNT equivalent. A measure of the energy release from the detonation of a nuclear weapon, or from the explosion of a given quantity of fissionable or fusionable material, in terms of the amount of TNT (trinitrotoluene) which would release the same amount of energy when exploded.

tolerance. The allowable variation in measurements within which the dimensions of an item are judged acceptable.

toll enrichment. A proposed arrangement whereby privately owned uranium could be enriched in uranium-235 content in government facilities upon payment of a service charge by the owners. (See **isotopic enrichment, uranium.**)

topple. Of a gyro, the vertical component of **procession** or **wander,** or the algebraic sum of the two.

topple axis. That horizontal axis, perpendicular to the (horizontal) **spin axis** of a gyroscope, around which **topple** occurs.

topographic map. A map which presents the vertical position of features in measurable form as well as their horizontal positions. See also **map.**

top secret. See **defense classification.**

torching. The burning of fuel at the end of an exhaust pipe or stack of a reciprocating aircraft engine, the result of an excessive richness in the fuel air mixture.

tornado. A violently rotating column of air attended by a funnel-shaped or tubular cloud hanging beneath a **cumulonimbus cloud.**

torque. About an axis, the product of a **force** and the distance of its line of action from the **axis** such as the rolling force imposed on an airplane by the engine in turning the propeller.

torquer. In a gyro, a device which produces **torque** about an **axis of freedom** in response to a signal input.

torquing. In a gyro, the application of **torque** to a gimbal about an **axis of freedom** for the following purposes: **processing, capturing, slaving, caging,** or **slewing.**

torr. Provisional international standard term to replace the English term *millimeter of mercury* and its abbreviation *mm of Hg* (or the French *mm de Hg).* The torr is defined as 1/760 of a standard atmosphere, or 1,013,250/760 dynes per square centimeter. This is equivalent to defining the torr as 1333.22 microbars and differs by only one part in 7 million from the International Standard millimeter of mercury. The prefixes *milli* and *micro* are attached without hyphenation.

toss bombing. A method of bombing where an aircraft flies on a line towards the target, pulls up in a vertical plane, releasing the bomb at an angle that will compensate for the effect of gravity drop on the bomb. Similar to loft bombing; unrestricted as to altitude. See also **loft bombing; over-the-shoulder bombing.**

total energy equation. In meteorology, an expression relating all forms of energy obtained by combining the **thermodynamic energy equation** with the **mechanical energy equation.** This equation expresses the fact that the combined internal, kinetic, and potential energy in a given volume of the atmosphere can vary only as a result of: (**a**) the transport of these forms of energy across the boundaries of the volume; (**b**) the work done by pressure forces on the boundary; (**c**) the addition or removal of heat; and (**d**) the dissipational effect of friction.

total head. Same as **total pressure,** sense **3.**

total pressure. 1. stagnation pressure. **2.** impact pressure. **3.** The pressure a moving fluid would have if it were brought to rest without losses. **4.** The pressure determined by all the molecular species crossing the imaginary surface.

touch and go/touch and go landing. An operation by an aircraft that lands and departs on a runway without stopping or exiting the runway.

touchdown RVR. The RVR readout values obtained from RVR equipment serving the runway threshold.

touchdown zone. The first 3,000 feet of the runway beginning at the threshold. The area is used for determination of Touchdown Zone Elevation in the development of straight-in landing minimums for instrument approaches.

toughness. The ability of a metal to absorb energy and deform plastically before fracturing. Toughness is usually measured by the energy absorbed in a notch impact test, but the area under the stress-strain curve in tensile testing is also a measure of toughness.

tower. See **airport traffic control tower.**

townsend discharge. A type of **direct-current discharge** between two **electrodes** immersed in a gas and requiring electron emission from the cathode.

toxic chemical, biological, or radiological attack. An attack directed at man, animals, or crops, using injurious agents of radiological, biological, or chemical origin.

trace. The line appearing on the face of a **cathode-ray tube** when the visible dot repeatedly sweeps across the face of the tube as a result of deflections of the **electron beam.** See **sweep.** The path of the dot from the end of one sweep to the start of the next sweep is called a *retrace.* If more than one trace is shown on the same scope, the traces may be called *A-trace, B-trace,* etc.

tracer, isotopic. As isotope of an element, a small amount of which may be incorporated into a sample of material (the carrier) in order to follow (trace) the course of that element through a chemical, biological or physical process, and thus also follow the larger sample. The tracer may be radioactive, in which case observations are made by measuring the radioactivity. If the tracer is stable, mass spectrometers, density measurement, or neutron activation anaylsis may be employed to determine isotopic composition. Tracers also are called *labels* or *tags,* and materials are said to be labeled or tagged when radioactive tracers are incorporated in them.

track. 1. The path or actual line of movement of an aircraft, rocket, etc., over the surface of the earth. It is the projection of the **flight-path** on the surface. 2. To observe or plot the path of something moving, such as an aircraft or rocket, by one means or another, as by telescope or by radar—said of persons or of electronic equipment, as *the observer, or the radar, tracked the satellite.* 3. To follow a desired track.

track crossing angle. In air interception, the angular difference between interceptor track and target track at the time of intercept.

tracking. 1. The process of following the movements of an object. This may be done by keeping the **reticle** of an optical system or a **radar beam** on the object, by plotting its **bearing** and distance at frequent intervals, or by a combination of the two. 2. A motion given to the major **lobe** of an **antenna** so that a preassigned moving target in space remains in the lobe's field as long as it is within viewing range.

tracking antenna. A **directional antenna** system which changes in position, or characteristics, automatically or manually to follow the motions of a moving **signal** source.

tracking radar. A radar used for following a **target.**

tracking rate. The rate at which an operator or a system follows a **target.** Usually expressed in terms of the rate of change of the parameter being measured.

tracking station. A station set up to **track** an object moving through the atmosphere or space, usually by means of **radar** or radio. See **minitrack.**

track made good. The actual path of an aircraft over the surface of the earth, or its graphic representation.

tractor propeller. A propeller designed normally to produce tension in the propeller shaft.

trade winds. Two belts of winds, one on either side of the equatorial doldrums, where the winds blow almost constantly from easterly quadrants.

traffic information (radar). Information issued to alert an aircraft to any radar targets observed on the radar display which may be in such proximity to its position or intended route of flight to warrant its attention.

traffic pattern. The traffic flow that is prescribed for air craft landing at, taxiing on, and taking off from an airport. The usual components of a traffic pattern are upwind leg, crosswind leg, downwind leg, base leg, and final approach.

trailer. (Military) Aircraft which are following and keeping under surveillance a designated airborne contact.

trail formation. Aircraft flying singly or in elements in such manner that each aircraft or element is in line behind the preceding aircraft or element.

trailing edge. The rear edge of an airfoil or other body moving through the air.

train bombing. A method of bombing in which two or more bombs are released at a predetermined interval from one aircraft as the result of a single actuation of the bomb release mechanism.

trainer (T), (Military designation). Aircraft designed for training personnel in the operation of aircraft and/or related equipment, and having provisions for instructor personnel. See **military aircraft types.**

training aids. Any item which is developed and/or procured with the primary intent that it shall assist in training and the process of learning.

trajectory. In general, the path traced by any body moving as a result of an externally applied force, considered in three dimensions. *Trajectory* is sometimes used to mean *flight path* or *orbit,* but *orbit* usually means a closed path and *trajectory,* a path which is not closed.

trajectory-measuring system. A **system** used to provide information on the spatial postion of an object at discrete time interval throughout a portion of the **trajectory** or flightpath.

transceiver. A combination **transmitter** and **receiver** in a single housing, with some components being used by both units.

transcribed weather broadcast equipment (TWEB). A transcribed broadcast providing continuous recorded aeronautical and meteorological information over a low frequency or VOR.

transducer. A device capable of being actuated by **energy** from one or more **transmission** systems or media and of supplying related energy to one or more other transmission systems or media, as a microphone, a thermocouple, etc. The energy in input and output may be of the same or different types (e.g., electric, mechanical, or acoustic).

transducer gain. The ratio of the **power** that a **transducer** delivers to a specified **load** under specified operating conditions to the available power of a specified source. If the input and/or output power consist of more than one component, such as multifrequency signal or noise, then the particular components used and their weighting must be specified. This gain is usually expressed in decibels.

transfer ellipse. See **transfer orbit.**

*Jet airliners are certificated by the Federal Aviation Administration under the **transport** category.*

transfer of control. (air traffic control) That action whereby the responsibility for the provision of separation to an aircraft is transferred from one controller to another.

transfer orbit. In interplanetary travel, an elliptical **trajectory** tangent to tne **orbits** oi botn the departure planet and the target planet. Also called *transfer ellipse*.

transistor. An active **semiconductor** device with three or more **electrodes**.

transit. 1. The passage of a **celestial body** across a **celestial meridian,** usually called *meridian transit*. **2.** The apparent passage of a celestial body across the face of another celestial body or across any point, area, or line. **3.** An instrument used by an astronomer to determine the exact instant of meridian transit of a celestial body. **4.** A reversing instrument used by surveyors for accurately measuring horizontal and vertical angles; a theodolite which can be reversed in its supports without being lifted from them.

transition maneuver. In lifting flight, a maneuver required to fly smoothly from one **equilibrium glidepath** to another, performed by changing **attitude** in some manner.

transition point. In aerodynamics, the point of change from **laminar** to **turbulent flow.**

transition temperature. 1. An arbitrarily defined temperature within the temperature range in which metal fracture characteristics determined usually by notched tests are changing rapidly such as from primarily fibrous (shear) to primarily crystalline (cleavage) fracture. **2.** The arbitrarily defined temperature in a range in which the ductility of a material changes rapidly with temperature.

translation. Movement in a straight line without **rotation.**

translator. A **network** or system having a number of inputs and outputs and so connected that **signals** representing information expressed in a certain **code,** when applied to the inputs, cause output signals to appear which are a representation of the input information in a different code. Sometimes called *matrix*.

translunar. Outside the moon's orbit about the earth.

translunar space. As seen from the earth at any moment, **space** lying beyond the **orbit** of the moon.

transmission. 1. The process by which **radiant flux** is propagated through a medium or body. **2. transmittance.**

transmission loss. The reduction in the **magnitude** of some characteristic of a **signal** between two stated points in a **transmission system.** Also called *loss*. The characteristic is often some kind of level, such as power level or voltage level; in acoustics, the characteristic that is commonly measured is sound pressure level.Thus,if the levels are expressed in decibels, the transmission level loss is likewise in decibels. It is imperative that the characteristic concerned (such as the sound pressure level) be clearly identified because in all transmission systems more than one characteristic is propagated.

transmission system. A **system** which propagates or transmits **signals.**

transmission time. The time interval between dispatch and reception of a **signal** in a particular transmission system.

transmissometer. An instrument for measuring the **extinction coefficient** of the atmosphere and for the determination of **visual range.** Also called *telephotometer, transmittance meter, hazemeter.* See **photoelectric**

transmittance meter, visibility meter.

transmissometry. The technique of determining the **extinction** characteristics of a medium by measuring the **transmittance** of a light beam of known initial intensity directed into that medium.

transmitted power. The **power** which is radiated from an **antenna.** Compare **received power.**

transmitter. A device used for the generation of **signals** of any type and form which are to be transmitted. See **receiver.** In radio and radar, it is that portion of the equipment which includes electronic circuits designed to generate, amplify, and shape the radiofrequency energy which is delivered to the antenna where it is radiated out into space.

transmutation. The transformation of one element into another by a *nuclear reaction* or series of reactions. Example: the transmutation of uranium-238 into plutonium-239 by absorption of a neutron.

transponder. A combined **receiver** and **transmitter** whose function is to transmit signals automatically when triggered by an **interrogator.** See **transceiver, air traffic control radar beacon system, radar, responder, IFF.**

transponder beacon. A **beacon** having a **transponder.** Also called *responder beacon*.

transponder india. International civil aviation organization/secondary surveillance radar.

transponder sierra. (Military) **Identification Friend or Foe** mark X (selective identification feature).

transponder tango. (Military) **Identification Friend or Foe** mark X basic.

transonic. Pertaining to that which occurs or is occurring within the range of speed in which flow patterns change from subsonic to supersonic or vice versa, about Mach 0.8 to 1.2, as in *transonic flight, transonic flutter;* that operates within this regime, as in *transonic aircraft, transonic wing;* characterized by transonic flow or transonic speed, as in *transonic region, transonic zone.*

transonic flow. In aerodynamics, flow of a fluid over a body in the range just above and just below the **acoustic velocity.** Transonic flow presents a special problem in aerodynamics in that neither the equations describing subsonic flow nor the equations describing supersonic flow can be applied in the transonic range.

transonic speed. The speed of a body relative to the surrounding **fluid** at which the **flow** is in some places on the body **subsonic** and in other places **supersonic.**

transparent plasma. A **plasma** through which an electromagnetic wave can **propagate.** In general, a plasma is transparent for frequencies higher than the **plasma frequency.**

transpiration. The passage of gas or liquid through a porous solid (usually under conditions of **molecular flow**).

transpiration cooling. A process by which a body having a porous surface is cooled by forced flow of **coolant** fluid through the surface from the interior. Compare **film cooling.**

transport category. As applied to aircraft **type certificates** issued by the **FAA,** transport **category** applies to aircraft over 12,500 pounds gross weight and/or a passenger seat configuration of 10 or more.

trap. A part of a **solid-propellant rocket engine** used to prevent the loss of unburned **propellant** through the **nozzle.**

trapping. The process by which radiation **particles** are caught and held in a **radiation belt.**

traveling-wave tube *(abbr* **TWT** *).* An **electron tube** in which a stream of electrons interacts continuously or repeatedly with a guided **electromagnetic wave** moving substantially in synchronism with it, and in such a way that there is a net transfer of energy from the stream to the wave.

triage. The process of determining which casualties (from a large number of persons exposed to heavy radiation) need urgent treatment, which ones are well enough to go untreated, and which ones are beyond hope of benefit from treatment. Used in medical aspects of civil defense.

triangle of velocities. The fundamental triangle associated with dead-reckoning. It is composed of the following vectors; **a.** Heading and true airspeed. **b.** Track and groundspeed. **c.** Wind speed and wind direction.

tri-camera photography. Photography obtained by simultaneous exposure of three cameras systemactically disposed in the air vehicle at fixed overlapping angles relative to each other in order to cover a wide field.

tricycle landing gear. A three-wheeled landing gear in which the third wheel is placed well forward under the nose of the airplane. In this type of landing gear the two rear wheels are located a short distance behind the c.g. The airplane rides approximately level on all three wheels, when taking off and landing.

trim. The condition of static balance in pitch of an airplane.

trimming. To set or adjust the flying controls and/or trimming devices so that the aircraft will maintain a desired attitude in steady flight. An airplane is normally trimmed by the pilot so that control forces are reduced to zero.

trim tab. A **tab** used to **trim** an airplane. See **trimming.**

triple ejection rack. A device designed to carry up to three bombs or other munitions on a single station of an aircraft.

triple point. The thermodynamic state at which three phases of a substance exist in equilibrium. The triple point of water occurs at a saturation vapor pressure of 6.11 millibar and at a temperature of 273.16° K.

triplexer. A dual-**duplexer** which permits the use of two receivers simultaneously and independently in a **radar** system by disconnecting the receivers during the transmitted **pulse.**

tropical. Of or pertaining to the **vernal equinox.** See **sidereal.**

tropical air. (Metgy). Warm air having its source in the low latitudes, chiefly in the regions of the subtropical high pressure systems.

tropical month. The average period of the **revolution** of the moon about the earth with respect to the **vernal equinox,** a period of 27 days 7 hours 43 minutes 4.7 seconds, or approximately 27 1/3 days.

tropical storm. A tropical cyclone in which the surface wind speed is at least 34, but not more than 63 knots.

tropical year. The period of one **revolution** of the earth around the sun, with respect to the **vernal equinox.** Because of precession of the equinoxes, the tropical year is not 360° with respect to the stars, but 50 minutes 0.3 second less. A tropical year is about 20 minutes shorter than a sidereal year, averaging 365 days 5 hours 48 minutes 45.68 seconds in 1955 and is increasing at the rate of 0.005305 second anually. Also called *astronomical, equinoctial, natural,* or *solar year.*

tropic of Cancer. The northern **parallel of declination,** ap-proximately 23°27' from the celestial equator, reached by the sun at its maximum declination, or the corresponding parallel on the earth. It is named for the sign of the zodiac in which the sun reached its maximum northerly declination at the time the parallel was so named.

tropic of Capricorn. The southern **parallel of declination,** approximately 23°72'; from the celestial equator, reached by the sun at its maximum declination, or the corresponding parallel on the earth. It is named for the sign of the zodiac in which the sun reached its maximum southerly declination at the time the parallel was so named.

tropopause. The boundary between the **troposphere** and **stratosphere,** usually characterized by an abrupt change of **lapse rate.** The change is in the direction of increased atmospheric stability from regions below to regions above the tropopause. Its height varies from 15 to 20 kilometers (9 to 12 miles) in the tropics to about 10 kilometers (6 miles) in polar regions. In polar regions in winter it is often difficult or impossible to determine just where the tropopause lies, since under some conditions there is no abrupt change in lapse rate at any height.

troposphere. That portion of the **atmosphere** from the earth's surface to the **stratosphere;** that is, the lowest 10 to 20 kilometers (6 to 12 miles) of the atmosphere. The troposphere is characterized by decreasing temperature with height, appreciable vertical wind motion, appreciable water-vapor content, and weather. Dynamically, the troposphere can be divided into the following layers: **surface boundary layer, Ekman layer,** and **free atmosphere.** See **atmospheric shell.**

tropospheric scatter. The propogation of radio waves by scattering as a result of irregularities or discontinuities in the physical properties of the troposphere.

tropospheric wave. A **radio wave** that is **propagated** by reflection from a place of abrupt change in the **dielectric constant** or its gradient in the **troposphere.** In some cases the ground wave may be so altered that new components appear to arise from reflections in regions of rapidly changing dielectric constants; when these components are distinguishable from the other components, they are called tropospheric waves.

trouble shooting. The process of investigating and detecting the cause of aircraft or equipment malfunctioning.

trough (metgy). An elongated area of low atmospheric pressure, usually extending from the center of a low pressure system.

true. 1. Related to or measured from true north. **2.** Actual, as contrasted with fictitious, as *true sun.* **3.** Related to a fixed point, either on the earth or in space, as *true wind;* in contrast with *relative.* **4.** Corrected, as *true altitude.*

true airspeed. See **airspeed.**

true altitude. 1. Actual height above sea level; calibrated altitude corrected for air temperature. **2.** The actual **altitude** of a celestial body above the celestial horizon. Usually called *observed altitude.*

true bearing. The horizontal angle at a given point measured from true north clockwise to the great circle passing through the point and the object.

true course. navig. The horizontal angle measured clockwise from true north to the line representing the intended path of the aircraft.

true indicated airspeed. See **airspeed.**

true meridian. A **great circle** through the geographical

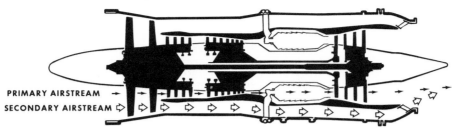

Air flow diagram of a **turbofan engine.** *The secondary air flow bypasses all but the first few compressor stages.*

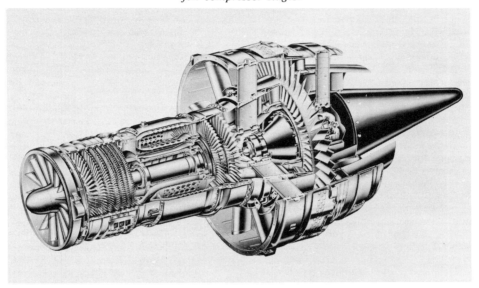

Cutaway view of the General Electric CF700 **turbofan engine** *which incorporates an aft fan.*

The Pratt & Whitney JT9D **turbofan engine.**

poles, distinguished from **magnetic meridian, grid meridian,** etc.

true north. The direction from an observer's position to the geographical north pole. The north direction of any geographic meridian.

true position. The position of a **celestial body** (or space vehicle) on the **celestial sphere** as computed directly from the elements of the **orbit** of the earth and the body concerned without allowance for **light time.** Also called *geometric position.*

true sun. The actual sun as it appears in the sky. Usually called *apparent sun.* See **mean sun, dynamical mean sun.**

T-time. Any specific time, minus or plus as referenced to zero or launch time, during a **countdown** sequence that is intended to result in the firing of a rocket propulsion unit that launches a rocket vehicle.

tumble. 1. To rotate end over end—said of a rocket, of an ejection capsule, etc. **2.** Of a gyro, to **precess** suddenly and to an extreme extent as a result of exceeding its operating limits of bank or pitch.

tumbling. An attitude situation in which the vehicle continues on its flight, but turns end over end about its center of mass.

tunnel. 1. A structure, installation, or facility incorporating apparatus to simulate flight conditions in one way or another, specially designed for testing or experimenting with power plants, or with aircraft rockets, or other aerodynamically designed bodies, engine installations, or models; specifically a **wind tunnel. 2.** A longitudinal protuberance on a rocket body used to house wiring, piping, etc., so as to not route the wiring through the propellant tanks.

turbidity. In meteorology, any condition of the **atmosphere** which reduces its transparency to **radiation,** especially to visible radiation. Ordinarily, this is applied to a cloud-free portion of the atmosphere that owes its turbidity to air molecules and suspensoids such as smoke, dust, and haze, and to scintillation effects.

turbine. 1. machine consisting principally of one or more **turbine wheels** and a **stator. 2.** A turbine wheel. **2.** A turbine engine. See **blowdown turbine, explosion turbine, free turbine, gas turbine, impulse turbine, partial-admission turbine, reaction turbine, single-stage turbine.**

turbine blade. Any one of the **blades** of a **turbine wheel.**

turbine engine. An engine incorporating a **turbine** as a principal component; especially, a **gas-turbine engine.**

turbine wheel. A multivaned wheel or **rotor,** especially in a **gas-turbine engine,** rotated by the impulse from or reaction to a **fluid** passing across the vanes. Often called a *turbine.*

turbofan. A **turbojet engine** in which additional propulsive thrust is gained by extending a portion of the **compressor** or **turbine** blades outside the inner engine case. The extended blades propel bypass air which flow along the engine axis but between the inner and outer engine casing. This air is not combusted but does provide additional thrust caused by the propulsive effect imparted to it by the extended compressor blading.

turbojet. 1. turbojet engine. 2. A craft propelled by a turbojet engine. See **jet engine.**

turbojet engine. A **jet engine** incorporating a turbine-driven air **compressor** to take in and compress the air for the combustion of fuel (or for heating by a nuclear reactor), the gases of combustion (or the heated air) being used both to rotate the **turbine** and to create a thrust-producing jet. Often called a *turbojet.* See **jet**

engine, sense **2.**

turboprop. A gas turbine engine in which a proportion of the net energy is used to drive a propeller.

turbo-starter. A starter incorporating a small turbine energized by compressed air or other gas with or without combustion.

turbulence. A state of fluid **flow** in which the instantaneous velocities exhibit irregular and apparently random fluctuations so that in practice only statistical properties can be recognized and subjected to analysis. Compare **laminar flow.** These fluctuations constitute major deformations of the flow and are capable of transporting mementum, energy, and suspended matter at rates far in excess of the rate of transport by the molecular processes of diffusion and conduction in a nonturbulent or laminar flow.

turbulence. (meteorology) Irregular motion of the atmosphere produced when air flows over a comparatively uneven surface, such as the surface of the earth, or when two currents of air flow past or over each other in different directions or at different speeds. Also see **clear air turbulence (CAT), wake turbulence.**

turbulent boundary layer. The layer in which the **Reynolds stresses** are much larger than the **viscous stresses.** When the **Reynolds number** is sufficiently high, there is a turbulent layer adjacent to the **laminar boundary layer.**

turbulent flow. Fluid motion in which random motions of parts of the **fluid** are superimposed upon a simple pattern of **flow.** All or nearly all fluid flow displays some degree of turbulence. The opposite is **laminar flow.**

turn and bank indicator. A gyroscopic instrument for indicating the rate of turning and the quality of coordination. Also called **turn and slip indicator.**

turn and slip indicator. Same as **turn and bank indicator.**

turnaround. The length of time between arriving at a point and departing from that point. It is used in this sense for the turnaround of shipping in ports, and for aircraft refueling and rearming. See also **turnaround cycle.**

turnaround cycle. Used in conjunction with vehicles, ships, and aircraft and comprises the following: loading time at home, time to and from destination; unloading and loading time at destination; unloading time at home; planned maintenance time and, where applicable, time awaiting facilities. See also **turnaround.**

turn error. Any error in **gyro** output due to **cross-coupling** and **acceleration** encountered during vehicle turns.

turn indicator. See **turn-and-bank indicator.**

turnstile antenna. An **antenna** composed of two **dipole antennas,** normal to each other, with their axes intersecting at their midpoints. Usually, the currents are equal and in phase quadrature.

twenty-four hour satellite. A synchronous **satellite** of the earth.

twilight. The periods of incomplete darkness following sunset (evening twilight) or preceding sunrise (morning twilight). Twilight is designated as *civil, nautical,* or *astronomical,* as the darker limit occurs when the center of the sun is at zenith distances of 96°, 102°, and 108°, respectively.

two-body problem. That problem in classical **celestial mechanics** which treats of the relative motion of two point masses under their mutual **gravitational** attraction.

two minute turn. See **standard rate turn.**

type. 1. As used with respect to the FAA certification,

ratings, privileges, and limitations of airmen, means a specific make and basic model of aircraft, including modifications thereto that do not change its handling or flight characteristics. Example include: DC-7, 1049, and F-27; and **2.** As used with respect to the FAA certification of aircraft, means those aircraft which are similar in design. Examples include: DC-7 and DC-7C; 1049G and 1049H; and F-27 and F-27F.

type certificate. A certificate issued by the **FAA** to a manufacturer meeting the **airworthiness** requirements of the applicable **Federal Aviation Regulations** for an **aircraft, aircraft engine, appliance** or **propellor.** See **category, class.**

typhoon. A severe tropical cyclone in the western Pacific.

U

U. Military mission designation for **utility** aircraft.

U-235. Uranium-235. (See **uranium**.)

UAM. Underwater-to-Air Missile.

ullage. The amount that a container, such as a fuel tank, lacks of being full.

ullage rocket. A small **rocket** used in space to impart an **acceleration** to a tank system to insure that the liquid propellants collect in the tank in such a manner as to flow properly into the pumps or thrust chamber.

ultimate load. The load which will, or is computed to, cause failure in any structural member. An ultimate load is a limit load multiplied by the appropriate factor of safety. In aircraft design the factor of safety is usually 1.5 unless otherwise specified.

ultimate strength. The maximum conventional **stress** (tensile, compressive, or shear) that a material can withstand.

ultrahigh frequency *(abbr* **UHF**). See **frequency bands.**

ultrasonic. In acoustics, of or pertaining to **frequencies** above those that affect the human ear, i.e., more than 20,000 vibrations per second. The term *ultrasonic* may be used as a modifier to indicate a device or system intended to operate at an ultrasonic frequency, as an *ultrasonic vibrator. Supersonic* was formerly used in acoustics synonymously with *ultrasonic;* this usage is now rare.

ultrasonic frequency. A frequency lying above the **audiofrequency** range. The term is commonly applied to elastic waves propagated in gases, liquids, or solids. See **sound.**

ultraviolet *(abbr* **UV**). Pertaining to or same as **ultraviolet radiation.**

ultraviolet imagery. That imagery produced as a result of sensing ultraviolet radiations reflected from a given target surface.

ultraviolet radiation. Electromagnetic radiation of shorter wavelength than **visible radiation;** roughly, radiation in the wavelength interval from 100 to 4000 angstroms. Also called *ultraviolet.* See **X-ray.** Ultraviolet radiation from the sun is responsible for many complex photochemical reactions characteristic of the upper atmosphere e.g., the formation of the ozone layer through ultraviolet dissociation of oxygen molecules followed by recombination to form ozone.

umbilical. Short for **umbilical cord.** Often used in the plural, *umbilicals.*

umbilical cord. Any of the servicing electrical or fluid lines between the ground or a tower and an uprighted **rocket vehicle** before the **launch.** Often shortened to *umbilical.*

umbilical tower. A verical structure supporting the **umbilical cords** running into a **rocket** in launching position.

umbra. 1. The darkest part of a shadow in which light is completely cut off by an intervening object. A lighter part surrounding the umbra, in which the light is only partly cut off, is called the *penumbra.* **2.** The darker central portion of a **sun spot,** surrounded by the lighter penumbra.

unclassified matter. Official matter which does not require the application of security safeguards, but the disclosure of which may be subject to control for other reasons. See also **classified matter.**

uncontrolled airspace. Uncontrolled airspace is that portion of the airspace that has not been designated as continental control area, control area, control zone, terminal control area, or transition area and within which ATC has neither the authority nor the responsibility for exercising control over air traffic. (See **controlled airspace**)

unconventional warfare. Includes the three interrelated fields of guerrilla warfare, evasion and escape, and subversion. Unconventional warfare operations are conducted within enemy or enemy-controlled territory by predominantly indigenous personnel, usually supported and directed in varying degrees by an external source.

uncoupled mode. A **mode** of **vibration** that can exist in a system concurrently with and independently of other modes.

undamped natural frequency. Of a mechanical system, the frequency of free vibration resulting from only elastic and **inertial forces** of the system.

undershoot. To land or to follow an approach path which would cause an aircraft to land, short of the intended area.

under the hood. Indicates that the pilot is using a hood to restrict visibility outside the cockpit while simulating instrument flight. An appropriately rated pilot is required in the other control seat while this operation is being conducted.

unicom. A privately owned radio communications station located at airports. Its use is limited to the necessities of safe and expeditious operation of private aircraft pertaining to runway and wind conditions and other advisory information. It is not used for air traffic control purposes.

unidirectional antenna. An **antenna** which has a single well-defined direction of maximum **gain.**

unified command. A command with a broad continuing mission under a single commander and composed of significant assigned components of two or more Services, and which is established and so designated by the President, through the Secretary of Defense with the advice and assistance of the Joint Chiefs of Staff, or, when so authorized by the Joint Chiefs of Staff, by a commander of an existing unified command established by the President.

unipole. A hypothetical **antenna** radiating or receiving equally in all directions. Also called *isotropic antenna.* A pulsating sphere is a unipole for sound waves. In the case of electromagnetic waves unipoles do not exist physically but represent convenient reference antennas for expressing directive properties of actual antennas.

unit. 1. Any military element whose structure is prescribed by competent authority, such as a table of organization and equipment; specifically, part of an organization. **2.** An organizational title of a subdivision of a group in a task force. **3.** A standard of basic quantity into which an item of supply is divided, issued, or used.

*The Piper Archer (upper) and Cessna Skyhawk (lower) are type certificated by the Federal Aviation Administration (FAA) in both the normal and **utility categories**. At a lower gross weight they can be flown in the utility category and limited acrobatics are permissible.*

In this meaning, also called *"unit of issue."*

unit aircraft. Those aircraft provided an aircraft unit for the performance of a flying mission. See also **aircraft.**

United States. In a geographical sense, (1) the States, the District of Columbia, Puerto Rico, and the possessions, including the territorial waters, and (2) the airspace of those areas.

United States air carrier. A citizen of the United States who undertakes directly by lease, or other arrangement, to engage in air transportation.

United Sates Air Force (USAF). See **Armed Forces of the United States.**

United States Air Force Reserve. All reserves of the Air Force except those units, organizations, and members of the Air National Guard of the United States. See also **Air Reserve Forces.**

United States Armed Forces. Used to denote collectively only the regular components of the Army, Navy, Air Force, Marine Corps, and Coast Guard. See also **Armed Forces of the United States.**

United States Army (USA). See **Armed Forces of the United States.**

United States Coast Guard (USCG). See **Armed Forces of the United States.**

United States Navy (USN). See **Armed Forces of the United States.**

universal time *(abbr* **UT). Time** defined by the rotational motion of the earth and determined from the apparent diurnal motions which reflect this **rotation;** because of variations in the rate of rotation, universal time is not rigorously uniform. Also called *Greenwich mean time.* Compare **ephemeris time.** In the years preceding 1960 the arguments of the ephemerides in the American Ephemeris and Nautical Alamanac were designated as *universal time.*

universal transmission function. A mathematical relationship that attempts to describe quantitatively the complex infrared propagation (including absorption and reradiation) in the atmosphere.

unsatisfactory report. A report of a material deficiency on items or equipments not covered by time change items. See also **time change items.**

unscheduled maintenance. Those unpredictable maintenance requirements that had not been previously planned or programmed but require prompt attention and must be added to, integrated with or substituted for previously scheduled workloads.

unstable air. (meteorology) An **air layer** in which the temperature decreases upward at a rate greater than the **adiabatic lapse rate,** or if saturated, greater than the **saturated adiabatic lapse rate.** In unstable air, if a particle is given a small upward or downward impulse, it will continue to move with increasing velocity, thus setting up a vertical current. This is because the vertical temperature distribution is such that ascending air becomes lighter than its surroundings and descending air heavier than its surroundings.

upper air. In **synoptic meteorology** and in weather observing, that portion of the **atmosphere** which is above the lower troposphere. Compare **upper atmosphere.** No distinct lower limit is set but the term is generally applied to the levels above 850 millibars.

upper air observation. A measurement of atmospheric conditions aloft, above the effective range of a surface weather observation. Also called *sounding, upper air sounding.* See **radiosonde.** This is a general term, but is usually applied to those observations which are used in the analysis of upper air charts (as opposed to measurements of upper atmospheric quantities primarily for research).

upper air sounding. Same as **upper air observation.**

upper atmosphere. The general term applied to the **atmosphere** above the **troposphere.** Compare **upper air.** For subdivisions of the upper atmosphere, see **atmospheric shell.**

upper branch. That half of a **meridian** or **celestial meridian** from pole to pole which passes through a place or its **zenith.**

upper limb. That half of the outer edge of a **celestial body,** especially the moon, having the greatest **altitude** in contrast with the lower limb, that half having the least altitude.

upper stage. A second or later **stage** in a **multistage rocket.**

upwind leg. A flight path parallel to the landing runway in the direction of landing.

uranium. (Symbol U) A radioactive element with the atomic number 92 and, as found in natural ores, an average atomic weight of approximately 238. The two principal natural isotopes are uranium-235 (0.7% of *natural uranium*), which is fissionable, and uranium-238 (99.3% of natural uranium) which is fertile. Natural uranium also includes a minute amount of uranium-234. Uranium is the basic raw material of nuclear energy. (See **fertile material, fissionable material.**)

uranium enrichment. (See **isotopic enrichment.**)

uranium series (sequence). The series of nuclides resulting from the radioactive decay of uranium-238, also known as the uranium-radium series. The end product of the series is lead-206. Many man-made nuclides decay into this sequence. (See **decay, radioactive; radioactive series.**)

Uranus. See **planet.**

USA. United States Army.

USAF. United States Air Force.

USCG. United States Coast Guard.

useful load. In airplanes, the difference, in pounds, between the empty weight and the maximum authorized gross weight.

USM. Underwater-to-Surface Missile.

USN. United States Navy.

UT *(abbr).* **universal time.**

utility (U), (military designation). Aircraft used for miscellaneous missions such as carrying cargo and/or passengers, towing targets, etc. These aircraft include those having a small payload. See **military aircraft types.**

utility category. As applied to an aircraft **type certificate** issued by the **FAA,** utility **category** is limited to aircraft intended for **limited acrobatic** operations.

utilization rate. (Military) 1. *Normal rate* — The flying rate produced in a 40-hour work week during normal, nonemergency conditions previously considered as *peacetime.* 2. *Emergency rate* — The maximum sustained flying rate achievable by operating the assigned human resources on an extended work week (6 days). 3. *Wartime rate* — The maximum sustained flying rate achievable by operating the assigned human resources initially on a 7-day week and by applying wartime crew, maintenance, and safety criteria.

UV *(abbr).* **ultraviolet radiation.**

V

V. Military mission designation for **staff** aircraft.

V. Military mission designation for **VTOL and STOL** aircraft.

vacua. Sometimes used as the plural of **vacuum.**

vacuum. 1. A given space filled with **gas** at pressures below **atmospheric pressure.** Various approximate ranges are:

low vacuum, torr . . . . 760 to 25
medium vacuum, torr . . . . 25 to 10^{-3}
high vacuum, torr . . . .10^{-3} to 10^{-6}
very high vacuum, torr . . . . .10^{-6} to 10^{-9}
ultrahigh vacuum, torr . . . . 10^{-9} and below

2. In reference to **satellite** orbital parameters, without consideration of the perturbing effects of an atmosphere, as in *vacuum perigee, vacuum apogee.*

vacuum gage. An instrument for measuring pressure below **atmospheric pressure.** Some of the more common types of vacuum gages listed in order of descending pressure range of use are: (**a**) **Manometer,** usually consists of a column of liquid supported by the pressure to be measured, the determination of which is a matter of measuring the colum height. (**b**) **Thermal conductivity gage,** consisting of a heated surface. The heat transported by the gas molecules from the surface is related to gas pressure. The heat transfer is reflected in changes in surface temperature (or in the heating power required to maintain constant temperature). Various types of thermal conductivity gages are distinguished according to the method of indicating the surface temperature. The most common types are **Pirani gage** and **thermocouple gage.** (**c**) **Knudsen gage,** which measures pressure in terms of the net rate of transfer of momentum by molecules between two surfaces maintained at different temperatures and separated by a distance smaller than the mean free path of the gas molecule. Also called radiometer vacuum gage. (**d**) **McLeod gage,** in which a known volume of the gas, at the pressure to be measured, is compressed by the movement of a liquid column to a much smaller known volume, at which the resulting higher pressure is measured. (**e**) **Ionization gage,** comprising a means of ionizing the gas molecules and a means of correlating the number and type of ions produced with the pressure of the gas. Various types of ionization gages are distinguished according to the method of producing the ionization. The common types are **hot-cathode ionization gage, cold-cathode ionization gage, radioactive ionization gage.**

vacuum pump. A device which sets up a flow of gas in a **vacuum system.** Some of the more common types are mechanical pump, vapor or diffusion pump, cryopump.

vacuum system. A chamber, or chambers, having walls capable of withstanding atmospheric pressure and having an opening through which the gas can be removed through a pipe or manifold to a pumping system. The pumping system may or may not be considered as part of the vacuum system. A complete vacuum system contains all necessary pumps, gages, valves, work-holding fixtures, and other components necessary to carry out some particular process; such a system is referred to in England as a *vacuum plant.*

vacuum tube. An **electron tube** evacuated to such a degree that its electrical characteristics are essentially unaffected by the presence of residual gas or vapor.

value engineering. An organized effort directed at analyzing the function of systems, equipment, facilities, procedures and supplies for the purpose of achieving the required function at the lowest total cost of effective ownership, consistent with requirements for performance, reliability, quality, and maintainability.

Van Allen belt, Van Allen radiation belt. (For James A. Van Allen) The zone of high-intensity **particulate radiation** surrounding the earth beginning at altitudes of approximately 1000 kilometers. The radiation of the Van Allen belt is composed of protons and electrons temporarily trapped in the earth's magnetic field. The intensity of radiation varies with the distance from the earth.

Van de Graaff generator. An electrostatic generator which employs a system of conveyor belt and spray points to charge an insulated **electrode** to a high potential.

Van der Waal equation. The best known of the many laws which have been proposed to describe the **thermodynamic** behavior of real gases and their departures from the ideal **gas laws.**

vane. 1. A thin and more-or-less flat object intended to aline itself with a stream or **flow** in a manner similar to that of the common weathercock, as: (**a**) a device that projects ahead of an aircraft to sense gusts or other actions of the air so as to create impulses or signals that are transmitted to the control system to stabilize the aircraft; (**b**) a fixed or movable surface used to control or give stability to a rocket. See **control vane. 2.** A blade or paddle-like object, often fashioned like an **airfoil** and usually one of several, that rotates about an axis, either being moved by a flow or creating a flow itself, such as the blade of a **turbine,** of a fan, or a rotary pump or **air turbine,** of a fan, or a rotary pump or air compressor, etc. See **impeller vane. 3.** Any of certain stationary blades, plates, or the like that serve to guide or direct a flow, or to create a special kind of flow, as: (**a**) any of the blades in the **nozzle ring** of a gas-turbine engine; (**b**) any of the plates or slatlike objects that guide the flow in a wind tunnel; (**c**) a plate or fence projecting from a wing to prevent spanwise flow. See **contravane.** See **airfoil,** note.

vapor. A gas whose temperature is below its **critical temperature,** so that it can be condensed to the liquid or solid state by increase of pressure alone.

vapor pressure. 1. The **pressure** exerted by the molecules of a given **vapor.** For a pure confined vapor, it is that vapor's pressure on the walls of its containing vessel; and for a vapor mixed with other vapors or gases, it is that vapor's contribution to the total pressure (i.e., its **partial pressure**). Also called *vapor tension.* In meteorology, *vapor pressure* is used almost exclusively to denote the partial pressure of water vapor in the atmosphere. See **saturation vapor pressure, equilibrium vapor pressure. 2.** The sum of the partial pressures of all

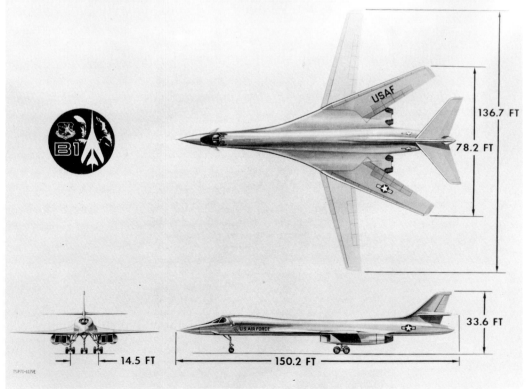

The F-111D fighter-bomber (upper) and the B-1 bomber (lower) are **variable geometry** *aircraft. Wings are extended forward (like the F-111D shown) for slow speed and take-off and landing. At two and one-half times the speed of sound, the swing wings are in the swept position.*

the vapors in a system.

vapor tension. The maximum possible vapor pressure that can be exerted, at a given temperature, by a system composed of a plane surface of a liquid or solid substance in contact with that substance's **vapor.**

vapor trail. See **condensation trail.**

variable-area exhaust nozzle. On a **jet engine,** an exhaust **nozzle** of which the exhaust exit opening can be varied in area by means of some mechanical device, permitting variation in the jet velocity. Compare **fixed-area exhaust nozzle.**

variable cycle engine. A **jet engine** with variable airflow capability to match the varying requirements with Mach number. This engine operates much like a turbojet at supersonic cruise and more like a turbofan for takeoff and subsonic operation. A significant reduction in both specific fuel consumption and noise results from the variable flow capability.

variable geometry aircraft. Aircraft with variable profile geometry, such as variable sweep wings.

variable geometry. (Air inlet). An air intake whose area or shape can be varied in flight. A device usually a part of the **turbojet engine** air inlet duct of a **supersonic** airplane.

variation. The angle between the **magnetic** and **geographical meridians** at any place, expressed in degrees east or west to indicate the direction of magnetic north from true north. Called *magnetic variation* when a specificity is needed to prevent possible ambiguity. Also called *magnetic declination.* The angle between the magnetic and **grid meridians** is called *grid variation* or *grivation.*

variometer. An instrument for comparing magnetic **forces,** especially of the earth's **magnetic field.**

VASI. Visual approach slope indicator.

V-band. A **frequency band** used in **radar** extending approximately from 46 to 56 gigacycles per second. See **frequency bands.**

vector. **1.** Any quantity, such as force, velocity, or acceleration, which has both magnitude and direction at each point in space, as opposed to a **scalar** which has magnitude only. Such a quantity may be represented geometrically by an arrow of length proportional to its magnitude, pointing in the assigned direction. **2.** A heading issued to an aircraft to provide navigational guidance by radar. For air Intercept and Close Air Support and Air Interdiction usage, alter heading to magnetic heading indicated. Heading ordered must be in three digits; e.g., "vector" zero six zero (for homing use "steer").

vector product. A **vector** whose **magnitude** is equal to the product of the magnitudes of any two given vectors and the sine of the angle between their positive directions. Also called *cross product, outer product.* See **scalar product.**

vector steering. A steering method for **rockets** and **spacecraft** wherein one or more **thrust chambers** are gimbal mounted so that the direction of the thrust force (thrust vector) may be tilted in relation to the center of gravity of the vehicle to produce a turning movement.

vehicle. Specifically, a structure, machine, or device, such as an aircraft or **rocket,** designed to carry a burden through air or space; more restrictively, a **rocket vehicle.** This word has acquired its specific meaning owing to the need for a term to embrace aircraft, rockets, and all other flying craft, and has more currency than other words used in this meaning. See **launch vehicle.**

vehicle control system. A system, incorporating control surfaces or other devices, which adjusts and maintains the altitude and heading, and sometimes speed, of a **vehicle** in accordance with signals received from a guidance system. The essential difference between a control system and a **guidance** system is that the control system points the vehicle and the guidance system give the commands which tell the control system where to point. However, the control system maintains the instantaneous orientation of the vehicle without specific commands from the guidance system.

velocimeter. A continuous-wave reflection **Doppler** system used to measure the radial velocity of an object.

velocity *(symbol V).* **1.** speed. See note. **2.** A **vector** quantity equal to speed in a given direction. In sense **1,** *velocity* is often used synonymously with *speed,* as in *the velocity of the airplane,* but in such contexts *speed* is properly the preferred term; except in the compound *airspeed, velocity* is preferred to *speed* in reference to motion of air or other fluid.

velocity head. **1.** Same as **velocity pressure.** **2.** The unit energy of a **fluid** stream owing to its motion.

velocity of escape. The initial speed an object, particularly a molecule of **gas,** must have at the surface of a **celestial body** to overcome the **gravitational** pull and proceed out into space without returning to the celestial body. Also called *escape velocity, escape speed.* The velocity of escape determines a body's ability to retain an atmosphere. The velocity of escape on the surface of the earth is nearly 7 miles per second, neglecting air resistance.

velocity of light. Same as **speed of light.**

velocity of propagation. Rate of flow of **electromagnetic radiation,** including: **(a)** Phase velocity. The velocity of propagation of surfaces of constant phase. Strictly, this definition is applicable only to space periodic fields of infinite length. **(b)** Group velocity. The velocity of propagation of electromagnetic radiant energy in a nondispersive or normally dispersive medium. For a complex waveform, *group velocity* refers to the velocity of propagation of the *beats* between the component frequencies of the waveform. **(c)** Signal velocity. The velocity of propagation of a signal. In a nondispersive or normally dispersive medium, *signal* and *group velocity* are the same. For pure CW (continuous-wave) systems, utilizing no modulation, phase velocity is applicable. For systems utilizing modulated CW, signal velocity is applicable.

velocity of sound. Same as **speed of sound.**

velocity pressure. The difference between **dynamic** (or total) **pressure** and **static pressure.** Also called *velocity head.*

ventilation garment. A lightweight, specially designed garment that is integrated with the **pressure suit** for providing adequate evaporation and heat dissipation from the surface of the body, by circulating dry air through the porous material.

ventral. Pertaining to the belly or the underside of a vehicle, as *ventral camera.*

Venturi tube. A short tube of smaller diameter in the middle than at the ends. When a **fluid** flows through such a tube, the pressure decreases as the diameter becomes smaller, the amount of the decrease being proportional to the speed of **flow** and the amount of restriction.

Venus. See **planet.**

verify. To insure that the meaning and phraseology of the transmitted message conveys the exact intention of the originator.

vernal equinox. 1. That point of intersection of the **ecliptic** and the **celestial equator,** occupied by the sun as it changes from south to north **declination,** on or about March 21. Also called *March equinox, first point of Aries.* **2.** That instant the sun reaches the point of zero declination when crossing the celestial equator from south to north.

vernier. A scale or control used for fine adjustment to obtain a more precise reading of an instrument or closer adjustment of any equipment.

vernier engine. A rocket engine of small thrust used primarily to obtain a fine adjustment in the **velocity** and **trajectory** of a **rocket vehicle** just after the thrust cutoff of the last **sustainer** engine, and used secondarily to add thrust to a booster or sustainer engine. Also called *vernier rocket.*

versus. As a function of, as *temperature versus time.*

vertex. 1. The highest point of a **trajectory** or other curve, as the vertexes of a **great circle,** the points nearest the poles. **2. node,** sense 3.

vertical air photograph. An air photograph taken with the optical axis of the camera perpendicular to the earth's surface.

vertical and short take-off and landing capability. The capability of an aircraft to meet both vertical take-off and landing and short take-off and landing requirements. See **VTOL and STOL, short take-off and landing, vertical take-off and landing.**

vertical circle. A **great circle** of the **celestial sphere,** through the **zenith** and **nadir.** Vertical circles are perpendicular to the horizon. The *prime vertical circle* or *prime vertical* passes through the east and west points of the horizon. The *principal vertical circle* passes through the north and south points of the horizon and coincides with the celestial meridian.

vertical development clouds. Vertical development clouds can vary from bases at 500 feet to tops at 40,000 feet or higher. These are the **cumulus** and **cumulonimbus** clouds. Also see **cloud.**

vertical gyro. A two-degree-of-freedom **gyro** with provision for maintaining its **spin axis** vertical. In this gyro, output signals are produced by gimbal angular displacements which correspond to components of the angular displacements of the base about two orthogonal axes.

vertical launch. A launch in which the missile or other vehicle starts from a vertical position.

vertical scanning. See **scanning.**

vertical separation. A specified vertical distance measured in terms of space between aircraft in flight at different altitude or flight levels.

vertical speed indicator. An instrument which indicates the rate of ascent or descent of an airplane. Sometimes called **rate-of-climb indicator.**

vertical stabilizer. See **stabilizer.**

vertical strip. A single flightline of overlapping photos. Photography of this type is normally taken of long, narrow targets such as beaches or roads.

vertical takeoff and landing (VTOL). The capability of an aircraft to take off and land vertically and to transfer to or from forward motion at heights required to clear surrounding obstacles.

vertigo. The sensation that the outer world is revolving about the patient (*objective vertigo*) or that he himself is moving in space (*subjective vertigo*).

The word frequently is used erroneously as a synonym for dizziness or giddiness to indicate an unpleasant sensation of disturbed relations to surrounding objects in space.

very high. (military) A height above fifty thousand feet.

very high frequency (*abbr* **VHF**). See **frequency bands.**

very-high-speed motion-picture photography. Picture taking at a frequency range from 500 to 10,000 pictures per second.

very low. (military). A height below five hundred feet.

very low frequency (*abbr* **VLF**). See **frequency bands.**

vestigial sideband. The transmitted portion of the **sideband** which has been largely suppressed by a **transducer** having a gradual cutoff in the neighborhood of the **carrier frequency,** the other sideband being transmitted without much suppression.

VFR. Visual Flight Rules. See **VFR Conditions, IFR Conditions.**

VFR Conditions. Basic weather conditions prescribed for flight under **Visual Flight Rules** in accordance with the applicable **Federal Aviation Regulations.**

VFR over-the-top. With respect to the operation of aircraft, the operation of an aircraft over-the-top under VFR when it is not being operated on an IFR flight plan.

VHF *(abbr).* See **frequency band.**

VHF omnidirectional range (VOR). A radio navigation aid which provides 360 courses radiating like spokes on wheels. With the proper airborne equipment, a pilot can fly any selected course to or from the station or determine a bearing to or from the station. Omniranges operate within the 108.0-117.95 MHz frequency band and have a power output of approximately 200 watts. The equipment is VHF, thus, it is subject to line-of-sight restriction, and its range varies proportionally to the altitude of the receiving equipment. Ground to air communications is normally available through the navigational frequencies.

VHF omnirange. See **omnirange, VHF omnidirectional range (VOR).**

vibration. 1. Motion due to a continuous change in the magnitude of a given force which reverses its direction with time. Vibration is generally interpreted as the cyclical (symmetrical or nonsymmetrical) fluctuations in the rate at which an object accelerates. In *longitudinal vibration* the direction of motion of the particles is the same as the direction of advance of the vibratory motion; in *transverse vibration* is is perpendicular to the direction of advance. **2.** The motion of an oscillating body during one complete **cycle;** two **oscillations.**

vibration isolator. A resilient support that tends to isolate a system from steady-state excitation. Also called *isolator.*

video. Pertaining to the picture signals in a television system or to the information-carrying signals which are eventually presented on the **cathode-ray tubes** of a **radar.**

videofrequency. Any **frequency** used in transmission images, as by television.

video signal. See **target signal.**

vidicon. A television pickup tube utilizing a photoconductor as the sensing element. In conjunction with a telescope this is known as a *vidicon telescope.*

view factor. The fraction of the total energy emitted by one surface that is directly incident on another surface. Also called *geometric factor, configuration factor, shape factor.*

virga. Wisps or streaks of water or ice particles falling out of a **cloud** but evaporating before reaching the earth's surface.

virtual height. The apparent height of an ionized atmospheric **layer** determined from the time interval between the transmitted signal and the ionospheric **echo** at vertical incidence, assuming that the velocity of propagation is the velocity of light in a vacuum over the entire path.

viscosity. That molecular property of a **fluid** which enables it to support tangential **stresses** for a finite time and thus to resist deformation; the ratio of shear stress divided by shearing strain. See **viscosity coefficient.**

viscosity coefficient. The ratio of the shearing component of **stress** to the velocity **gradient** in a fluid where the stress acts across a plane perpendicular to the direction of the velocity gradient. Also called *viscosity.* See also **dynamic viscosity, kinematic viscosity.**

viscous. Pertaining to **viscosity,** as a *viscous fluid.*

viscous damping. The dissipation of energy that occurs when a **particle** in a vibrating system is resisted by a **force** that has a magnitude proportional to the magnitude of the velocity of the particle and direction opposite to the direction of the particle.

viscous flow. The flow of **fluid** through a duct under conditions such that the **mean free path** is very small in comparison with the smallest dimension of a transverse section of the duct. This flow may be either laminar or turbulent.

viscous fluid. A fluid whose molecular **viscosity** is sufficiently large to make the **viscous forces** a significant part of the total force field in the fluid. Compare **inviscid fluid.**

viscous force. The force per unit volume or per unit mass arising from the action of tangential **stresses** in a moving **viscous fluid.** This force may then be introduced as a term in the **equations of motion.**

visibility meter. The general term for instruments used to make direct measurements of **visual range** in the **atmosphere** or of the physical characteristics of the atmosphere which determine the visual range. Visibility meters may be classified according to the quantities that they measure. Telephotometers and transmissometers measure the **transmissivity** or alternatively, the **extinction coefficient** of the atmosphere. Nephelometers measure the **scattering function** of the atmospheric **suspensoids.** A third category of visibility meters makes use of an artificial haze of variable density which is used to obscure a marker at a fixed distance from the meter.

visibility, prevailing. The horizontal distance at which targets of known distance are visible over at least half of the horizon. It is normally determined by an observer on or close to the ground viewing buildings or other similar objects during the day and ordinary city lights at night. Under low visibility conditions the observations are usually made at the control tower. Visibility is REPORTED IN MILES AND FRACTIONS OF MILES in the Aviation Weather Report. If a single value does not adequately describe the visibility, additional information is reported in the "Remarks" section of the report.

visibility, runway visibility by observer (RVO). The horizontal distance that an observer near the end of the runway can see an ordinary light (about 25 candlepower) at night or a dark object against the horizon sky in the day time. Visibility is reported in miles and fractions of miles. RVO is used in place of prevailing visibility in determining minimums for a particular runway.

visibility, runway visual range (RVR). An instrumentally derived value, based on standard calibrations, that represents the horizontal distance a pilot will see down the runway from the approach end; it is based on the sighting of either high intensity runway lights or on the visual contrast of other targets— whichever yields the greater visual range. RVR, in contrast to prevailing or runway visilbility, is based on what a pilot in a moving aircraft should see looking down the runway. RVR is horizontal, AND NOT SLANT, visual range. It is based on the measurement of a transmissometer made near the touchdown point of the instrument runway and is REPORTED IN HUNDREDS OF FEET. RVR provides an additional operating minimum at fields equipped with specified navigational aids.

visible radiation. Electromagnetic radiation lying within the wavelength interval to which the human eye is sensitive, the spectral interval from approximately 0.4 to 0.7 micron (4000 to 7000 angstroms). The term is without reference to the variable response of the human eye in its reception of radiation.

visible spectrum. That portion of the **electromagnetic spectrum** occupied by the wavelengths of **visible radiation,** roughly 4000 to 7000 angstroms. This portion of the electromagnetic spectrum is bounded on the short-wavelength end by ultraviolet radiation, and on the long-wavelength end by infrared radiation.

visual approach. An approach wherein an aircraft on an IFR flight plan, operating in VFR conditions and having received an air traffic control authorization, may deviate from the prescribed instrument approach procedures and proceed to the airport of destination by visual reference to the surface.

visual approach slope indicator (VASI). An airport lighting facility in the terminal area navigation system used primarily under VFR conditions. It provides vertical visual guidance to aircraft during approach and landing, by radiating a directional pattern of high intensity red and white focused light beams which indicate to the pilot that he is "on path" if he sees red/white, "above path" if white/white, and "below path" if red/red.

visual/aural range, *abbr* **VAR.** A very-high-frequency radio range. In the aircraft, two tracks are identified by visuals and two by aural indication. This is an obsolete system and was replaced by VOR.

visual flight rules (VFR). See **VFR.**

visual photometry. A subjective approach to the problem of **photometry,** wherein the human eye is used as the sensing element; to be distinguished from **photoelectric photometry.**

visual range. The distance, under daylight conditions, at which the apparent contrast between a specified type of target and its background becomes just equal to the **threshold** contrast of an observer; to be distinguished from the **night visual range.** Also called *daytime visual range.*

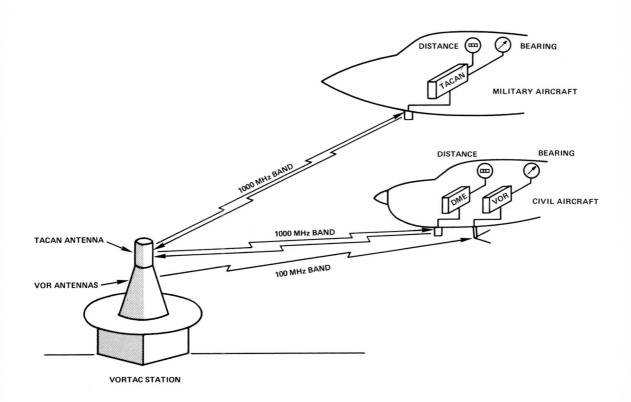

DISTANCE
BEARING
MILITARY AIRCRAFT
TACAN

1000 MHz BAND

DISTANCE
BEARING
CIVIL AIRCRAFT
DME
VOR

1000 MHz BAND

100 MHz BAND

TACAN ANTENNA

VOR ANTENNAS

VORTAC STATION

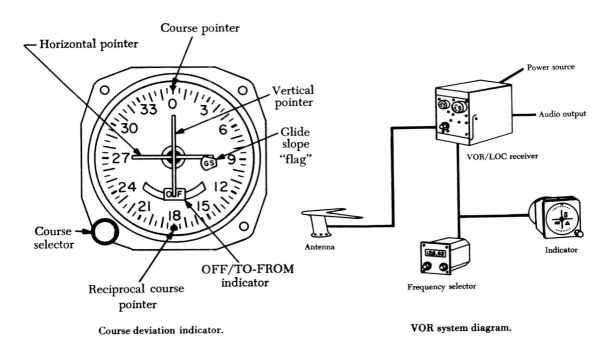

Course pointer

Horizontal pointer

Vertical pointer

Glide slope "flag"

Course selector

OFF/TO-FROM indicator

Reciprocal course pointer

Course deviation indicator.

Power source

Audio output

VOR/LOC receiver

Antenna

Frequency selector

Indicator

VOR system diagram.

A **VORTAC** *navigation facility combines* **VOR** *and Tacan.*

vital area. A designated area or installation to be defended by air defense units.

vitrifying tendency. Tendency of the crystalline phase of a **ceramic** to transform into an amorphous or glassy phase when subjected to aging or temperature cycling.

VLF *(abbr).* See **frequency band.**

volt. The unit of electric potential difference and electromotive force, equal to the difference of electric potential between two points of a conductor carrying a constant current of 1 ampere when the power dissipated between these points equals 1 watt.

VOR *(abbr).* Same as **VHF omnirange.**

VOR/DME. A VOR to which a specific kind of distance measuring device (DME) has been added. (See **VORTAC.**)

VOR-Doppler. A navaid essentially the same as VOR but with an electronically scanned antenna system used to produce the equivalent of a rotating antenna.

VOR receiver. An airborne equipment for receiving the signals of VOR stations and ILS localizers. VOR receivers provide a visual indication of deviation from a course line. They operate in the radio frequency band 108.0 to 117.95 MHz. The VOR receiver consists of three groups of circuits: **1.** Frequency control circuits. **2.** Conventional superheterodyne circuits. **3.** Navigational circuits. The basic components of a VOR receiver are: **a.** Frequency or station selector. **b.** Course or bearing selector (OBS). **c.** Course Deviation indicator. **d.** To/From indicator, also called sense indicator to show whether the selected bearing is from the airplane TO the station or FROM the station to the airplane.

VOR test signal. See **VOT.**

vortex. 1. Any **flow** possessing **vorticity. 2.** Specifically a flow with closed **streamlines** or the idealized case in which all vorticity is concentrated in a **vortex filament.**

vortex filament. A line along which an infinite **vorticity** in a **fluid** motion is concentrated, the surrounding fluid being free of vorticity.

vortex generator. A movable surface on an aerodynamic body that may be used as a spoiler to break down the airflow.

vortices. As pertaining to aircraft, circular patterns of air created by the movement of an airfoil through the atmosphere. As an airfoil moves through the atmosphere in sustained flight, an area of high pressure is created beneath it and an area of low pressure is created above it. The air flowing from the high pressure area to the low pressure area around and about the tips of the air foil tends to roll up into two rapidly rotating vortices, cylindrical in shape. These vortices are the most predominant parts of aircraft **wake turbulence** and their rotational force is dependent upon the wing loading, gross weight, and speed of the generating aircraft.

VOT (VOR Test Signal). A ground facility which emits a test signal to check VOR receiver accuracy. System is limited to ground use only.

V/STOL. vertical/short takeoff, landing. An aircraft capable of both vertical and short takeoff and landing. A VTOL (vertical takeoff, landing) aircraft can normally carry a heavier payload by accomplishing a short takeoff run.

VTOL and STOL (V), (Military designation). Aircraft designed for vertical takeoff or landing with no takeoff or landing roll, or aircraft capable of takeoff and landing in a minimum prescribed distance. See **military aircraft types.**

vulnerability. The characteristics of a system which causes it to suffer degradation (incapability to perform the designated mission) as a result of having been subjected to a certain level of effects in unnatural (manmade) hostile environment.

VORTAC. A navigation facility consisting of co-located **VOR** and **Tacan** equipment. VORTAC is a facility consisting of two components, VOR and TACAN, which provides three individual services: VOR azimuth, TACAN azimuth and TACAN distance (DME) at one site. Although consisting of more than one component, incorporating more than one operating frequency, and using more than one antenna system, a VORTAC is considered to be a unified navigational aid. Both components of a VORTAC are envisioned as operating simultaneously and providing the three services at all times.

W

wake. The region of fluid behind a body in which the total pressure has been changed by the presence of the body.

wake turbulence. The **turbulence** associated with the pair of counter-rotating vortices trailing from the wing tips of an airplane in flight. See **vortex.**

walk-around bottle. A personal supply of oxygen for the use of crewmembers when temporarily disconnected from the craft's system.

waning moon. The moon between full and new when its visible part is decreasing. See **phases of the moon.**

war air service program. The program designed to provide for the maintenance of essential civil air routes and services, and to provide for the distribution and redistribution of air carrier aircraft among civil air transport carriers after withdrawal of aircraft allocated to the Civil Reserve Air Fleet.

war consumables. Expendable items directly related, and absolutely necessary to, a weapon/support system or combat/combat support activity, for which the expenditure factors are indicated in the United States Air Force War and Mobilization Plan. Examples of these items are: auxiliary fuel tanks, pylons, petroleum, oil lubricants, chaff, aircraft guns and gun barrels, airmunition, bombs, rockets, air-to-ground and Air-to-air missiles, in-flight food packets, dropsondes and related flight expendables, racks, adaptors, launchers, film, etc.

war game. A simulation, by whatever means, of a military operation involving two or more opposing forces, using rules, data, and procedures designed to depict an actual or assumed real life situation.

war gas. Any chemical agent (liquid, solid, or vapor), used in war, which produces poisonous or irritant effects on the human body. See also **chemical agent.**

warhead. That part of a missile, projectile, torpedo, rocket, or other munition which contains either the nuclear or thermonuclear system, high explosive system, chemical or biological agents or inert materials intended to inflict damage.

warhead mating. The act of attaching a warhead section to the rocket, or missile body, torpedo, airframe, motor or guidance section.

warm front. The discontinuity at the forward edge of an advancing current of relatively warm air which is displacing a retreating colder air mass. Also see **front.**

warmup time. The time interval required for a **gyro** to reach specified performance from the instant that it is energized.

warning area. Airspace which may contain hazards to nonparticipating aircraft in international airspace.

warning receiver (electronic warfare). A receiver with the primary function of warning the user that his unit is being illuminated by an electromagnetic signal of interest.

warning red. See **air defense warning conditions.**

warning white. See **air defense warning conditions.**

warning yellow. See **air defense warning conditions.**

war readiness materiel. Materiel required in addition to peacetime assets, to support forces, missions, and activities which have been approved in the United States Air Force War and Mobilization Plan.

war readiness spares kit. An air transportable package of spares and repair parts required to sustain planned wartime or contingency operations of a weapon system for a specified period of time pending resupply. War readiness spares kits will include spares and repair parts for aircraft, vehicles, and other equipment, as appropriate. War readiness spares kits are normally prepositioned with the using unit.

war reserve (nuclear) Nuclear weapons material stockpiled in the custody of the Atomic Energy Commission or transferred to the custody of the Department of Defense and intended for employment in the event of war.

war reserves. War reserves are stocks of material amassed in peacetime to meet the increase in military requirements consequent upon an outbreak of war. War reserves are intended to provide the interim support essential to sustain operations until resupply can be effected.

wartime rate. The maximum sustained flying rate achievable by operating the assigned human resources initially on a seven-day week, and by applying wartime crew, maintenance, and safety criteria.

wash. The disturbed air in the wake of an airplane, particularly behind its propeller.

washin. A greater angle of incidence (and attack) in one wing, or part of a wing, to provide more lift; usually used to overcome torque effects. Also see **washout, aerodynamic twist.**

washout. A lesser angle of incidence to decrease lift. Also see **washin, aerodynamic twist.**

waste, radioactive. Equipment and materials (from nuclear operations) which are radioactive and for which there is no further use. Wastes are generally classified as highlevel (having radioactivity concentrations of hundreds to thousands of curies per gallon or cubic foot), lowlevel (in the range of 1 microcurie per gallon or cubic foot), or intermediate (between these extremes).

water. Dihydrogen oxide (molecular formula H_2O). The word is used ambiguously to refer to the chemical compound in general and to its liquid **phase;** when the former is meant, the term *water substance* is often used. Water is distinguished from other common terrestrial substances in existing in all three phases at atmospheric temperatures and pressures (see **ice, water vapor**).The phase changes, with the consequent latent-heat energy changes, are of great significance in many geophysical processes. The same is true of the large specific heat of liquid water and ice relative to both land surface and atmosphere. Water's complex absorption spectrum gives rise to the **greenhouse effect.**

waterspout. A tornado occurring over water; rarely, a lesser whirlwind over water, comparable in intensity to a dust devil over land.

water suit. A G-suit in which water is used in the interlining thereby automatically approximating the required hydrostatic pressure-gradient under G forces. See also **pressure suit.**

water vapor. Water (H_2O) in gaseous form. Also called *aqueous vapor.* See **vapor.** The amount of water vapor present in a given gas sample may be expressed in a number of ways. See **absolute humidity, mixing ratio, dewpoint, relative humidity, specific humidity, vapor pressure.**

watt *(abbr* **w, W**). The unit of power in the **MKSA system;** that power which produces energy at the rate of 1 **joule** per second.

wave. A disturbance which is propagated in a medium in such a manner that at any point in the medium the **quantity** serving as measure of disturbance is a function of the time, while at any instant the **displacement** at a point is a function of the position of the point. Any physical quantity that has the same relationship to some independent variable (usually time) that a propagated disturbance has, at a particular instant, with respect to space, may be called a *wave.*

wave. (Military) A formation of forces, landing ships, craft, amphibious vehicles or aircraft, required to beach or land about the same time. Can be classified as to type, function , or order as shown: **a.** assault wave; **b.** boat wave; **c.** helicopter wave; **d.** numbered wave; **e.** on-call wave; and **f.** scheduled wave.

wave filter. A **transducer** for separating **waves** on the basis of their **frequency.** It introduces relatively small loss to waves in one or more frequency bands and relatively large loss to waves of other frequencies. Also called *filter.*

waveform. The graphical representation of a **wave,** showing variation of **amplitude** with time.

waveguide. A hollow pipe, usually of rectangular cross section, used to transmit or conduct RF energy. The dimensions of the pipe are determined by the wavelength of the conducted signal.

waveguide localizer. A modified type of ILS localizer which produces a highly directional signal along the approach corridor. The course radiated is of exceptional straightness and stability.

wave interference. The phenomenon which results when **waves** of the same or nearly the same **frequency** are superposed; characterized by a spatial or temporal distribution of **amplitude** of some specified characteristic differing from that of the individual superposed waves. Also called *interference.*

wavelength In general, the mean distance between maximums (or minimums) of a roughly **periodic** pattern. Specifically, the least distance between particles moving in the same **phase** of **oscillation** in a **wave** disturbance. The wavelength is measured along the direction of propagation of the wave, usually from the midpoint of a crest (or trough) to the midpoint of the next adjoining crest (or trough).

wave motion. The oscillatory motion of the particles of a medium caused by the passage of a **wave,** produced by forces external to the medium, but propagated through the medium by internal forces. Wave motion per se involves no net translation of the medium. Various types of **oscillation** are found in natural wave motions. Among the simplest are the linear oscillation parallel to the direction of propagation of a longitudinal wave, the linear oscillation perpendicular to the direction of propagation of a transverse wave, and the orbital motion produced by the passage of a progressive gravity wave.

wave of translation. A **wave** in which the individual particles of the medium are shifted in the direction of wave travel, as ocean waves in shoal waters; in contrast with an **oscillatory wave,** in which only the form advances, the individual particles moving in closed orbits, as ocean waves in deep water.

waypoint (w/p). A predetermined geographical position used for route definition and/or progress reporting purposes that is defined relative to a VORTAC station position. Two subsequently related waypoints define a route segment.

weapons state of readiness. The degree of readiness of air defense weapons which can become airborne or be launched to carry out an assigned task. The states of readiness are expressed in numbers of weapons and numbers of minutes. Weapons states of readiness are defined as follows: **a.** 2 minutes—Weapons can be launched within two minutes. **b.** 5 minutes—Weapons can be launched within five minutes. **c.** 15 minutes—Weapons can be launched within fifteen minutes. **d.** 30 minutes—Weapons can be launched within thirty minutes. **e.** 1 hour—Weapons can be launched within one hour. **f.** 3 hours—Weapons can be launched within three hours. **g.** released—Weapons are released from defense commitment for a specified period of time.

weapon system. A weapon and those components required for its operation. (The term is not precise unless specific parameters are established.) (*note;* The Air Force definition indicates that the Air Force parameters are: a composite of equipment, skills, and techniques that form an instrument of combat which usually, but not necessarily, has an aerospace vehicle as its major operational element. The complete weapon system includes all related facilities, equipment, materiel, services, and personnel required solely for the operation of the aerospace vehicle, or other major elements of the system, so that the instrument of combat becomes a self-sufficient unit of striking power in its intended operational environment.)

weather. The short-term variations of the atmosphere in terms of temperature, pressure, wind, moisture, cloudiness, precipitation, and visibility. See **National Weather Service.**

weather adivsory. In aviation weather forecast practice, an expression of anticipated hazardous weather conditions as they affect the operation of air traffic and as prepared by the **National Weather Service.**

Weather Bureau. See **National Weather Service.**

weather forecast. A prediction of weather conditions at a point, along a route, or within an area for a specified period of time. See **area forecast, terminal forecasts, National Weather Service.**

weather map. A map showing the weather conditions prevailing, or predicted to prevail, over a considerable area. Usually, the map is based upon weather observations taken at the same time at a number of stations. See also **map, National Weather Service.**

weather minimum. The worst weather conditions under which aviation operations may be conducted under either **visual** or **instrument flight rules.** Usually prescribed by directives and standing operating procedures in terms of minimum **ceiling, visibility** or specific hazards to flight.

weather radar. An airborne or ground based weather **radar** system used to detect heavy rainfall which is usually associated with **thunderstorms** and **turbulence.**

weather report. See **sequence report.**

weathervane. The tendency of an airplane on the ground or water to face into the wind, due to its effect on the vertical surfaces of the tail group.

wedge inlet. A **variable geometry air inlet** usually of rectangular form, whose area and shape are defined by the position of one or more variable ramps. A device usually a part of the **turbojet engine** air inlet duct of a **supersonic airplane.**

weight *(symbol w.).* **1.** The **force** with which a body is attracted toward the earth. **2.** The product of the **mass** of a body and the **acceleration** acting on a body. In a dynamic situation, the weight can be a multiple of that under resting conditions. Weight also varies on other planets in accordance with their gravity.

weight and balance sheet. A sheet which records the distribution of weight in an aircraft and shows the center of gravity of an aircraft at takeoff and landing.

weight flow rate. Mass flow rate multiplied by gravity.

weightlessness. 1. A condition in which no acceleration, whether of **gravity** or other force, can be detected by an observer within the **system** in question. Any object falling freely in a vacuum is weightless, thus an unaccelerated statellite orbiting the earth is *weightless* although gravity affects its orbit. Weightlessness can be produced within the atmosphere in aircraft flying a parabolic flightpath. **2.** A condition in which gravitational and other external forces acting on a body produce no stress, either internal or external, in the body.

welding. Joining two or more pieces of metal by applying heat, pressure, or both, with or without filler material to produce a localized union through fusion or recrystallization across the interface. The thickness of the filler material is much greater than the capillary dimensions encountered in **brazing.**

wet. To come in contact with, and flow across (a surface, body, or area)—said of air or other **fluid.**

wet-bulb potential temperature. (Meteorology). The temperature an air parcel would have if cooled from its initial state adiabatically to saturation, and thence brought moist adiabatically to the 1,000-millibar level. This temperature is conservative with respect to reversible adiabatic changes.

wet emplacement. A **launch emplacement** that provides a deluge of water for cooling the **flame bucket,** the **rocket engines,** and other equipment during the launch of a missile. See **flame deflector, dry emplacement.**

what luck. (Military) What are/were the results of assigned mission?

what state. (Military). Report amount of fuel, ammunition, and oxygen remaining.

what's up. (Military) Is anything the matter?

wheel load capacity. The capacity of airfield runways, taxiways, parking areas, or roadways to bear the pressures exerted by aircraft or vehicles in a gross weight static configuration.

which transponder. A code meaning report type of transponder fitted—**Identification Friend** or **Foe, Air Traffic Control Radar Beacon System,** or Secondary Surveillance Radar.

whistler. A **radiofrequency** electromagnetic **signal** generated by some lightning discharges. This signal apparently propagates along a geomagnetic line of force and often *bounces* several times between the Northern and Southern Hemispheres. Its name derives from the sound heard on radio receivers.

white body. A hypothetical body whose surface absorbs no **electromagnetic radiation** of any wavelength, i.e., one which exhibits zero **absorptivity** for all wavelengths; an idealization exactly opposite to that of the black body. See **gray body.** In nature, no true white bodies are known. Most white pigments possessing high reflectivity for visible radiation are farily good absorbers in the infrared; hence, they are not *white bodies* in the sense of the radiation theory. However, the term *white body* is used for physical objects with respect to a particular wavelength interval.

whiteout. Loss or orientation with respect to the horizon caused by sun reflecting on snow and overcast sky.

white room. A clean and dust-free room used for assembly and repair of precise mechanisms such as gyros.

whole body counter. Nucleonics. A device used to identify and measure the radiation in the body (body burden) of human beings and animals; it uses heavy shielding to keep out background radiation and ultrasensitive scintillation detectors and electronic equipment. (See **body burden.**)

Wien law. One of the **radiation laws** which states that the wavelength of maximum radiation **intensity** for a **black body** is inversely proportional to the **absolute temperature** of the radiating black body.

wind axis. Any one of a system of mutually perpendicular reference axes established with respect to the undisturbed wind direction about an aircraft or similar body. See **axis,** sense **2.**

wind direction indicator. See **windsock, wind tee, wind tetrahedron.**

windmilling. A propeller is "windmilling" when it is delivering power to the propeller shaft. On multi-engine aircraft, the excessive drag of a windmilling propeller is reduced by **feathering** the propeller of an inoperative engine.

window. 1. Any device introduced into the atmosphere for producing an appreciable radar **echo,** usually for tracking some airborne device or as a tracer of wind. **2.** A World War II code name for a type of **radar**-jamming device employed to confuse the operators of enemy radars (also referred to by the code names of *rope, chaff,* and *clutter).* One type of window consists of packages containing thousands of small strips of paper-backed tinfoil which may be dropped from aircraft and balloons, ejected from rockets, and carried within balloons. The packages burst open upon ejection, scattering the tinfoil widely, producing a radar echo which looks like a small shower of a tight formation of aircraft on plan-position-indicator scopes. **3.** Any gap in a linear **continuum,** as *atmospheric windows,* ranges of wavelengths in the electromagnetic spectrum to which the atmosphere is transparent, or *firing windows,* intervals of time during which conditions are favorable for launching a spacecraft on a specific mission.

wind shear. Change of wind velocity with distance along an axis at right angles to the wind direction (usually specified as vertical or horizontal).

wind speed. See **wind velocity.**

wind shift. (or wind shift line). An abrupt change in the direction or velocity, or both, of the wind. Usually associated with a front.

windsock. A cloth sleeve, mounted aloft at an airport to use for estimating wind direction and velocity.

windtee. An indicator for wind or takeoff and landing

*Special models are tested in a **wind tunnel** to develop desired aerodynamic characteristics.*

direction at an airport. It can be freely pivoted to indicate wind direction or manually set to indicate takeoff and landing direction.

wind tetrahedron. Same as a **wind tee** except shaped in the form of a tetrahedron.

wind triangle. See **triangle of velocities.**

wind tunnel. A tubelike structure or passage, sometimes continuous, together with its adjuncts, in which a high-speed movement of air or other gas is produced, as by a fan, and within which objects such as engines or aircraft, airfoils, rockets (or models of these objects), etc., are placed to investigate the **airflow** about them and the **aerodynamic forces** acting upon them. Tunnels are designated by the means used to produce the gas flow, as *hot shot tunnel, arc tunnel, blow down tunnel;* by the speed range, as *supersonic tunnel, hypersonic tunnel;* or by the medium used, as *plasma tunnel, light gas tunnel.*

wind-tunnel balance. A device or apparatus that measures the **aerodynamic forces** and **moments** acting upon a body tested in a **wind tunnel.**

wind velocity. Wind direction and speed. Wind direction is the direction from which the wind is blowing. The direction is expressed as an angle measured clockwise from true north. Wind speed is generally expressed in nautical miles per hour (knots).

wing. (Military. **1.** An Air Force unit composed normally of one primary mission group and the necessary supporting organizations, i.e., organizations designed to render supply, maintenance, hospitalization, and other services required by the primary mission groups. Primary mission groups may be functional, such as combat, training, transport, or service. **2.** A fleet air wing is the basic organizational and administrative unit for naval land and tender-based aviation. Such wings are mobile units to which are assigned aircraft squadrons and tenders for administrative control. **3.** A balanced Marine Corps task organization of aircraft groups/squadrons together with appropriate command, air control, administrative, service, and maintenance units. A standard Marine Corps aircraft wing contains the aviation elements normally required for the air support of a Marine division. **4.** A flank unit; that part of a military force to the right or left of the main body.

wing. A general term applied to the airfoil, or one of the airfoils, designed to develop a major part of the lift of a heavier-than-air craft.

wing flap. See **flap.**

winglets. Small, nearly vertical aerodynamic surfaces mounted at the tips of airplane wings. Winglets produce large side forces even at low angles of attack, which

combined with the perpendicular local flow, provide a forward thrust component. This reduces the airplane **induced drag.**

wing loading. Gross weight of an airplane divided by gross wing area.

wing panel. A section of the wing which is constructed separately from the adjoining structure, such as the center panel or outer panel. On smaller aircraft the wing is often assembled in one integral panel.

wingover. An airplane flight maneuver consisting of a climbing turn followed by a diving turn.

wing root. The end of a wing which joints the fuselage, or the opposite wing.

wing slot. See **slotted airfoil.**

wingspan. The distance from one wingtip of an aircraft to the opposite wingtip.

winter solstice. 1. That point on the **ecliptic** occupied by the sun at maximum southerly **declination.** Sometimes called *December solstice, first point of Capricornus.* **2.** That instant at which the sun reaches the point of maximum southerly declination, about December 22.

wireless. Sometimes used as the equivalent of **radio,** particularly in British terminology.

wire link telemetry. Telemetry in which no **radio** link is used. Also called *hard wire telemetry.*

word. In electronic computers, an ordered set of **characters** which is the normal unit in which **information** may be stored, transmitted, or operated upon within a **computer.**

word rate. In **computer** operations, the **frequency** derived from the elapsed period between the beginning of transmission of one **word** and the beginning of transmission of the next word.

work *(symbol W).***Energy** resulting from the motion of a system against a **force** and existing only during the process of energy conversion.

working fluid. A **fluid** (gas or liquid) used as the medium for the transfer of **energy** from one part of a **system** to another part.

World Geographic Reference System. A geographic reference system for the world, used in the Air Force for aircraft position reports and target designation, and for the control and direction of air units engaged in air defense, air-sea rescue, and tactical air operations. The short title for this system is *georef.*

World Meteorological Organization (WMO). An organization of the United Nations whose objective is to develop the principles and techniques of international meteorology.

write. In computer terminology, record.

X

X. Military mission designation for **research** aircraft.

X-axis. A horizontal axis in a system of rectangular coordinates; that line on which distances to the right or left (east or west) of the reference line are marked, especially on a map, chart, or graph.

X-band. A **frequency band** used in **radar** extending approximately from 5.2 to 10.9 kilomegacycles per second.

X-ray. **Nonnuclear electromagnetic radiation** of very short **wavelength,** lying within the interval of 0.1 to 100 angstroms (between gamma rays and ultraviolet radiation). Also called *X-radiation, Roentgen ray.* X-rays penetrate various thicknesses of all solids and they act upon photographic plates in the same manner as light. Secondary X-rays are produced whenever X-rays are absorbed by a substance; in the case of absorption by a gas, this results in ionization.

Y

yard. (International). Exactly 0.9144 **meter.** The U.S. yard before 1 July 1959 was 0.91440183 meter.

yaw. 1. The rotational or oscillatory movement of an aircraft, rocket, or the like about a vertical axis. **2.** The amount of this movement, i.e., the angle axis. **3.** To cause to **rotate** about a vertical axis. **4.** To **rotate** or **oscillate** about a vertical axis.

yaw angle. See **angle of yaw.**

yaw axis. A vertical **axis** through an aircraft, rocket, or similar body, about which the body **yaws.** It may be a body, wind, or stability axis. Also called a *yawing axis.*

yawing moment. A **moment** that tends to **rotate** an aircraft, an airfoil, a rocket, etc., about a vertical **axis.** This moment is considered positive when it rotates clockwise.

Y-axis. A vertical axis in a system of rectangular coordinates; that line on which distances above or below (north or south) the reference line are marked, especially on a map, chart or graph.

year. A period of one revolution of the earth around the sun. The period of one revolution with respect to the vernal equinox, averaging 365 days 5 hours 48 minutes 45.68 seconds in 1955, is called a *tropical, astronomical, equinoctial, natural,* or *solar* year. The period with respect to the stars, averaging 365 days 6 hours 9 minutes 9.55 seconds in 1955, is called a *sidereal* year. The period of revolution from perihelion to perihelion, averaging 365 days 6 hours 13 minutes 53.16 seconds in 1955, is an *anomalistic year.* The period between successive returns of the sun to a sidereal hour angle of 80° is called a *fictitious* or *Besselian* year. A *civil year* is the calender year of 365 days in common years, or 366 days in leap years. A *light year* is a unit of length equal to the distance light travels in one year, 9.460×10^{12} kilometers. The term *year* is occasionally applied to other intervals such as an *eclipse year,* the interval between two successive conjunctions of the sun with the same node of the moon's orbit, a period averaging 346 days 14 hours 52 minutes 52.23 seconds in 1955, or a *great* or *Platonic year,* the period of one complete cycle of the equinoxes around the ecliptic, about 25,800 years.

yield. Nucleonics. The total energy released in a nuclear explosion. It is usually expressed in equivalent tons of TNT (the quantity of TNT required to produce a corresponding amount of energy). Low yield is generally considered to be less than 20 kilotons; low intermediate yield from 20 to 200 kilotons; intermediate yield from 200 kilotons to 1 megaton. There is no standardized term to cover yields from 1 megaton upward. (Compare **fission yield;** see **TNT equivalent.**)

Young modulus *(symbol E).* The ratio of normal **stress** within the proportional limit to the corresponding normal **strain.**

Z

Z. Military mission designation for **airship.**

zenith. That point of the **celestial sphere** vertically overhead. The point 180° from the zenith is called the *nadir.*

zenith distance. Angular distance from the **zenith;** the arc of a **vertical circle** between the zenith and a point on the **celestial sphere,** measured from the zenith through 90°, for bodies above the horizon. This is the same as coaltitude with reference to the celestial horizon.

zero fuel weight. The maximum weight authorized for an aircraft with payload, with no unuseable fuel.

zero gravity. The complete absence of gravitational effects, existing when the gravitation attraction of a primary is exactly nullified or counterbalanced by inertial force. For example, during the proper parabolic, flightpath of high-performance aircraft or an orbiting satellite. See also **agravic, weightlessness.**

zero-g. See **weightlessness, zero gravity.**

zero launch. The launch of a rocket or aircraft by a **zero-length launcher.**

zero-length launcher. A **launcher** that holds a vehicle in position and releases the rocket simultaneously at two points so that the buildup of **thrust,** normally rocket thrust, is sufficient to take the missile or vehicle directly into the air without need of a take-off run and without imposing a pitch rate release. The term is not normally applied to a pad used for a vertical launch.

zero-length rocket. A **rocket** with a sufficient thrust to launch a vehicle directly into the air. Said especially of a rocket used to launch an aerodynamic vehicle.

zero-lift chord. A **chord** taken through the trailing edge of an **airfoil** in the direction of the relative wind when the airfoil is at a zero-lift angle of attack.

zip fuel. A boron-base high-energy **liquid propellant.**

zippers. Target dawn and dusk combat air patrol.

Z marker beacon. Equipment identical with the fan marker except that it is installed as part of a four-course radio range at the intersection of the four range legs, and radiates vertically to indicate to aircraft when they pass directly over the range station. It is usually not keyed for identification. (Also known as **cone of silence marker.**) See also **beacon, LF/MF four course radio range.**

zodiac. The band of the sky extending 8° either side of the **ecliptic.** The Sun, Moon, and navigational planets are always within this band, with the occasional exception of Venus. The zodiac is divided into 12 equal parts, called *signs,* each part being named for the principal constellation originally within it.

zodiacal counterglow. See **gegenschein.**

zooming. Utilizing kinetic energy of an airplane to gain altitude.

zone time. See **time.**

z-time. Same as **Greenwich mean time.**

zulu time. An expression indicating Greenwich mean time.

APPENDIX

ABBREVIATIONS, SYMBOLS AND ACRONYMS

Abbreviations, symbols and acronyms are entered in normal alphabetical order in the main part of this dictionary; however for convenience, the most commonly used acronyms and a few symbols and abbreviations are listed separately here in the Appendix. The use of capital and lower case letters may appear inconsistent, however an attempt was made to include the most commonly used form for each individual term.

A. attack (military designation) aircraft
A & P. aircraft and powerplant (mechanic)
AAM. Air-to-Air Missile
ADC. Aerospace Defense Command
ADF. Automatic Direction Finder
ADP. Automatic Data Processing
AGE. aerospace ground equipment
agl. above ground level
AIRMET. weather advisory service
AM. amplitude modulation
AMR. Atlantic Missile Range
APU. auxiliary power unit·
ARSR. Air Route Surveillance Radar
ARTCC. Air Route Traffic Control Center
ASDE. Airport Surface Detection Equipment
ASM. Air-to-Surface Missile
ASR. Airport Surveillance Radar
ATA. Actual Time of Arrival
ATC. Air Traffic Control
ATCRBS. Air Traffic Control Radar Beacon System
ATIS. Automatic Terminal Information Service
AU. astronomical unit
AUM. Air-to-Underwater Missile
avgas. aviation gasoline
B. bomber (military designation) aircraft
C. cargo/transport (military designation) aircraft
CAP. Civil Air Patrol
CAS. Calibrated Air Speed
CAT. clear air turbulence
CGS. centimeter-gram-second
D. director (military designation) aircraft
DH. Decision height
DME. distance measuring equipment
DR. dead reckoning
DSIF. Deep Space Instrumentation Facility
E. special electronic installation (military designation) aircraft
EAS. equivalent air speed
ECM electronic counter measures
EDP. electronic data processing
EHF. extremely high frequency
ELF. extremely low frequency

ELT. emergency locator transmitter
EMA. electronic missile acquisition
ETA. Estimated Time of Arrival
EVA. Extra Vehicular Activity
F. fighter (military designation) aircraft
FAA. Federal Aviation Administration
FAR. Federal Aviation Regulations
FM. frequency modulation
FM/AM. alternate FM/AM operation
FSS. Flight Service Station
G or g. acceleration of gravity, gravity, G-force
GCA. ground controlled approach
GCI. ground controlled intercept
georf. World Geographic Reference System
GMT. Greenwich mean time
gox. gaseous oxygen
GSE. ground-support equipment
H. search/rescue (military designation) aircraft, helicopter
HAT. height above touchdown
Hz. hertz
IAS. Indicated Airspeed
ICAO. International Civil Aviation Organization
ICBM. intercontinental ballistic missile
ICSU. International Council of Scientific Unions
IFF. Identification Friend or Foe
IFR. Instrument Flight Rules
Igor. Intercept ground optical recorder
IGY. International Geophysical Year
ILS. instrument landing system
IRAN. inspect and repair as necessary
IR. infrared or infrared radiation
IRBM. Intermediate range ballistic missile
JATO. jet-assisted take-off
JP. jet propellant
K. tanker (military designation) aircraft
kHz. kilohertz
LMM. compass locater combined with middle marker
LOM. compass locater combined with outer marker
Lorac. long-range accuracy
loran. long-range navigation
lox. liquid oxygen

loz. liquid ozone
LP. liquid propellant
M. Missile Carrier (military designation) aircraft
MCA. minimum crossing altitude
MDA. minimum descent altitude
MEA. minimum enroute altitude
MHz. megahertz
MOCA. minimum obstruction clearance altitude
MRA. minimum reception altitude
MSFN. Manned Spaceflight Network
MSL or msl. mean sea level
MTI. moving target indicator
NACA. National Advisory Committee for Aeronautics
NASA. National Aeronautics and Space Administration
NATO. North Atlantic Treaty Organization
NAVAID. Air Navigation Facility
NDB. non-directional radio beacon
NOTAM. notice to airmen
NRC. Nuclear Regulatory Commission
O. observation (military designation) aircraft
OMNI. See VOR
P. patrol (military designation) aircraft
PAM. pulse amplitude modulation
PAR. Precision Approach Radar
PCM. pulse code modulation
PDM. pulse duration modulation
PFM. pulse frequency modulation
PM. phase modulation
PMR. Pacific Missile Range
PPI. plan position indicator
Q. drone (military designation) aircraft
q. dynamic pressure
R. reconnaissance (military designation) aircraft
racon. radar beacon
rad. radiation absorbed dose
radar. radio detection and ranging
raob. radiosonde observation
RBN. radio beacon
RATO, Rato or rato. rocket assisted takeoff
RF. radio frequency
RMI. Radio Magnetic Indicator
Rnav. area navigation
RP. rocket propellant
R-T unit. receiver-transmitter unit

RVR. runway visual range
S. antisubmarine (military designation) aircraft
SAC. Strategic Air Command
SAM. Surface-to-Air Missile
SAR. Search and Rescue
Sarah. search and rescue homing
Secor/DME. sequential collation of range/distance measuring equipment.
Shoran. short range navigation
SID. Standard Instrument Departure
SIGMET. weather advisory service
SNAP. Systems for Nuclear Auxiliary Power
SONAR. sound navigation and ranging
Spadats. space detection tracking system
SSB. single sideband
SSM. Surface-to-Surface Missile
STAR. Standard Terminal Arrival Route
STOL. Short Take-Off and Landing
SUM. Surface-to-Underwater Missile
T. trainer (military designation) aircraft
TAC. Tactical Air Command
Tacan. tactical air navigation
TAS. True Air Speed
TIAS. True Indicated Air Speed
U. utility (military designation) aircraft
UAM. Underwater-to-Air Missile
UHF. ultra high frequency
USA. United States Army
USAF. United States Air Force
USCG. United States Coast Guard
USM. Underwater-to-Surface Missile
USN. United States Navy
UT. universal time
UV. ultraviolet radiation
V. staff (military designation) aircraft
VAS1. visual approach slope indicator
V. VTOL and STOL (military designation) aircraft
VFR. Visual Flight Rules
VHF. very high frequency
VLF. very low frequency
VOR. VHF omnirange
VOT. VOR test signal
VTOL. Vertical Take-Off and Landing
VORTAC. VOR and Tacan (combined)

REFERENCE   G26683

OVERSIZE $18.95
TL509   Aviation and space dictionary.
A8
1980